Chemistry for Sustainable Technologies
A Foundation

To Susan

Chemistry for Sustainable Technologies
A Foundation

Neil Winterton
Department of Chemistry, University of Liverpool, Liverpool, UK

RSC Publishing

ISBN: 978-1-84755-813-8

A catalogue record for this book is available from the British Library

Published by The Royal Society of Chemistry,
Thomas Graham House, Science Park, Milton Road,
Cambridge CB4 0WF, UK

Registered Charity Number 207890

For further information see our web site at www.rsc.org

Foreword

We live in an age of instantaneous access to information and instantaneous ability to comment. This may not always be good when the need to check facts and ensure reliability is a vitally necessary condition for the validity of arguments. Furthermore, the perception that can be created seems now to be more the aim of people who seek influence than honest opinion-forming discussion based on reliable evidence. It is thus hardly surprising that most 'true' scientists, whose intrinsic raison d'etre is the disinterested quest for the truth about the way the Universe works, are reluctant to get too closely involved with the instantaneous communication process and often feel uncomfortable about the results when they do.

Scientists, along with experts in other fields, are often intimately aware of the limitations which are intrinsic to the doubt-based understanding that surrounds many complex issues of importance at the science–society interface. Consequently many scientists can appear diffident and hesitant when involved in open public debate on scientific and related matters. By comparison professional media commentators often portray their ignorance in the communications arena and seem truly unaware of their limitations. Unfortunately the uninhibited aura of their commentary tends to propagate misunderstanding as having a scientific basis. In this age of the 'sound bite' almost no issue is dealt with in adequate depth. It was for these reasons that some time ago I decided to create an effective platform for professional scientists with state-of-the-art expertise and understanding to communicate on major issues via TV and the Internet. The original Vega Science Trust initiative with an open access website (www.vega.org.uk) streaming free material has now been joined by the Global Educational Outreach for Science Engineering and Technology (GEOSET) sites (www.geoset.info, www.geoset.fsu.edu).

So many of the benefits are now taken for granted that we have all but lost an appreciation of the overwhelming impact on everyday life and importance of science for our continued well-being. We are presently struggling to respond to our apprehensions about the implications of climate change and a plethora of equally disturbing global challenges. Thus, more than ever before, there is a critical need for a deeper understanding of the contributions of science and technology. Scientists must become involved, as only they can have the intrinsic awareness of the complex issues involved, and must carefully articulate them. A new generation of young scientists is needed desperately which has the energy, deep insight and enthusiasm to take up the challenge of communicating with the general public if we are to have any chance of a sustainable future for the environment and society. One specific concern is the fact that the majority of our children receive a relatively poor education in the sciences in schools and universities, and do not develop the intrinsic critical thinking that only hands-on empirical experimentation can instil. There is not only a need for broad cross-disciplinary education across the arts/sciences/humanities spectrum but also for multi-disciplinarity training within the whole science/engineering/technology educational arena.

As a chemist I am acutely aware that, in the 21st century, my discipline overlaps with physics, engineering and biology more than ever before and, where that overlap is strongest, exciting advances are forthcoming. I am delighted to support the major effort that Neil Winterton has made in writing this book to elaborate, in detail, the wide-ranging implications of chemistry to everyday life and the future of our socio-economic environment. He has clarified the pros and cons of chemistry and highlighted the ways in which it can contribute to the development of sustainable technologies. His text fills in the gaps in standard chemistry teaching by highlighting how science is done and how the outcomes and applications of research in ground-breaking areas are often unpredictable and invariably lead to the most revolutionary advances. These are of particular importance for our decision-makers (government leaders, university administrators, science funding bureaucrats as well as those who direct industrial programmes) involved with the funding and exploitation of the sciences.

Those prepared to carefully assess the detailed arguments presented here will be rewarded with an enhanced insight into science and its contributions to society and its development. As the author points out, this book is for the reader with a major concern about modern science and keenly aware that understanding of complex issues requires a significant commitment to assess the arguments.

Neil Winterton's career and experience have given him the background necessary to create this text. Our first contact came when he called out-of-the-blue to discuss his (and ICI's) interest in C_{60}, the third form of carbon. This unique carbon cage molecule, which we had discoverd some years before, was finally available in sufficient quantities for its novel chemical behaviour and possible applications to be examined experimentally. He and his colleagues at ICI (and others at BP) were so interested in this that they were able to provide the support that my group at the University of Sussex desperately needed to exploit this molecule's exciting promise and that had not been forthcoming from academic funding agencies. ICI at that time was managed at the highest level by people who recognised the crucial importance of fundamental science and were prepared to support exciting new developments. The great bonus of such a perspective was the generation of initiatives which fostered synergistic collaborations between industrial scientists, engineers and technologists, and academic researchers. It was a revelation to me that ICI had the vision (at that time) to employ people like Neil, who had a PhD in inorganic reaction mechanisms, to work in extremely basic industrial areas (*e.g.* chlorine and its products) and also have a company-wide ancillary brief to explore innovative new research avenues. It is vital to our future that innovators in business-driven environments are encouraged to explore new topics and be able to fund them; this ICI initiative should be a blueprint for encouraging innovation today.

This book should be read by all chemists and chemical technologists as this new perspective will reveal topics and issues with which few are adequately familiar. Graduates, in particularly those who may no longer use their scientific knowledge directly, will benefit from its focus on contemporary questions and it will enable them to become acutely aware of the scientific issues involved. As is indicated by the title, in a globally-interconnected world in which complex technologies are evolving rapidly, this book provides an excellent foundation for involvement and succeeds in providing a starting point for anyone wanting to follow the general science–society interface topic further. In particular it enables the reader to track down original and reliable information.

Harry Kroto
Florida

Preface

As a chemist, I have often felt pressed to express an opinion when chemistry-related controversy erupts, though forming what I believe to be a robust and well-grounded view has been constrained by my ignorance of wider relevant factors. If today's chemists are anything like me in this respect, particularly relating to questions about the role of chemistry in sustainable development, they need help. This book is for them, as few chemistry courses (and texts associated with them) place, as I hope to do, an equivalent emphasis on both the subject matter and the context in which it should be considered.

Chemistry for Sustainable Technologies: a Foundation seeks to meet this need by combining two key requirements: first, **rigour**, appropriate to a course for undergraduates studying for a Chemistry BSc or MChem, and second, **relevance**, by placing chemistry into the widest of contexts. I hope that even a non-specialist general reader will understand the relevance of (and I would say the central importance of) chemistry to efforts to move to a more sustainable world. Even so, the complexity of the technological challenges is immense; confronting and overcoming these challenges needs an improved dialogue of experts, the public and government, the success of which requires, on an unprecedented scale, informed, pragmatic, enlightened and disinterested decision-making.

The book is not explicitly written to popularise science or chemistry (as there are many who have done this much more effectively than I could), though I hope it may promote the greater understanding of the role of science, and chemistry in particular, in seeking solutions to societal problems. I hope the reader will recognise and appreciate the motivation and practice of science, particularly its (sometimes imperfect) correction of error and its self-criticism. Indeed, in what other sphere would a proponent of a revolutionary idea provide a list of its weaknesses with which its opponents could challenge it, as Charles Darwin did in Chapter 6 of his *On the Origin of Species*?[1] Darwin lived in a less cynical age in which the expression of wonder was not viewed askance;

Chemistry for Sustainable Technologies: A Foundation
By Neil Winterton

Published by the Royal Society of Chemistry, www.rsc.org

the sort of wonder that was felt when it was first realised that we depended on plants to produce the oxygen we need to survive by converting the CO_2 waste product from our own respiration.[2] Wonder leads to curiosity and a wish to understand. Curiosity needs to be combined with scepticism so that ideas and evidence (and the tentative theories that arise) are continually scrutinised and tested. Richard Feynman said:[3] '*. . .each generation that discovers something from its experience must pass that on, but it must pass it on with a delicate balance of respect and disrespect, so that (. . .it...) does not inflict its errors too rigidly on its youth, but it does pass on the accumulated wisdom, plus the wisdom that it may not be wisdom*'. I thus believe that it is important for chemists (and others) to understand something of the ethos, methods and history of science and the ways scientists go about their business, none of which is a topic covered adequately in today's mainstream chemistry courses. This becomes of particular importance when a scientific issue becomes a matter of public controversy.

My own career has been one in which chemistry and chemistry research have been the central motivating interests. The difficulty (for me at least) has been to explore and understand all those related aspects of science, technology and society necessary for the successful and acceptable application of the subject (and its defence against those wishing to demonise it) while retaining expertise in my chosen specialist chemistry research areas. I did not want to become a specialist in sustainability science (important as the topic may be) or a science communicator; I wanted to remain a mainstream chemist, with an awareness of what contribution chemistry might make to sustainability. This is why I hope graduate and professional chemists may also find this book of value.

I have written this book with the aim of equipping those studying chemistry to degree level with some insights, starting points and tools to put their learning into a wider context, particularly relating to the challenges of sustainability. It is not a book that is written primarily for those concerned with the other important disciplines (such as economics, sociology, anthropology, political science) the understanding of which is necessary for sustainable development. However, it is hoped that they, too, might find something of value in its coverage of topics other than those in which they specialise. The book introduces several other disciplines that are important to the way chemistry is applied and to its impact, both social and environmental. I hope the specialists in these disciplines will not be too critical of the treatment of their specialism and my temerity in encroaching into areas where I am not expert. The purpose is twofold: first, to identify domains that play a role in putting chemistry and science into a wider context, and second, by providing a basic introduction as a starting point, to highlight many chemists' ignorance of these domains. One thing my 25 years in industry and the subsequent 10 years in the academic world have taught me is that any such sensitivity is misplaced. The importance of interdisciplinarity and multi-disciplinarity in ICI where I worked was assumed, and one was expected to search for ideas and solutions to problems irrespective of from where they might have originated.

I hope that this book can begin to equip young chemists in this way.

As the topics covered are fast moving and ever-developing, it is not the purpose of this book to provide the latest work in any particular area (though I hope the book is illustrated with useful contemporary examples). Rather, its purpose is to provide the tools and the starting points to enable the reader to explore the body of scientific, technical and other knowledge and, independently, to seek out current and future perspectives and developments. The book should be seen very much like the tip of an iceberg, with its contents a reflection of how much more one would need to know for even a sketchy appreciation of sustainable development. Where specific sources are quoted a full citation is provided at the end of the relevant chapter. These are often early landmark contributions that are of particular value as they do not assume an understanding of some of the basic ideas that later treatments can do. More general sources are collected in the Bibliographies. Increasingly, the internet can be a valuable source, when used with appropriate care. Webliographies provided a personal selection of websites.

The reader for which I have written this book will need to bring an active mind to its reading and a preparedness to explore further on his or her own. For this reason, there are no set questions at the end of each chapter: readers should identify their own. It is, thus, not for the 'shallow learner'. As space does not permit full explanations and even, on occasion, definitions, it will be up to the reader to seek out (at least) the sources I point to (and preferably additional sources that I don't). Appendix 1 summarises some basic principles of searching out information (including web-based sources). I will also assume that the reader is able (and will be willing) to access a dictionary for unfamiliar terms that will arise from time to time.

This book is not designed to fit with the modular model of degree courses (though it grew out of a short optional module at the University of Liverpool). In my view, the modular treatment does not, sufficiently, emphasise connections between, nor stimulate thinking on the wider relevance of, diverse aspects of chemistry, particularly those that are important to recall when addressing the complexities of sustainable development. These connections are also necessary to appreciate and learn of (and from) the fundamental insights from past generations of scientists that are both the bedrock on which all science is based and a pre-requisite for the developing of new insights. (I do not apologise, therefore, for referring to some historic and classic texts: I just wish I had come across them earlier!) There is, I believe, a need for mainstream chemistry courses to have their specialist content, where appropriate, more explicitly linked to the challenges of sustainability. I hope that this book may stimulate my teaching colleagues to develop the content of their courses in this way.

While we have certainly moved from a time when it was possible for a leading scientist of his day to say:[4] '*A serious chemist would begin with an elementary knowledge of mathematics, general physics, languages, natural history and literature. His*[i] *imagination must be active and brilliant in seeking analogies... The*

[i]This is the only time that I will point out that I am, of course, conscious of these sorts of anachronisms of expression but believe they do not invalidate the intended sentiment of this, and other, quotations that I use.

memory must be extensive and profound', the ever-constricting narrowness of our specialisations has made communication even within disciplines ever more difficult. Bearing in mind that the timescale on which we must learn the interdisciplinary means to find solutions to the challenges of sustainability, we must take note of inspiring efforts to get us to climb out of our disciplinary bunkers.[5] This should provide the sobering experience of realising the level of our own ignorance, a state that, paradoxically, should get ever deeper the more we know. The need for us to live with this ignorance and, most importantly, not to be borne down by it nor delay making judgements in the belief they may be easier with more information, is something that scientists can teach the world. We should also argue for the importance of expertise in guiding us to those areas of ignorance most important to explore.

The front cover of the book, therefore, includes an image of an ivory bridge (in the sense of both a link and a crossing), a choice that requires some explanation, not least because there are those who might choose to misconstrue its meaning and imply I am promoting the illegal trade in ivory. The idea of an ivory bridge as opposed to an ivory tower was embodied in a title of a book,[6] *Ivory Bridges: Connecting Science and Society* by Gerhard Sonnert and Gerald Holton. This is the central motivation of my book, an attempt to widen the horizons of young (and perhaps not so young) chemists (and possibly others, too) to take an interest in, and seek a better understanding of, a whole range of other disciplines and domains, ideas and perspectives that I believe the specialist should be aware of (at the very least) to be respected and listened to beyond his or her specialism. The particular choice was motivated by its depiction of a family of elephants, symbols in Japan of prosperity and peace. The use of ivory is of course problematic, but does highlight the dilemma (or, as readers of the book will discover in Section 2.2, the 'trilemma') that humankind faces. The use of ivory from the tusks of animals that die naturally is one thing (and, if it is appropriate to use such a term in such circumstances, would be considered 'sustainable'), whereas the purposeful killing of such creatures for their ivory is rightly regarded as criminal and despicable. Whether we are talking about ivory, the installation of new nuclear electricity-generating capacity, the growth of transgenic crops or the use of the pesticide DDT in controlling malaria, we seem increasingly to be faced, in our debates about the way forward, with a discourse that is characterised more by polarisation and heated disputation between extremes rather than the cool-headed rational pragmatism that is required. Conflict may have many origins, including ignorance and lack of understanding of scientists' work and scientists' own ignorance and lack of understanding about the way their work is perceived. I would like to believe that this book may make a small contribution to changing this situation.

Finally, I would like to thank all those who have helped me in the preparation of this book, particularly my friends and colleagues John Winfield, Jim Swindall and John Satherley who read the whole of the book and Graham Eastham who commented helpfully on particular sections. My thanks also to the many colleagues over the years who have inspired me and have written and spoken eruditely on the very many topics I discuss (with less expertise) in my book. Any errors are my own. Finally, words are inadequate to express my appreciation to my wife, Susan, who has put up, cheerfully and almost uncomplainingly, with my preoccupation with the preparation and completion of this book.

REFERENCES

1. C. R. Darwin, *On the Origin of Species by Means of Natural Selection*, John Murray, London, 1859.
2. Michael Faraday, *On the Chemical History of a Candle*, 1861; Faber Book of Science, ed. J. Carey, 2003, p. 90.
3. Richard. P. Feynman, *'What is Science?'*, in *The Pleasure of Finding Things Out: The Best Short Works of Richard P. Feynman*, ed., J. Robbins, Penguin Books, London, p. 188.
4. Humphry Davy, *Consolations in Travel, or the Last Days of a Philosopher*, Charles Press Pubs, October 2007, Dialogue the Fifth, The Chemical Philosopher, pp. 223–255.
5. E. O. Wilson, *Consilience: The Unity of Knowledge*, Alfred A Knopf, New York, 1998.
6. G. Sonnert, with the assistance of G. Holton, *Ivory Bridges: Connecting Science and Society*, The MIT Press, Cambridge, MA, 2002.

Contents

Chemistry for Sustainable Technologies: A Foundation
By Neil Winterton

Published by the Royal Society of Chemistry, www.rsc.org

CHAPTER 1

Scope of the Book

'For every human problem, there is a neat simple solution. And it is always wrong!'

H. L. Mencken

Chemistry for Sustainable Technologies: A Foundation is intended to be a different type of book. While the treatment attempts to be rigorous, it treats the chemical fundamentals quite broadly, connecting material found in a range of more specialist courses and books. Aspects of other disciplines, relevant to sustainability, are introduced, considered and explored.

The book is also designed to help the reader, particularly students of chemistry, understand the scientific method (and, as a consequence, more consciously to think as scientists)—something not formally taught in undergraduate chemistry courses. In addition to encouraging the use of these tools to get to (and to interpret) the underlying evidence behind the images and headlines we see in the media and on the internet, the book more conventionally does the following:

- **explains** the concepts and terminology of sustainability and sustainable development and the associated complexity, inter-relatedness and uncertainty;
- **highlights** the necessary role of science and technology in the transition towards sustainable development;
- **exemplifies** new approaches to chemistry driven by the need for more sustainable chemical technologies; and
- **illustrates** the central role of metrics in the critical and comparative assessment of the sustainability of technologies.

Chemistry for Sustainable Technologies: A Foundation
By Neil Winterton

Published by the Royal Society of Chemistry, www.rsc.org

The aim is to equip the reader to:

- **understand** the basic terminology of sustainable development and chemistry for sustainable technologies (also known as 'green' chemistry);
- **appreciate** the non-rigorous nature of this terminology and its consequences;
- **place** chemistry and chemical technology in a wider societal context;
- **recognise** the importance of thermodynamic principles in judgements about what may be considered sustainable;
- **recognise** the strengths and weaknesses of green chemistry; and
- **appreciate** the importance of catalysis and the use of renewable feedstocks in developing sustainable chemical technologies and the challenges associated with their implementation.

The topic of sustainable development, the factors driving it and efforts being made to bring it about continue to change over time. To maintain currency, I refer to websites and weblinks that may be of use as starting points to supplement the material to be found in the academic peer-reviewed literature. The extra care needed when using such web-based material is discussed in Appendix 1.

To keep the scope of this treatment manageable but while meeting the book's prime purpose to provide a foundation to the topic of chemistry for sustainable development, there will be some matters that are not explored in the detail to satisfy every reader. In these instances, I point to accessible and peer-reviewed sources of additional information which readers could profitably explore further.

The selection of topics addressed and the examples used to illustrate them are governed, to a large extent, by the fact that this book is aimed primarily at chemists and chemical technologists. The selection I have made is different from that which those with other specialisms and interests might have made. That this is so is a reflection of the complexity and inter-relatedness of sustainable development and sustainability (Glaze[1] called this 'hyperdisciplinarity'), something that it is important, at the outset, to recognise. The role of chemistry, and of science itself, is shown to be critically important. While absolutely necessary, however, neither is sufficient.

The main themes covered by this book include:

- Sustainability and sustainable development: the impact of climate change
- Science: what is it and what is its role?
- Carrying capacity of the Earth; the 'master' equation and our reliance on technology; ecological footprints; can humankind survive?
- The 'Gaia' principle (or Earth systems science); environmental chemistry
- Waste and its minimisation; pollution and its prevention: historical and modern perspectives

- Metrics, life-cycle analysis and chemical technology: the process and product chain; technological integration and industrial ecology
- Importance of the Second Law of Thermodynamics: the concept of exergy
- Green chemistry: principles and pitfalls; contributions from new chemistry
- Central importance of catalysis
- Renewable feedstocks: the transition from fossil sources; what are the constraints; biotechnology; the 'biorefinery'
- Energy production: prospects and timescales
- The chemist as citizen: a statement of the challenges.

The theme running through the book is chemistry's central importance both to our attempts to understand the environment and the lifeforms that populate it, as well as to our efforts to develop ways to make the demands of the human population on the planet's resources (and its associated impact) more sustainable. The technological application of chemistry requires some basic understanding of process engineering and process economics and these are introduced as part of the foundation that represents the purpose of this book. Furthermore, this foundation also encompasses the economic and social context (and associated political ramifications) of technological development, particularly relating to the challenge of climate change.

The book tries not to be polemical: it takes no position in areas of controversy, but seeks to reflect my best personal assessment of the consensus position. It is worth pointing out that a consensus view (such as prevailed when it was believed that the Earth was the centre of the solar system) can be wrong and those seeking to change this can appear to be outlandish, even dangerous, mavericks[2] who challenge established authority. My own view on the anthropogenic (*i.e.* man-made) contribution to climate change has moved over the last 10 years or so, in the light of the evidence, from a point where I accepted the evidence for climate change with a lack of conviction concerning the role of anthropogenic emissions to a position now where I accept that there is a contribution to climate change arising from our emissions of greenhouse gases. While there may be a developing general acceptance that action needs to be taken to ameliorate the situation, it is less likely that there will be a consensus on what form (or forms) this action might take. However, my hope is that readers will be assisted in arriving at a rational view based on the scientific evidence as to what the current position is that may help to inform their judgements about how to proceed.

We are, in historical terms, at the beginning of the road to sustainability, so it is possible only to frame the very hard questions to which we must find answers. However, because it is possible that we are close to a point of no return (and in no position to judge whether or not how close we are), there is a sense of urgency in our search for answers to these questions. I hope that this book will enable those of the new generation who must exercise judgement to develop 'a deeper kind of prudence'[3] and a 'capacity to worry intelligently'.[4]

REFERENCES

1. W. H. Glaze, *Environ. Sci. Technol.*, 2001, **35**, 471A.
2. N. Winterton, *Clean Technol. Environ. Policy*, 2007, **9**, 153.
3. W. R. Freudenburg, *Science*, 1988, **242**, 44.
4. R. W. Kates, *Ambio*, 1977, **6**, 247.

CHAPTER 2

Setting the Scene

'... out of this nettle, danger, we pluck this flower, safety.'

William Shakespeare (Henry IV Part 1, Act 2, Scene 3).

2.1 THE STATE OF THE PLANET

Our immediate perceptions of the world about us are governed by what we see directly with our own eyes and the images that the media select for us to see. Both, in their different ways, are incomplete pictures. We attempt to fill in the gaps by seeking out more information and by exercising judgement based on our experience, knowledge and attitudes. This will be supplemented by additional information and insights from other sources, usually of varying reliability. However this may be done, it is true to say that many of us, while living longer and more comfortably largely as a result of improvements to our health and well-being arising from the benefits of technology, are increasingly concerned about the Earth's continuing ability to support us all. More recently, we have been exposed to more apocalyptic visions for the future of humankind relating to the consequences of the human contribution to climate change associated with increased emissions of greenhouse gases such as carbon dioxide. This has served to stimulate debate about the need for political action (and what form this should take), which has brought with it further questions about the nature of society, the economy and the environment, and their future. These components of our world are interconnected and overlapping, complex and dynamic, reflecting the diversity of the ways people have come to live with one another and with the natural world. Any approach to solving the problems of sustainable development needs to reflect on this reality. No-one has dealt with

Chemistry for Sustainable Technologies: A Foundation
By Neil Winterton

Published by the Royal Society of Chemistry, www.rsc.org

the broader aspects of this issue as well as Mike Hulme in his recent book, *Why We Disagree About Climate Change.*[1]

Because of the impact that technological development has had on the environment in seeking to meet the needs of a growing population (as illustrated in Figures 2.1–2.3), there is a perhaps understandable view in some quarters that we should not look to science and technology to help map out a more sustainable future. Such a view discounts the undoubted benefits of technology. It also ignores the somewhat paradoxical fact that science and technology, increasingly, have enabled us to observe—and thereby understand—the environment and to assess the nature, extent and consequences of our impact. It is technological developments, such as rocketry and satellite construction, associated with science-driven advances in computation, communication and spectroscopy, which now allow us to see the whole planet from a space platform. We can also 'see' at different wavelengths, as the two images in Figure 2.4 from NASA's Earth Observing System satellite show: one, the reflected short-wave radiation and the other, the emitted long-wave radiation, allow climatologists, earth scientists and others to gain a more precise, global and detailed understanding of the energy received and given out by the Earth (and to do so over time).

So science and technology can, with some justification, be seen as both the source of the problem of anthropogenic environmental impact as well as the

Figure 2.1 Compactor at domestic waste landfill site. Image copyright SergioZ 2010. Used under license from Shutterstock.com.

Figure 2.2 Ice collapsing from the Perito Moreno Glacier, Patagonia. Image copyright PSD Photography 2010. Used under license from Shutterstock.com.

Figure 2.3 Emissions from cooling towers and smoke stacks of a coal-fired power station. Image copyright Mark Smith 2010. Used under license from Shutterstock.com.

Figure 2.4 Two images from the NASA Earth Observing System satellite showing reflected and emitted radiation. Source: NASA Earth Observatory (Web 1).

means of detecting what some of its consequences are. Can it also provide the basis for solutions to these problems? My quotation from Henry IV at the beginning of this chapter expresses my own, more hopeful view, and it is the purpose of this book to make this case—particularly from the point of view of chemistry and chemical technology.

Satellite observations also provide evidence (if such evidence was really needed) of a particular feature of our quest for sustainable development, *i.e.* the disparity in prosperity across the globe from the relatively well-off north (Europe, North America, Japan) and the less prosperous south. Figure 2.5 shows visible light emitted at night around the world. The contrast between North America, Europe and Japan and the countries of the southern hemisphere shows the difference in the number of population centres with public lighting infrastructure and domestic light sources that contribute to light emission. This disparity is a major factor in how acceptable, in different parts of the world, might be different policy prescriptions to address the origins of anthropogenic climate change and to bring about more sustainable development. How would car users in Europe or North America view severe restrictions on the types of cars they might be permitted to buy and on the extent to which they might use them? How acceptable to those in the developing world, seeking to reduce malnutrition, would be constraints on the use of genetically modified (GM) crops driven by the concerns of well-fed environmental activists in the developed world? It is a matter of perspective or, rather, of many perspectives.

Whatever position one takes on the role of technology, it is inescapable that the physical and chemical processes taking place in the environment are governed by the same laws that control all of chemistry. Understanding the

Figure 2.5 Composite image showing visible light emitted at night, contrasting the centres of population in the northern hemisphere with public lighting infrastructure and those in the southern hemisphere largely without. Source: NASA/Goddard Space Flight Center (Web 2).

environmental chemistry perspective, therefore, is important in addressing questions of sustainable development. I introduce environmental chemistry and the associated topic of climate change in Chapter 5.

2.2 THE 'TRILEMMA'

Seeking ways forward for society, the economy and the environment that are, at the same time, equitable, acceptable and practical is something that has been termed the 'trilemma'.[2] Resolving the trilemma will involve reconciling the consequences of meeting the societal and developmental needs of a growing world population with the associated deleterious effects on the environment. The degree to which environmental degradation can be limited through the control of emissions is an important question addressed in Chapter 6, particularly from the perspective of thermodynamics.

The urgency expressed by some on the matter of climate change arises from their judgement of just how close we are to some point or state at which changes to the Earth's systems that support human life might become irreversible. There is a high degree of uncertainty about this: there are some who say we are at or, irretrievably, beyond this point; others say we are close to it.[i] A few (a declining number) believe the problem is not as serious as has been

[i] The term 'tipping point' has been coined to describe critical thresholds at which small changes can abruptly alter the condition or state of a system. Lenton *et al.*[3a] list nine components of the Earth system ('tipping elements') that may reach the 'tipping point' as a consequence of humankind's activities sometime this century. More recently, Rockström *et al.*[3b] have attempted to define the boundaries of 10 global processes and how close they believe we are to them (or beyond them), eliciting a series of responses.[3c–i]

made out. However, in truth, bearing in mind how limited our understanding of the complex behaviour of global climate systems actually is, it is impossible to provide scientific certainty of the quality that might allow a consensus to be arrived at and (more to the point) a way forward that might be agreed upon. This apparent absence of quantitative certainty may encourage those who are unconvinced to doubt the need for special or urgent action. Furthermore, uncertainty may encourage the simple belief that more knowledge will resolve the matter. While greater understanding is certainly needed, the degree of complexity is such that uncertainty will always remain. In these circumstances, we must rely on subjective assessments of probability—anathema to most physical scientists—to guide action under conditions of uncertainty, particularly when the time needed to reduce uncertainty sufficiently is longer than the time by which a decision needs to be made. On the other hand, scientists should be able to come to judgments about their confidence in these assessments of probabilities, underpinned by their scientific insight, knowledge and expertise.

Furthermore, it is clear that changes we are already fully aware of such as the depletion of resources, the loss of wilderness and its impact on biodiversity, constrain the options that will be considered acceptable. It is also evident that an understandable preoccupation with local problems can deflect attention from considerations of more global questions whose local impact is perhaps seen as less immediate. Indeed, it is paradoxical that action against such a threat as climate change, which is perceived by many to be invisible and intangible, may only arise when it is too late.[4]

The nature of sustainable development is addressed in Chapter 3.

2.3 HUMAN POPULATION AND ITS GROWTH

The potentially fateful consequences of the geometric growth of a population and the mere linear growth of the resources available to sustain it were first commented on in the late 18th century by Thomas Malthus (1766–1834)[5] when the population of the planet was *ca.* 0.9–1.0 billion. So, any consideration of sustainable development needs to begin with a consideration of the Earth's human population, its size and rate of growth, and what has brought this about. It is the consequence of greater understanding (and control) of the world about us that has improved average human health and longevity and allowed many of us to live lives of greater comfort and convenience brought about by technological development. A further question relates to the difficult job of estimating how many of us the Earth can reasonably sustain indefinitely.[6] The number is believed not to be much greater than about 10 billion.

Figure 2.6 plots the growth of the population of human beings (*homo sapiens*), particularly since modern man emerged about 200 000 years ago and since agriculture began to be practised about 10 000 years ago.[7] Our unique ability to control, and more directly exploit, our environment was critical in enabling a more rapid, near exponential, growth in population, leading to its current value of about six billion. It is particularly significant that about 80% of the increase in population numbers has occurred during the last 200–300 years

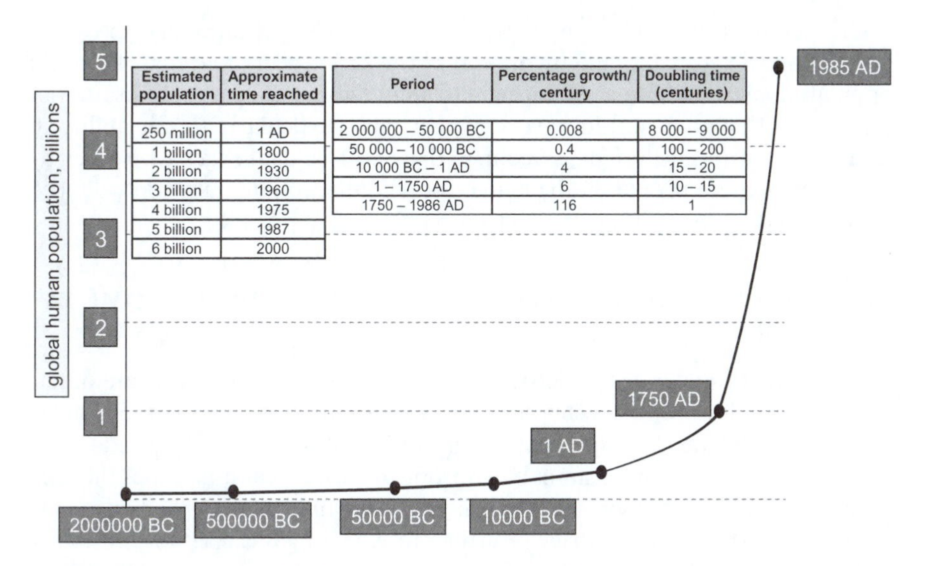

Estimated population	Approximate time reached
250 million	1 AD
1 billion	1800
2 billion	1930
3 billion	1960
4 billion	1975
5 billion	1987
6 billion	2000

Period	Percentage growth/ century	Doubling time (centuries)
2 000 000 – 50 000 BC	0.008	8 000 – 9 000
50 000 – 10 000 BC	0.4	100 – 200
10 000 BC – 1 AD	4	15 – 20
1 – 1750 AD	6	10 – 15
1750 – 1986 AD	116	1

Figure 2.6 Growth in world human population[7] showing accelerated growth since agriculture began to be practised and since the beginning of the industrial revolution.

(0.01–0.02% of man's history), *i.e.* since the industrial revolution, when increased mechanisation and the exploitation of power sources arising from the use of fossil energy resources amplified many-fold what unaided human labour could achieve. This period is seen as representing the beginning of a new geological period,[ii] the Anthropocene, a term first coined by the Nobel Prize winning atmospheric chemist, Paul Crutzen.[8a]

Critical to the proper management of our future impact on the planet is an understanding of how many of us there will be for the available resources to support (and over what timescale). The forecasting of future population growth made by specialists is itself a complex process which carries with it uncertainties associated with which factors are taken into account (and which ignored), the weight given to them and the models[iii] set up to calculate the size of populations in the future. While recognising these uncertainties, there appears to be a view among specialists that there is a slowing in the rate of increase in population (*i.e.* future population growth will not simply be exponential, as one might have at first thought). Fertility rates fall as prosperity (and associated health) improves, though recent work[9] suggests the decline is reversed among the most highly developed countries.

[ii] Geologists have begun to consider[8b] whether or not the Anthropocene should be formally included in the geological timescale and whether it should be denoted an 'epoch' or an 'era'.

[iii] As we do not know how future populations will grow, specialists make mathematical projections using simulations containing what are thought to be the key variables and influences (assigning numbers and weights to these factors). The simulation uses a mathematical model to calculate these projections.

One authoritative study[10] suggests that there is an 85% chance that the world population will stop growing by the end of the 21st century and that the world population may stabilise at about nine billion in about 2070. A stabilisation of the world population would have some benefits, particularly to our ability to feed ourselves (even though, of course, there will be challenges to be met by having 50% more people on the planet for whom adequate nutrition will need to be provided through secure food supplies).

2.4 OUR ATTITUDES TO TECHNOLOGY AND HOW WE COME BY THEM

At the heart of much of the debate about sustainable development are differences in attitudes towards technology and whose purposes it serves and benefits (a historical debate that goes back to the 19th century). Some believe technology is at the heart of the sustainability problem, others that it provides the only practical means of resolving it. Yet others believe that there is probably some truth in both these points of view. Furthermore, teasing out what responsibility technology might be said to have (and the way it may be deployed in the future to aid the transition to a more sustainable society) is complicated by the close linkage between technology and science. While science and technology are intimately related (and some hold science responsible for the failings of its industrial and technological application), it is more constructive to see them as distinct, with different motivations, objectives and ways of working. The fear of the consequences of technology[11] has some justification when considering some of its historical failings (and, in some cases, perceptions about its failings). However, this has to be set against many technological triumphs that certainly have improved the health, life expectancy and quality of life for many (though clearly not universally so).

There is no doubt that a mutual incomprehension exists between the lay public and scientists (and other experts) that arises from a fear of technology (for which, in most cases, scientists have little or no direct responsibility) and from the mismatch between general 'common sense' experience and the less familiar ways of thinking and working of science (and its sometimes counter-intuitive outcomes[12]). The lay public, themselves, of course are not monolithic in their views, with a diversity of attitudes arising from individual education, experience and belief. Despite much effort to improve public understanding and awareness of science, acceptance of the decisions to move to a more sustainable society does depend on an improved dialogue between experts (including scientists), government and population.

What science is and what scientists do, and science's relationship with technology, and other matters, are all covered in Chapter 4. In Chapter 9, I explore aspects of chemical technology (including some basic ideas of chemical and process engineering) having, in Chapter 6, introduced the thermodynamic concept of exergy and the myth of zero waste processes—an appreciation of which are necessary to understand the contribution (and its limitations) chemical technology can make towards cleaner chemical products and processing.

2.5 SCIENCE, CONTROVERSY AND THE MEDIA

Constructive public debate on scientific and related matters depends, to an important degree, on an understanding of the processes by which information is processed and transmitted and a recognition that the treatment of scientific and technological controversies (both by protagonists and the media) can lead to scepticism, suspicion and cynicism towards experts and expertise. Public attitudes can often be influenced by what is read in newspapers and on the web, and what is seen and heard on TV or radio. There is no doubt that attitudes to science (and even the attitudes of scientists themselves) can be influenced by the way scientific evidence, speculation and uncertainty are presented (and re-presented) in the media. This underlines the necessity (including for scientists) to avoid falling into the trap of forming judgements based on newspaper articles and instead, to seek out the original sources of information to find out the factual basis behind news items. Some examples of how this may be done are provided in Appendix 1.

This is particularly relevant to public perceptions about chemistry and the chemical industry that have been formed from its history of poor emissions control allied with a small number of high impact events that are seen as being typical of the performance of today's chemical industry. The reality (I believe) is somewhat different.[13] Without implying that control of pollution (including that produced by the chemical industry) is unimportant, nevertheless it is important to focus on those critical problems identified by the evidence rather than those that arise through media campaigns. We should note the words of the co-founder of Greenpeace, Paul Wilson, who characterised the cynical view—held in some quarters—that equates to: *'It doesn't matter what is true; it only matters what people believe is true* . . . ' (quoted in ref. 14). Neither the chemical industry nor its critics (nor even the media) can claim to hold the moral high ground when it comes to providing the public with a balanced and objective picture of the impact of chemicals in the environment.

Such perceptions are often allied with an ignorance of, or a discounting of, the benefits of technology such as clean water, improved health and well-being, and reliable availability of food, power, transport and communication. They are also associated with a heightened aversion to risk.[15] Added to this mix is the influence of some social theorists who have suggested that scientific knowledge is culturally determined and has no special authority over other sorts of knowledge. This has been a major area of discourse and academic debate,[iv] and is another topic relevant to sustainable development and the shaping of attitudes to the role of science and technology that I will return to in Chapter 4.

[iv] The heated exchanges became known as the 'Science Wars'. Physical and social scientists (as well as philosophers of science) learnt much in these exchanges about the way they work and how their work is perceived beyond science. The more virulent expression of the views of some social theorists, however, had some amusing consequences. Suffice it to say that the paradox at the heart of the social theorists' view (that no one sort of knowledge has any special authority over another) was used against them. Those wishing to explore this topic further should begin with the texts listed in the Bibliography at the end of Chapter 4.

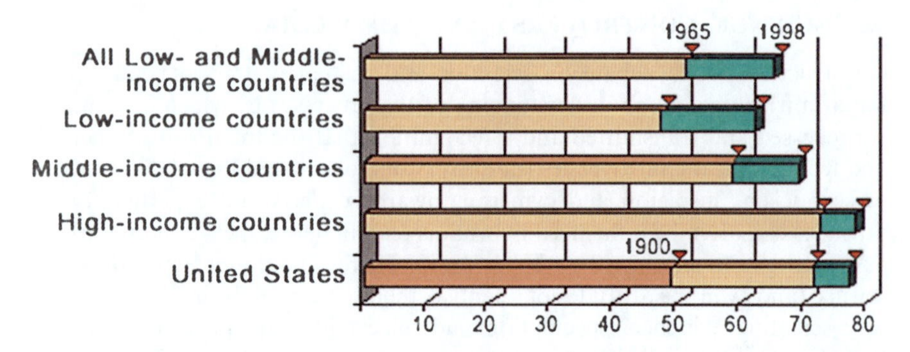

Figure 2.7 Increases in estimates at 1900 (US only), 1965 and 1998 of average life expectancy for low-, middle- and high-income countries. Source: UNESCO (Web 3).

It is not too long ago that projections of population growth were such as to lead commentators such as Paul Ehrlich to foresee an apocalypse. Writing in 1968, he said: '*The battle to feed all of humanity is over. In the 1970s and 1980s hundreds of millions of people will starve to death* . . . '.[16] Such tragic outcomes did not come about on anything like the scale Ehrlich forecasted (though, of course, famine did occur and malnutrition remains widespread). The fact is that, during the 1960s, the so-called 'green revolution'[v] took place in farming and agricultural practices that led to higher agricultural productivity. As a consequence, the per-person calorie intake has increased[18] globally by 24% and in the developing world by 38% since 1961 (despite the population having doubled since then). Such technical developments have led to better nutrition and health. Figure 2.7 shows how the average life expectancy of the human population has changed over time, with increases being seen in high-, middle- and low-income countries. Many of these improvements stem from developments in chemical technology and from the application of scientific and chemical understanding, though this has brought with it direct and indirect environmental impact through land-use changes, urban encroachment, pollution of watercourses and emissions of greenhouse gases.

The crucial question now is whether technology will find the answers to the challenging problems that are currently faced by humankind. A further question is whether technology will be permitted to do so based on the view—particularly in countries of the developed world—that technology is the originator of the problems with which the Earth is faced and that non-technological solutions are more appropriate.

This is a matter discussed in more detail in Chapter 13 (The Chemist as Citizen).

[v] Norman Borlaug, an agronomist who died on 12 September 2009, was awarded the 1970 Nobel Peace Prize for his contributions to the green revolution. It was Borlaug's contention that land-use intensification arising from improved crop yields would help preserve forests and native lands. The associated benefits in respect of net carbon emissions have recently been quantified.[17]

Figure 2.8 Skyline of Widnes in the late 19th century showing pollution arising from industrial activity (particularly relating to the production of soda ash) (Web 4).

2.6 CHEMISTRY AND THE CHEMICAL INDUSTRY

Despite the fact that the image is over 100 years old, Figure 2.8 graphically encapsulates the enduring widespread negative perception of the chemical industry. I suggest that the true picture is somewhat different; indeed, that the performance of the chemical industry with respect to the efficiency with which it converts raw materials into useful products (and the reduction of its emissions in doing so) has improved markedly over the last 50–60 years. While it is widely believed that synthetic chemicals which find their way into the environment are a dominant health hazard[19] (and notwithstanding tragic events such as that at Bhopal or other historic instances of pollution that have lead to significant and unacceptable suffering), the reality concluded by the eminent epidemiologists,[vi] Doll and Peto, in their classic study published in 1981[20a] is that the proportion of (US) cancers attributable to various factors show pollution (at 2%) to be less important than diet, tobacco, infection, sex, sun and radon (in that order) (Figure 2.9).[vii] More direct evidence can be seen in the marked decline in levels of the insecticide, DDT,[viii] found in human milk and fat during the period 1963–1997 (Figure 2.10). DDT was used in the 1950s and 1960s to control insects—and overused, particularly in the USA where its impact on bird populations led Rachel Carson to produce her classic polemic, *Silent Spring*.[21]

[vi] Epidemiologists study factors affecting the health and illnesses of populations (Web 5).

[vii] Doll and Peto suggest [20a] that occupational cancers represent 4 (2–8)% of total cancers in the USA. A recent study for the UK,[20b] which brings the topic of work-place cancer up to date, gives a figure of 5.3%.

[viii] DDT is the insecticide, **d**ichloro**d**iphenyl**t**richloroethane [1,1-bis(4-chlorophenyl)-2,2,2-trichlorethane].

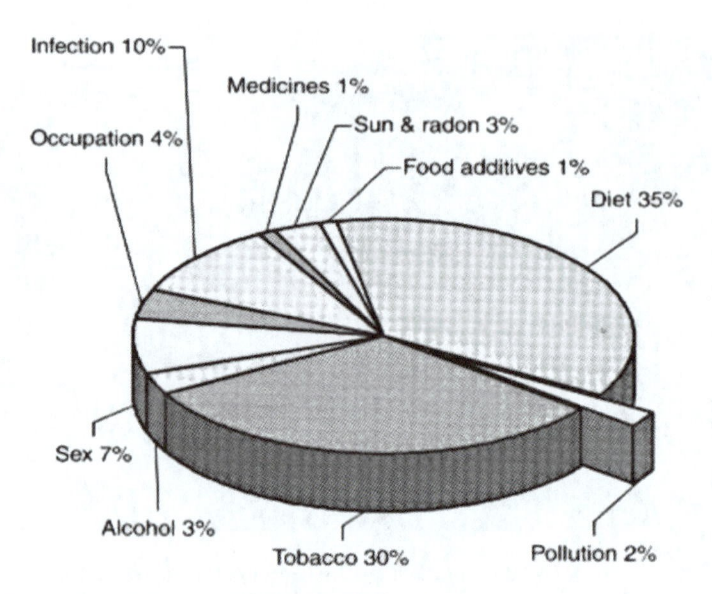

Figure 2.9 Proportion of US cancers attributable to various factors. Data from Doll and Peto[20a] presented graphically by Lomborg.[18] Reproduced with permission of Cambridge University Press.

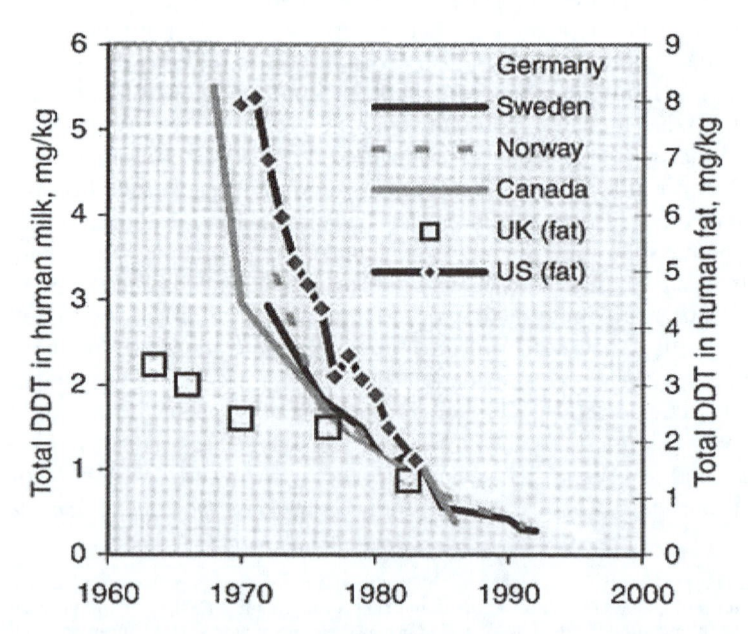

Figure 2.10 Trends in DDT content (1963–1997) of human fat and human milk. Data from various sources presented graphically by Lomborg.[18] Reproduced with permission of Cambridge University Press.

2.7 WHY WE CANNOT TURN THE CLOCK BACK

One understandable response to the situation we find ourselves in might be to seek to return to some pre-industrial past when we all lived in harmony with nature. Unfortunately, not only does this notion represent a somewhat romantic idealisation of life (and life expectancy) before the 18th century, there are also just too many of us currently inhabiting the earth for this to be possible. Indeed, the negative impact on the environment began even before the 18th century. A statute issued in 1398 by Richard II (1367–1400), anticipating 'polluter pays' regulations, required:

> '. . . *all they which do cast and lay such annoyances, dung, garbage, entrails, and other ordure in ditches, rivers, waters and other places shall cause them utterly to be removed, avoided and carried away*'.[ix]

The population of the UK[22] in 1750 was less than six million suggesting that, without industrial technology, the country could support a population of no more than about ten million—or about 50 million less than today. To quote the famous American engineer, Norbert Wiener who recognised our dependence on technology over 50 years ago,[23] '*we live only by the grace of invention*'. A policy of de-industrialisation would be a political prescription that no responsible government could seek to enact on which to base a future society, though of course there could be no objection to such an approach as an individual lifestyle choice.[x] If we cannot go back, how then can we move forward? I address this question in later chapters.

That we are dependent on technology (and that we are unaware of this fact or that we take such dependence for granted) can be illustrated by the simple example presented in Box 2.1 which demonstrates how far we have gone in intensifying land use compared with an equivalent, but non-industrial, activity.[24] The example compares two processes for obtaining fibre for use in clothing: the first uses chemical technology to make the polymer, poly(ethylene terephthalate) (PET) (Figure 2.11); the second uses agriculture to farm sheep to make wool (Figure 2.12). If we go back 50 or 60 years, a sweater would be made of wool. So, why cannot we go back to making fibre from wool instead of consuming fossil feedstocks for PET? The answer is to be found in Box 2.1. (I come back to the question of more renewable feedstocks for such products in Chapter 11).

2.8 SYNTHETIC BAD, NATURAL GOOD?

The history of DDT serves to illustrate the benefits and problems that arise in the application of a new chemical development. DDT proved its worth during

[ix] A readable introduction to sustainable development (see Bibliography) points to an even earlier law, this time relating to water pollution, propounded by Plato sometime during the 4th century BC.

[x] Follow an interesting, if rather disturbing, exchange on this topic in *The Guardian* of 18 August 2009 between George Monbiot and Paul Kingsnorth (Web 7).

BOX 2.1 LAND-USE INTENSIFICATION IN MANUFACTURING FIBRE

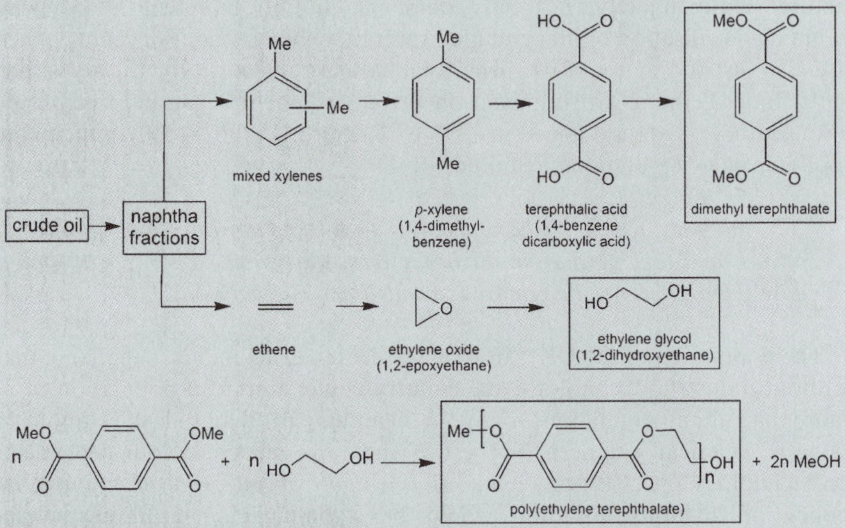

Scheme 2.1 Production route to PET from catalysed transesterification using petrochemically-derived dimethyl terephthalate and ethylene glycol.

Poly(ethylene terephthalate) is a condensation polymer (Scheme 2.1) made by the transesterification of dimethyl terephthalate (dimethyl 1,4-benzene-dicarboxylate) with ethylene glycol (1,2-dihydroxyethane).[xi] Both these materials are made in huge tonnages in well-established processes from basic organic feedstocks derived from oil or natural gas. In the case of dimethyl terephthalate, mixed xylenes (1,2-, 1,3- and 1,4-dimethylbenzenes) are fractionated to give the 1,4-isomer. This is then oxidised to the dicarboxylic acid and esterified with methanol. The other monomer may be obtained by the hydrolysis of epoxyethane, itself obtained by the epoxidation of ethene. There are now many plants making PET to meet the annual worldwide demand, for textile use alone, of 39 million tonnes (Mt)[xii] in 2008.

In making this quite crude comparison, we need to make certain assumptions:

- The areas needed for infrastructure (providing raw materials, transport, processing) for chemicals production and sheep-rearing are broadly similar.
- All grazing can sustain the number of sheep as the most productive pasture.

[xi] Details of the process and product technologies of these and other important industrial chemicals can be found in sources listed in the Bibliography and at Web 8 and 9.

[xii] In this book the abbreviation ‘t’ is used to refer to units of ‘tonne’, where 1 tonne = 1 metric ton = 10^3 kg. Elsewhere, however, ‘t’ may also be used to represent ‘short’ ton (2000 lb, 907 kg) or ‘long’ ton (2240 lb, 1016 kg). Turn-over number, a measure of catalyst performance we meet in Section 10.3, may also be abbreviated to ‘ton’.

- There is a technical equivalence (weight for weight) of wool and PET when used as fibre in the production of clothes.

For the purposes of this calculation, I assume that a world-scale installation might produce about 500 000 tonnes of PET per year. I estimate roughly such a plant would occupy about four hectares.

If an average fleece weighs 2–4 kg (say *ca.* 5 kg), we would need the fleece from approximately 100 million sheep to obtain 500 000 tonnes of wool, *i.e.* equivalent to the production from a single world-scale PET plant.

If we assume that the best pasture can support around 25 sheep per hectare, then to farm 100 million sheep would need more than 4×10^6 ha of prime pasture. The land area, just for pasture, would be 40 000 km^2 (100 ha = 1 km^2). This area, for just one plant, is equivalent to the size of The Netherlands; 78 such plants would be needed to meet global demand, which would require pasture equivalent in area to the whole of India.

An approximate land-use intensification factor can now be calculated. This is the ratio of the land area needed to grow wool to that needed to produce PET fibre, *i.e. ca.* $4\times10^6/4$, or bearing in mind the assumptions made, *ca.* $10^{(6\pm1)}$. This establishes a largely unacknowledged benefit of chemical technology, *viz.*, the land not used in natural fibre production[xiii] can be given over to some other use or can be preserved as wilderness in the protection of species diversity.[xiv]

There are some additional specific aspects of sheep-rearing we need to be aware of which might make the use of the chemical technological approach more or less favourable. We would need, for instance, to be able to dispose of 8–10 million sheep carcasses annually (based on the life expectancy of sheep of 10–12 years). While wool is a source of lanolin, the processing of fleece to make wool is an especially waste-producing activity. Whether this is more of less problematic than the wastes associated with PET manufacture would require more detailed analysis (though we do know that sheep, like other ruminants, are responsible for the emission of significant quantities of the greenhouse gas, methane). Furthermore, as they are raised out of doors, sheep would be subject to a range of climatic, seasonal and disease factors that would not impact similarly on the chemicals process (though of course there may be other factors that might).[xv]

Therefore, there are intrinsic, and possibly counter-intuitive, benefits that have arisen from the development of chemical technology—though

[xiii] By the foregoing calculation, sheep produce *ca.* 0.13 t wool ha^{-1} y^{-1}. Ramie, a vegetable fibre blended into fabrics for indoor use, can be produced at 1.3 t fibre ha^{-1} y^{-1} from *ca.* 4 t ha y (dry weight) of harvested stems (Web 10). Even so, the land-use intensification factor for chemical technology would still be $>10^4$.

[xiv] The benefits of the 'green' revolution have been discussed in similar terms. See footnote, p. 14.

[xv] Such 'back-of-the-envelope' calculations as these can often provide useful 'broad brush' insights. For precise comparisons, however, more detailed analyses are needed such as those provided by life-cycle methods (introduced in Section 6.9).

Figure 2.11 Fleece made from recycled synthetic PET fibre. Image copyright O. Akhøj.

Figure 2.12 A sheep (in Swaledale, Cumbria), the source of the natural fibre, wool. Image copyright David Iliff 2010. Used under license from Shutterstock.com.

this is not necessarily evident in the way the wider public has come to perceive it.[xvi]

the Second World War and its aftermath in controlling the parasite-borne disease, typhus, to the extent that its developer (it was first synthesised in 1894), the Swiss chemist Paul Müller, was awarded the 1948 Nobel Prize in Medicine (Web 11). However, its later use (due to its effectiveness and low toxicity to humans) and misuse resulted in deleterious effects on birdlife. Because of its environmental persistence (and its storage in the fatty tissues of living things), DDT is even detected in fish and marine mammals in the Arctic and Antarctic. However, DDT was also credited with bringing malaria under control through its effect on the mosquito responsible for transmitting the disease. For example, there were millions of cases of malaria every year in Sri Lanka prior to DDT use; by the early 1960s, case numbers were down to double figures. However, the campaign against DDT, particularly in the developed world, led to its withdrawal from use in Sri Lanka. As a consequence, the incidence of malaria was, by 1969, back up to millions of cases.[25] There followed a search for alternatives to DDT (of which more in Chapter 9), but nothing as cheap nor as effective has so far been found for controlling malaria. It was only recently that the World Health Organization (WHO) (Web 12) was able to gain acceptance for the controlled re-introduction of DDT for the preventative spraying of houses in countries where alternative strategies are unacceptably expensive.[xvii] This example serves to highlight the dilemma that can be faced when seeking to balance human and environmental welfare.

Another perception in the general population about chemicals is the belief that because something is natural it is intrinsically better than something that is synthetic. This point of view neglects the fact that some of the most toxic chemicals are produced naturally [*e.g.* aflatoxin B_1 (Figure 2.13), the most toxic of a group of 13 fungal metabolites that can be present at low levels in some foodstuffs].

Many pharmaceuticals in widespread use today such as aspirin or quinine were identified following investigations[27] of the active ingredient in folk remedies such as plant extracts. While, drinking an infusion of St John's wort (*Hypericum perforatum*) is a known and effective herbal remedy against depression, it may not be without potentially serious side effects arising from the presence of other components. Infusions of plant material containing varying amounts of the useful ingredient, as well as a variety of other substances, cannot deliver the pure or controlled dose that a synthetic pharmaceutical can.

[xvi] It is worth noting here that some success has been achieved in commercialising polyesters made from terephthalic acid and 1,3-propanediol, the latter manufactured from non-fossil raw materials. Related polymers are under investigation in which terephthalic acid is substituted by a sugar-derived di-acid. The biodegradable polyester, poly(hydroxybutyrate), has been manufactured on an industrial scale in a biotechnological process (though one that uses petrochemical-derived methanol as the carbon source). These are discussed further in Chapter 11.

[xvii] See ref. 26 for a recent perspective.

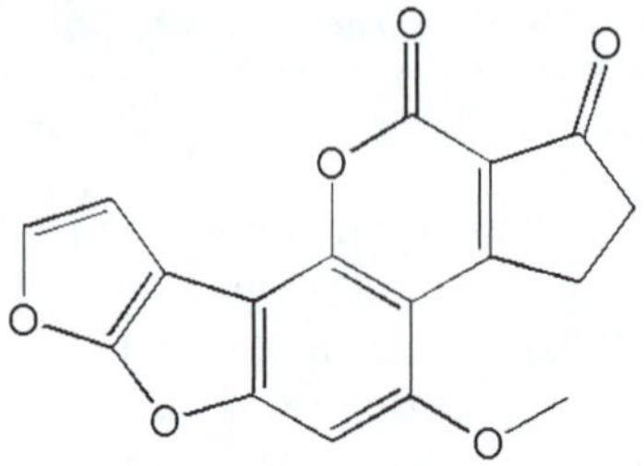

Figure 2.13 Aflatoxin B_1.

The regulations surrounding the use of pharmaceuticals (but not of herbal remedies[xviii]) require that the pharmacological[xix] effects of all components are known. In addition, demand for a herbal medicine may be so great as to threaten the extinction of its source.[28] Therefore, the synthetic production of a chemical found in nature can satisfy demand for it which may otherwise threaten the survival of the species from which it is isolated. In Chapters 10 and 11, I examine two examples of this when we look at the availability of the flavouring, vanilla, and the anticancer treatment, taxol.

That being said, the production of such ‘nature-identical’ products will not satisfy those who still believe that the material isolated from a plant or other living thing contains some residue of the ‘life-force’ that animated the organism. It may not have occurred to those holding such views (and wishing to persuade others to follow them) that they may be doing harm to the natural world. The belief that ‘natural’ is synonymous with ‘good’ and ‘synthetic’ with ‘bad’ is thus simplistic, erroneous and potentially dangerous. The benefits and dangers of each material, whether natural or synthetic, need to be evaluated in the overall context of the circumstances of its source and its use.

There is no doubt that the campaign against DDT led to the more general view, propounded vigorously by some environmental groups, that anything containing chlorine must, *ipso facto*, be bad. TV programmes about chlorine with titles such as *The Devil’s Element* (broadcast by the BBC in 1991) not too surprisingly reinforced a fear of anything containing chlorine. While clearly there are hazardous chemical compounds containing chlorine, there are plenty just as hazardous that contain no chlorine at all. However, the impression was fostered that nature does not make chlorine-containing compounds, so that anything found in the environment that might contain chlorine must be man-made. This is not so.[29] Examples (among thousands) include the chlorinated anisyl alcohol and related compounds shown in Figure 2.14 that are present naturally in the biomass produced by certain common species of fungus. Ironically, the concentrations of chlorinated metabolites found in this natural

[xviii] Aristolochic acid {8-methoxy-6-nitrophenanthro[3,4-d][1,3]dioxole 5-carboxylic acid (CAS no. 313-67-7)} continues to be provided in herbal remedies (Web 13) and to cause serious adverse health effects despite the fact that its toxicological hazards are well-known.

[xix] Pharmacology is the study of drugs, their composition, action on and interaction with the body.

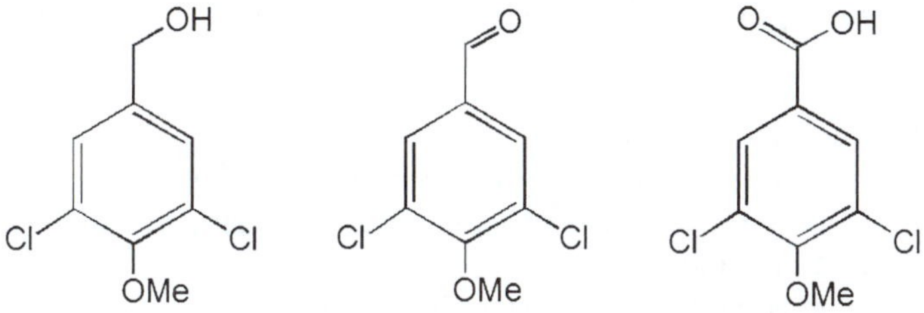

Figure 2.14 Chlorinated anisyl metabolites present in fungal biomass.[30]

material can be so high that, if released into the environment by a chemical company rather than by a fungus, it would have to be labelled as hazardous waste.[30]

Such perceptions can have unforeseen and sometimes tragic consequences. The use of chlorine to treat potable water was first carried out on a large scale in the early part of the 20th century. Water chlorination is seen as one of the most important developments in enhancing public health and life expectancy. Because many millions of people are exposed to drinking water treated this way, there is obviously a need to monitor whether there are any deleterious consequences arising from such exposure. A review of epidemiological studies (that investigate illnesses and diseases in the human population as a means to identifying possible causes) by the US Environmental Protection Agency (EPA) in the late 1980s discovered a weak statistical association between drinking water that had been chlorinated and the incidence of a type of bladder cancer.[31] While this obviously would be a cause for concern, the study did not conclude that exposure to chlorinated drinking water caused the small number of extra cancers but that such a possibility warranted further investigation. As a subsequent article in *Nature*[32] revealed, uncertainties in balancing risks between various courses of action—and perhaps a flawed application of something called the precautionary principle (see Chapter 6)—appears to have led the authorities responsible for public health in Peru to cease water chlorination and to do so without a full consideration of the consequences. Unfortunately an outbreak of cholera then occurred—a disease, ironically, that had essentially been wiped out where sources of potentially contaminated water had been chlorinated. The resulting epidemic is believed to have killed 3000 people.[32]

Real harm can arise from precipitate action by public officials when driven by anxiety felt when confronted by a public health issue that appears to arise from the use of chemicals.

2.9 DECISION-MAKING AND 'WICKED' PROBLEMS

Achieving sustainable development is a so-called 'wicked' problem involving the domains of economics, law, culture and politics, as well as science and technology, in the process of decision-making. Problems are considered 'tame' or 'wicked'[33] according to their complexity and amenability to defined solutions (and with no ethical or moral judgement being made). A 'tame' problem is one:

- that doesn't change over time
- for which a solution is usually self-evident
- that may be amenable to technically-based solution that would be broadly considered acceptable by a non-expert
- with low uncertainty and indeterminacy, its resolution being achieved by working towards goals likely to be non-controversial and having clear success criteria.

Among such problems might be an effective cancer treatment, reducing famine, better animal welfare, safer road transport and better air quality. On the other hand, a 'wicked' problem is one:

- that is complex and can change with time,
- that is intractable and interconnected, with many overlapping social and political factors.
- for which there will be difficulty in defining a solution with a predictable outcome and, often, a serious lack of consensus on goals to be aimed at.

Solutions to problems raised by GM crops, terrorism, climate change, access to water resources or a third runway at Heathrow Airport may all be recognised as being 'wicked'.

'Tame' problems can become 'wicked' ones. For instance, agriculture has seen the role of science move from being largely uncontested (during the so-called 'green revolution' of the 1960s and 1970s referred to in the previous chapter) to one in which the role of science and its application, through technology, has become highly controversial. The consequences can be bizarre as was the case, reported in *The Independent* of 30 August 2002, in which the government of Zambia, having been influenced by single-interest pressure groups, was persuaded not to release GM-containing food aid to the hungry and the starving.

2.10 SUSTAINABLE DEVELOPMENT AND HYPERDISCIPLINARITY

The multidisciplinary nature of sustainable development and the need to include intellectual domains other than science in its consideration are illustrated in two diagrams (Figures 2.15 and 2.16) based on ref. 34.

There is no doubt that introductory texts of equal relevance and importance could be written on the political, economic, sociological, religious or ethical dimension of sustainable development with science not being mentioned until well into the book. This would not be a criticism; rather a reflection of the difficulty of delivering in written form contributions to sustainable development from any perspective with sufficient background to ensure that they are not

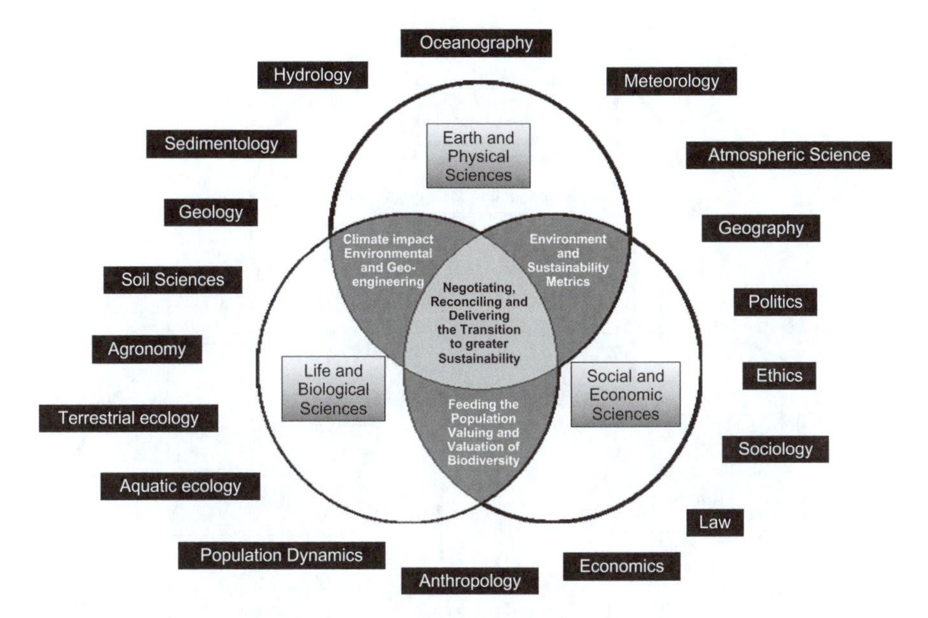

Figure 2.15 A selection of key academic disciplines relevant to an understanding of the environment, human impact on it, and means of living more sustainably. Adapted from Figure 1.3.1, ref. 34.

narrowly and meaninglessly parochial. Success might be assessed by the degree to which those within the domain broadly accept the contribution as valid and the degree to which those from outside read it and take note.

2.10.1 Economics

The review for the UK government of the economics of climate change (Web 14, 15) by Nicholas Stern, a former Chief Economist and Senior Vice President of the World Bank, may (at some point in the future) be seen to be one such influential text, though Stern's review initiated a vigorous debate amongst economists. A subsequent book by Stern[35] and earlier contributions (for example by Jonathon Porritt[36]) address the challenges to the world economic system.

The brief introduction to economics that follows considers, quite pragmatically, only what is currently so (rather than what should or might be) and avoids political analysis or prescription. Meeting the challenges associated with the transition to more sustainable technologies will inevitably bring about social, political and, possibly, cultural change. For the purposes of this book, I therefore assume (begging a number of questions in so doing) that the technological transition will, in general, involve companies and other entities acting within a global social–market framework subject to governmental regulation under political control. So, while the analysis of the domains relevant to individual and collective responses to these challenges lies wholly outside the

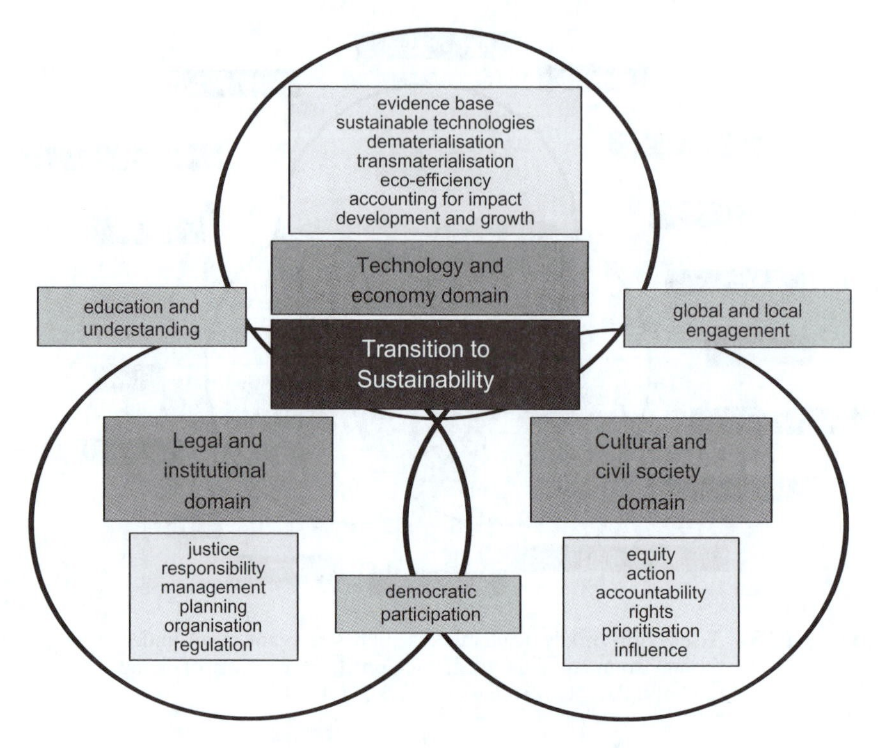

Figure 2.16 A selection of key intellectual, professional and social domains relevant to determining the transition to more sustainable living. Adapted from Figure 2.4, ref. 34.

scope of this book, a brief sketch must be drawn of some of the economic background (dealt with more completely and rigorously in texts listed in the Bibliography) to matters dealt with later.

The term 'economy' has the same Greek root [οἶκος (oikos, or house)] as 'ecology' [-λογία (-logia, or study of)] with economics being concerned, originally, with the management of a household. A useful definition suggests economics is the study of human behaviour that relates human needs (which can be seen as limitless) to the means of meeting them. Because resources [human (labour), physical (land, raw materials), financial (capital)] are not limitless, personal and political choices are made between their alternative uses.

There are three basic systems for the allocation, distribution and utilisation of resources to meet human needs. The market ('capitalist') economy allocates resources using the price mechanism, based on the decisions of individual consumers and producers in markets governed by supply and demand. In a planned ('communist') economy, the government decides what is produced and how this is allocated. Today, most economies are 'mixed', having features of both the market and the planned system, in which governments allow markets to operate, intervening and regulating when the market fails to address inequalities (through taxation and government spending) or to deal with so-called 'externalities' (of particular relevance to the question of sustainable development).

The millions of decisions made by individuals every day in buying what they want or need (and the prices they are willing to pay) determines, in principle, what suppliers and producers (in seeking to profit by such supply and production) choose to invest in. The employment generated by such investment provides those employed—through wages—with the economic means (money) to satisfy their needs and wants, some of which is taxed for public spending purposes. The profit rewards investors, encouraging further investment. Such, in basic terms, is the process of 'wealth creation',[xx] itself a vague, value-laden term. The imbalances between needs and wants (and differences in view as what constitutes the one or the other) as well as between the private, individual, public and collective good provides the stuff of political argument.

The mixed economies of North America, Western Europe and Japan have (broadly) delivered economic growth and increased prosperity particularly since the Second World War, doing so in the belief that natural resources would continue to be discovered to meet demand (as technology utilised them more efficiently) and that the environment could continue to receive the associated wastes (suitably controlled by regulation-driven technology). Such technological developments and the decline in the rate of population growth associated with increased prosperity encouraged the view that Malthus had been wrong when he suggested that the (geometric) increase in population would inevitably outstrip the (linearly increasing) supply of food. However, it is now no longer at all clear that the resources and services of the planet will be adequate indefinitely to meet the future needs of the human population.

Part of the problem (that cannot be decoupled from our individual role as consumers who materially benefit from the use of the output of industrial activity) is that the true total cost of the environmental impact of manufacture is not accounted for in (is 'external' to) companies' balance sheet and accounts. The latter normally include (for a defined period) income from sales, money spent on wages, dividends, taxes and material needed for their operations and statements of their assets, which will include physical assets such as manufacturing plant, unsold stocks of product, financial assets such as cash, investments and commitments such as loans. At the national level, at which economic activity is aggregated in metrics such as gross domestic product (GDP),[xxi] 'externalities' such as taking account of declining stocks of natural resources or the costs of pollution damage (or even our ultimate dependency on a 'service'—our survival—provided by ecosystems that contain oxygen-generating organisms) are not taken into account. Indeed, the reverse can be true, in that the costs associated with the clean-up of an oil spillage will be added

[xx] For a recent discussion of the various meanings of the term wealth creation and its links with globalisation and the question of corporate responsibilities, see ref. 37.

[xxi] GDP is one estimate of national income that measures the value of output produced and income received by companies, organisations and households located within a country's economy.

positively into GDP. Seeking the inclusion of such externalities into the accounts of companies and other operations is the province of environmental accounting. More generally, environmental economics and ecological economics (with different philosophical bases) address these so-called market failures and seek a basis for the valuation of environmental goods and services with an appreciation of the limits imposed on resource use that arise from a consideration of the laws of thermodynamics.[38]

Bearing in mind that cost avoidance or reduction is a major stimulus in the profit (or surplus) motive, the internalising of environmental costs should force companies to seek to reduce them. Unfortunately, avoidance could be achieved by moving operations to jurisdictions in which such internalising was not required. For any company or industrial sector to account, unilaterally, for the full environmental costs of its operations and products would put itself at a competitive disadvantage. Only by ensuring all business and industrial activity is subject to these costs (by providing a 'level playing field') will the matter be addressed. The Stern Review (Web 14) uses (with many qualifications) an economic model to provide (inevitably imprecise) estimates of the costs (in reduced consumption per person) of climate change if nothing is done ('business as usual'); these estimates are based on International Panel on Climate Change (IPCC) projections of average global temperature that are believed to be realistic. The review also estimates the costs of mitigation (M) to avoid much of the loss of economic activity—seeing the expenditure on mitigation as an 'investment' with a return equivalent to the climate change costs (C) thus avoided. Furthermore, stabilising greenhouse gases at 500–550 parts per million (ppm) CO_2e[xxii] by 2050 would cost about 1% of global GDP per year, with any delay leading to an increase in annual costs. Stern concludes that, as $M < C$, mitigation would be a highly productive investment.

2.11 THE ROLE OF THE EXPERT

Even when assembling factual evidence about the state of the planet and the factors and influences that are bringing about change (whether for good or ill), there is a very large number of academic disciplines (*e.g.* earth sciences, life sciences and social sciences) that will have something relevant to say. Because of the very large number of disciplines involved, a new term, 'hyperdisciplinarity', has been coined.[39] A new discipline has also arisen called 'sustainability science',[40] now with its own journal (Web 16).

An examination of the disciplines shown in Figure 2.15 reveals only one that explicitly includes the term 'chemistry' in its name. However, it should be understood that all the other earth and life sciences are underpinned, to varying degrees, by the fundamental understanding provided by physics, chemistry and biology (as well as mathematics).

[xxii] As explained in Chapter 5, there is a number of greenhouse gases with different potentials to contribute to global warming. It is, therefore, convenient to express their combined effect in terms of an equivalent concentration of carbon dioxide, CO_2e.

Addressing sustainable development (with its wide range of cultural, religious, political and social attitudes and influences) in the wider public domain must take account of the fact that attitudes and influences will differ from region to region and from society to society, despite any globalising trends that may be evident. So, even if there was agreement among the scientists about what was needed to bring about sustainable development (which, for reasons discussed below, is most unlikely), an agreed outcome would still depend on discussion and debate in the political and social arena.

The unsurprising conclusion from a consideration of Figures 2.15 and 2.16 is that different disciplines will have different perspectives on what the key problems are. Neither are individual disciplines themselves monolithic: there will be range of conclusions that practitioners in those disciplines will arrive at and seek to articulate, based upon their research. Once these conclusions reach the public domain, those that are most directly affected (or interested) will seek to deploy them to their own ends, or to neutralise them if they are believed to run counter to their interests. So, the conclusion of the economist might be taken up by the company or an institution, that of the ecologist by an environmentalist group, of an anthropologist by an indigenous people, of a lawyer by a litigant, of a sociologist by social group, of a scientist by a technologist or a consumer, of a political scientist by a politician, interest group, voter or journalist, and so on. Overlaying this complex web of information and response, there will be further reactions based on differing perspectives:

- the present *vs.* the future
- north *vs.* south
- developed *vs.* developing
- rich *vs.* poor
- consumer *vs.* producer
- government *vs.* populace
- urban *vs.* rural
- radical *vs.* status quo
- expert *vs.* non-expert

and many other possible combinations of equal significance.

The foregoing highlights the difficulty faced by someone with what they believe to be expert knowledge in approaching the complex questions of sustainability. How relevant is their expertise? How can it be put into a wider context? What if it is ignored or challenged? You, the reader, might wish to give some thought to how you might approach a controversy in which the public had been persuaded to a view in contradiction to what you knew from your position as an expert to be reliable and relevant evidence. This is not an academic point and may arise for a range of reasons. It is quite possible that during your career as a scientist in industry, in education or in public service, you may be (or even have been) confronted with such a dilemma. How would you approach such a problem?

In addressing such controversies, it is advisable not to rely on the force of advocacy and intensity of belief of the protagonists as they do not provide any

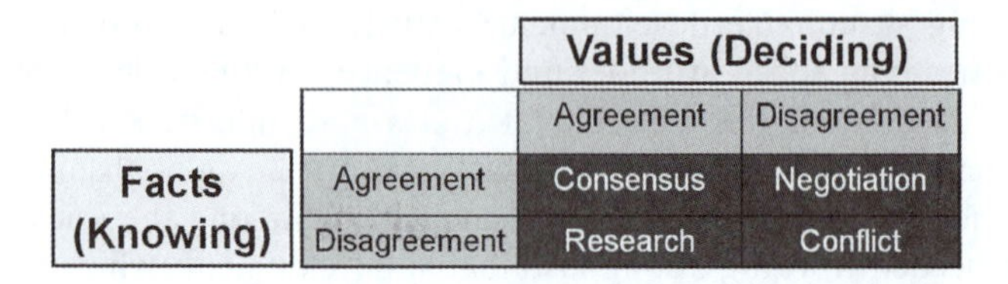

Figure 2.17 Possible outcomes from attempts to reconcile attitudes and evidence. Adapted from King *et al.*[41]

measure of the reliability or otherwise of the factual basis for the controversy. There is always a need (particularly for scientists) in the middle of a heated debate to assess all the evidence (and its likely implications) in a dispassionate and pragmatic manner. This underlines the importance of knowledge 'unmediated' [*i.e.* information from the original (preferably peer-reviewed[xxiii]) sources], as the media stand between you and the original information and, in their selection of what to present and the way in which they present it, they can influence how it is perceived.

In simple terms, the process of decision-making will depend on what we know to be the facts of the case (often, but not invariably, in the case of environmental controversies provided by scientists) and on the attitudes, perspectives, experiences, emotions and values that are used and considered in coming to a decision. Figure 2.17 shows four possible scenarios that may arise depending on whether or not there is agreement about the underlying facts and the values that are used in seeking to come to a decision. In the event that there is an acceptance of the facts and the values employed in decision-making are shared, then a consensus can be said to exist. If we agree on the facts, but the attitudes and priorities driving the outcome that we seek are in conflict, then some form of negotiation is likely to resolve the question. If we have shared values, but disagree about the factual basis of the controversy, then we might support the idea of further research or fact-finding to find a way out of the impasse. However, if we cannot agree on the facts and differ in the values that are uppermost in attempting to secure an outcome, then we face a potentially irreconcilable conflict. The latter situation often arises in dealing with the 'wicked' problems mentioned previously.

It is my belief that in generating reliable information of relevance to an area of political conflict, particularly concerning the environment and sustainable development, science can both minimise conflict and guide good decision-making. However poorly this might work it is usually better than the alternative of acting without reference to the evidence.

The distinctive nature of scientific knowledge and some insights into the practices of science, and how they have changed, are covered in Chapter 4.

[xxiii] Peer-review is an important aspect of quality control in the publication of scientific information, something discussed in more detail in Section 4.4.

REFERENCES

1. M. Hulme, *Why We Disagree About Climate Change: Understanding Controversy, Inaction and Opportunity*, Cambridge University Press, 2009.
2. Y. Nitta and S. Yoda, *Technol. Forecast. Soc. Change*, 1995, **49**, 175.
3. (a) T. M. Lenton, H. Held, E. Kriegler, J. W. Hall, W. Lucht, S. Rahmstorf and H. J. Schellnhuber, *Proc. Natl. Acad. Sci. U.S.A.*, 2008, **105**, 1786; (b) J. Rockström, W. Steffen, K. Noone, Å. Persson, F. S. Chapin, E. F. Lambin, T. M. Lenton, M. Scheffer, C. Folke, H. J. Schellnhuber, B. Nykvist, C. A. de Wit, T. Hughes, S. van der Leeuw, H. Rodhe, S. Sörlin, P. K. Snyder, R. Costanza, U. Svedin, M. Falkenmark, L. Karlberg, R. W. Corell, V. J. Fabry, J. Hansen, B. Walker, D. Liverman, K. Richardson, P. Crutzen and J. A. Foley, *Nature*, 2009, **461**, 472; (c) W. H. Schlesinger, *Nature Rep. Clim. Change*, 2009, **3**, 112; (d) S. Bass, *Nature Rep. Clim. Change*, 2009, **3**, 113; (e) M. Allen, *Nature Rep. Clim. Change*, 2009, **3**, 114; (f) M. J. Molina, *Nature Rep. Clim. Change*, 2009, **3**, 115; (g) D. Molden, *Nature Rep. Clim. Change*, 2009, **3**, 116; (h) P. Brewer, *Nature Rep. Clim. Change*, 2009, **3**, 117; (i) C. Samper, *Nature Rep. Clim. Change*, 2009, **3**, 118.
4. A. Giddens, *The Politics of Climate Change*, Polity Press, Cambridge, 2009.
5. T. R. Malthus, *An Essay on the Principle of Population*, 1798, Oxford World's Classics reprint.
6. (a) J. E. Cohen, *Science*, 1995, **269**, 341; (b) J. E. Cohen, *How Many People Can the Earth Support?*, W.W. Norton, New York, 1995.
7. K. Davis, *Bull. At. Sci.*, 1986, **42** (April), 20.
8. (a) P. J. Crutzen, *Nature*, 2002, **415**, 23; (b) J. Zalasiewicz, M. Williams, W. Steffen and P. Crutzen, *Environ. Sci. Technol.*, 2010, **44**, 2228.
9. M. Myrskylä, H.-P. Kohler and F. C. Billari, *Nature*, 2009, **460**, 741.
10. W. Lutz, W. Sanderson and S. Scherbov, *Nature*, 2001, **412**, 543.
11. U. Beck, *Risk Society: Towards a New Modernity*, Sage, London, 1992.
12. L. Wolpert, *The Unnatural Nature of Science*, Faber and Faber, London, 2000.
13. J. F. Jenck, F. Agterberg and M. J. Droescher, *Green Chem.*, 2004, **6**, 544.
14. L. Spencer, J. Bollwerk and R.C. Morais, 'The not so peaceful world of Greenpeace', *Forbes*, 11 November 1991, 174.
15. F. Furedi, *Risk-taking and the Morality of Low Expectation*, Cassell, London, 1997.
16. P. Ehrlich, *The Population Bomb*, Ballantine Books, New York, 1968, p. xi.
17. J. A. Burney, S. J. Davis and D. B. Lobell, *Proc. Natl. Acad. Sci. U.S.A.*, 2010, **107**, 12052.
18. B. Lomborg, *The Skeptical Environmentalist*, Cambridge University Press, Cambridge, 2001, pp. 60–67.
19. B. N. Ames and L. S. Gold, *Angew. Chem., Int. Ed. Engl.*, 1990, **29**, 1197.
20. (a) R. Doll and R. Peto, *J. Natl. Cancer Inst.*, 1981, **66**, 1191; (b) L. Rushton, S. Bagga, R. Bevan, T. P. Brown, J. W. Cherrie, P. Holmes, L.

Fortunato, R. Slack, M. Van Tongeren, C. Young and S. J. Hutchings, *Br. J. Cancer*, 2010, **102**, 1428.
21. R. Carson, *Silent Spring*, Houghton Mifflin, Boston, 1962.
22. E. A. Wrigley and R. S. Schofield, *The Population History of England 1541-1871: a Reconstruction*, Edward Arnold, London, 1981, quoted by J. Jeffries, *Focus on Migration: The UK Population Past, Present and Future* (Web 6).
23. N. Wiener, *Invention: The Care and Feeding of Ideas*, MIT Press, Cambridge, MA, 1993, p. 3.
24. N. Winterton, *Clean Technol. Environ. Policy*, 2003, **5**, 8.
25. E. M. Whelan, *Toxic Terror: The Truth Behind the Cancer Scares*, Prometheus Books, Buffalo, NY, 1993, p. 101.
26. H. Van den Berg, *Environ. Health Perspect.*, 2009, **117**, 1656.
27. J. W.-H. Li and J. C. Vederas, *Science*, 2009, **325**, 161.
28. C. Delvaux, B. Sinsin, F. Darchambeau and P. Van Damme, *J. Appl. Ecol.*, 2009, **46**, 703.
29. N. Winterton, *Green Chem.*, 2000, **2**, 173.
30. J. A. Field, J. M. Verhagen and E. de Jong, *Trends Biotechnol.*, 1995, **13**, 451.
31. P. A. Murphy and G. F. Craun, in *Water Chlorination: Chemistry, Environmental Impact and Health Effects*, ed. R. L. Jolley, Lewis Publishers, Chelsea, MI, 1990, **Vol. 6**, pp. 361–372.
32. C. Anderson, *Nature*, 1991, **354**, 255.
33. (a) H. W. J. Rittel and M. M. Webber, *Policy Sci.*, 1973, **4**, 155;31.(b) S. S. Batie, *Am. J. Agric. Econ.*, 2008, **90**, 1176.
34. T. O'Riordan, *Environmental Science for Environmental Management*, Prentice Hall, London, 2000, pp.13, 53.
35. N. Stern, *A Blueprint for a Safer Planet: How to Manage Climate Change and Create a New Era of Progress and Prosperity*, The Bodley Head, London, 2009.
36. J. Porritt, *Capitalism as if the World Matters*, Earthscan, London, 2005.
37. G. Enderle, *J. Bus. Ethics*, 2009, **84**, 281.
38. D. G. Ockwell, *Energy Policy*, 2008, **36**, 4600.
39. W. H. Glaze, *Environ. Sci. Technol.*, 2001, **35**, 471A.
40. (a) R. W. Kates, W. C. Clark, R. Corell, J. M. Hall, C. C. Jaeger, I. Lowe, J. J. McCarthy, H. J. Schellnhuber, B. Bolin, N. M. Dickson, S. Faucheux, G. C. Gallopin, A. Grübler, B. Huntley, J. Jäger, N. S. Jodha, R. E. Kasperson, A. Mabogunje, P. Matson, H. Mooney, B. Moore III, T. O'Riordan and U. Svedin, *Science*, 2001, **292**, 641;38.(b) W. C. Clark and N. M. Dickson, *Proc. Natl. Acad. Sci., U.S.A.*, 2003, **100**, 8059; (c) J. R. Mihelcic, J. C. Crittenden, M. J. Small, D. R. Shonnard, D. R. Hokanson, Q. Zhang, H. Chen, S. A. Sorby, V. U. James, J. W. Sutherland and J. L. Schnoor, *Environ. Sci. Technol.*, 2003, **37**, 5314.
41. D. M. King, in *Investing in Natural Capital The Ecological Economics Approach to Sustainability*, ed. A. M. Jansson, M. Hammer, C. Folke and R. Costanza, Island Press, Washington, 1994, p. 325.

BIBLIOGRAPHY[xxiv]

Ullmann's Encyclopedia of Industrial Chemistry, Wiley-VCH Verlag, Weinheim, Germany, 7th edn, 2009 (see Web 7).

Kirk-Othmer Encyclopedia of Chemical Technology, John Wiley & Sons, Chichester, 5th edn, 2007 (27-volume set) (see Web 8).

P. P. Rogers, K. F. Jalal and J. A. Boyd, *An Introduction to Sustainable Development*, Earthscan, London, 2008.

S. Ison and S. Wall, *Economics*, Prentice Hall/Pearson Education, Harlow, UK, 4th edn, 2007.

N. Hanley, J. F. Shogren and B. White, *Environmental Economics in Theory and Practice*, Macmillan Press, Basingstoke, 1997.

I. Goldin and L. A. Winters, *The Economics of Sustainable Development*, Cambridge University Press, Cambridge, 1995.

D. Pearce and E. B. Barbier, *Blueprint for a Sustainable Economy*, Earthscan, London, 2000.

D. Pearce, R. K. Turner, T. O'Riordan, N. Adger, G. Atkinson, I. Brisson, K. Brown, R. Dubourg, S. Frankhauser, A. Jordan, D. Maddison, D. Moran and J. Powell, Blueprint 3: *Measuring Sustainable Development*, Earthscan, London, 1993.

WEBLIOGRAPHY

1. www.earthobservatory.nasa.gov/IOTD/view.php?id = 2984
2. http://en.wikipedia.org/wiki/File:Flat_earth_night.png
3. www.unesco.org/education/tlsf/TLSF/theme_c/mod13/www.worldbank.org/depweb/english/modules/social/life/chart1.htm
4. http://commons.wikimedia.org/wiki/file:Widnes_Smoke.jpg
5. http://en.wikipedia.org/wiki/Epidemiology
6. www.statistics.gov.uk/downloads/theme_compendia/fom2005/01_FOPM_Population.pdf
7. www.guardian.co.uk/commentisfree/cif-green/2009/aug/17/environment-climate-change
8. http://mrw.interscience.wiley.com/emrw/9783527306732/home/
9. http://eu.wiley.com/WileyCDA/WileyTitle/productCd-0471484946.html
10. www.swicofil.com/products/007ramie.html#Background%20information
11. http://nobelprize.org/nobel_prizes/medicine/laureates/1948/index.html
12. www.who.int/mediacentre/news/releases/2006/pr50/en/index.html
13. http://news.bbc.co.uk/1/hi/uk/8520171.stm
14. http://webarchive.nationalarchives.gov.uk/ + /www.hm-treasury.gov.uk/media/4/3/Executive_Summary.pdf (executive summary); http://webarchive.nationalarchives.gov.uk/ + /www.hm-treasury.gov.uk/

[xxiv] These texts supplement those already cited as references.

independent_reviews/stern_review_economics_climate_change/stern_review_report.cfm
15. http://en.wikipedia.org/wiki/Stern_review
16. www.springer.com/environment/environmental + management/journal/11625

All the web pages listed in this Webliography were accessed in May 2010.

CHAPTER 3

Sustainability and Sustainable Development

' . . . wise consumption is a far more difficult art than wise production.'

John Ruskin, 1860

During the past 100 years, the world's population has multiplied four-fold and global economic output has increased more than 20-fold. This and the approximately 50 billion tons of material we extract each year[1] (Figures 3.1 and 3.2) have driven dramatic changes in the way the human population interacts with the Earth and its ecosystems, shifting from a dominantly rural–agrarian to a dominantly urban–industrial society.[i] The consequences of this pace of consumption challenge government, industry and individuals alike and motivate moves to find a more sustainable means for humanity and the planet to co-exist. This chapter explores what this may mean.

3.1 WHAT IS SUSTAINABILITY? AND IS IT DIFFERENT FROM SUSTAINABLE DEVELOPMENT?

This book is about 'sustainability' and 'sustainable development'. These terms are often used interchangeably, so we need to define them carefully (or, at least, try to).

[i] Indeed, a recent UN report, *State of the World Cities 2008/2009* (Web 1), concludes that just over half the world's population now lives in cities and, by 2050, over 70% will be urban dwellers.

Chemistry for Sustainable Technologies: A Foundation
By Neil Winterton

Published by the Royal Society of Chemistry, www.rsc.org

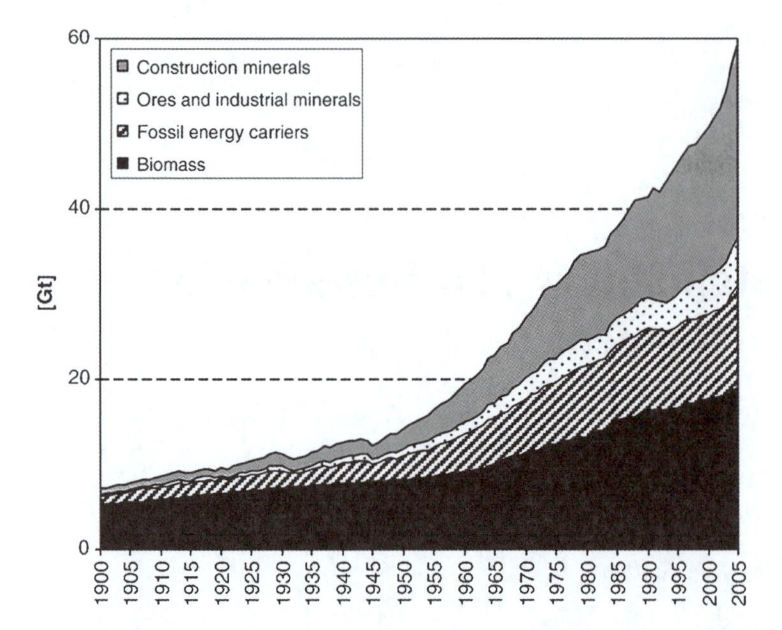

Figure 3.1 Changes with time (1900–2005) in extraction of various classes of material [expressed in gigatons ($1\,Gt = 10^9\,t$) per year]. Reproduced from ref. 1, with permission from Elsevier.

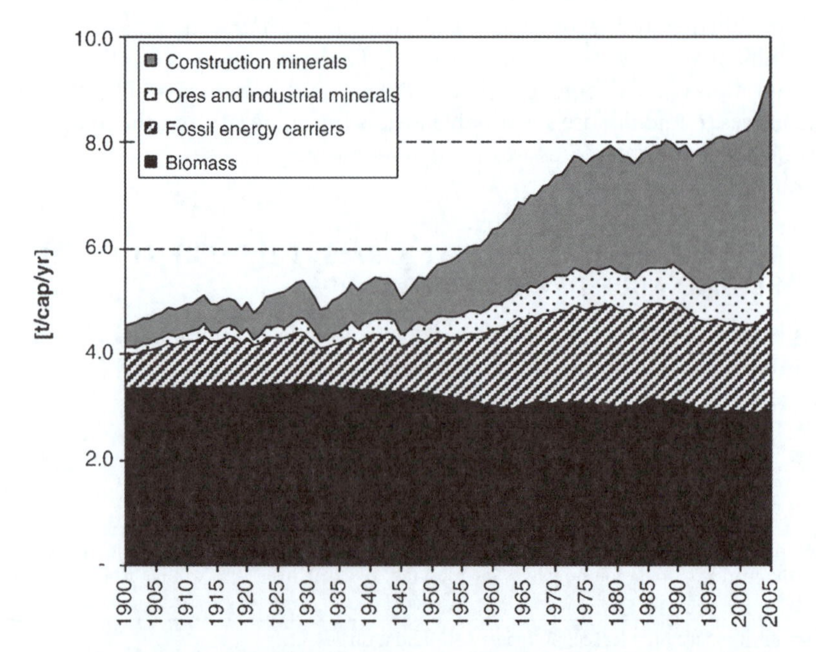

Figure 3.2 Changes with time (1900–2005) in extraction of various classes of material (expressed in tons per person per year). Reproduced from ref. 1, with permission from Elsevier.

Roland Clift[2] gives the following definition of sustainability:

> '*Sustainability is a . . . state of existence, in which humanity's techno-economic skills are deployed, within the long-term ecological constraints imposed by the planet, to provide resources and absorb emissions and to provide the welfare on which human society relies for an acceptable quality of life*'.

Sustainability, the perfect balance between man and nature, is thus best seen as an ideal (like justice) that, intuitively, can readily be agreed upon in the abstract but whose precise definition and the processes of bringing it into realisation cannot.

Sustainable development, on the other hand, is the means by which we move towards this ideal. There have been several hundred definitions, or attempts at definitions, with many books (*e.g.* ref. 3) having been written on the various interpretations designed to elicit a robust, objective and universally accepted definition (without success). The most famous, and most oft-quoted, definition is from the Brundtland Commission,[4] which in 1987 said:

> ' . . . *Sustainable development meets the needs of the present without compromising the ability of future generations to meet their own needs.*'

In 1998, Angela Merkel wrote in the international journal, *Science*, that sustainable development means:

> ' . . . *using resources no faster than they can regenerate themselves and releases pollutants to no greater extent than natural resources can assimilate them.*'[5]

Pictorial representations (such as the widely reproduced Figure 3.3) are used to convey the interactions between society, the economy and the environment, and some of the preferred characteristics of sustainable development—particularly that any outcome should be bearable, equitable and viable.

As definitions, those of Brundtland and Merkel are imprecise and unclear. Even as descriptions, they are uninformative and hardly provide robust guidance for decision-makers and others responsible for making society and technologies more sustainable. Indeed, one observer has described the management rules for sustainable development as '*empirically-empty categorical imperatives*'.[6] Definitions also use terms that are heavily 'value-laden', *i.e.* how one interprets what they might mean depends subjectively on one's attitudes and beliefs.[ii] Some would say that there is an internal contradiction, even a

[ii] In the characterisation of the required outcomes for sustainable development in Section 2.2 I used 'acceptable' instead of 'bearable' and 'practical' instead of 'viable'. Think about the differences these changes might convey and reflect on which adjective you prefer. Then, ask yourself: 'Why'?

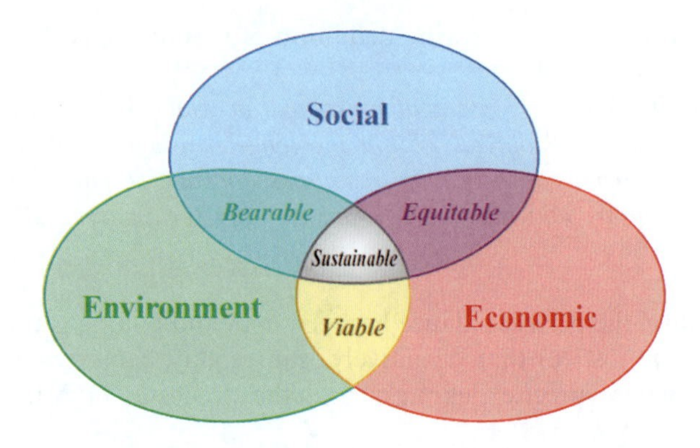

Figure 3.3 Representation by Johann Dréo of the characteristics of the ideal social, environmental and economic changes necessary to bring about an optimal transition to sustainable development.

paradox, at the heart of sustainable development in that it is impossible simultaneously both to protect the environment and bring about development.

Sustainable development is, undoubtedly, the ultimate 'wicked' problem. Giddens, in his recent book,[7] urges that more time be spent on solving problems that are more manageable and less 'wicked' (while still being relevant) rather than seeking a global over-arching consensus that can never be achieved. The view I adopt in this book is that it is important (indeed necessary in the light of current evidence) to make a start on the transition to more sustainable technology. Indeed, some changes (ref. 8, discussed further in Chapter 12) do not necessarily need a significant emphasis on research and could be implemented relatively quickly, once agreed.

There are many associated terms widely used in discussions on sustainable development that, likewise, are held to mean different things by different people. A selection of the most relevant to considerations of sustainable technology is given below and is taken from ref. 9, where more detailed definitions are given:

- cleaner production
- degradation
- environmental accounting
- eco-design
- environmental engineering
- ethical investment
- environmental legalisation
- environmental management strategy
- environmental technology
- ecoefficiency
- factor *X*
- green chemistry

- health and safety
- industrial ecology
- integrated pollution prevention and control
- life-cycle assessment
- mutualism
- minimisation of resource usage
- pollution control
- 'polluter pays' principle
- product service system
- pollution prevention
- responsible care
- recycling
- remanufacturing
- regeneration
- repair
- reuse
- recovery
- renewable resources
- sustainable consumption
- supply chain management
- sustainable development
- sustainable production
- source reduction
- social responsibility
- waste minimisation
- zero waste.

To which I add the following:

- back-casting
- ecobalances
- ecological modernisation
- ecothermodynamics
- environmental justice
- intergenerational equity
- precautionary principle
- triple bottom line.

When coming across these (and related) terms, it is important always to seek to identify the precise meaning that is intended or assumed when they are used. During the course of this book, I focus in some detail on the following:

- carrying capacity (Section 3.2)
- environmental chemistry (Chapter 5)
- dematerialisation and transmaterialisation (Chapter 6)
- Factor 4 and Factor 10, ecological footprints, life-cycle analysis (Section 6.9)

- exergy (Section 6.13)
- green chemistry (Chapter 8)
- industrial ecology (Section 9.12).

3.2 ENVIRONMENTAL BURDEN OR CARRYING CAPACITY

We start by considering the ability of the Earth's ecosystems to deal with the impact of humankind, to provide its needs and assimilate its wastes in ways that are not terminally damaging to those ecosystems. This is a fundamental question[10] which, until quite recently, was largely ignored, as it was thought that the Earth would have essentially an unlimited capacity to 'carry' the human race. So, we need now to ask these questions:

- Is there, in fact, a limit to the Earth's carrying capacity or to the environmental burden that we humans can, indefinitely, place upon it?

If there is a limit (as Malthus suggested 200 years ago), there are then further crucial questions:

- How close are we to it? Are there parts of the world or parts of the natural world that are currently at this critical point? Or, indeed, are we (or they) beyond it?

There is an increasing number of credible and expert scientists who take the view that we are beyond the limit, particularly in terms of greenhouse gas emissions. In 2006, James Lovelock, an independent and widely respected scientist who propounded Gaia theory (Section 5.10) (now associated with the academic disciplines of geophysiology and Earth systems science), produced a polemical text: *The Revenge of Gaia.*[11] The book was launched with an article in *The Independent* of 16 January 2006 headlined: 'The Earth is about to catch a morbid fever that may last as long as 100 000 years'. The paper highlighted this on its front page with the banner headline: 'Green guru says: we are past the point of no return'. It is sufficient to note that this view is at the more pessimistic end of a wide spectrum of views held by expert scientists, more soberly expressed and gathered together in the latest IPCC[iii] report.[12] However, scientific projections (Web 5) prepared for the meeting of governments to discuss the updating of the Kyoto Protocol (Copenhagen, December 2009) appear to reinforce Lovelock's view (one updated more forcefully in his new book[13]). All these judgements are based on informed assessments of the facts, as understood by their proponents. They form a growing consensus that has become more generally held over the last 10 years. As the forecasts cannot be proven (and

[iii] The Intergovernmental Panel on Climate Change (Web 2) is an influential body of scientists and others whose periodic reports on aspects of climate science we will meet in Chapter 5. The IPCC shared the 2007 Nobel Peace Prize with Al Gore *'for their efforts to build up and disseminate greater knowledge about man-made climate change and to lay the foundations for the measures that are needed to counteract such change'* (Web 3).

anyway the methodology used to generate them has been challenged by Lovelock and others), it is possible, but increasingly unlikely, that the consensus view may turn out to be wrong. The question to be answered is what should be done about the consequences of climate change, particularly to define the investment needed (and where and how it should be deployed) that might succeed in minimising humankind's impact on the rest of the natural world. A selection of additional and very readable perspectives is provided in the Bibliography.

The environmental burden or carrying capacity is related to three (non-independent) factors expressed in the so-called 'master' equation,[14] also known as the IPAT equation. Eqn (3.1) should not be seen as a precise mathematical relationship, more a simple (possibly simplistic) way of representing a complex idea.

$$EB = P \times W_P \times B_W \tag{3.1}$$

The environmental burden on the Earth (EB) arises from a combination of the following:

- the number of us on the planet
- the amount of the Earth's resources each of us consumes (and have consumed), directly and indirectly
- the efficiency with which we process those resources (which affects what we throw away and the impact this has on the environment).

P represents the population of humankind and W_P the average wealth per person (or, as some prefer, the amount of economic activity or consumption per person). B_W is a measure of the burden that we expect the earth to accept per unit of such economic activity. These, individually, are complex interacting factors.

We have already seen that P is increasing, though the growth rate may well be declining. P may well stabilise during this century; it is unlikely to decline significantly unless something cataclysmic occurs that we cannot foretell.

Our average prosperity, or economic activity/person, W_P, is also unlikely to decline (notwithstanding the small and temporary slowing the global economy is currently experiencing as I write). The aspirations of those in the developing world to see their prosperity improve are unlikely to be prevented. Nor should we seek to prevent attempts to alleviate inequalities evident elsewhere.

One of the interactions that eqn (3.1) does not take account of is the effect that increasing prosperity tends to have on the birth rate of populations. As economic and associated development occurs in less developed parts of the world, improving average standards of life tend to be associated with a decline in the birth rate—though, of course, life expectancy (see Figure 2.7) also increases, leading to changes in patterns of resource consumption as the proportion of the elderly increases.[15]

If P and W_P both continue to grow, and we believe (or accept)[iv] that we may be close to the limit of the natural world to provide resource for our welfare and to assimilate wastes that we produce, then bringing these into balance will

[iv] See footnote, p. 9.

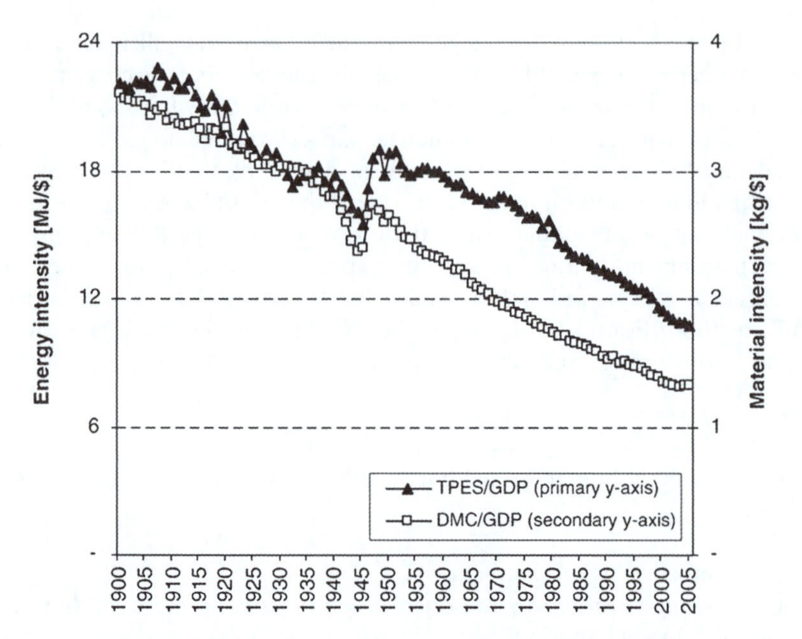

Figure 3.4 Changes with time (1900–2005) in global energy intensity (MJ $\$^{-1}$) and material intensity (kg $\$^{-1}$). Reproduced from ref. 1, with permission from Elsevier.

require an offsetting decrease in B_W, the burden per unit of wealth. This decrease in B_W can be brought about by doing what we currently do, but doing it while consuming less resource, emitting less waste and emitting waste with less impact. Clearly this can only be done through the contributions of technology: in fact, this can be seen as a continuation of an ongoing historical process, as shown in Figures 3.4–3.7.

Figure 3.4 shows the changes occurring between 1900 and 2005 in the energy [in MJ ($1\,MJ = 10^6$ Joules)] consumed to produce one US dollar (of constant value) of gross domestic product (GDP). The value of this parameter ('energy intensity'; Web 6) declined during this period by 0.7% per year, a reflection of improvements in the overall efficiency of global energy generation and use. A similar global improvement can be seen in so-called 'material intensity'. However, such plots tend to mask the overall increases in consumption reflected in Figures 3.1 and 3.2, as well as increases in per capita (or per person) use of energy and materials as shown in Figure 3.5. (We will come back to this in Section 3.3.)

The plots also mask important national and regional variations, as revealed in Figures 3.6 and 3.7. Figure 3.6 shows the relative decline in resources consumed to create a unit of economic value (normalised to US$ of 1980) in different countries around the world. The economic value is represented by the abscissa (x axis) in units of industrial value-added (an overall measure of economic activity or wealth) per person. This is achieved through the use of resources (ordinate or y axis), expressed as carbon intensity, in units of kg of

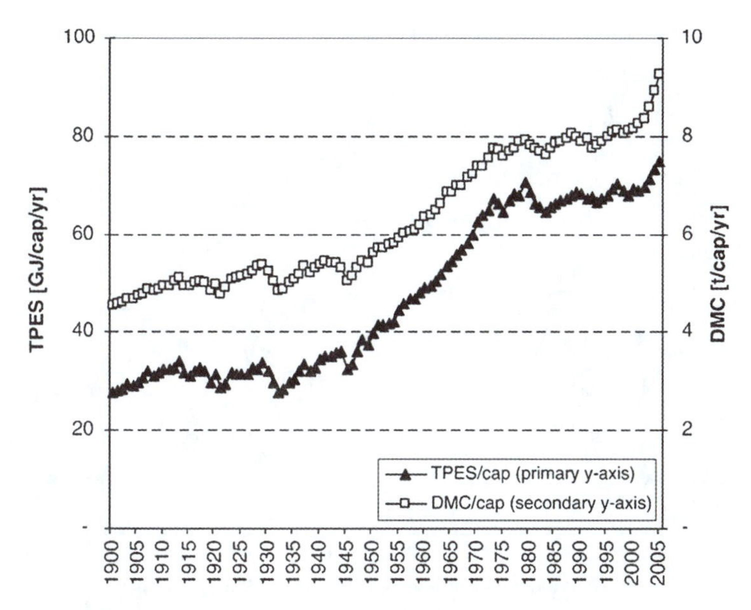

Figure 3.5 Changes with time (1900–2005) in mean individual energy usage (GJ person^{-1} year^{-1}) and mean individual material usage (t person^{-1} year^{-1}). Reproduced from ref. 1, with permission from Elsevier.

carbon consumed per unit of economic value. This introduces a term, 'carbon intensity' (Web 6), which aggregates resource consumption into a single metric. The carbon intensity declines over time as average industrial value-added increases, meaning that between 1958 and 1988, carbon intensity reduced from about 1 to about 0.4 in the US (*ca.* 2.5-fold), with similar declines in the developed world. South Korea, an emerging economy, shows an even steeper decline (Figure 3.6). This confirms that, for a relatively long period, the efficiency with which we transform raw materials into manufactured products has been steadily increasing.

A similar picture of energy usage is seen in Figure 3.7. The relative change in energy usage [expressed as energy intensity (Web 6) in units of MJ used to create a unit of economic value (normalised to US$ of 1980)] is plotted as a function of time. During the period shown, energy intensity decreased for the developed economies but, surprisingly, increased for some other countries. According to the plots in Figure 3.7, in developed countries, the energy used to create a unit of wealth decreased by a factor of *ca.* 2–2.5 over a period of about 30 years. Globally, the plots from ref. 1 suggest this decrease took about 100 years, reflecting the regional variations in technological development, highlighting the benefits to improvements in the efficiency of resource consumption that might arise from purposeful technology transfer.

Will it be sufficient simply to extrapolate such a trend into the future on a 'business-as-usual' basis, or will it be necessary to improve the trends of the

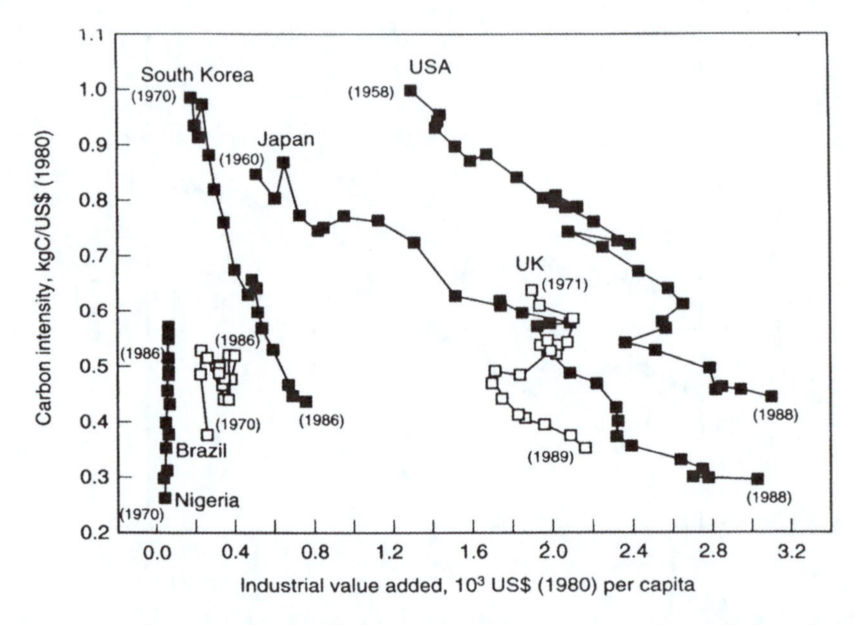

Figure 3.6 Plot for a series of national economies showing the changes in carbon intensity [expressed as kg of carbon consumed per unit of wealth created (in US$ of 1980)] as a function of industrial development (expressed in value added in US$ of 1980 per person).[16a,16b] Reproduced with permission from the International Institute for Applied Systems Analysis (IIASA) and from Cambridge University Press.

badly performing economies to those of the well-performing economies? If so, by how much? Would even this be sufficient?

Before we move on to these questions, there is a caveat that should be entered concerning an unintended consequence of improvements in production efficiency—the so-called 'rebound' effect.[v] If efficiency reduces costs, a benefit (to the environment) only accrues if the money saved is not spent on something with greater environmental impact. While it is possible to see financial capital circulating indefinitely to bring such change about, a constraint on economic activity is provided by the limits on the availability of natural capital (itself determined ultimately by the first and second laws of thermodynamics). In addition, if we reduce waste we are then increasing the efficiency with which we transform material. Such increased efficiency should lead to reduced costs and, possibly, to reduced prices. Price reductions, in turn, are likely to increase demand, resulting in increased turnover. A consequence of increased turnover is an increase in resource use and the possibility of an increase in the *absolute* levels of waste produced.

[v] This effect has long been known; indeed, it was discussed by W. S. Jevons in his book, *The Coal Question*, first published in 1865 (Web 7) to counter the idea that technological progress would lead to a reduction in future coal consumption. It has become known as the Jevons paradox.

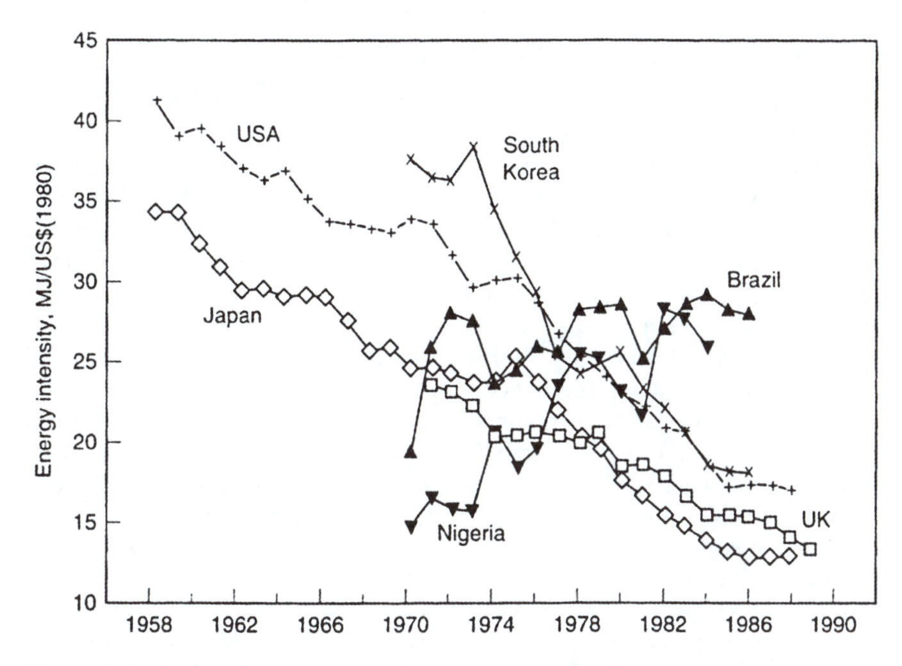

Figure 3.7 Plot for a series of national economies showing the changes in energy intensity [expressed as MJ of energy consumed per unit of wealth created (in US$ of 1980)] as a function of time (1958–1988).[16b] Reproduced with permission from the International Institute for Applied Systems Analysis (IIASA).

So efficiency, taken on its own, may not be the complete answer to balancing the environmental burden. This can only be offset by a sufficient increase in efficiency to take account of such likely increases in demand. This means that B_W must decline at a rate significantly greater than the historic trend combined with a focus on the minimisation of its impact. The difficulty is, of course, knowing how much greater the future trend must be. Decreases by factors of at least four and as much as 10 have been proposed.[17] Bringing about such change represents a major challenge to our economic and industrial systems, and an even greater one to our political and social systems.

Success in controlling the environmental burden cannot arise from a single approach. There is no doubt that the introduction of new technology could occur at a faster rate, though processes of industrialisation of a truly innovative technology take decades[18] (dealt with in Chapter 9). The two general strategies, key to sustainable development, can be summarised by the terms '**dematerialisation**' (*i.e.* 'do more with less') and '**transmaterialisation**' (or 'change to using something better' such as the use of renewables, solar energy, *etc.*). Both are addressed further in Chapters 9–11. As we shall see, chemistry and chemical technology will play an important, if not central, role.

3.3 FOOTPRINTS: ECOLOGICAL, CARBON AND WATER

Material and energy intensities relate consumption to a unit of cost and provide a measure of economic efficiency. Other metrics have attempted to represent disparate and individually complex measures of environmental performance and sustainable development in composite indicators such as the World Economic Forum's Environmental Sustainability Index (ESI),[19] for benchmarking, performance monitoring and policy development. Surveys suggest that none of these measures is universally accepted both because of confusion in, and lack of agreement about, terminology and methodology[20a] combined with subjective and inconsistent information[20b] and difficulties in aggregating environmental indicators together.[20c] The metrics considered so far are difficult to relate to any upper limit of material or energy availability. Indeed, past indications of such limits have simply motivated searches for new sources which, in large measure, subsequently appeared. Whether or not new sources can continue to be discovered (or new ways of processing them) in the future has given rise to vigorous argument.

A quite different measure of our impact on the planet (individually or collectively), originally developed by Wackernagel and Rees,[21] estimates our use of the products and services (such as providing a sink for our emissions and wastes) of ecosystems. This is the 'ecological footprint' (Web 9 provides a footprint atlas where data for individual countries can be obtained). The footprint identifies the area needed to sustain a population through the growing of crops, grazing of animals, forestry, fishing, the generation of energy and that taken up by buildings. This has the particular advantage that there is a more readily definable limit to the total available area to support our needs.

The footprint concept is used to manage resources [now extended to focus explicitly on carbon (ref. 22, Web 10) and water usage[23]] by measuring (or estimating approximately) how much land area and water a human population (or even an individual) requires to produce the resources consumed and to absorb wastes using current technology. These measures reveal the much larger land area or volume of water needed to sustain individuals in the developed world compared with that for someone in the developing world. For instance, the global water footprint (estimated for the period 1997–2001) was 7450 $Gm^3\ year^{-1}$ with an average of 1240 $m^3\ person^{-1}\ year^{-1}$.[23] Values ranged from $>2300\ m^3\ person^{-1}\ year^{-1}$ for Italy, Spain, Greece and the USA to *ca.* 700 $m^3\ person^{-1}\ year^{-1}$ for China.

Some everyday products assigned an individual 'virtual' water content are listed in Table 3.1. The largest water footprints tend to be associated with agricultural production. Such considerations and differences in water use in delivering energy by different means now influence discussions of the relative merits of biofuels.[24] Indeed, it is not simply (!) a question of how we get fuel from biomass. We must now also ask: where do we get the water to grow the biomass? And, we need (quickly) to develop crops that can grow with less water with no loss of productivity.

Table 3.1 Estimates of the average quantity of water required to produce everyday products.[23]

Product	*Unit*	*Average virtual water content (L)*
1 A4 sheet of paper	80 g m^{-2}	10
1 tomato	70 g	13
1 potato	100 g	25
1 slice of bread	30 g	40
1 cup of coffee	125 ml	140
1 bag of crisps	200 g	185
1 cotton T-shirt	250 g	2000
1 hamburger	150 g	2400
Leather shoes	1 pair	8000

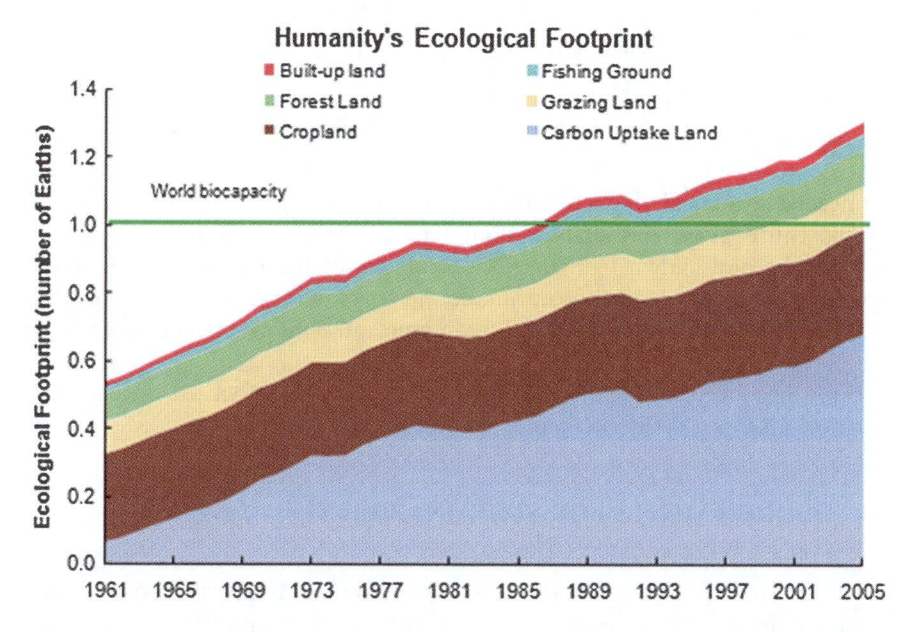

Figure 3.8 Changes in ecological footprint as a function of time (1961–2005) related to an estimate of world biocapacity. Reproduced courtesy of the Global Footprint Network (Web 9).

Some suggest (see Figure 3.8, Web 9) that humanity's ecological footprint overshot the Earth's capacity by 23% in 2002. Put dramatically, we currently need more than 1.2 Earths to sustain ourselves. This might excite the thought: 'Hold on! We still only have one Earth!' How come we can get from the Earth more than one Earth can supply? However, it is probably sterile to argue about the basis of this form of accounting and the accuracy of the estimates; that the number is close to unity and the degree to which the developed world is dependent on the productive capacity of the developing world and what would

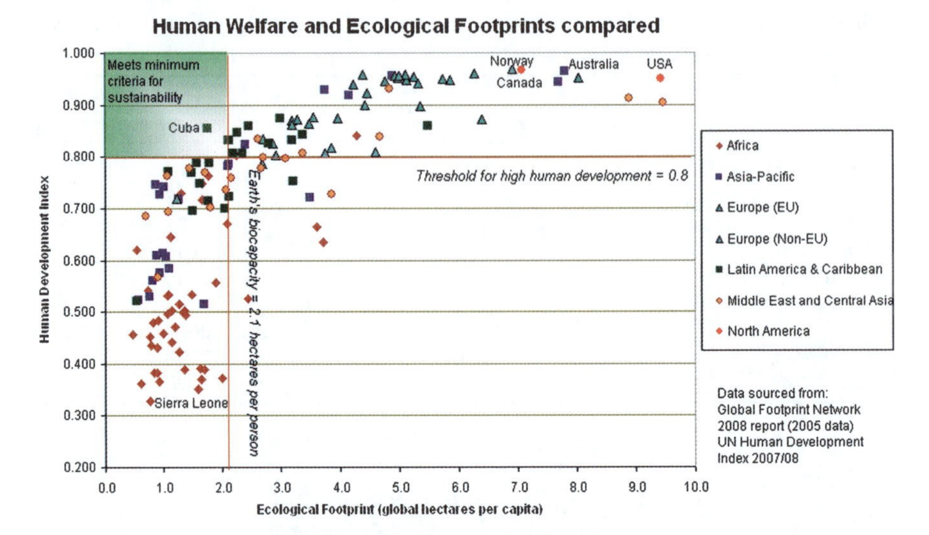

Figure 3.9 Plot of Human Development Index *vs.* ecological footprint. Reproduced courtesy of the Global Footprint Network (Web 11).

be needed to extend the standard of living of those in the developed world to the entire global population should be sufficient to gain our attention. Indeed, the plot shown in Figure 3.9 (Web 11) of the Human Development Index *versus* the ecological footprint, country by country, provides a sobering graphic representation of the social, economic and political challenges both for the 'have' countries and the 'have not' countries.

We will come back again to the question of metrics focused on chemical reactions and processes in Chapter 7 (particularly Section 7.9).

3.4 REQUIREMENTS FOR SUSTAINABILITY

From the forgoing discussion it is possible to set down the general conditions for sustainability. The requirements and constraints that any technology must meet for the generation and supply of food, materials and energy to be sustainable are as follows:[25]

- Resources to produce the goods the human population needs should not run out. This may be achieved by minimising use through greater efficiency, recycling, dematerialisation, transmaterialisation and increased use of renewable resources.
- Emissions must not endanger the Earth's ecological systems. This may be achieved by reducing emissions, improved process efficiency and extended product life, and abatement of emissions by converting them to harmless wastes or sequestration and isolation (after assessing fully the energy costs).

Approaches to these objectives are the concern of Chapters 9–11.

REFERENCES

1. F. Krausmann, S. Gingrich, N. Eisenmenger, K.-H. Erb, H. Haberl and M. Fischer-Kowalski, *Ecol. Econ.*, 2009, **68**, 2696.
2. R. Clift, *Clean Prod. Proc.*, 2000, **2**, 67.
3. K. Lee, A. Holland and D. McNeill, *Global Sustainable Development in the Twenty-first Century*, Edinburgh University Press, Edinburgh, 2000.
4. World Commission on Environment and Development, *Our Common Future*, Oxford University Press, Oxford, 1987.
5. A. Merkel, *Science*, 1998, **281**, 336.
6. J. Huber, Towards industrial ecology: sustainable development as a concept of ecological modernization, *J. Environ. Policy Plann.*, 2000, **2**, 269.
7. A. Giddens, *The Politics of Climate Change*, Polity Press, Cambridge, 2009.
8. S. Pacala and R. Socolow, *Science*, 2004, **305**, 968.
9. P. Glavič and R. Lukman, *J. Cleaner Prod.*, 2007, **15**, 1875.
10. K. Arrow, B. Bolin, R. Costanza, P. Dasgupta, C. Folke, C. S. Holling, B. O. Jansson, S. Levin, K.-G. Mäler, C. Perrings and D. Pimentel, *Science*, 1995, **268**, 520.
11. J. Lovelock, *The Revenge of Gaia*, Allen Lane: Penguin Books, London, 2006.
12. (a) Core Writing Team, R. K. Pachauri and A. Reisinger, *Climate Change 2007: Synthesis Report: Contribution of Working Groups I, II and III to the Fourth Assessment Report of the Intergovernmental Panel on Climate Change*, IPCC, Geneva, Switzerland, 2007; (b) for a briefer overview, read the Summary for Policymakers 'approved in detail at IPCC Plenary XXVII (Valencia, Spain, 12–17 November 2007), represents the formally agreed statement of the IPCC concerning key findings and uncertainties contained in the Working Group contributions to the Fourth Assessment Report', Web 4.
13. J. Lovelock, *The Vanishing Face of Gaia: a Final Warning*, Allen Lane: Penguin Books, London, 2009.
14. P. R. Ehrlich and J. P. Holdren, *Science*, 1971, **171**, 1212.
15. (a) K. Christensen, G. Doblhammer, R. Rau and J. W. Vaupel, *Lancet*, 2009, **374**, 1196; (b) W. Lutz, W. Sanderson and S. Scherbov, *Nature*, 2008, **451**, 716.
16. (a) T. E. Graedel and B. R. Allenby, *Industrial Ecology*, Prentice-Hall, Eaglewood Cliffs, NJ, 1995; (b) A. Grübler, *Industrialization as a Historical Phenomenon*, International Institute for Applied Systems Analysis, Laxenburg, Austria, 1995, Working Paper WP-95-29; (c) A. Grübler, Industrialization as a Historical Phenomenon, in *Industrial Ecology and Global Change*, ed. R. H. Socolow, C. Andrews, F. Berjout and V. Thomas, Cambridge University Press, Cambridge, 1994.
17. K.-H. Robèrt, B. Schmidt-Bleek, J. Aloisi de Larderel, G. Basile, J. L. Jansen, R. Kuehr, P. Price Thomas, M. Suzuki, P. Hawken and M. Wackernagel, *J. Cleaner Prod.*, 2002, **10**, 197.

18. M. Hirooka, *Innovation Dynamism and Economic Growth: A Nonlinear Perspective*, Edward Elgar Publishing Ltd, Cheltenham, 2006 [cited in G. Huppes and M. Ishikawa, *Ecol. Econ.*, 2009, **68**, 1687].
19. D. C. Esty, M. Levy, T. Srebotnjak and A. de Sherbinin, *2005 Environmental Sustainability Index: Benchmarking National Environmental Stewardship*, Yale Center for Environmental Law and Policy, New Haven CT, 2005 (see Web 8).
20. (a) T. M. Parris and R. W. Kates, *Annu. Rev. Environ. Resour.*, 2003, **28**, 559; (b) B. F. Giannetti, S. H. Bonilla, C. C. Solva and C. M. V. b. Almeida, *J. Environ. Manage.*, 2009, **90**, 2448; (c) G. H. Orians and D. Policansky, *Annu. Rev. Environ. Resour.*, 2009, **34**, 375.
21. M. Wackernagel and W. Rees, Ecological Footprints and Appropriated Carrying Capacity, in *Investing in Natural Capital: The Ecological Economics to Sustainability*, ed. A.-M. Jansson, M. Hammer, C. Folke and R. Costanza, Island Press, 1998, pp. 362–390.
22. Parliamentary Office of Science and Technology, *Carbon Footprint of Electricity Generation*, POST, London, 2006, POSTnote 268.
23. A. Y. Hoekstra and A. K. Chapagain, *Water Resour. Manage.*, 2007, **21**, 35.
24. See Table 1 in R. Dominguez-Faus, S. E. Powers, J. G. Burken and P. J. Alvarez, *Environ. Sci. Technol.*, 2009, **43**, 3005.
25. J. Dewulf, H. Van Langenhove, J. Mulder, M. M. D. van den Berg, H. J. van der Kooi and J. de Swaan Arons, *Green Chem.*, 2000, **2**, 108.

BIBLIOGRAPHY[vi]

M. Hulme, *Why We Disagree about Climate Change: Understanding Controversy, Inaction and Opportunity*, Cambridge University Press, Cambridge, 2009.

B. Lomborg, *Cool It: The Skeptical Environmentalist's Guide to Global Warming*, Alfred A Knopf, New York, 2007.

I. Plimer, *Heaven and Earth. Global Warming: the Missing Science*, Quartet Books, London, 2009.

G. Hardin and J. Baden, *Managing the Commons*, W. H. Freeman and Co., San Francisco, 1977.

G. Walker and D. King, *The Hot Topic: What we can do about global warming*, Harcourt, London, 2008.

N. Stern, *A Blueprint for a Safer Planet: How to manage climate change and create a new era of progress and prosperity*, The Bodley Head, London, 2009.

T. Jackson, *Prosperity without Growth: Economics for a Finite Planet*, Earthscan, London, 2009.

J. R. Ehrenfeld, *Sustainability by Design*, Yale University Press, New Haven, CT, 2008.

[vi] These texts supplement those already cited as references..

M. Giampetro and K. Mayumi, *The Biofuels Delusion: The fallacy of large agro-biofuel production*, Earthscan, London, 2009.
W. Visser, *The Top 50 Sustainability Books*, Greenleaf Publishing Ltd., Sheffield, 2009.

WEBLIOGRAPHY

1. www.unhabitat.org/pmss/listItemDetails.aspx?publicationID = 2562
2. www.ipcc.ch
3. http://nobelprize.org/nobel_prizes/peace/laureates/2007/
4. www.ipcc.ch/pdf/assessment-report/ar4/syr/ar4_syr_spm.pdf
5. http://climatecongress.ku.dk
6. (a) http://en.wikipedia.org/wiki/Carbon_intensity
   (b) http://en.wikipedia.org/wiki/Energy_intensity
7. www.econlib.org/library/YPDBooks/Jevons/jvnCQ.html
8. www.yale.edu/esi/ESI2005_Main_Report.pdf
9. www.footprintnetwork.org/images/uploads/Ecological_Footprint_Atlas_2009.pdf
10. www.footprintnetwork.org/en/index.php/GFN/page/carbon_footprint/
11. (a) www.footprintnetwork.org/download.php?id = 506
    (b) http://en.wikipedia.org/wiki/File:Human_welfare_and_ecological_footprint_sustainability.jpg

All the web pages listed in this Webliography were accessed in May 2010.

CHAPTER 4

Science and its Importance

'Men love to wonder, and that is the seed of our science.'

Ralph Waldo Emerson, 1883

'Science begets knowledge; opinion, ignorance.'

Hippocrates

Modern chemistry books tend not to begin with a description of what science is and does. The methods of science are not, in general, explicitly taught; they tend, if at all, to be picked up in an *ad hoc* manner. It is therefore important, in my view, for those who will have to face the challenges of sustainable development to have an understanding of the core ideas of science, how it is currently practised and the ethos that underpins it. This should not only provide an appreciation of its global cultural significance, but should also equip the reader sensibly to defend it when it comes under intellectual attack. In such, and other, circumstances it is wise to follow the advice, quoted by Hulme in the frontispiece to his book on climate change[1] from Haidt's book, *The Happiness Hypothesis*:[2] '*A good place to look for wisdom ... is where you least expect to find it: in the minds of your opponents*'. The benefit to be gained by a reading of both E. O. Wilson's plea for a unity of knowledge built on the foundations of science, *Consilience,*[3] and Wendell Berry's riposte, *Life is a Miracle*,[4] makes Haidt's point perfectly.

4.1 WHAT IS SCIENCE?

It is of the first importance (for scientists at the very least) in considering sustainable development, and the role of chemistry and chemical technology in

Chemistry for Sustainable Technologies: A Foundation
By Neil Winterton

Published by the Royal Society of Chemistry, www.rsc.org

seeking to achieve it, to start with an up-to-date understanding of what science is and to be able to defend its role in public discourse. The text by Ziman combines erudition, cogency and readability in delivering an eminently sensible contemporary picture.[5]

In essence, science is, simply, shared, systematic and formulated knowledge,[i] based on observation and measurement,[ii] which attempts to understand the natural and physical world. The term 'scientist' was first coined in the early 19th century and took over the description of practitioners of the intellectual pursuit previously known as 'natural philosophy'. John Herschel, son of the famous astronomer, William Herschel, and himself an influential 19th century scientist, said in an historically important text:[6a] '*science is the knowledge of many, orderly and methodically digested and arranged, so as to become attainable by one*'. It was possible for Herschel, in his book, to survey all the important areas of scientific enquiry in a manner that would not be possible today because of modern specialisation and compartmentalisation, even within individual disciplines such as chemistry. Nevertheless, the principle of universal availability of scientific information broadly remains (notwithstanding constraints associated with commercial and institutional interests) and the requirements of sustainable development demand an effort to be made to climb out of our individual sub-discipline silos and to view science (and everything else) broadly (as well as deeply).

Science is also a way of abstract reasoning characterised by certain traits. Wolpert's book, *The Unnatural Nature of Science*,[7] does justice to this topic, particularly emphasising the difference between this way of thinking and common sense. Wolpert repeats Galileo's wonderful thought experiment that destroyed (after 1800 years!) Aristotle's view that the rate of fall of a body is proportional to its weight. Galileo argued as follows: if two bodies whose natural speeds of fall are different are combined, then the more rapidly falling one will be retarded by the slower and the slower will be hastened by the swifter. However, the combined body is heavier than either of the individual bodies, so it should fall faster than them both. This cannot be so, thus the original supposition must be incorrect. Aristotle's view was wrong and the rate of fall is not proportional to weight.

Scientists, in their scientific work, seek consciously and routinely to be sceptical, open-minded, heterodox, creative, conceptual, objective and rigorous, and to be bound by reason and rationality. One must immediately say that these traits are not the exclusive province of the scientist nor are they always displayed by all scientists all the time. Science is also a creative process in which intuition, guesswork and chance[iii] play a part. Scientific discovery can, more

[i] The study of what knowledge is is known as 'epistemology'.

[ii] The term 'empirical' is used to describe data or information obtained from experience, observation or experiment.

[iii] While chance plays a significant part in discovery (and the term 'serendipity' is used to describe finding what you are not looking for), scientists have long known that chance favours the prepared mind. Acuity of observation lies at the heart of scientific enquiry—always taking note of, and seeking an explanation for, what may arise unexpectedly.

often than not, come from an intellectual jump and not by logical inference. Indeed, P. B. Medawar[8] suggested that '*the "spontaneity" of an idea signifies nothing more than our unawareness of what preceded its irruption into conscious thought*'.

Scientists are also human, so science is also a social institution though one that is rather loosely organised. It has its own ethos, rules and methods, which make the activities and motivations of scientists an interesting area of study for social scientists. Science not only has a sociology, but also a history and a philosophy. Is the scientist motivated altruistically, by curiosity and a wish to understand phenomena at the fundamental level? Or is he or she motivated by discovery and its possible exploitation, and the wealth or recognition that might result? Some areas of science, therefore, can be closely linked with its application, through technology.

The history of science is a fascinating topic in its own right,[9] and helps us to understand how it has developed and changed.[iv] One can imagine science's earliest origins in the behaviour of the first farmers who, observing the changing length of the day and sensing the changing temperature (and maybe linking these with observations of the stars or the height of the Sun at its maximum), would decide when best to plant crops and to harvest them. Its further development can be traced through the influences of early cultures and civilisations.[10]

Probably the most important development in the way modern science is carried out is associated with the name of Francis Bacon (1561–1626) (Web 1). Bacon is credited with formalising the role of experiment (though Bacon's was not necessarily the modern meaning of the term) in acquiring knowledge (and the principle of reductionism,[v] by which we seek to understand the whole by the study of the individual parts). This marked a radical departure from the approach of Aristotle (384–322 BC) (Web 2) (who believed experiment was an interference with nature, rendering invalid the reliability of any associated observation), which had influenced enquiry for nearly two millennia.

Philosophy of science addresses some very deep questions about the nature of knowledge and questions of truth and belief (and whether and how they are different) and the existence, status and validity of the theoretical entities and concepts that science deals with. There are some very accessible texts (see Bibliography) that address these questions and will enrich readers' understanding of the science with which they are engaged. The most influential of these are the contributions by Karl Popper and Thomas Kuhn. To the former, we owe the clear exposition of the 'hypothetico-deductive' model of scientific reasoning (to which most physical scientists probably subscribe); to the latter,

[iv] The inclusion throughout this book of some older, but still relevant, references and sources should be seen in this light.

[v] Reductionism is like taking a watch apart to see how it works by examining its component parts. Its antithesis, 'holism', by observing the watch complete, can tell us what its function is. It should be self-evident, when looking at Earth systems in the context of sustainable development, that both reductionist and holistic approaches are needed.

we owe the notion that science does not progress *via* a steady process of accumulating knowledge but undergoes periodic revolutions.[vi]

Traditionally, the sociology of science involved the study of how scientists work together. Social scientists sought to understand the nature of the process by which ideas were shared and developed, the methods used to arrive at scientific theories and the means by which they were tested and could be said to be successful or otherwise. This led Robert Merton (1910–2003) to define the key precepts or norms of academic science (CUDOS) as:

- Communalism—science is a social institution, a community following certain rules of behaviour
- Universalism—science is a universal means of enquiry, in which contributions are not to be excluded on grounds of gender, race, *etc.*
- Disinterestedness—science has a common style of presentation of work characterised by honesty and acknowledgement of the contributions of others
- Originality—which motivates new work; grants priority to those that originate new ideas; abhors plagiarism
- Scepticism—scrutiny by peers; reproducibility and verification of results.

While most physical scientists would say, if asked, that they continue to subscribe to these norms, Ziman[5] identified recent changes that affect the character of science (and which may in some ways be responsible for changes in the way it is perceived). He suggests adding to (but not replacing) the Mertonian norms to include trends towards science's collectivisation, its closer links to science policy and technology and to its bureaucratisation.

The focus of sociology of science has shifted in recent years to encompass the actual process of idea construction and, more controversially, whether or not scientific knowledge can be said to be different from other sorts of knowledge and whether or not it can be said to be superior, or 'privileged'. This has not gone uncontested, leading to vigorous exchanges, best summarised in a text edited by Segerstråle.[11] However, I suspect most physical scientists would agree with the physicist, Steve Weinberg, who said:[12] '*physical scientists generally do not speculate why they believe something to be true, but only why it* <u>*is*</u> *true*'.

Those who use the arguments of science in approaching questions of sustainable development now have to have in mind the need sometimes to convince sceptics of the importance of reproducible and testable evidence in providing a foundation for any collective decision-making. In the view of Ziman,[5] scientists should not claim that science has 'transcendental' authority but, better, should argue pragmatically from its historic and contemporary successes as a '*knowledge-generating institution trading publicly in credibility and criticism*'.

[vi] This is associated with the (now much misused) idea of the 'paradigm shift' (a term that he did not, in fact, coin).

Physical scientists accept science's universal applicability without question as well as its universalism (not seeing science as allied with any religious creed or political philosophy). Nevertheless, scientific experts have still felt able to organise themselves into activist groups concerned with social or political questions. The challenge for each scientist involved is to reconcile the ethos of science with the selective argumentation that can go with political activism. I return to this question in Chapter 13.

4.2 THE SCIENTIFIC METHOD

So, what is the scientific method? Much has been written on the ways scientists do science (including two very readable personal accounts by Medawar[13] and Braben[14]), though much of this has probably passed the practicing physical scientist by. Suffice it to say that, while there is no single method, there are some basic elements that are broadly accepted and followed by practitioners of science.

Science essentially involves making purposeful and repeated observations of some natural or physical phenomenon that may be of interest. This can involve the simple act of counting (*e.g.* the annual RSPB bird counting exercise in which large numbers of measurements are aggregated together that allow estimates to be made of bird populations in the UK and how they are changing). Or an experiment can be set up, in which some factors are controlled and other characteristics monitored and measured, with the purpose of seeing and learning from what happens (particularly if an idea or theory may stand or fall by the outcome).

For such an exercise to be 'good' science, someone else must be able later, quite independently, to reproduce the experiment and get the same results. They must be able to **verify** what the experimenter claims to have observed. The observations must be **reproducible**. As experiments have become ever more refined and the phenomena being studied ever more subtle, the means by which the relevant data are extracted from the raw observations and the methods then used to process them and establish their significance (in a statistical sense) can come under challenge. Evaluation of data, therefore, requires an understanding of statistical methods (and an appreciation of the appropriate method to use in particular circumstances).

Data from a single observation such as from one experiment (or from the results of one individual's observations of the number and type of bird seen in a garden during one hour over a certain weekend) are not a great deal of use on their own. However, when aggregated together and particularly if they have been verified, the sum of data can provide information that is the raw material of knowledge. Extensive and reliable knowledge can enable new insights to be gained. The assembly of information to produce new knowledge requires original thinking, and rigorous and objective analysis of the information that makes it up (as well as the other characteristics listed above).

Based upon a series of observations and experiments (sometimes over a lifetime, sometimes fortuitously), it may be possible for a scientist to propose a new theory. For a theory to be considered scientific, it must be couched in terms that enable it to be tested as to its **falsifiability**. The basic idea is that no scientific theory is safe from such tests; even if a set of observations from one study are consistent with the theory, no one can be certain that the theory will survive the next test.[vii] The classic example of such challenges to theories is the displacement of the Newtonian idea of the mechanics of interacting bodies by that based upon quantum mechanics, associated with figures such as Heisenberg, Pauli, Planck, Schrödinger and others. The key point here is that Newton's laws are still a good approximation on the macroscopic scale but they fail at the atomic scale. This is how science has progressed. Some theories turn out to be more provisional than others. Despite their best efforts, for instance, creationists have been unsuccessful in having the theory of so-called 'intelligent design'[viii] accepted as having the same standing as the now well-established theory of evolution by natural selection. To be scientific, a theory must be capable of being tested experimentally as to its validity (or otherwise). This is not the case for intelligent design.

4.3 HYPOTHESES, MODELS, THEORIES AND LAWS

We have begun to use terms that, because of their importance, we should define with care. The word 'hypothesis' comes from the Greek 'to suppose' and is a proposed explanation for an observed phenomenon. A 'theory' is designed to explain a set of empirical observations by making general statements about a deeper underlying reality.[ix] Darwin's theory of evolution by natural selection is a good example. A 'scientific law', on the other hand, is a law of nature based on repeated observations and experiment that has become universally accepted and believed to be universally valid. Such laws would include, for instance, the laws of thermodynamics. Scientists use 'models' (*e.g.* the Bohr model of the atom) to provide a simplified conceptual representation of a complex physical or chemical system. An everyday example of a model might be the map of the London underground. Few would imagine that the tube stations are distributed

[vii] Scientific and other research is, therefore, better seen in terms of uncertainty reduction than in certainty creation.

[viii] The proponents of intelligent design claim that the present structure of certain organisms has arisen through the intervention of a 'designer' and not as a consequence of the processes of Darwinian evolution. The existence of such a supernatural entity may be part of a belief system that many may adhere to, but it is not science. Science has nothing to say about such belief systems one way or the other (though individual scientists, speaking as individuals, may). In an important opinion (Web 3) handed down on 20 December 2005, a United States District Court decided that intelligent design is not a scientific theory in that it is not open to **falsification**, a necessary property of any scientific theory.

[ix] We can immediately see that, when we attempt such definitions, we are forced to use other terms that themselves require explanation. What does 'reality' mean? Physical scientists tend to leave such 'metaphysical' or 'ontological' considerations to others. However, while we consult the dictionary, we do need to recognise that our ideas are not quite as solidly based as we like to think (and as others will certainly tell us).

this way in reality, but the simplified form of the tube system enables us to understand and use it effectively.

Science, therefore, collects information and seeks to understand the natural world by setting up hypotheses and models that allow predictions to be made and general theories to be enunciated. We must always distinguish between observable and verifiable facts and the theories that may seek to interpret or explain them.

4.4 EXCHANGE OF SCIENTIFIC KNOWLEDGE: PEER REVIEW

Scientific information (including the critical analysis of the data that make it up and the basis of the conclusions drawn) to be added to the body of knowledge must be available to others both to scrutinise and to test, as well as to provide the basis of further scientific research. This sharing of knowledge, and its availability to all in the common scientific enterprise, is the cornerstone of science. Indeed, it is not surprising that the development of science followed closely the invention of the printing press and movable type that facilitated communication through wider dissemination of the printed word.

The processes of communicating scientific knowledge are varied (including *via* the internet), but the essential method is *via* publication in scientific journals (whether hardcopy or online; whether open access or not). Proper communication of scientific information, sufficient to meet the criteria of the scientific method (*i.e.* disclosure of sufficient information about how the study was carried out and what was observed for another scientist to be able to repeat it) is never achieved solely *via* a newspaper or other journalistic source. So, when some scientific claim is made in the media [*e.g.* by the BBC about the recent report from the Food Standards Agency about the nutritional benefits or otherwise of 'organically' grown food (Web 4)], it is crucial to access the original publications (Web 5, ref. 15) linked with the news item, and not to rely on some digest of it that appears in a newspaper or arguments about it that appear in the blogosphere (or even *via* Twitter) (as interesting and stimulating as these might be).

The most reliable sources of scientific information are journals or periodicals that contain so-called 'peer-reviewed' publications. Peer review involves several quality control steps. The authors of a piece of work they want to share with the wider world submit their manuscript for publication in a scientific journal. The editors of the journal (after checking the subject matter is appropriate for the journal) then send it, in confidence, to two or three experts in the field covered by the submitted manuscript. These experts (the 'peers' of the authors) examine the material submitted to test whether it is new and original, and whether the experimental method is described in sufficient detail to enable the work to be reproduced. They also examine whether the data presented and the rigour of the authors' analysis are satisfactory, and whether their conclusions are reasonable in the light of the results presented and of what others may have done before. This process of peer review prevents many of the more obvious errors from being published. Nevertheless, it is by no means foolproof. However, the scrutiny by other scientists that follows publication can ferret out further errors, often quite quickly.[x] This further scrutiny is fundamental to the scientific process. It is always important to recognise that material found on the web (other than that from journals whose papers are subject to peer review) will not have been subjected to this scrutiny. Its reliability should not be assumed and it should be approached with deliberate caution.

Peer review also ensures that a check is made to see whether related work done by others has been acknowledged properly. This both places the work in the context of what is already known but, importantly, also delineates what is new in the work being reported. The full acknowledgement of the prior contributions made by others forms part of the basic ethic of science. Indeed, plagiarism (*i.e.* passing off the work of others as your own) is a cardinal sin. It is a form of intellectual theft and is unacceptable under any circumstances. For scientists who plagiarise, it can mean dismissal and the end of their career.

4.5 SCIENCE AND AUTHORITY

The astronomical observations of Galileo (1564–1642) (Web 7) confirmed the Copernican[xi] heliocentric view of the solar system. This ran counter to the teaching of the Catholic Church that had elevated to doctrine the belief that the Earth was the centre of universe ('geocentrism') and that 'heliocentrism' was 'false and contrary to scripture'. Galileo was subject to two inquisitions, the second of which found him to be a heretic. The historical importance of science in challenging views that so fundamentally ran counter to observable fact that people were expected to accept them simply because an established authority (however well-meaning or benign) said so cannot be overstated. It is no accident, therefore, that the Royal Society–founded in 1660 and possibly the oldest scientific society still in existence—should recognise the firmness of purpose[xii] scientists sometimes need to show in seeking out such truths by having as its motto '*nullius in verba*': take nothing on authority (even scientific authorities).

4.6 SCIENCE AND TECHNOLOGY

Science is closely linked to its technological application. Karl Popper is quoted as having said:

[x] How quickly can be seen from the blog (Web 6) that reported data from experiments in real time in response to some remarkable claims[16a] made in a leading chemistry publication, *Journal of the American Chemical Society*, that sodium hydride could act as an oxidant. The paper has recently been retracted.[16b]

[xi] The Polish astronomer Nicolaus Copernicus (1473–1543) has been immortalised by the discoverers of element 112 who have proposed the name 'Copernicium' to replace the systematic name Ununbium (Latin for one-one-two-um). This choice (and the symbol Cn) has been formalised by the International Union of Pure and Applied Chemistry.[17]

[xii] An interesting contemporary insight into where 'authority' currently lies is to be found in the fate of Michael Reiss, the Royal Society's Director of Education. In 2008, Reiss proposed that teachers should, if the matter was raised, take time to explain why creationism has no scientific basis rather than simply dismissing it. Some media reports misrepresented his proposal, claiming he supported the teaching of creationism. Reiss subsequently resigned.

BOX 4.1 NOTABLE HISTORIC SUCCESSES.

Some of the rather abstract ideas covered so far are illustrated in the following very notable historic successes:

Observation, Generalisation and Prediction: The Work of Kepler, Newton, Halley and Others

Johannes Kepler (1571–1630) is known for his empirical laws of planetary motion which provided one of the foundations for the laws of motion and theory of gravitation, generalisations first propounded by Isaac Newton (1643–1727). A good test of a scientific theory or mathematic model is whether or not it can predict the outcome of future experiments, observations or events. Such was the case with the prediction, based on the laws of motion, by Edmund Halley (1656–1742) of the return in 1758 of the only periodic comet visible to the naked eye (which now bears Halley's name). Interestingly, Halley was not precisely correct in his calculations, the return being slowed by 618 days due to the gravitational influence of Jupiter and Saturn as calculated by three French mathematicians—Alexis Clairault, Joseph Lalande and Nicole-Reine Lepaute (the latter a rare example of an 18th century female scientist). Figure 4.1

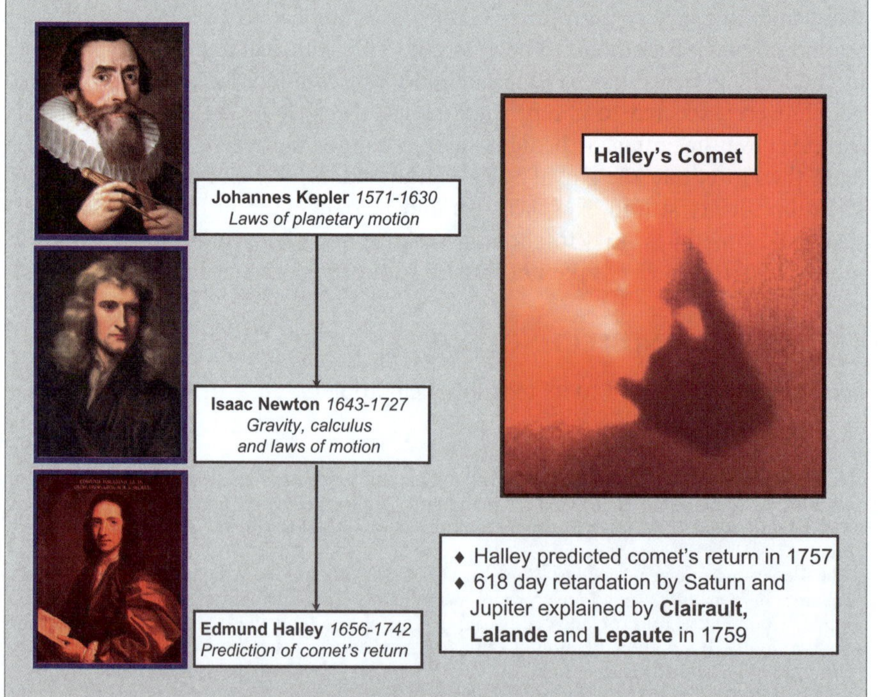

Figure 4.1 Some historically important contributors (and their contributions) to the development of our understanding of astronomy, exemplified by predictions of the close approach of Halley's Comet to Earth.

Systematisation and Theory: The Work of Linnaeus and Darwin

The beginning of any scientific venture is the observation of some phenomenon, possibly using some experimental approach to aid this process. The observation and description of the different types of living things required a logical system of classification that could be used universally (using the then universal language of science, Latin). The Swedish botanist, Carl Linnaeus (1707–1778), developed the Linnaean system that is still used today. Such systematisation and classification allowed similarities and differences between species to be seen, described and shared and was instrumental in leading Charles Darwin (1809–1882) to write his famous book, *On the Origin of*

Figure 4.2 The importance of systematisation as a precursor to the development of soundly based theories, exemplified by the contributions of the widely used method of classification developed by Carl Linnaeus that enabled Charles Darwin (and others) to formulate evolutionary theories.

Species (see Web 8 for an online version), which set down his theory of evolution by natural selection. While the possibility of such a process had occurred to others (indeed, Alfred Russell Wallace, a contemporary of Darwin's produced a similar theory at the same time as Darwin), it is largely as a consequence of Darwin's own extensive observations—and not simply those he made on his voyage on HMS *Beagle* to the Galapagos Islands[18]—that we give the larger part of the credit to Darwin (Figure 4.2).

Periodicity and the Periodic Table: Mendeleev

The value of assembling observations made by others and searching for underlying patterns within those observations (from which more fundamental generalisations might then arise) is exemplified by the Periodic Table. This is now seen as a major milestone in human culture and the development of our understanding of the natural world (see Bibliography). There were many

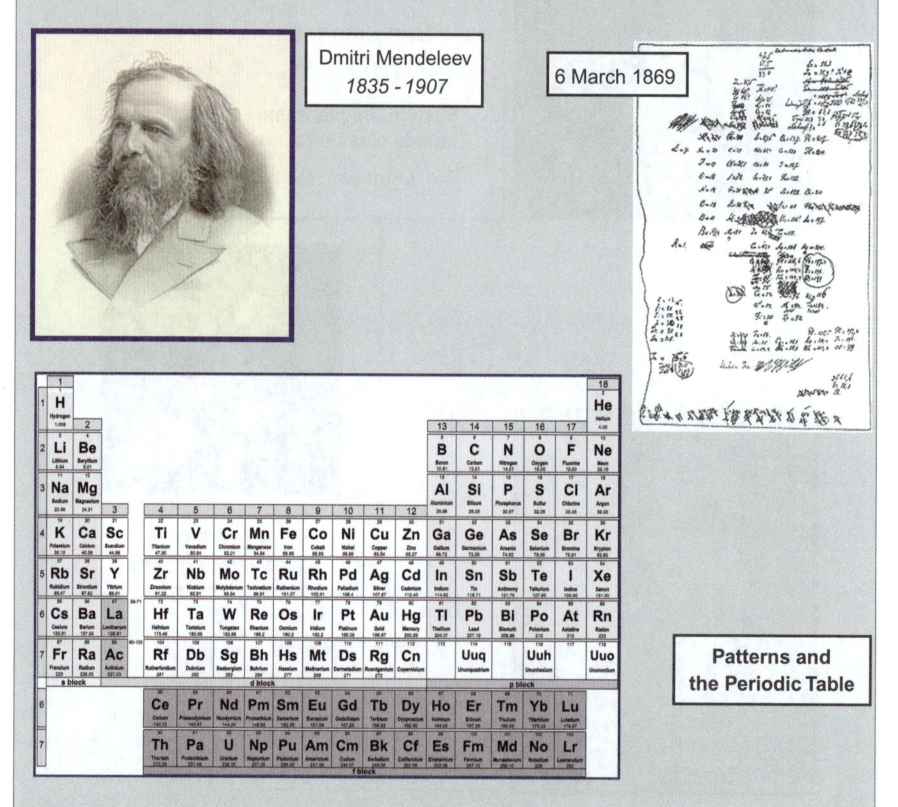

Figure 4.3 The seeking of patterns from one's own observations and those of others can lead to relationships whose further study can reveal the most profound and fundamental characteristics in the nature of matter. The Periodic Table, now associated with Dmitri Mendeleev, has become an icon of the development of human culture and scientific thought.

scientists (most notably Johann Dobereiner, John Newlands and Lothar Meyer) who sought to systematise the disparate observations made about the properties of the chemical elements and their compounds. The most successful turned out to be Dmitri Mendeleev, who made his first documented attempt in March 1869. Not only did this first periodic table reveal patterns that were later to be explained from developments in our understanding of the structure of the atom and its theoretical underpinning, but it also displayed gaps which focussed chemists in their efforts to discover new elements based on well-found predictions of their properties. The successful subsequent discovery of previously unknown elements such as germanium, gallium and scandium provided powerful support for the validity of periodicity and drove later efforts to understand how this periodicity arose (Figure 4.3).

'We advance like a person walking through a swamp, first painfully on one leg and advancing it, and then on the other. One leg is labelled 'technology' and the other 'science'.[19]

Often, in the public mind, science and technology are held to be one and the same and it is true that there are many examples in which a new technology has grown up based on the exploitation of scientific knowledge.[xiii] There are also examples in which a technology has grown to be successful by more trial and error methods, and has itself been the stimulus of scientific enquiry. The interplay between astronomy, navigation and measurement of time has, variously, had political, commercial and purely intellectual motivations. Relevant to the topic covered in Chapter 6 is the fact that the science of thermodynamics grew out of the development of steam power in the late 18th century, stimulated by (and stimulating) the process of industrialisation.

There is a temptation, particularly prevalent among non-scientists, to believe that it will be possible to map out the pathway to sustainable development by identifying scientific problems and setting scientists to work on them, and to have their work linked with some societal and technological objective. While some work by scientists can be done this way and some progress will be made, it is more true to say that discovery, particularly of the most profound sort, cannot be planned and timetabled, nor can the technological benefits of a major scientific advance be foreseen.

[xiii] Sometimes there is a long gap between a scientific study (often carried out without any regard for its possible usefulness) and its technological application. For instance, the observations of Stern and Gerlach on the deflection of the path of streams of silver atoms by a magnetic field were later shown to have arisen from the phenomenon of nuclear spin.[20] This led to the observation of nuclear spin resonance (NMR), subsequently exploited to great effect by chemists. It was more than 50 years after the Stern–Gerlach experiments that the revolutionary non-invasive medical diagnostic technique of magnetic resonance imaging (MRI) became widely available. This was only made possible as a result of the development of injectable non-toxic MRI contrast agents for which chemists, particularly coordination chemists, were largely responsible. When in Chapter 12 we consider dihydrogen as a transportation fuel, we come across another, perhaps surprising, reason why knowing about nuclear spin is rather important.

John Herschel[6b] provides an interesting example of the serendipitous process of fundamental discovery that may arise from seeking to understand an apparently modest technological problem and that may subsequently lead to significant and wholly unanticipated societal benefits. He describes a problem faced by a 19th century soap manufacturer who noted that the residue from his process corroded his copper boiler. A chemist who looked into the problem discovered the hitherto unknown element, iodine. The origin of iodine was traced to sea plants (whose ashes were the principal ingredient of the soap being manufactured) and then to sea water. A medical practitioner, a Dr Coindet of Geneva, learning of these observations and remembering that an unpleasant deformity (goitre) had previously been cured using the ashes from burnt sponge, tried a compound of iodine directly and found that it rapidly effected a complete cure.

A more recent example highlights the unplanned (and unanticipated) outcome of exploratory research in industry that arose from the contributions of Stoddart to the current 'hot' area of molecular machines.[21] His seminal work on host-guest chemistry grew out of piece of speculative product development undertaken during a period at ICI's Corporate Laboratory in the early 1980s when he was asked to find a safener (an additive designed to modify or reduce the toxicity of another component) for the herbicide, diquat.

4.7 GOOD SCIENCE, BAD SCIENCE AND THE MEDIA

Openness to the consideration of opposing ideas and the associated evidence is the hallmark of the true scientist. This characteristic (which need not be restricted to professional scientists nor even demonstrated by all of them!) reflects one of Merton's norms, that scientists should show 'a willingness to change their opinion on the basis of new evidence'. In his famous and historical text, *On the Origin of Species by Means of Natural Selection*, Darwin includes a whole chapter (Chapter 6 entitled 'Difficulties on Theory') identifying and addressing the weaknesses he, and others, perceived in his theory to prompt them to follow up the flaws in his theory or absences in the evidence. If such new material was consistent with the theory, then the theory would become stronger; if it did not support the theory, then the theory would be undermined and would require revision, or in the limit, abandonment. This is 'good' science and the basis of the scientific method.

'Good' science should be distinguished from 'bad' science. This is not simply a means of distinguishing between work that might lead to a Nobel prize and work that may not. Unfortunately, some work is published and presented in ways that claim to be scientific, but which fails to follow the basic precepts of objectivity, rigour and publication covered earlier. Here we are not focussing on the failings of journalists—though their gullibility (associated with wishful thinking or the absence of open-minded scepticism) is often exposed. Rather, we are focussing on stories in the press that are based on research (or, more often, work that purports to be research) that is poor in its execution, weak in its analysis as well as being unavailable for independent scrutiny. Much of this work has the intention (and certainly the result) of misleading the public in ways that can cause injury.

Until very recently, it was very difficult to challenge such stories, largely because few journalists have a scientific training. So, we are indebted to the GP, Ben Goldacre, for his weekly column, 'Bad Science', in *The Guardian*, which exposes errors and lies perpetrated on the public by those claiming to argue from the point of science. For those wishing to understand what scientists are up against, they might want to listen (*via* Ben Goldacre's website, Web 9) to a recent discussion on LBC Radio on the triple vaccine for measles, mumps and rubella (MMR) and autism. His book, also called *Bad Science*,[22] exposes chicanery, arrogance and venality dressed up as science in a way my writing skills and the space available cannot. It is an enjoyable and lively read. A more scholarly work (on MMR) should also be consulted where the evidence is reviewed and assessed.[23] You can find in Ben Goldacre's book good examples of the failure to follow the scientific method, particularly:

- about failing to make results available to others (*e.g.* homoeopathy; see Chapter 4 of *Bad Science*[22])
- of the fabrication and selective use of data
- of the making of unsupportable claims or questionable claims based on poorly controlled experiments and data analysis.

The claimed benefits for homoeopathic treatment are likely to arise either from the operation of the 'placebo' effect (the perceived benefit from a treatment that, in fact, contains no active ingredient whatsoever) or by getting better naturally (known technically as 'regressing to the mean'). What this bad science has done is to stimulate a proper scientific interest in what actually gives rise to the placebo effect. Interesting as these concepts are, they are beyond the scope of this book; so see Chapters 5 and 13 of *Bad Science*.[22]

A delicate and often difficult-to-perceive line does, however, need to be drawn between such pseudo-science on the one hand, and ground-breaking and unconventional thinking on the other, which might lead to new ideas about the natural world and our understanding of it. This is why scientists should always be open to new ideas, while being persistently critical when the methods used do not conform to proper scientific standards and norms or where there is a disinclination to provide details of such methods for the scrutiny of others.

4.8 CARE IN WHAT WE SAY AND HOW WE SAY IT

At the heart of all that has gone before is the use we make of language and data in communicating science and technology. Some concepts are vague and do not lend themselves to precise, or even agreed, definitions. Some ideas are complex and difficult to understand. Furthermore, they may be poorly explained and interpreted, either by the original author, by the reporter or editor discussing them, or by a non-expert seeking to use them. Therefore it is always crucial to track down the original, to establish precisely what was said in it and to distinguish this from what was said about it. While these steps should help in the approach to scientific controversies and to science in general, particularly when seeking to assess the impact of events and ideas on you as a citizen (something I return to in the final chapter), disputation (see, for example, ref. 24) will inevitably remain.

The care we need to exercise in our use of words[xiv] is illustrated by the following example. In his book, *Damned Lies and Statistics*,[25] Joel Best provides two statements that are identical except that two phrases, 'each year' and 'has doubled', appear in a different order: The first reads: The number of children killed by guns each year has doubled since 1950. The second: The number of children killed by guns has doubled each year since 1950. Does this matter? If there is a different meaning for each of these statements, is it big enough to worry about? We can look at this by asking what the values are in the two cases, assuming one child only was killed by a gun in 1950. In the first case, the answer in 2010 would be 2. In the second case, it would be 2^{60} or 1 152 921 504 606 846 976. The difference between these two statements is quite clearly of some significance.

An additional concern, highlighted by J. Boardman,[26] relates to a different use, or misuse, of numbers that also underlines the central importance of going back to the very first source of data (even if later publications that quote this source are by authoritative and well-respected authors in journals that are subject to the peer-review process). This is often a tedious and time-consuming process, though it can become necessary if the numbers take on a particular importance. Boardman describes as '*number laundering*' the process by which data initially reported for a particular and limited set of circumstances are quoted and re-quoted by authors each citing the immediately preceding, but not the original, source. In the end, the data are believed to be relevant to circumstances wholly different from the ones initially intended.

Boardman's example relates to the important societal and economic question of soil erosion. In 1982 Bollinne reported, in his PhD thesis in Belgium, a study which gave a range for the average rate of soil erosion of 12.8–15.6 t ha^{-1} y^{-1} for three plots totalling 0.11 ha. To his credit, Bollinne warned against extrapolating these results to larger areas. *Via* a chain of citation involving five other studies, each citing the preceding report but not Bollinne's original, this number became (in the internationally renowned journal, *Science*[27a]) an average erosion rate for Europe of 17 t ha^{-1} y^{-1}. Europe has a land area of *ca.* 10.36 million km^2 (or 1 036 million ha), an inflation factor of *ca.* 10^{10}! The problem is that the *Science* paper, although criticised shortly after its publication,[27b] has nevertheless become widely cited and used in land management policy development.

4.9 IGNORANCE, UNCERTAINTY AND INDETERMINANCY

The difficulty of reaching a consensus on how to move towards a more sustainable society is not helped by our ignorance about the problems we face. Some concern things we do not know or are uncertain about. This is something

---

[xiv] Sometimes, of course, ambiguity can be deliberate and designed to provoke thought: consider what James Watson (of double helix fame) meant when he entitled his recent book, *Avoid Boring People* (see Bibliography).

that can be rectified by finding out. However, some areas of ignorance are indeterminate (*i.e.* for a variety of reasons are impossible to know with certainty), regardless of how much research is done. Complex interacting physical and biological systems may be deterministic, *i.e.* their behaviour can be described by a mathematical expression, the variables of which arise from what we have learned about the system under study. Unfortunately, the future behaviour of this system, which may follow this precisely expressed deterministic law, is heavily dependent on the definition of the starting state. As we can never know the starting state precisely enough, we are unable to predict accurately the system's behaviour into the indefinite future.

Ehrlich's population projections and, according to James Lovelock, the projections of the impact of climate change made by the Intergovernmental Panel on Climate Change, combine a mixture of weakness in the model and such indeterminacy. The combination of uncertainty and indeterminacy requires a different approach to forecasting to be adopted. Certain types of uncertainty (*e.g.* those that arise from random or systematic errors in experimental or instrumental observations) can be evaluated using standard statistical methods (see Bibliography).

In epidemiological studies, practitioners are faced with the difficulty of establishing whether or not a relationship between an observed health effect and an environmental factor is statistically associated or causally related.[xv] Epidemiologists have therefore developed additional criteria[29] to guide them in forming their expert judgements (as getting them wrong can have serious consequences). These factors include the strength and consistency of the association, its specificity and time dependence, dose–response relationship, plausibility, analogy and experiment.[xvi] Where we have few or no quantitative data, we have to find ways of assessing probability using techniques that may be unfamiliar to the physical scientist such as Bayesian methods (Web 10).

There will also be differences in the priority we each may attach to different aspects of sustainable development. Which should take precedence—economic development or protection of the environment? Should we be more concerned with what may happen tomorrow in preference to events or outcomes next year or in 100 years' time? Should we be concerned only with global effects that concern everyone, or should we address local concerns as well? How do we approach decisions that require us to compare things that, strictly, are impossible to compare (*i.e.* they are 'incommensurable')? For instance, how do

[xv] A cautionary (if amusing) example[28] highlights the danger of finding implausible associations if data analysis is undertaken without first specifying the hypothesis to be tested. Demographic data for the residents of Ontario between the ages of 18 and 100 in the year 2000 were analysed for 223 common reasons for hospitalisation. For just two of these 223 conditions, a statistically significant higher probability of hospitalisation was found for individuals born under one astrological sign compared with those born under all the remaining signs. This statistical association is not evidence for a zodiacal effect.

[xvi] In a famous example, the physician John Snow (1813–1858) on analysing the death rates from a cholera epidemic in London in the 1850s noted that those drawing water from a pump served by one water company suffered 71 deaths per 10 000 houses, 14 times the figure (five deaths per 10 000 houses) of those served by another. He removed the handle from the former water pump and the epidemic subsided.

we assess whether it is wiser to spend limited resources on further eliminating ozone-depleting chemicals from the environment or to protect biodiversity by limiting encroachment into wilderness areas? These are all 'wicked' questions for which there are no simple answers. All that realistically can be done is to provide individuals—scientists certainly, citizens hopefully—with an idea of the basic tools needed to approach the process of identifying, in a timely fashion, the least worst answer.

REFERENCES

1. M. Hulme, *Why We Disagree about Climate Change: Understanding Controversy, Inaction and Opportunity*, Cambridge University Press, Cambridge, 2009.
2. J. Haidt, *The Happiness Hypothesis: Finding Modern Truth in Ancient Wisdom*, Basic Books, London, 2006, p. 242.
3. E. O. Wilson, *Consilience: The Unity of Knowledge*, Alfred A Knopf, New York, 1998.
4. W. Berry, *Life is a Miracle: An Essay Against Modern Superstition*, Counterpoint, Washington DC, 2000.
5. J. Ziman, *Real Science: What it is, and what it means*, Cambridge University Press, Cambridge, 2000.
6. (a) J. F. W. Herschel, *Preliminary Discourse on the Study of Natural Philosophy*, Longmans, 1846, p.18; (b) J. F. W. Herschel, *Preliminary Discourse on the Study of Natural Philosophy*, Longmans, 1846, p. 50.
7. L. Wolpert, *The Unnatural Nature of Science*, Faber and Faber, London, 1992.
8. P. B. Medawar, *The Art of the Soluble*, Methuen & Co., London, 1967, p. 124.
9. J. L. Heilbron, *The Oxford Companion to the History of Modern Science*, Oxford University Press, Oxford, 2003.
10. P. Whitfield, *Landmarks in Western Science: From Prehistory to the Atomic Age*, The British Library, London, 1999.
11. U. Segerstråle, *Beyond the Science Wars: The Missing Discourse about Science and Society*, State University of New York Press, Albany, NY, 2000.
12. S. Weinberg, *Facing Up: Science and its Cultural Adversaries*, Harvard University Press, Cambridge, MA, 2001.
13. P. B. Medawar, *Advice to a Young Scientist*, Basic Books, London, 1979.
14. D. Braben, *To be a Scientist*, Oxford University Press, Oxford, 1994.
15. (a) A. Dangour, A. Aikenhead, A. Hayter, E. Allen, K. Lock and R. Uauy, *Comparison of Putative Health Effects of Organically and Conventionally Produced Foodstuffs: A Systematic Review*, Food Standards Agency, London, 2009, Report prepared under contract number PAU221 (see Web 5); (b) A. Dangour, S. Dodhia, A. Hayter, A. Aikenhead, E. Allen, K. Lock and R. Uauy, *Comparison of Composition (Nutrients and Other Substances) of Organically and Conventionally Produced Foodstuffs: A Systematic Review of the Available Literature*, Food Standards Agency, London, 2009, Report prepared under contract number PAU221 (see Web 5).

16. (a) X. Wang, B. Zhang and D. Z. Wang, *J. Am. Chem. Soc.*, doi: 10.1021/ja904224y, 21 July 2009; (b) Additions and corrections, *J. Am. Chem. Soc.*, 2010, **132**, 890.
17. K. Tatsumi and J. Corish, *Pure Appl. Chem.*, 2010, **82**, 753.
18. S. Jones, *Darwin's Island: the Galapagos in the Garden of England*, Little Brown, London, 2009.
19. H. Bondi, *Nature*, 1991, **358**, 363.
20. B. Friedrich and D. Herschbach, *Phys. Today*, 2003, **56** (December), 53.
21. N. Winterton, *Clean Technol. Environ. Policy*, 2008, **10**, 309.
22. B. Goldacre, *Bad Science*, Harper Collins, London, 2008.
23. K. M. Madsen, A. Hviid, M. Vestergaard, D. Schendel, J. Wohlfahrt, P. Thorsen, J. Olsen and M. A. Melbye, *N. Engl. J. Med.*, 2002, **347**, 1477.
24. H. Ledford, *Nature*, 2008, **455**, 1023.
25. J. Best, *Damned Lies and Statistics: Untangling Numbers from the Media, Politicians and Activists*, University of California Press, Berkeley, 2001.
26. J. Boardman, *J. Soil Water Conserv.*, 1998, **53**, 46.
27. (a) D. Pimentel, C. Harvey, P. Resosudarmo, K. Sinclair, D. Kurz, M. McNair, S. Crist, L. Shpritz, L. Fitton, R. Saffouri and R. Blair, *Science*, 1995, **267**, 1117; (b) P. Crosson, *Science*, 1995, **269**, 461.
28. P. C. Austin, M. M. Mamdani, D. N. Juurlink and J. E. Hux, *J. Clin. Epidemiol.*, 2006, **59**, 964.
29. A. B. Hill, *Proc. R. Soc. Med.*, 1965, **58**, 295.

BIBLIOGRAPHY[xvii]

K. Popper, *The Logic of Scientific Discovery*, Routledge, London 2002 [first published (in German) in 1935, first English edition published in 1959].

T. S. Kuhn, *The Structure of Scientific Revolutions*, University of Chicago Press, Chicago, 2nd edn, 1970.

A. F. Chalmers, *What is this Thing called Science?* Open University Press, Milton Keynes, 3rd edn, 1999.

S. Fuller, *Science*, Open University Press, Milton Keynes, 1997.

G. Holton, *Science and Anti-Science*, Harvard University Press, Cambridge, MA, 1993.

R. G. Newton, *The Truth of Science: Physical Theories and Reality*, Harvard University Press, Cambridge, MA, 1997.

E. R. Scerri, *The Periodic Table: Its Story and Significance*, Oxford University Press, Oxford, 2007.

R. G. Brereton, *Applied Chemometrics for Scientists*, J. Wiley and Sons, Chichester, 2007.

J. D. Watson, *Avoid Boring People: and other lessons from a life in science*, Oxford University Press, Oxford, 2008.

D. Beach, *The Responsible Conduct of Research*, VCH, Weinheim, 1996.

[xvii] These texts supplement those already cited as references.

WEBLIOGRAPHY

1. http://en.wikipedia.org/wiki/Francis_Bacon
2. http://en.wikipedia.org/wiki/Aristotle
3. www.pamd.uscourts.gov/kitzmiller/kitzmiller_342.pdf
4. http://news.bbc.co.uk/1/hi/health/8174482.stm
5. www.food.gov.uk/news/newsarchive/2009/jul/organic
6. http://totallysynthetic.com/blog/?p = 1903
7. http://en.wikipedia.org/wiki/Galileo; http://en.wikipedia.org/wiki/Dialogue_Concerning_the_Two_Chief_World_Systems
8. http://darwin-online.org.uk/content/frameset?itemID = F373&viewtype = text&pageseq = 1
9. www.badscience.net
10. http://en.wikipedia.org/wiki/Bayesian_probability

All the web pages listed in this Webliography were accessed in May 2010.

CHAPTER 5

Chemistry of the Environment

'Never does nature say one thing and wisdom another'

Juvenal

Before moving on to consider chemical approaches to more sustainable technologies we need to take a closer look at those features of the planet relevant to our discussion. 'Green' chemistry (of which more in Chapter 8) represents an approach to chemistry that consciously takes account of its potential for environmental impact (to a greater extent than its proponents believe chemists have in the past) and that seeks to prevent (or at least lessen) this impact though the application of its principles.

We need to begin, therefore, with some basic appreciation of the natural world itself and how humankind interacts with it. Such interactions take many forms, with evidence for them being found in the most remote parts of the planet and not just in areas that have become urbanised and industrialised. Man's impact on the landscape has been profound and extensive. The clearing of much of the forests of Europe in the Middle Ages shows that such impact began well before industrialisation, though its consequences have grown as the Earth's population increased. While we are particularly concerned in this book with the impact of chemicals, the understanding of the broader impact of humanity begins with what the disciplines of environmental science can tell us. For instance, the study of geochemical and biogeochemical cycles can reveal (amongst other things) how the production and flows of nutrients affect the ability of an ecosystem to support life. From an understanding of trophic levels (where in a food chain an organism sits, what it eats and what eats it), we can follow processes of biomagnification (by which an environmental component becomes more concentrated towards organisms at the top of a food chain). As a consequence, we can detect and measure the impact of a pollutant and follow

Chemistry for Sustainable Technologies: A Foundation
By Neil Winterton

Published by the Royal Society of Chemistry, www.rsc.org

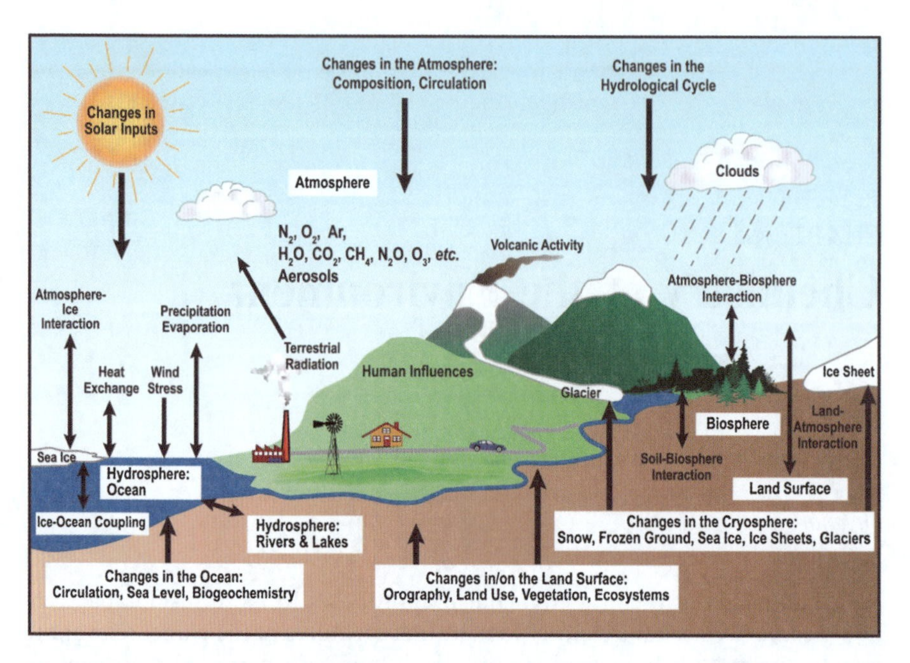

Figure 5.1 Schematic view of the components of the climate system, their processes and interactions. Source: IPCC 4th Assessment Report (2007).[3a]

its migration (as well as gain an appreciation of the inter-connectedness and inter-dependence of all living things).

5.1 ENVIRONMENTAL SCIENCE AND ENVIRONMENTAL CHEMISTRY

Environmental science[1,2] is the interdisciplinary study of the Earth's soil, air and water and associated biosphere (made up of a series of inter-dependent ecosystems) and the effects of human activity (Figure 5.1). Environmental chemistry[4,5] encompasses the study of the chemical processes that occur in nature, their complexity and dynamic character, and the way humankind's activities have, may have or could perturb them. Only from such a perspective will it be possible to establish ways of using ecosystems to support ourselves so as not, seriously and irreversibly, to damage them. Environmental chemistry, therefore, is the study of the sources, reactions, transport, effects and fates of chemical species in soil, air and water, whether from a biotic, geological or anthropogenic origin. Environmental chemistry (along with geology and geochemistry, atmospheric chemistry, chemical oceanography and associated disciplines) is concerned with observing, characterising and understanding these basic processes and the materials involved in them.

The fundamental principles embodied within the discipline of chemistry apply equally to natural processes. So, chemical processes occurring naturally in the environment will follow the principles, laws and relationships familiar to chemists, though of course their scale, scope and complexity will be different. Understanding them is central to the protection of the environment. (It may appear ironic to some that the very discipline that is held responsible for pollution and environmental degradation is also responsible for developing the means to detect, reveal and understand such impact and its origin.) Chemistry enables the characterisation and understanding of the environment's pristine state (or as near to pristine as humankind's activities now permit) and the identification of where and how humankind has impacted on it.[i] Among the more important aspects of chemistry that will aid the observation and understanding of the environment are the following:

- **Chemical analysis, spectroscopy and characterisation:** to detect, isolate and identify compounds, natural or man-made, that are present in an environmental sample, to quantify how much is present and to monitor changes with time and location
- **Phase behaviour of chemical compounds and their adsorption/desorption characteristics:** to understand where an environmental component (whether a pollutant or a compound naturally present, or both) comes from, where it might be going, how it might move about and where it might end up. Such transport phenomena will be dependent on meteorological, atmospheric, geological, oceanographic and hydrological phenomena, highlighting their multi-disciplinarity
- **Kinetics:** to establish how long an environmental component takes to be converted into something else (and the factors that might affect these processes).

Environmental chemistry, therefore, covers all aspects of chemistry essential to an understanding of the natural world. Figure 5.2 illustrates human impact on the environment and displays some of the processes responsible for the dispersion, transport and transformation of emissions from human activity, domestic and industrial, how these may be washed out of the atmosphere by precipitation (locally or after travelling great distances) and deposited on the soil, and how they get into watercourses and ultimately to the sea. Much of the sodium chloride to be found on the surface at the heart of the continental United States, for instance, comes from the entirely natural process of deposition of sea salt aerosol generated from the breaking of waves at the surface of the oceans. Other phenomena will be responsible for the transport of pollutants to remote regions, *via* migration, predation or *via* physical processes of volatilisation or transport following absorption onto aerosols or dust particles. By such mechanisms, persistent pollutants (*e.g.* DDT whose use was banned in the 1960s) have found their way into the fatty tissues of marine animals in remote environments.

[i] This is not always immediately obvious, so it is important to monitor changes believed to provide early warning of possible human impact. Biologists and ecologists use so-called 'biomarkers' or 'sentinel' species to aid this process.

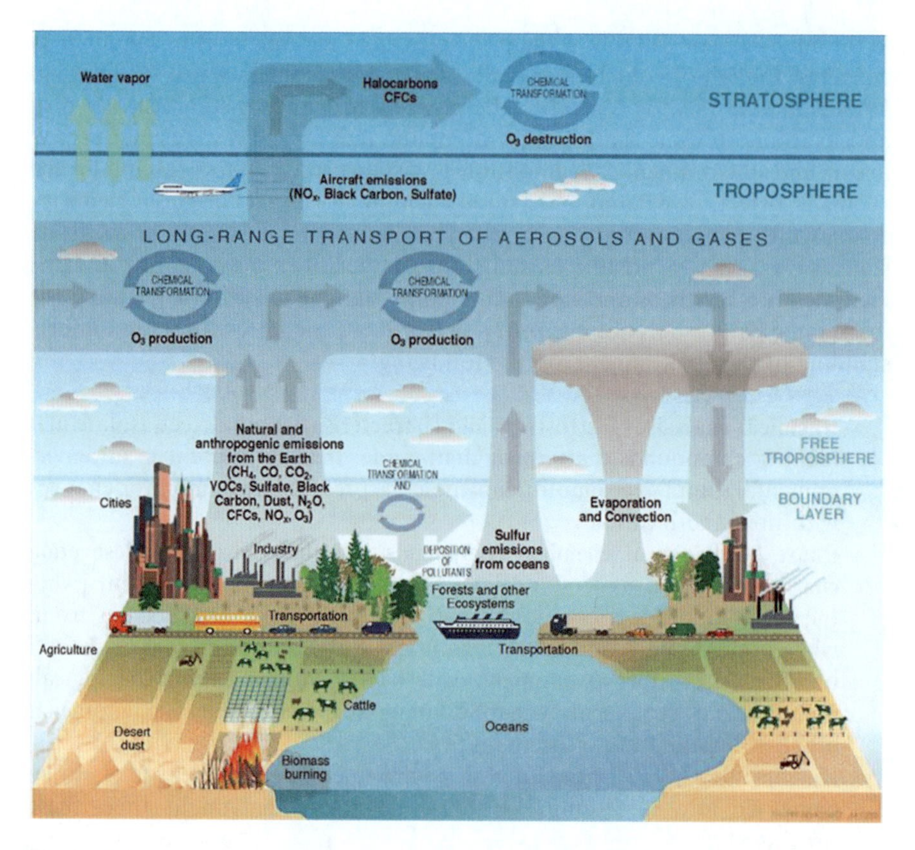

Figure 5.2 Simplified graphic representation of the main global environmental compartments and processes and some of the ways human activity impacts on them. Source: US Climate Change Science Program: illustration prepared by P. Rekacewicz, ref. 6.

Once a material believed to be a pollutant is detected in the environment we can then begin to answer a series of related questions:

- What is it that is present?
- How much of it is there?
- In what form?
- Where does it come from?
- Where does it end up?
- How does it get there?
- What happens to it?
- How quickly?
- What effects does it have?
- How do we stop (or reduce) those that are harmful?
- How do we know when we have?
- Does this meet environmental standards?

- How much is it costing?
- Who is responsible for its presence?
- What are the consequences to those who may be exposed to it?

It is also important to be able to produce a picture of what the unperturbed environment might look like. This is more difficult than it might appear. How long into the past should we go to arrive at what might be a good approximation to the environment without the imprint of humankind? The difficulty of such a question is illustrated in Box 5.1.

5.2 GEOCHEMISTRY

Geology is another historically important scientific discipline that has, at various times, been driven by the search for commercial advantage or the furthering of strategic political interests, in addition to intellectual curiosity. Its pioneers also came into conflict with established religious beliefs, particularly over the age of the Earth and the processes of its formation.

Those that study geochemical and other processes in the environment find it useful to divide the Earth up into compartments (Figure 5.3). Such divisions, while highly oversimplified (indeed, they represent good examples of the scientific models discussed in Chapter 4), aid the study of Earth processes, the chemical aspects of which are the province of geochemistry. It is easy to distinguish:

- the **atmosphere**—made up of gaseous components, condensed water vapour and other aerosols, and particulates
- the **hydrosphere**—the oceans, lakes, rivers, groundwater (and the **cryosphere** representing glaciers and polar regions)
- the **lithosphere**—the soils (the **pedosphere**) and the Earth's crust.

These are the parts of the Earth that also define the biosphere or ecosphere in which the majority of life forms, including humankind, live. We also readily recognise that these spheres act as reservoirs for storing enormous[ii] quantities of material that make up the Earth. In addition these spheres are highly dynamic, with processes occurring within them and exchanges between them, many on the gigantic scale. While most living things occupy a relatively small part of the Earth, represented by its surface layers, nevertheless the development of living things (and their survival) is also linked to what goes on in the core of the Earth and its mantle (through the generation of heat and the emission of gases) and in outer space, from which we receive energy (most importantly, solar energy) and matter (in the form of meteorites) and to which we lose gases that permeate to the outer reaches of the atmosphere. The topic of

[ii] We will meet a number of unfamiliar abbreviations denoting very large numbers. For instance, 1 Gt, one gigatonne, is 10^9 t, or 10^{15} g. A list of important constants, units and abbreviations is given in Appendix 2.

BOX 5.1 THE HISTORY OF COPPER POLLUTION IN GREENLAND

It may be thought that the ice fields of Greenland would be a good place to start to find a pristine environment. However, the impact of man is evident even here, illustrated in some revealing studies by Hong and colleagues[7a] who have investigated the history of copper pollution. They have taken ice core samples from the ice fields of Greenland (these can be many hundreds of metres thick). This ice arises from the accumulation of snow precipitated over long periods and provides a historical record of deposition, with the deepest being the oldest. By sampling such cores and analysing them for trace components, in this case for copper, it is possible to estimate the amount of copper trapped by the snow. It is known that copper smelting began 3000 years ago, though the early processes were exceedingly primitive,[7b] with losses up to 15% of the available copper which were dispersed into the atmosphere, mostly in the northern hemisphere. As far as we know, very little copper processing occurred in Greenland itself.

Table 5.1 Estimates of copper emissions to atmosphere linked with copper production. Based on Hong *et al.*[7a]

Years ago[a]	*Estimated*[b] *copper emission to the atmosphere* ($t\ y^{-1}$)
5000	5
3000	20
2500	300
2000[c]	2300
1250	300
900[d]	2100
500	800
250	1500
150	1500
50	16 000
5	23 000

[a]From *ca.* 1996.
[b]See ref. 2a for assumptions made.
[c]Peak of Roman copper production.
[d]Peak of Chinese copper production (Northern Sung dynasty).

Hong's data in Table 5.1 provide a record of copper pollution covering the last 5000 years and reveal a number of interesting points: copper concentrations greater than those evident before copper smelting began (and what is reasonable to assume are the natural 'geochemical' background levels) appear in the record *ca.* 2500 years ago. Using the copper content measurements, it is possible to make an estimate of annual copper emissions and how these varied with time. Interestingly, peaks may be seen some 2000 and 900 years ago, thought to arise from the rise and fall of Roman and early

Chinese civilisations. Emissions have grown in recent times, though the amounts lost as a proportion of useful product have declined steeply as process efficiency improved as a consequence of technological developments in the means of smelting copper. Surprisingly, Hong *et al.* concluded that the cumulative deposition on Greenland from copper smelted before the Industrial Revolution was greater than that from the period following the Industrial Revolution (up to 1996 when his paper was written), despite the very much greater production in the latter period.

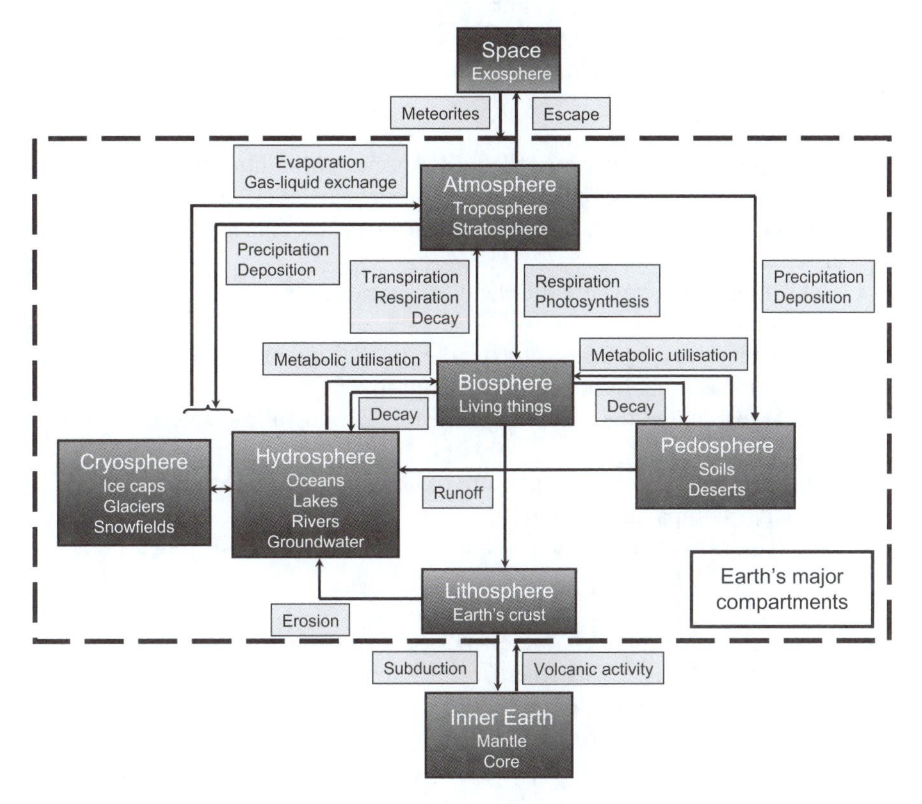

Figure 5.3 Representation of the main global environmental compartments and exchanges and flows between them. Based on Figure 6.1 of ref. 8.

renewables (covered in Chapter 11) and our dependence on the energy from the Sun that the Earth receives and absorbs puts the Sun at the centre of this discussion.

5.3 GLOBAL GEOCHEMICAL CYCLING OF THE ELEMENTS

From the study of environmental chemistry, particularly geochemistry, the more important pathways by which chemical elements and compounds move in the environment have been identified and characterised. These geochemical and biogeochemical '**cycles**', involving the most abundant chemical elements and the most important of their compounds, include those of oxygen (the hydrological cycle, involving water, is sufficiently important to be considered a separate cycle), nitrogen, carbon, phosphorus and sulfur. These analyses allow atmospheric scientists and chemists to track changes (spatially or temporally) of individual environmental components. The Bibliography provides a starting point for those wishing to explore these aspects in more detail.

These dynamic cycles have important characteristics that are used to define them and which highlight the importance of the chemical and physical properties of environmental components. These characteristics include:

- **Carrier species:** the various chemical species containing a particular element involved in a geochemical or biogeochemical cycle
- **Sources:** these specify where the chemical species comes from and how it arises or is emitted
- **Sinks:** these specify where the chemical species ends up and how it is stored or is consumed
- **Budget/inventory/burden:** equivalent terms that quantify how much of each chemical species is found in each compartment
- **Fluxes:** estimate the rate of transfer of chemical species between compartments
- **Residence times:** provide a measure of how long a carrier species stays in a particular compartment; and
- **Lifetimes:** provide a measure of reactivity to estimate how long it takes for a chemical species to react into something else.

An understanding of these characteristics provides some basis for comparison when we attempt to assess whether or not humankind is having an impact on global processes, and if so, to what degree. Being aware of the massive quantities involved in global processes, their complexity, interconnectedness and variability, enables us better to put the effect we may be having, and the difficulty of detecting it, into some sort of context. Assessing any anthropogenic perturbation requires us to have some idea of the characteristics that might define an unperturbed environment, such as the concentration of some important component at the point in the past at which human activity's contribution could be said to be insignificant.

The data on copper emissions discussed in Box 5.1 are a good example. However, such historic data are more usually not available or are questionable (or at least have been questioned). On the other hand, where a material is known to have no natural sources, the baseline concentration in the environment will, of course, be zero. Not all man-made chemicals fall into this category. Indeed, a surprising number of materials we emit to the environment (initially thought to have no natural sources) are now known to be produced naturally, sometimes in surprising quantities. For instance, chloromethane is produced naturally on a scale of 3–5 million tonnes per year (far in excess of industrial emissions) and plays a significant role in natural ozone depletion.

The scale of the inventories and fluxes of materials involved in geochemical cycles are substantial. For example, Figure 5.4 shows the planetary water, or hydrological, cycle in which the key reservoirs, sources and sinks are defined and the flows between them (the fluxes) estimated. The units of the latter are $10^{12}\,L\,d^{-1}$. (As an exercise, work out the amount of energy required to distil the water that is transferred each day from the oceans to the atmosphere. How many 1 GW power stations would be needed to provide this energy?)

5.4 THE CARBON CYCLE I: THE ROLE OF CARBON DIOXIDE

Figure 5.5 [taken from the Intergovernmental Panel on Climate Change (IPCC) 4th Assessment Report (4AR)[9a] and discussed in refs. 1, 2 and 5] shows a corresponding representation of the carbon cycle—both geochemical and biogeochemical including inorganic, biotic (*i.e.* produced in living processes) and xenobiotic (introduced from the technosphere) components—in units of GtC (1 GtC = 10^9 t carbon). Bearing in mind the difference in formula mass between carbon (12) and carbon dioxide (44), it is always important to check whether the estimates are tonnes of C or CO_2.

The cycles for the carbon carriers, carbon dioxide and methane, are explored in more detail below (Section 5.8). For carbon dioxide, the IPCC has produced estimates in its assessment reports (see Table 5.2) for the global sources and sinks of CO_2. The estimate published in 2007 in the 4th Assessment Report[9b] suggests that known sources are greater than known sinks. However, there is some statistical uncertainty in the imbalance, arising from the difficulties in making reliable estimates of the contributions of the component processes. Indeed, there are difficulties in identifying what all these processes might be.

Some key natural processes involving carbon dioxide[8,10] are shown in Figures 5.6 and 5.7. Carbon dioxide is cycled between the atmosphere, vegetation and soil (related processes take place in the oceans involving phytoplankton) through the key biological processes of photosynthesis, respiration and decay. As we cannot carry out photosynthesis ourselves, humankind is wholly dependent on terrestrial plants, algae and phytoplankton for food. In photosynthesis,[12] carbon dioxide is converted in a process driven by energy from the Sun and dependent on the enzyme ribulose bisphosphate carboxylase oxygenase ('rubisco') into carbohydrate—see Figure 5.8 and eqn (5.1). Each mole of carbon dioxide 'fixed' is associated with the consumption of three moles of ATP and two moles of NADPH.

$$n\,CO_2 + n\,H_2O + h\nu \rightarrow (CH_2O)_n + n\,O_2 \qquad (5.1)$$

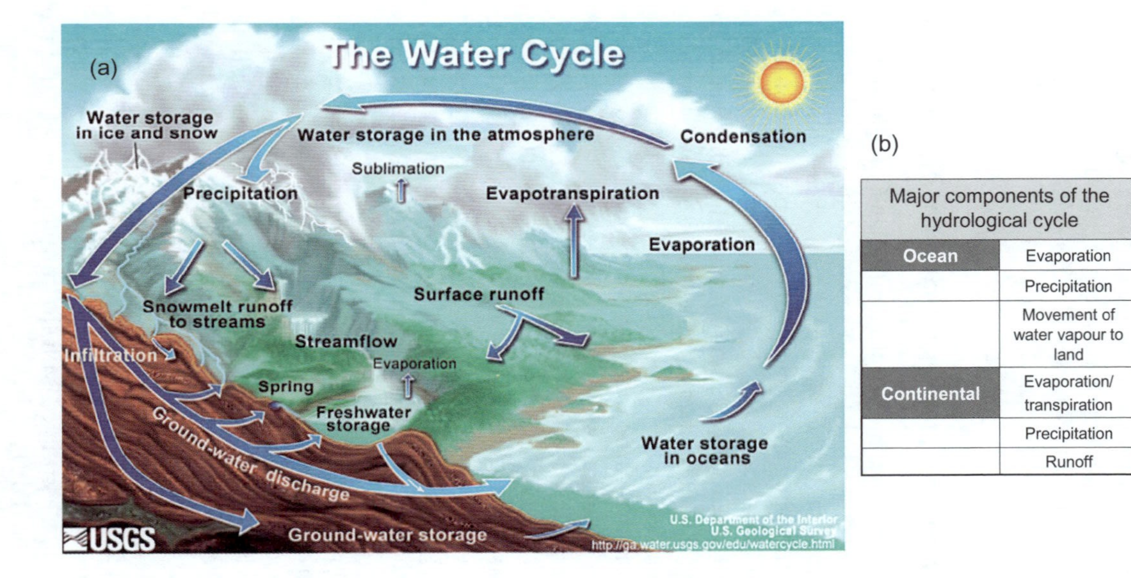

Major components of the hydrological cycle		Water Flux (10^{12}L day^{-1})
Ocean	Evaporation	1164
	Precipitation	1055
	Movement of water vapour to land	110
Continental	Evaporation/ transpiration	195
	Precipitation	304
	Runoff	110

Figure 5.4 Simplified graphic representation of the major reservoirs of the water, or hydrological, cycle and the fluxes between them. (a) Source: US Geological Survey: illustration prepared by John M. Evans (Web 2). (b) Fluxes shown in 10^{12} L d^{-1} (ref. 5).

Table 5.2 Estimates of some of the major global sources and sinks of carbon dioxide in GtC y^{-1} (±1 SD) taken from the most recent IPCC Assessment Reports. Data from ref. 9b.

Period	*1980s*	*1990s*	*1990s*	*2000–2005*
Assessment Report (AR)	Third (Revised)	Third	Fourth	Fourth
Source or sink:				
Atmospheric increase	3.3±0.1	3.2±0.1	3.2±0.1	4.1±0.1
Emissions (fossil fuel + cement)	5.4±0.3	6.4±0.4	6.4±0.4	7.2±0.3
Net ocean-to-atmosphere flux	−1.8±0.8	−1.7±0.5	−2.2±0.4	−2.2±0.5
Net land-to-atmosphere flux	−0.3±0.9	−1.4±0.7	−1.0±0.6	−0.9±0.6
Land-use change flux	(0.4–2.3)		(0.5–2.7)	
Residual terrestrial sink	(−3.4 to 0.2)		(−4.3 to −0.9)	

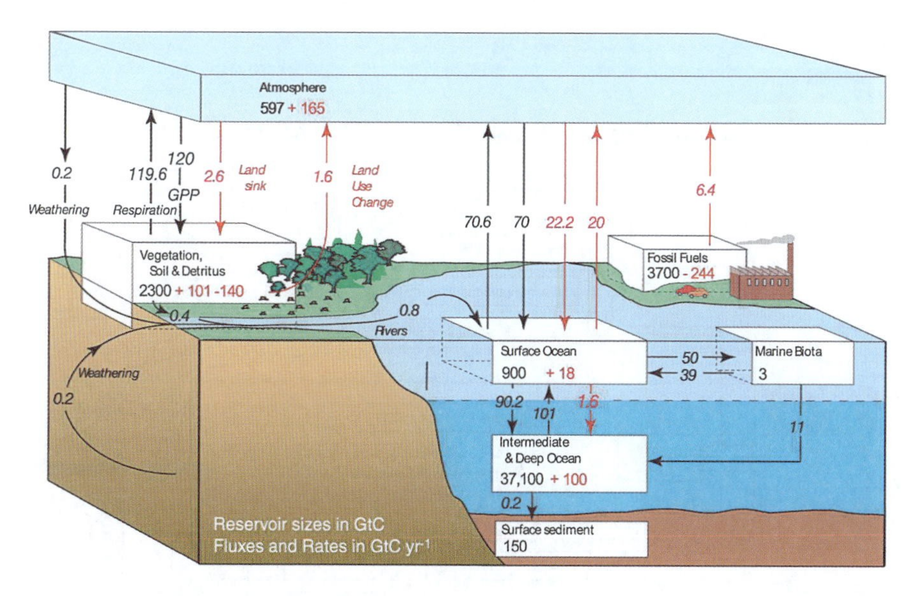

Figure 5.5 Representation of the major reservoirs of the global carbon cycle for the 1990s in GtC and the fluxes between them in GtC y^{-1}: pre-industrial 'natural' fluxes in black and 'anthropogenic' fluxes in red. Gross fluxes generally have uncertainties of more than ±20%. Source: IPCC 4th Assessment Report (2007).[9a] Original caption edited with the permission of the IPCC.

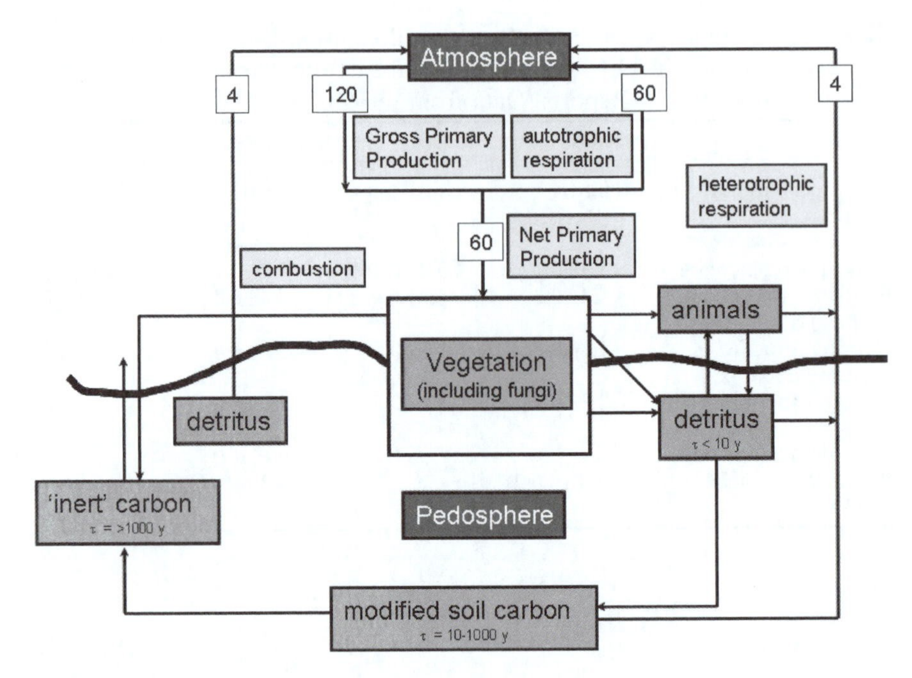

Figure 5.6 Representation of the key compartments, processes and fluxes in GtC y^{-1} for the carbon dioxide cycle on land. Redrawn from ref. 11.

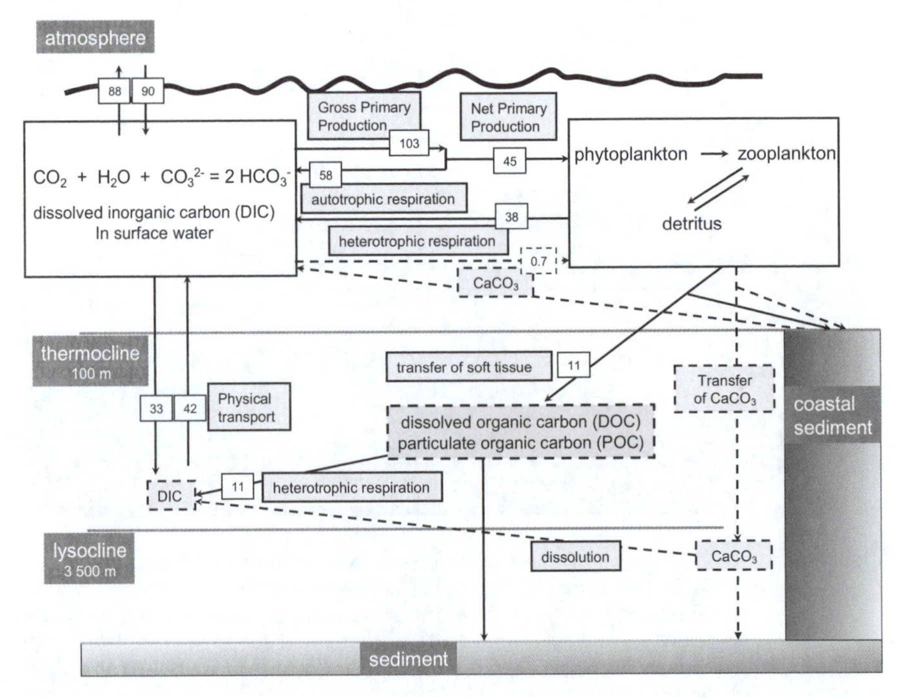

Figure 5.7 Representation of the key compartments, processes and fluxes in GtC y^{-1} for the oceanic carbon dioxide cycle. Redrawn from ref. 11.

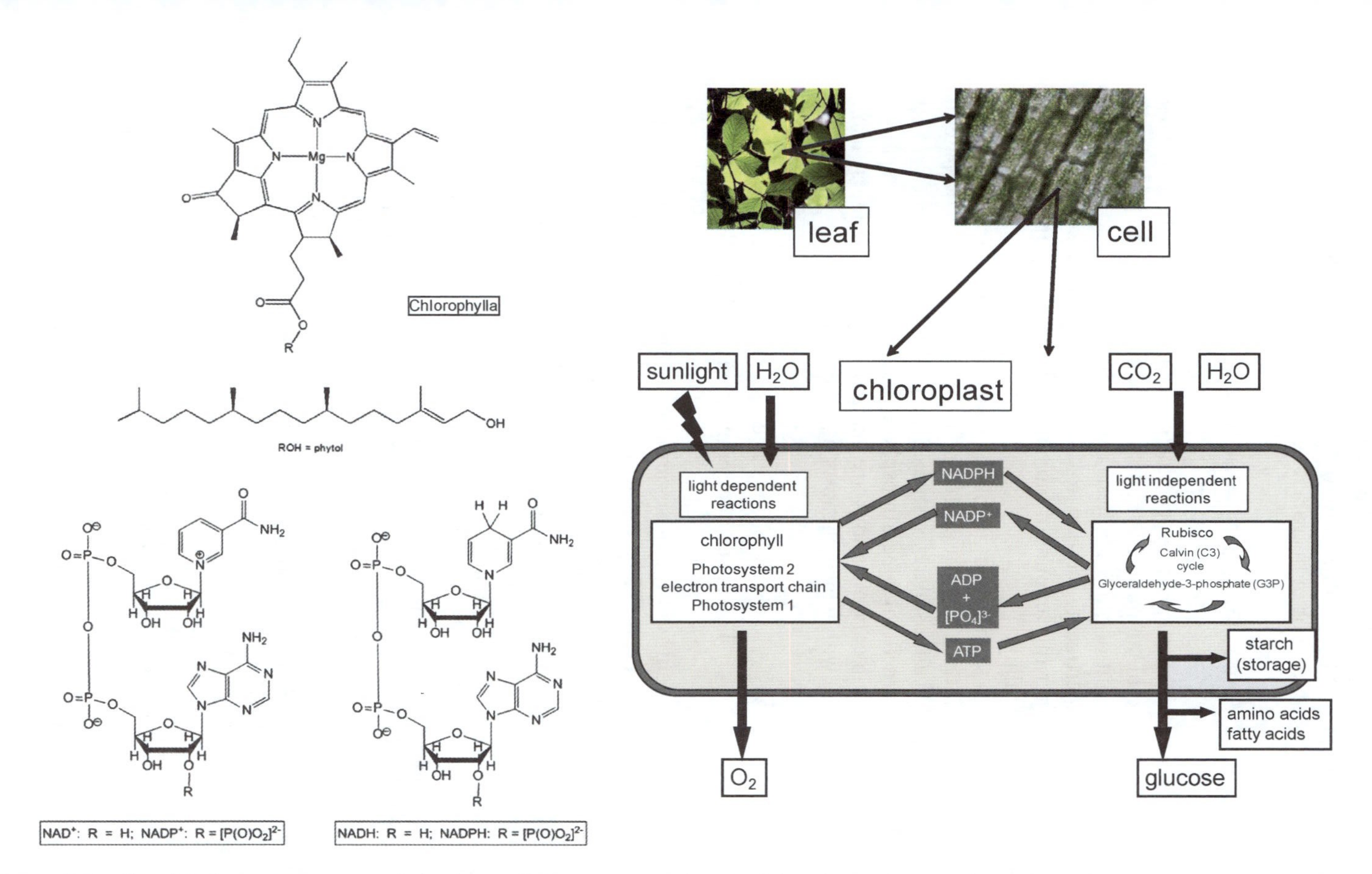

Figure 5.8 Simplified representation of the key cellular processes involved in plant photosynthesis including the chemical structure of chlorophyll a. ADP = adenosine diphosphate; ATP = adenosine triphosphate; G3P = glyceraldehyde-3-phosphate; $NADP^+$ and NADPH = oxidised and reduced forms of nicotinamide adenine dinucleotide phosphate respectively; rubisco = ribulose bisphosphate carboxylase oxygenase.

The carbohydrate generated by photosynthesis along with other natural products is then used by other life-forms using co-produced oxygen in respiration (as in eqn 5.2) or in microbial decay:

$$(CH_2O)_n + n\,O_2 \rightarrow n\,CO_2 + n\,H_2O \tag{5.2}$$

The global terrestrial net primary production (NPP) from photosynthesis (representing the total amount of 'biomass' produced on Earth[13a] taking account of that amount of photosynthesis needed for plants to do their job) is *ca.* 60 $GtC\,y^{-1}$. A pertinent question to ask (and one addressed briefly in Chapter 11) is whether this amount of biomass is sufficient to support a global population of 9–10 billion.[iii]

Geochemical cycles such as that involving carbon can, therefore, be seen to be highly dynamic, interdependent processes involving huge fluxes of material. Their impact on the atmosphere is studied in detail by atmospheric scientists.[8,10] Those concerned with the atmosphere and its components—particularly when atmospheric concentrations are being discussed—use units such as mixing ratio (eqn 5.3) with which chemists may be unfamiliar.

$$\text{Mixing ratio} = \frac{\text{Mass of substance in a given volume}}{\text{Total mass of all components in that volume}} \tag{5.3}$$

The atmospheric concentrations of gases are usually expressed in parts per unit volume as shown in Table 5.3 along with the more conventional chemical equivalent for comparison.

Figure 5.9[iv] illustrates how the concentration of CO_2 has changed over the past 1200 years. The data have been estimated for recent times from direct measurements of the CO_2 content of air and, for earlier times, from the CO_2 content of air bubbles trapped in columns of ice bored out of compressed snow in the polar regions. From 800 to about 1850, the concentration (as represented by the 100-year running mean (*i.e.* the mean value from 100 annual measurements) remained fairly constant at about 280 ppmv. After about 1850, a rapid increase is seen with the current average concentration in 2009 being 387 ppmv.

Table 5.3 Expression of atmospheric concentration of a gas.

Abbreviation	*Description*	*Number basis*	*Chemical equivalent*
ppmv	parts per million by volume	10^{-6}	$\mu mol\ mol^{-1}$
ppbv	parts per billion by volume	10^{-9}	$nmol\ mol^{-1}$
pptv	parts per trillion by volume	10^{-12}	$pmol\ mol^{-1}$

[iii] Human appropriation of net primary production (HANPP) is an important area of study in its own right. A series of articles that might be used as a starting point for finding out more is given in a recent special issue of the journal, *Ecological Economics*. The introductory paper is ref. 13b.

[iv] See Web 1 for the more comprehensive treatment contained in the most recent IPCC report.

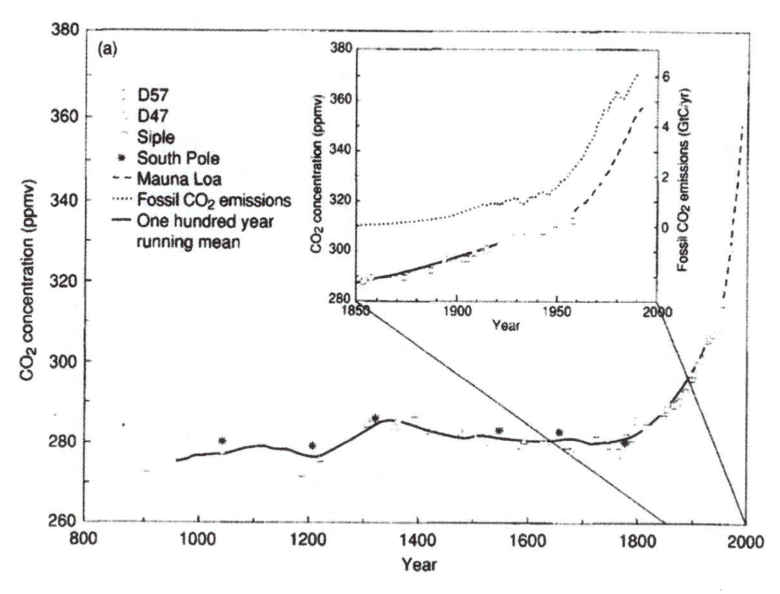

Figure 5.9 Changes in atmospheric carbon dioxide mixing ratio (ppm) for the period 800–2000 AD showing individual measurements and the 100-year running mean. Source: IPCC Second Assessment Report (1996).[14]

Figure 5.10 shows the trend in CO_2 concentrations from the 1950s to the present. This reveals some interesting effects. Figure 5.10a displays monthly data[15d] collected during the period 1960 to 2008 at the Mauna Loa Observatory, Hawaii. (The observatory is *ca.* 3400 m above sea level on the side of a volcano, well above local anthropogenic sources. The data presented have been corrected to take account of local CO_2 degassing from the volcano.) On top of the gradual increase that is clearly evident there is also a smaller oscillation. This arises as a consequence of the seasonal changes in photosynthesis (greater in summer) and vegetative decay (greater in the winter). Figure 5.10b provides similar data[15e] from the southern hemisphere, confirming the general trend, though with smaller seasonal cycles reflecting the smaller terrestrial biosphere.

While the amount of CO_2 in the atmosphere is clearly increasing with time (a growth rate of *ca.* 1 ppmv y^{-1}), it is also important to know whether the rate at which it is increasing is changing. Because of the great variability in these data, they have been 'smoothed' statistically (otherwise trends might be difficult to discern). Evidence that the rate of increase is changing is very difficult to pick out from data that are subject to large natural variations.

Before considering the origins of the increased amounts of atmospheric CO_2, we should look first at some of the natural inorganic cycles involving it. Carbon dioxide in the atmosphere may be taken up by oceans (eqn 5.4–5.6) where it is involved in a series of well-known equilibria:

$$CO_2 + H_2O \rightleftharpoons H_2O.CO_2 \quad (5.4)$$

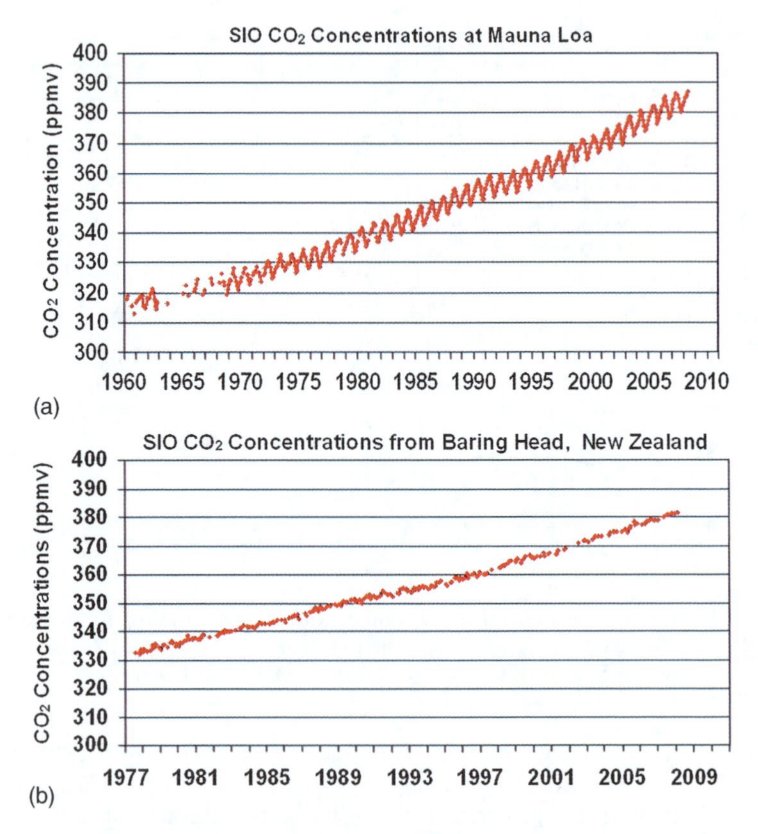

Figure 5.10 Changes in atmospheric carbon dioxide mixing ratio (ppm) (a) in the northern hemisphere between 1960 and 2008 measured at Mauna Loa, Hawaii (source: Carbon Dioxide Information Analysis Center, Oak Ridge National Laboratory,[15d] reproduced with permission) and (b) in the southern hemisphere, between 1977 and 2008 measured at Baring Head, New Zealand (source: Carbon Dioxide Information Analysis Center, Oak Ridge National Laboratory,[15e] reproduced with permission). See also IPCC 4th Assessment Report (2007).[15a]

$$H_2O.CO_2 \rightleftharpoons H^+(aq) + [HOCO_2]^- \; pK_a = 6.1 \tag{5.5}$$

$$[HOCO_2]^- \rightleftharpoons H^+(aq) + [CO_3]^{2-} \; pK_a = 9.2 \tag{5.6}$$

The oceans are slightly alkaline, with an average pH of 8.2, which results from the weathering of basic rocks from the lithosphere—particularly minerals based on Al_2O_3, SiO_2 and $CaCO_3$ (99% of planetary carbon is found in rocks and this is relatively immobile). Dissolution of alumina, for instance, leads to neutralisation of hydronium ions, as formally represented in eqn (5.7):

$$Al_2O_3 + 9\,H_2O + 4\,H^+(aq) \rightleftharpoons 2[Al(H_2O)_5(OH)]^{2+}(aq) \tag{5.7}$$

One consequence of the slight alkalinity of the oceans is that CO_2 dissolves mostly as bicarbonate, $[HOCO_2]^-$, providing a buffering effect on acidity that arises from CO_2 dissolution (eqn 5.8). This effect appears insufficient to prevent a decrease in the average surface ocean pH of *ca.* 0.07 pH units since 1700[16] (see Web 8 for the most recent IPCC summary).

$$H_2O.CO_2 + [CO_3]^{2-} \rightleftharpoons 2\,[HOCO_2]^- \tag{5.8}$$

Because of the huge volumes of water and the timescales of oceanic cycles (*e.g.* those that move ocean water from the depths to the surface or move water from the tropics to the colder high northern or southern latitudes), there is relatively slow mixing of ocean waters, such that its equilibration with atmospheric CO_2 is estimated to take *ca.* 200 years. In addition to these inorganic processes, CO_2 is captured during phytoplankton photosynthesis; 90% of this biomass carbon is returned to the CO_2 cycle by the processes of respiration and decay. The remaining 10% is sequestered to the deep ocean sediment when organisms die. At the sea floor, this sediment can be transferred to the Earth's mantle by a process called 'subduction', providing sources of carbon dioxide for re-emission from volcanoes and geysers and other geochemical sources.

5.5 THE SUN

Solar radiation is the result of nuclear fusion processes in the Sun's core and represents the most important physical influence (after gravity) affecting the Earth, driving most life processes and many physical processes (*e.g.* global atmospheric circulation and the evaporation of water). For completeness, we should note the contribution that other basic physical forces, gravitation and fission, make as sources of energy, tidal and geothermal, respectively. I examine in more detail in Chapter 12 the possibilities for making greater use of all these sources in shifting our dependence away from fossil energy sources.

There are clearly at least two basic aspects of the effect that the Sun's radiation has on the Earth: the first arises from the nature of the Sun's radiation (and that amount of it that is intercepted by the Earth); second is the way in which the Earth is affected by the solar radiation it intercepts. These factors are summarised graphically in Figures 5.11–5.13.

We can calculate the amount of solar energy incident on the Earth assuming that the Sun is a black body emitter and that we know the following fundamental dimensions:

The *total radiation flux* emitted by a black body is given by eqn (5.9), where σ is the Stefan–Boltzmann constant and T the absolute temperature:

$$\text{Flux} = \sigma T^4 \tag{5.9}$$

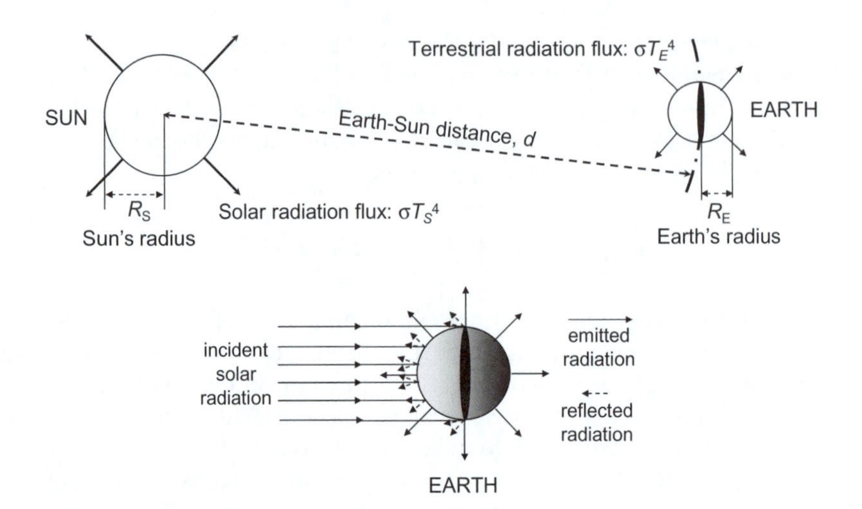

Figure 5.11 Solar irradiation of the Earth. Based on Figure 7.9 of ref. 8.

Total radiation emitted by the Sun per unit time (eqn 5.10) (the temperature of the Sun's surface, T_S, is 5800 K and R_S is the Sun's radius):

$$\text{Radiation} = 4\pi R_S^2 \sigma T_S^4 \tag{5.10}$$

The *solar radiation flux* (eqn 5.11) at Earth's distance from the Sun, d:

$$\begin{aligned}\text{Flux} &= \frac{4\pi R_S^2 \sigma T_S^4}{4\pi \text{d}^2} \\ &= 1370\ \text{W m}^{-2}\end{aligned} \tag{5.11}$$

This value (for a square metre of the Earth's surface facing the sun during daytime) is known as the **solar constant**. (This is not in fact a mathematical constant because, for reasons we discuss below, the energy from the Sun intercepted by the Earth has changed with time, and will continue to do so, on a range of timescales.)

The Earth intercepts this solar radiation flux over a disc of cross-section πR_E^2, where R_E is the Earth's radius. The amount of solar energy received at any one point on the Earth's surface will vary because of the curvature of the Earth and its rotation. The energy received, averaged over the Earth's surface, is 342 W m^{-2} —a value roughly 25% of the solar constant.

Figure 5.12 shows graphically that the Earth's atmosphere interacts with the incoming solar radiation as far out as *ca.* 500 km above its surface, where the most energetic radiation (with wavelengths <100 nm, in the high energy ultraviolet region) will ionise atmospheric species, raising the temperature to >1000 K. The 'thermosphere' extends down to about 85 km—the boundary

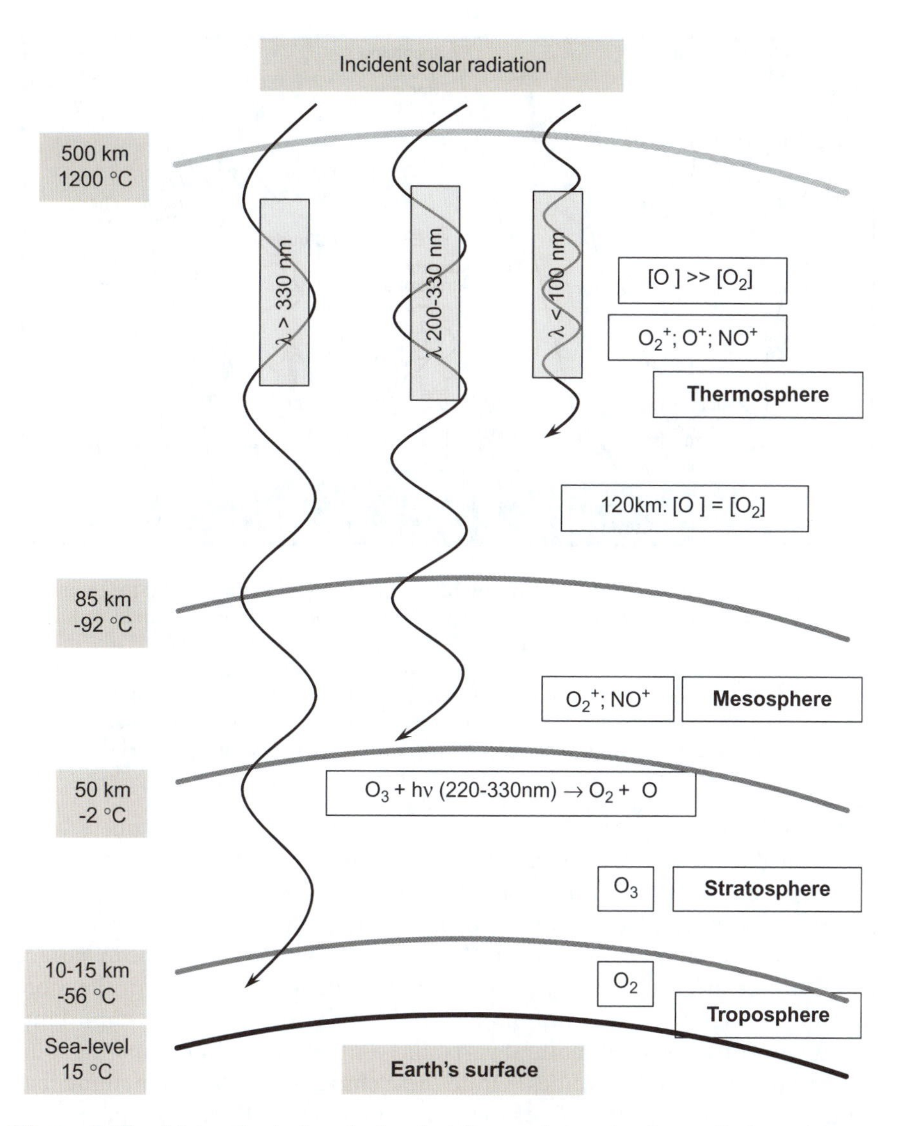

Figure 5.12 Main physical and chemical interactions of solar radiation with the terrestrial atmosphere as a function of distance above the Earth's surface. Also shown are the layers of the atmosphere, their approximate height and the temperatures (°C) at their boundaries. Based on Fig. 9.2, ref. 5.

with the 'mesosphere', which absorbs radiation in the range 200–330 nm. The stratosphere extends from about 50 km above the surface to about 10–15 km, and this is the region in which the concentrations of oxygen atoms (from photolysis of dioxygen) and dioxygen itself are high enough to form ozone, O_3, in a series of photocatalysed processes that prevent most of the high energy ultraviolet (UV) radiation from reaching the Earth's surface. The troposphere is the part of the atmosphere that extends from the boundary with the stratosphere

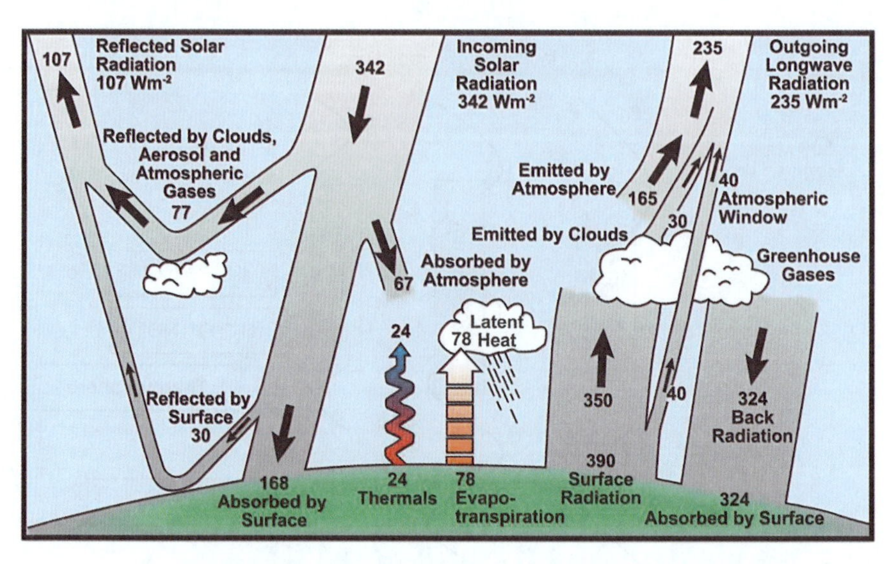

Figure 5.13 A simplified representation of the Earth's annual and global mean energy balance. Over the long term, the amount of incoming solar radiation absorbed by the Earth and atmosphere is balanced by the Earth and atmosphere releasing the same amount of outgoing longwave radiation. Source: IPCC 4th Assessment Report (2007)[3b] and Kiehl and Trenberth[3c] reproduced with the permission of the American Meteorological Society. Original caption edited with the permission of the IPCC.

down to the surface of the Earth and is that part of the atmosphere most controlling of the fate of living systems and most affected by them. Figure 5.13 illustrates how the 342 $W m^{-2}$ of incoming radiation is reflected, scattered, absorbed and re-emitted by various components of the Earth's surface.[3c]

It is now a matter of observation (Figure 5.14) that the global mean surface temperature[v] of Earth increased *ca.* 0.6 K during 20th century, with the more recent data suggesting that warming is accelerating. Why is this so important? Can it be a consequence of natural variability—after all, the Earth and the solar system are very complex entities? Or do these changes represent something that we should be concerned about or that, even, may pose a threat to humankind?

Figure 5.15 shows two graphs: the upper one is a record of CO_2 concentrations in the atmosphere estimated for the much longer period of 400 000

[v] I mention this parameter as it is critically important in the discussion of climate change. However, it is no easy matter to arrive at a precise value of something as apparently simple as the 'global mean temperature'. Because of this, it has proved difficult to delineate changes in it that might be ascribed to an anthropogenic effect (let alone estimate what the value of this parameter might be in 50–100 years' time). These difficulties arise because of the following questions: Is the number of measurements at any one time sufficiently representative of the temperature of the Earth's surface? Have the instruments that collect the data been correctly calibrated and properly maintained? Have individual sites been unduly affected by developments that now make them less representative than they were initially thought to be? Do measurements cover a long enough period for trends to be perceived? Are the methods of data processing and analysis robust? For further information, see refs. 17b–d.

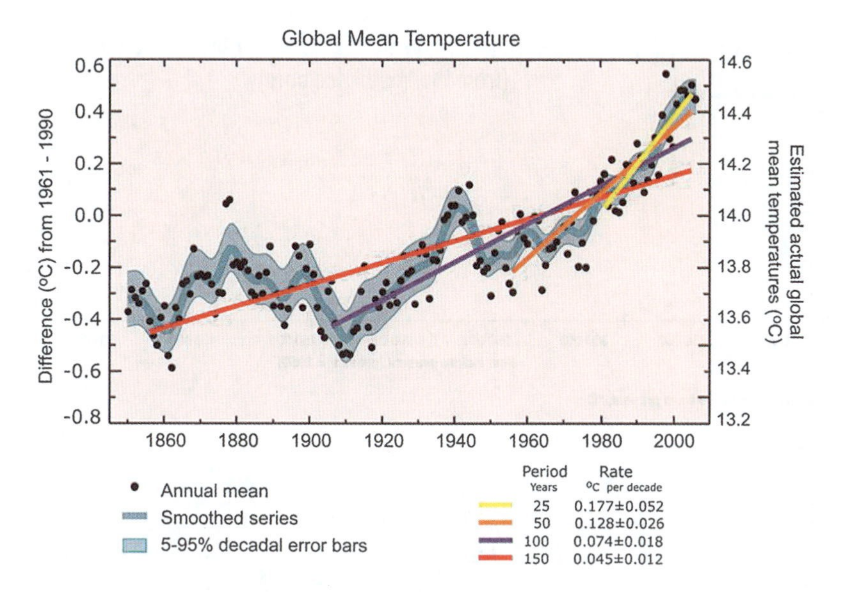

Figure 5.14 Annual global mean observed temperatures (black dots) 1850 to present, with the narrower blue curve (and the wider light grey curve) presenting smoothed fitted data (and 5–95% error bars). The left-hand axis shows anomalies relative to the 1961–1990 average and the right hand axis shows the estimated actual temperature. The increased slope of the linear trend fits for 25 (yellow), 50 (orange), 100 (purple) and 150 (red) years up to 2005, reveal accelerated warming. The blue curve is a smoothed depiction to capture decadal variations with (in light grey) 5–95% error ranges. Source: IPCC 4th Assessment Report (2007)[17a] (Web 9). Original caption edited with the permission of the IPCC.

years before the present; the lower one is the variation in global mean temperature compared with that of the present day. These two plots are striking because of their similarity and one may be tempted to argue that they are causally related, *i.e.* that one phenomenon is responsible directly or indirectly for the other. While this may be the case, we are unable to say (without additional evidence) which one is the cause and which the effect. Did the temperature change because of changes in CO_2, or did CO_2 concentrations change as a consequence of changes in temperature? Or indeed are they bound together in more complex processes of positive and negative feedback? In any event, the importance of the concentration of carbon dioxide in the atmosphere to understanding of the changes in global mean temperature is evident.

Natural phenomena that are believed to alter the mean surface temperature of the Earth include:

- long-term changes in solar flux
- sun-spot activity (11 year cycle)—see Figure 5.16 for changes in solar irradiance (W m^{-2}) 1980–2005

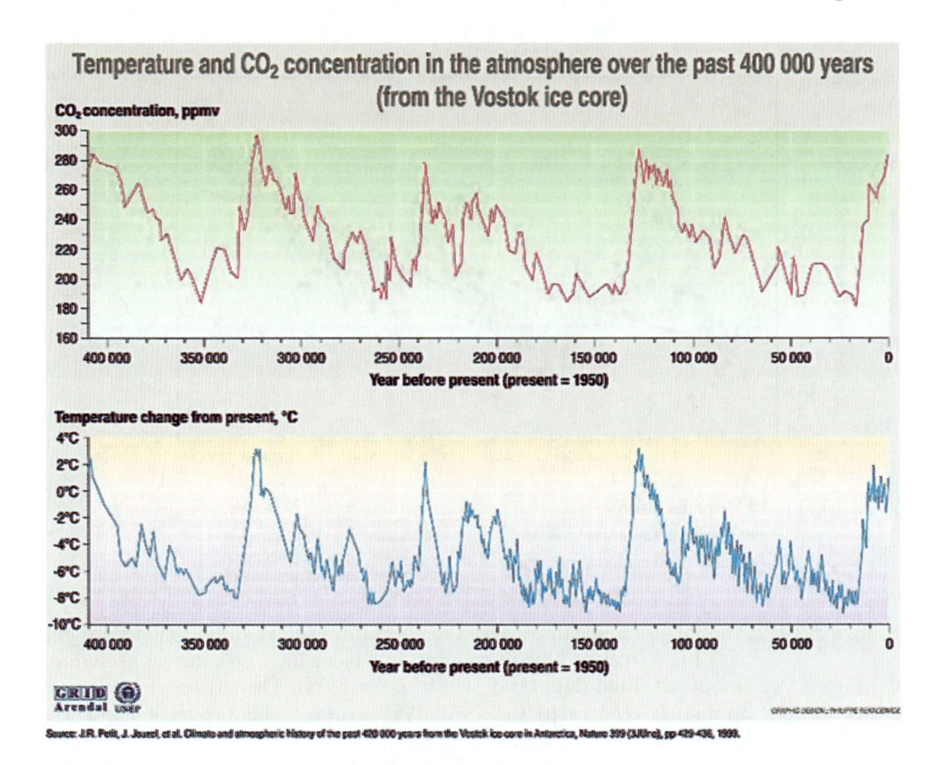

Figure 5.15 Changes with time of atmospheric CO_2 concentrations and temperatures over a period from *ca.* 400 000 years ago to 1950 from studies on an ice core sample. Source: Data from ref. 18 (adapted by permission of Macmillan publishers, copyright 1999); image courtesy of UNEP/GRID Arendel (Web 10).

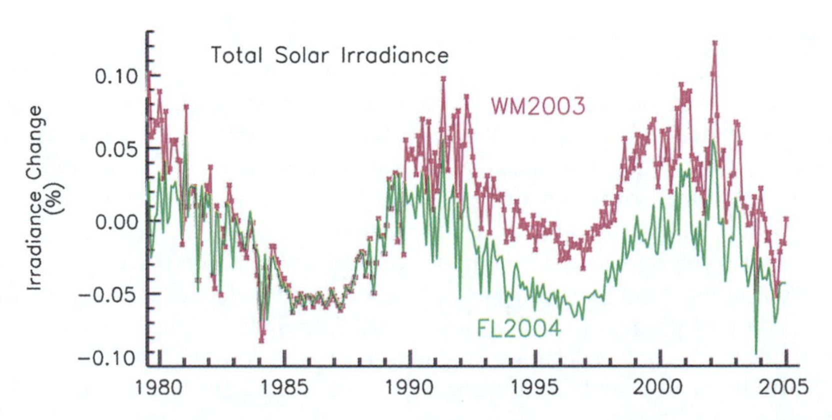

Figure 5.16 Percentage change in monthly values of total solar irradiance from two data series from 1980 [WM2003: Willson and Mordinov;[15f] FL2004: Frohlich and Lean (reproduced from ref. 15g with kind permission of Springer and Business Media)]. Source: IPCC 4th Assessment Report (2007)[15b] (Web 6).

- Earth's tilt angle (41 000 year cycle)
- month of closest approach (19 000–24 000 year cycle)
- shape of Earth's orbit (100 000 year cycle)
- changes in composition of atmosphere (irregular)
- emission of particulates (*e.g.* from volcanic eruptions).

A recent analysis[19] has attempted to separate out two of these factors (*i.e.* changes in solar irradiance and volcanic activity) to establish the significance of their contribution to recent changes in global temperature. This is shown in Figure 5.17, along with a representation of the wide range of estimates of past values of global mean temperature obtained from simulations. These highlight the difficulty of establishing whether a trend exists or not (let alone what might be responsible for it if it does).

5.6 THE GREENHOUSE EFFECT

The historical record suggests that the recent global warming trend began in about 1700 before the beginning of the Industrial Revolution and large-scale use of fossil fuels and associated emissions of CO_2. Indeed, there have been large swings over the last 150 000 years that were the consequence of natural phenomena.

In making projections of future changes in global mean surface temperature of the Earth—and particularly in gaining a proper appreciation of whether or not there is an anthropogenic contribution to such change (now largely accepted) and whether or not changes are dominated by such anthropogenic contributions (more controversial, but increasingly accepted)—has depended and will depend on taking full account of these many natural variations. The mathematical models on which such projections rely are assessed by the Intergovernmental Panel on Climate Change, whose 4th Assessment Report (4AR) (Web 1) appeared in 2007. Such projections are not without their critics.

The balance of radiation from the Sun impacting on the Earth and that reflected by it and emitted from it (Figure 5.13) will have a bearing on the global mean temperature at the surface. While this is a complicated parameter to calculate, the essential physical nature of the underlying processes can be readily understood using the basic ideas of optics and chemical spectroscopy. Of the radiation incident on the surface of the Earth from the Sun, some 28% is reflected back into space by clouds, snow and ice with high 'albedo'. Some penetrates through the clear atmosphere and impinges on the surface of the Earth (land or sea). Most of this is absorbed by the surface and warms it (driving movements of the atmosphere and evaporating water that directly lead to climatic and meteorological effects). The warmed Earth is an emitter in the infra-red (IR) region of the spectrum and, were there no absorbers of IR radiation in the atmosphere, this radiation would be lost into space. However, some of the IR radiation is intercepted by so-called 'greenhouse' gases (of which more later) that prevent the cooling that would result if all

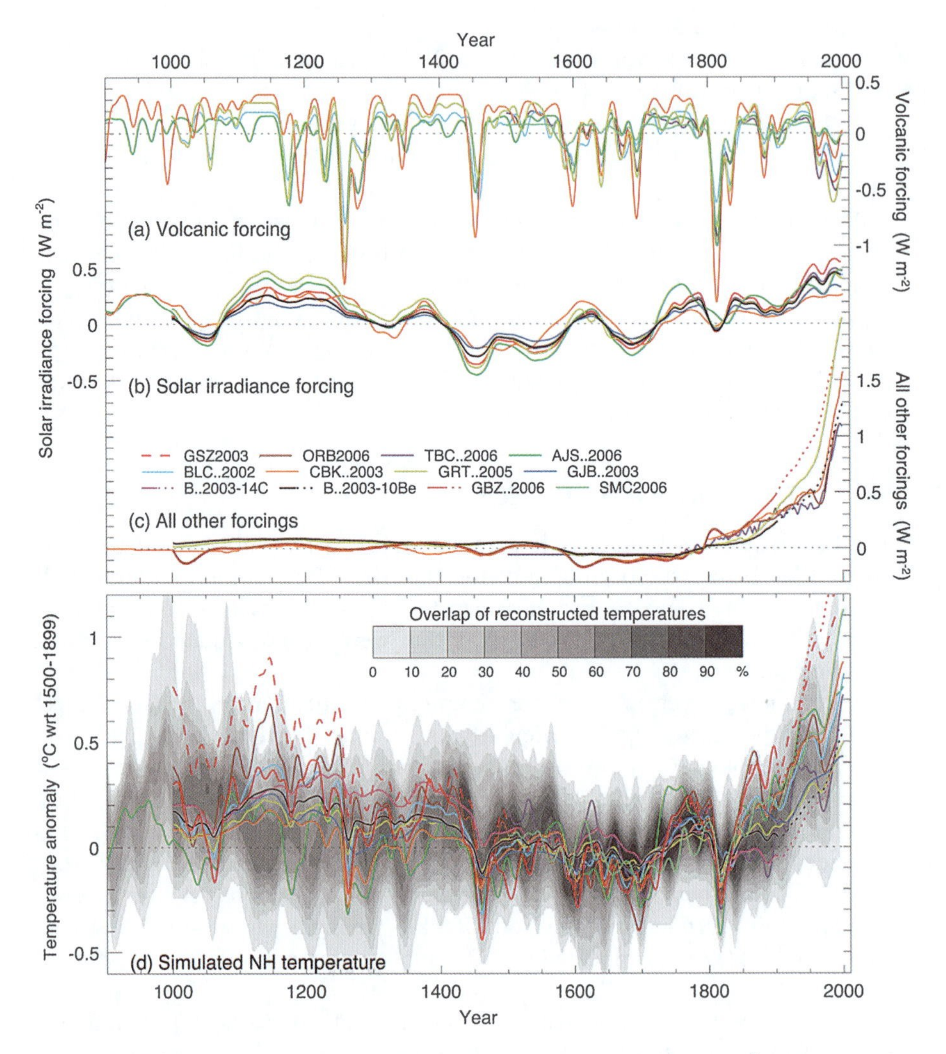

Figure 5.17 Top figure presents global mean radiative forcing (W m^{-2}) (Section 5.7) (estimated, for the last 1100 years, from climate model simulations) associated with (a) volcanic activity, (b) solar irradiance variations and (c) all other forcings (which vary between models). Bottom figure (d) data for annual mean northern hemisphere (NH) temperature (relative to the 1500–1899 mean) in grey, with coloured curves relating to the various individual simulations from (a) to (c) and smoothed to remove fluctuations on timescales <30 years. Source: IPCC 4th Assessment Report (1997)[19] (Web 11). Original caption edited with the permission of the IPCC.

such radiation was emitted to space. Note that this IR emission also occurs at night.

It is possible to envisage these interactions as a complex spectroscopic absorption phenomenon, from which absorption spectra can be obtained

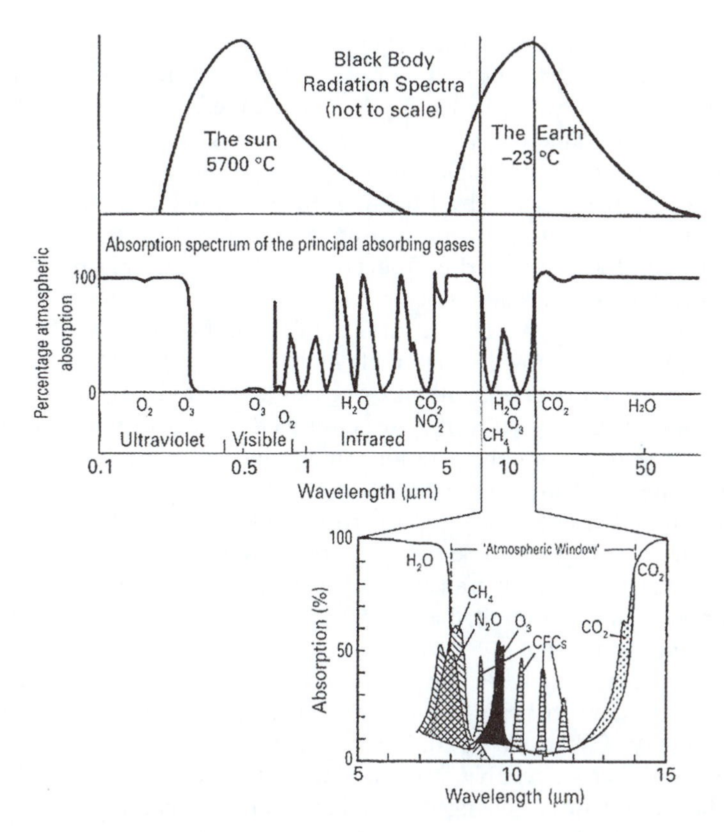

Figure 5.18 Origin of the 'greenhouse' effect. Radiation emitted from the Earth's surface (top graph) is in the range 5–50 μm (200–2000 cm^{-1}). The Earth's atmosphere only weakly absorbs in the range 8–13 μm (770–1250 cm^{-1}) (middle graph). Greenhouse gases absorb strongly in this 'window' (bottom graph). Reproduced from ref. 20a with permission of Wiley-Blackwell, based on Sawyer[20b] (reproduced with permission of Macmillan Publishers Ltd), Spedding[20c] and Turco[20d] (both reproduced with permission of Oxford University Press, Inc.).

(Figure 5.18). Solar radiation incident on the Earth is made up of components associated with electronic excitations and with molecular vibrations and rotations:

- ultra-violet (<0.4 μm, >25 000 cm^{-1})
- visible (0.4–0.7 μm, 25 000–14 000 cm^{-1})
- vibrational (0.7–20 μm, 14 000–500 cm^{-1})
- rotational (>20 μm, <500 cm^{-1}).

The radiation emitted by the Earth is predominantly in the region 5–50 μm or 200–2000 cm^{-1}. Interestingly, there is a 'window' in the absorption spectrum of the atmosphere (corresponding to energy in the range 8–13 μm or

770–1250 cm^{-1}) that allows much, but crucially not all, of the emitted radiation to escape. Were the atmosphere totally transparent to IR radiation, then the surface temperature would be *ca.* 255 K.[21] The difference between this value and the observed value of *ca.* 288 K is associated with the operation of the 'greenhouse' effect,[vi] a phenomenon first proposed by Fourier in 1827 and demonstrated experimentally by Tyndall in 1859. The first calculations relating to its effect on the Earth's temperature were reported by Arrhenius in 1896. The greenhouse effect, therefore, arises from components of the atmosphere that are not transparent to IR radiation in this critical region. It is a natural physico-chemical phenomenon vital for the survival of life on Earth in that it maintains the Earth's temperature within habitable limits.

The Earth's atmosphere is a mixture of gases (along with aerosols and particulates) whose contribution to the greenhouse effect is governed by their chemical structure and their concentration. Some components such as N_2, O_2, Ar, Ne and H_2 are IR transparent, whereas H_2O, CO_2, O_3, CH_4, N_2O (+SF_6 and the chlorofluorocarbons and fluorohydrocarbons), the greenhouse gases, are not. This arises from the simple spectroscopic selection rule (see Bibliography) that states that vibrational transitions associated with the absorption of IR radiation are allowed only if the change in the vibrational state results in a change in dipole moment.

Several of the IR-active components of the atmosphere are those whose concentrations are increasing, with recent increases primarily associated with human activity. Efforts involving many research groups have been directed towards the design of reliable models of the atmosphere so that meaningful predictions about future changes in global mean temperature can be made. These simulations show a gradual warming over the next 100 years.

However, the climate is a complex and a chaotic phenomenon, with many component factors (*e.g.* the role of clouds and the impact of aerosols) providing possibly important 'feedback' mechanisms that are only poorly modelled, so far. From the point of view of the Earth's climate and its heating and cooling arising from the absorption, reflection or re-emission of solar energy, feedback can be either negative or positive. Positive feedback may accelerate or amplify the consequences of an effect; negative feedback can inhibit or lessen the consequences of an effect. For example, positive feedback may result from the loss of reflectivity of solar radiation arising from the loss of ice sheets (ice is reflective, underlying soil less so). As a consequence, less solar energy is reflected and more is absorbed compared with the situation if no additional ice melted. Negative feedback may result from the greater reflectivity from additional cloud formation that is likely to result from increases in atmospheric water vapour. Some aspects such as the dynamic effects that living things have on the climate (*e.g.* marine algae which may affect ocean surface temperature and CO_2 fixation in surface layers) have not yet been taken into account. For

[vi] Some have pointed out that, strictly speaking, the precise mechanism by which heat is retained in the atmosphere is different from that occurring in a greenhouse. In a greenhouse, it is the prevention of convection that inhibits the movement of warmer air and not the absorption of IR radiation. Whatever, you get the picture.

this reason, some believe that little reliance should be placed on the projections generated by such models and that policy-makers should be much more influenced by direct observations as indicators of climate change such as sea level rise.[vii]

It would seem obvious that, as different molecules have different vibrational spectra, then their contribution to the greenhouse effect would also be different. Interestingly, water is probably the most important greenhouse gas and its involvement in the formation of clouds and ice fields of high reflectivity poses great problems to those seeking to model the atmosphere and future temperature changes. This can be seen qualitatively from the tendency of increased temperatures to put more water vapour into the atmosphere where it can absorb more infra-red (positive feedback), but which may also lead to greater formation of clouds, which being reflective can reduce incoming radiation thereby lessening warming (negative feedback). There are major uncertainties in assessing the net consequence of these factors. Similarly, the consequences of the presence of a component in the atmosphere can be affected by how long it spends in the atmosphere before being decomposed into something that may be less absorbing in the critical region of the electromagnetic spectrum or not absorbing at all.

Bearing in mind that we now believe that CO_2 from fossil sources, emitted anthropogenically, is contributing to an increase in the temperature of the Earth and that CO_2 mixing ratios may be close to or (as some now believe) have gone beyond the point at which irreversible damage will be done, we need to minimise such emissions. Ultimately, this can be achieved by shifting to sources of energy that avoid the use of fossil fuels. In the transition, ways need to be found to limit CO_2 emissions from fossil sources. The Kyoto Protocol (Web 12), which was agreed in 1997 but did not obtain the required number of ratifications for it to 'enter into force' until 2005, sought to limit CO_2 in the atmosphere to twice 'pre-industrial' levels (*i.e.* 2×280 or 560 ppm) and to reduce annual global emissions by 50% by 2050. While implementation of these provisions has been patchy, there are those (including James Hansen,[22] now an advisor on climate change to the US President) who have concluded that the Kyoto limit is much too high, believing we are already beyond an acceptable CO_2 mixing ratio. Governments are currently seeking to formulate a post-Kyoto agreement (that would operate from 2012) with preparatory meetings having taken place in December 2007 (Bali, Indonesia), December 2008 (Poznan, Poland) and in December 2009 in Copenhagen (Web 13).

5.7 GLOBAL WARMING POTENTIAL

Climate scientists use global warming potential (GWP) to estimate the impact of greenhouse gases and to put the contributions from different materials onto

[vii] Estimates of sea level increase are also difficult to measure bearing in mind, for example, the need to correct for the effects of barometric pressure, wave formation, tides, local gravitational effects and ocean currents. These aspects are discussed in Web 8–particularly Section 5.5 and Table 5.3.

a common scale. Understanding GWP requires us to introduce the concept of radiative forcing to assess the perturbations arising from an increase in the amount of a particular greenhouse gas in the atmosphere. Radiative forcing is the change in radiative flux (W m^{-2}) resulting from a change in the mass of a greenhouse gas in the atmosphere. All climate models show a linear relationship between radiative forcing and the ultimate change in surface temperature. So, radiative forcing is a useful quantitative metric to compare the potential effect of different greenhouse gases on climate.

$$\mathbf{GWP} = \frac{\int_{t_0}^{t_0+\Delta t} \Delta F_{1\,kg\,X}\,dt}{\int_{t_0}^{t_0+\Delta t} \Delta F_{1\,kg\,CO_2}\,dt} \tag{5.12}$$

The global warming potential of an atmospheric species, X, is the radiative forcing, ΔF, arising from the instantaneous injection of 1 kg of X into the atmosphere relative to that resulting from a similar injection of 1 kg of CO_2 integrated over time, from t_o to a time Δt, during which the atmospheric concentration of the injected gas decays through chemical transformation or sequestration. This time integration is important because individual atmospheric components have finite lifetimes before they are lost from the atmosphere and these lifetimes are different from component to component.

The formal equation for GWP is given in eqn (5.12), with some values for typical gases in Table 5.4. The second column gives the value of the atmospheric lifetime of X, with estimates varying from quite low (*e.g.* for chlorofluorohydrocarbon, HCFC-123) to very high (*e.g.* for sulfur hexafluoride). Clearly something that is added to the atmosphere that survives for a long period will produce a greater contribution to global warming (mole for mole,

Table 5.4 Global warming potentials and related data for selected greenhouse gases. Data from ref. 15c (Web 6).

Gas	*Atmospheric lifetime*	*Radiative efficiency*	*Global Warming Potential (GWP)* [a] *for given time horizon*		
	years	*W m^{-2} ppb^{-1}*	*20 years*	*100 years*	*500 years*
Carbon dioxide, CO_2	[b]	1.4×10^{-5}	1	1	1
Methane, CH_4	12	3.7×10^{-4}	72	25	7.6
Nitrous oxide, N_2O	114	3.0×10^{-3}	289	298	153
CFC-12, CF_2Cl_2	100	0.32	11 000	10 900	5200
HCFC-123, CF_3CHCl_2	1.3	0.14	273	77	24
HFC-134a, CF_3CH_2F	14	0.16	3830	1430	435
Sulfur hexafluoride, SF_6	3200	0.52	16 300	22 800	32 600

[a] These GWP values have an estimated uncertainty of ±35% 95 to 95% confidence range.
[b] The decay of a pulse of carbon dioxide in the atmosphere follows a complex relationship with time.

assuming similar absorbance characteristics in the critical region of the IR spectrum) than will one present in the atmosphere for a short period. Bearing in mind that these GWP values are normalised to CO_2 and that the decay of CO_2 introduced into the atmosphere is a complex function of time (though its lifetime is, very roughly, 100 years), then the value for GWP will change as the timescale (the so-called 'time horizon') over which the effect is estimated is changed. The values are estimated for time horizons of 20, 100 and 500 years. The dependency of the relative effects on the nature of the absorber and on its lifetime can be seen in the data.

5.8 THE CARBON CYCLE II: METHANE AND ITS ATMOSPHERIC LIFETIME

I want now to turn to another greenhouse gas, methane, which has a GWP of 25 times that of CO_2 on a 100-year time horizon. It might surprise many chemists to know how much methane can be detected in the atmosphere, bearing in mind that we know that hydrocarbons react readily, sometimes explosively, with oxygen. How is it possible for there to be detectable concentrations of methane in the atmosphere (and, indeed, of other hydrocarbons, some more reactive than methane)?

Atmospheric methane has also featured in a series of news items over the last three or so years, one of which (Web 14) reported that researchers had detected methane emissions from terrestrial vegetation.[23] Obtaining reliable estimates of the sources and sinks of methane and measures of its atmospheric concentration is vital because methane is one of a number of naturally occurring organic compounds that has shown steep increases in concentration from historical levels (currently *ca.* 1720 ppbv for methane) (Figure 5.19a),[24] though the more recent data (Figure 5.19b) show a reduced rate of increase. An inventory for methane (Table 5.5) shows some surprising sources. Some of these are clearly natural, some clearly anthropogenic and some that are more difficult to apportion. Global and seasonal variations in methane sources and sinks are not understood, with the global concentration of methane remained constant during the period 1997 to 2006.

Because methane is a greenhouse gas, it is important to estimate its atmospheric lifetime. Ideally, to do this we would need to know how fast it is lost for each of the processes that lead to its destruction. To gain an approximate value, we will concentrate on the main sink for methane in the atmosphere. The following treatment is based on that in ref. 8. Let us first imagine a volume element of the atmosphere (represented in Figure 5.20) into which a general species, X, may flow from an adjacent volume element and from which the species may flow out into a different volume element. Some of component X may be generated in the volume element itself and some of it may decay in this element into some other species. Some new emission of X may enter this volume element and material may be physically deposited from it. By integrating these volume elements (with due allowance for all of the important atmospheric

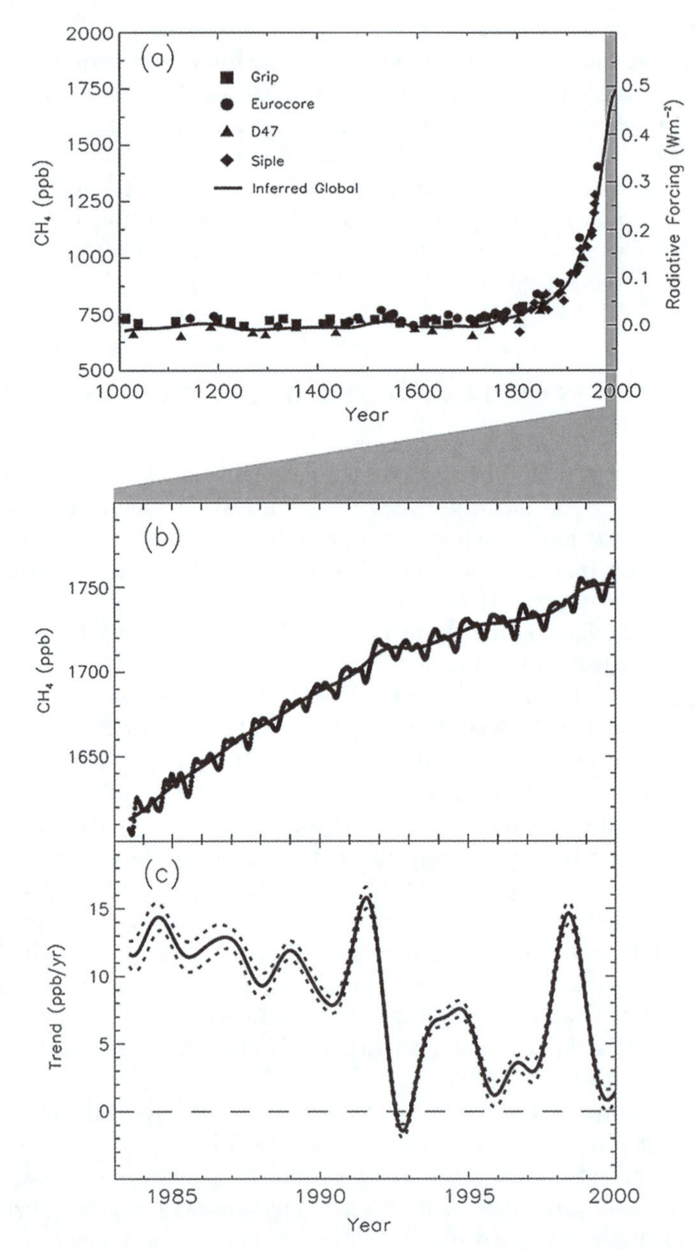

Figure 5.19 (a) Change in methane abundance (ppb) 1000–2000 AD determined from ice core, firn (compressed snow) and whole air samples. Radiative forcing, approximated by a linear scale since the pre-industrial era, is plotted on the right axis; (b) Globally averaged methane abundance (ppb) 1983–1999: monthly and 'deseasonalised' data (smooth line); (c) Instantaneous annual growth rate (ppb y^{-1}; ±1 standard deviation) of global atmospheric methane abundance 1983–1999. Source: IPCC 3rd Assessment Report (2001).[24] Original caption edited with the permission of the IPCC.

Table 5.5 Various inventories of natural and anthropogenic sources and sinks of atmospheric methane in $Tg(CH_4)\ y^{-1}$. [1 Tg = 10^{12} g = 10^{6} tonnes = 1 million tonne (Mt)]. Data from ref. 9c.

	Study 1	*Study 2*	*Study 3*	*Study 4*	*Study 5*	*4AR*
Anthropogenic sources						
Energy			77			
Coal mining		46		30	48	
Gas, oil, industry		60		52	36	
Landfill and waste		61	49	35		
Ruminants		81	83	91	189	
Rice agriculture		60	57	54	112	
Biomass burning		50	41	88	43	
Total anthropogenic sources		**358**	**307**	**350**	**428**	
Natural sources						
Wetlands	163	100	176	231	145	
Termites	20	20	20	29	23	
Oceans	15	4				
Hydrates		5	4			
Geological sources	4	14				
Wild animals	15					
Wildfires	5	2				
Total natural sources	**222**	**145**	**200**	**260**	**168**	
Total sources		**503**	**507**	**610**	**596**	**582**
Sinks						
Soils		30	34	30		
Tropospheric OH		445	428	507		
Stratospheric loss		40	30	40		
Total sinks		**515**	**492**	**577**		**581**

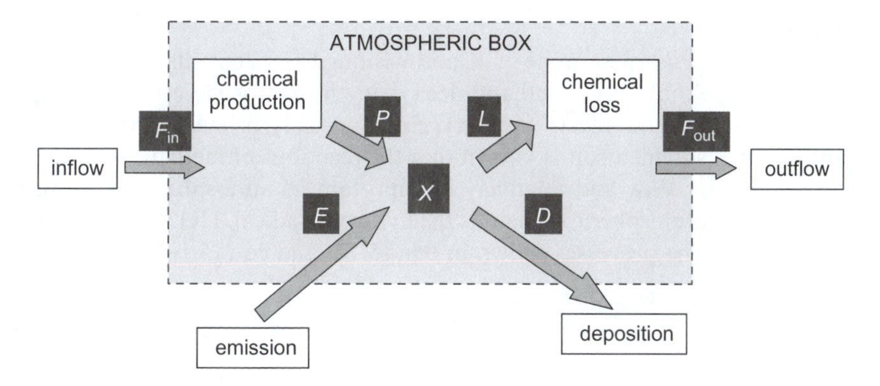

Figure 5.20 A volume element ('box') of the atmosphere showing the various sources, sinks and fluxes of atmospheric component *X*. Based on Figure 3.1 of ref. 8.

variables needed to make a credible model), atmospheric scientists can gain insights into the dynamic behaviour of important atmospheric components.

Let F represent transport of X from, or to, an adjacent atmospheric element or box; E and D represent emissions of X into the box and depositions of X from it; P and L represent chemical formation and decomposition of X within the box. F_{in}, E and P are 'sources'; F_{out}, D, L are 'sinks'. The mass of X is the 'inventory'. Let us assume the box or volume element is well-mixed.

If we see this box as a virtual reactor, our first job would be to establish a mass balance[viii] (eqn 5.13), representing the net contribution of sources and sinks.

$$\Sigma_{\text{sources}} - \Sigma_{\text{sinks}} = F_{\text{in}} + E + P - F_{\text{out}} - D - L \tag{5.13}$$

If we wish to determine the atmospheric lifetime, we proceed as follows:

If the concentration of X in a 1 m^3 box is A and the total source of X introduced into the box per unit time is S, then the lifetime of X, τ, is defined by eqn (5.14):

$$\tau = \frac{A(\text{kg m}^{-3})}{S(\text{kg m}^{-3}\,\text{s}^{-1})} \tag{5.14}$$

R, the rate of removal of X, is given by eqn (5.15), in which k is the rate constant for the loss of X.

$$R = \frac{\text{d[A]}}{\text{d}t} = k[\text{A}] \tag{5.15}$$

If the box is in balance, then $R = S$, and τ is related to k as follows (eqn 5.16):

$$\tau = \frac{[A]}{k[A]} = k^{-1} \tag{5.16}$$

The lifetime of X is, therefore, the inverse of the first order rate constant for its reaction.

Using methane as an example of X, and wishing to estimate its lifetime, τ, we would need to know how methane decays in the atmosphere. A first guess might be that methane reacts with oxygen or possibly ozone. However, from laboratory measurements, it is known that the reactions of methane with O_2 or O_3 are both very slow and relatively unimportant as atmospheric decay processes. The key atmospheric oxidant is the hydroxyl radical, $HO^•$, formed in the troposphere by the processes shown in eqn (5.17) and eqn (5.18):

$$O_3 + h\nu \rightarrow O_2 + O(^1D) \tag{5.17}$$

[viii] Mass balance is an important concept that is required by the First Law of Thermodynamics. This and related topics are considered in more detail in Chapter 6.

$$O(^1D) + H_2O \rightarrow 2\,HO^{\bullet} \quad (5.18)$$

Methane then reacts in the (simplified) Scheme 5.1:[8]

$CH_4 + HO^{\bullet} \rightarrow {}^{\bullet}CH_3 + H_2O$ (rate-determining step)

$M + {}^{\bullet}CH_3 + O_2 \rightarrow CH_3O_2^{\bullet} + M$

$CH_3O_2^{\bullet} + HO_2^{\bullet} \rightarrow CH_3O_2H + O_2$

$CH_3O_2H + h\nu \rightarrow CH_3O^{\bullet} + HO^{\bullet}$

$CH_3O^{\bullet} + O_2 \rightarrow H_2CO + HO_2^{\bullet}$

$H_2CO + HO^{\bullet} \rightarrow {}^{\bullet}CHO + H_2O$; $H_2CO + h\nu \rightarrow CO + H_2$

${}^{\bullet}CHO + O_2 \rightarrow CO + HO_2^{\bullet}$

$CO + OH^{\bullet} \rightarrow CO_2 + H^{\bullet}$; $H^{\bullet} + O_2 \rightarrow HO_2^{\bullet}$

Scheme 5.1 Reaction of methane with hydroxyl radicals.

The first process, hydrogen atom abstraction, is rate-determining (leading to formaldehyde, carbon monoxide or carbon dioxide) from which a rate equation (5.19) can be set up:

$$\frac{-d[CH_4]}{dt} = k_2[CH_4][HO^{\bullet}] \quad (5.19)$$

k_2 can be measured in the laboratory (and is found to be 8.4×10^{-15} cm^3 molec s^{-1}). From other measurements, the atmospheric concentration of the hydroxyl radical, $[HO^{\bullet}]$, is known and found to be approximately constant (10^6 molec cm^{-3}) (permitting a pseudo-first order rate constant, k_1', to be derived). A value for the atmospheric lifetime of methane can then be obtained (eqn 5.20):

$$k_2[CH_4][HO^{\bullet}] = k_1'[CH_4]; k_1' = k_2[HO^{\bullet}] = 8.4 \times 10^{-9}\ s^{-1} \quad (5.20)$$

Our estimated lifetime for methane, $\tau = k^{-1} = (8.4\times10^{-9})^{-1}$ s, is 3.8 years. (The currently accepted value is 12 years. As an exercise, list the assumptions we have made in arriving at our estimate.)

The value for the lifetime of methane indicates how rapid this compound's turnover is.[ix] From such estimates, it is possible to calculate an approximate measure of the total source of the methane needed to maintain the concentration at its measured level. This sort of calculation highlights the difficulty of separating out the impact of the anthropogenic emissions of compounds like methane that also occur naturally because of the relatively modest quantities of man-made emissions when compared with the huge sources, fluxes and inventories from natural sources. Even so, we are now beginning to understand that even these relatively modest emissions can have measurable, possibly significant, impact.

5.9 THE NITROGEN CYCLE, NITROUS OXIDE AND BIOMASS PRODUCTION

Nothing highlights the incompleteness of our understanding of the chemistry of the environment and humankind's perturbation of it than the increasing awareness of the role of nitrous oxide in both global warming and ozone depletion.[9c,26a–d]

Dinitrogen is the dominant atmospheric component of the global nitrogen geochemical and biogeochemical cycles, with NO, NO_2 and NH_3 also being carriers of the element, in addition to N_2O, with $[NO_3]^-$ and $[NH_4]^+$ important in the hydrosphere. The atmospheric mixing ratio for N_2O has increased from 270 ppb in about 1800 to 300 ppb in 1980, and 322 ppb at present.[9d] Its abundance, along with its long atmospheric lifetime (110–120 years for an emission to be reduced in a first-order exponential decay to $1/e$ of its initial level, or reduced by 63.2%), means that it is the fourth most important greenhouse gas after water, carbon dioxide and methane. Very recently,[26a] N_2O has also been suggested to be the single most important ozone-depleting compound emitted. Surprisingly, emissions of N_2O are not subject to the Montreal Protocol on Substances that Deplete the Ozone Layer which controls the production and use of chlorofluorocarbons (CFCs).

N_2O has both natural and anthropogenic sources, though there are large uncertainties in the estimates of the sources, fluxes and sinks (Table 5.6). N_2O is both emitted naturally by the oceans and by soil supporting the growth of vegetation and consumed there; 30% of the net emissions are believed to result from humankind's activities, particularly from agriculture. This raises immediate questions about the need to establish the net benefit associated with efforts to reduce our dependence on fossil sources of energy (and thereby reduce CO_2 emissions) by increasing our use of renewable sources such as biomass (and the associated increase in N_2O emissions). Indeed, this may affect the choices of the types of vegetation to be grown.[26c,d]

[ix] Methane has also very recently been detected on Mars,[25] though its origin is not known. Estimates were made of its Martian atmospheric lifetime (200 days) that were 600 times shorter than predicted from the expected photochemical routes to its decomposition. Methane is also a major primordial component of the atmosphere of Titan, a satellite of Saturn.

Table 5.6 Global sources of nitrous oxide in TgN y^{-1}. Data from ref. 9d.

	TgN y^{-1}	*(range)*	*TgN y^{-1}*	*(range)*
Anthropogenic sources			**6.7**	
Fossil fuel combustion/industrial processes	0.7	(0.2–1.8)		
Agriculture	2.8	(1.7–4.8)		
Biomass and biofuel burning	0.7	(0.2–1.0)		
Human excreta	0.2	(0.1–0.3)		
Rivers, estuaries and coastal zones	1.7	(0.5–2.9)		
Atmospheric deposition	0.6	(0.3–0.9)		
Natural sources			**11.0**	
Soils under natural vegetation	6.6	(3.3–9.0)		
Oceans	3.8	(1.8–5.8)		
Atmospheric chemistry	0.6	(0.3–1.2)		
TOTAL SOURCES			**17.7**	**(8.5–27.7)**

5.10 HUMAN IMPACT ON THE ENVIRONMENT

There is no doubt that human activity has led to substantial increases in emissions of greenhouse gases (Figure 5.21).[27] While long-term consequences based on computer simulations are less generally accepted, direct observations of critical parameters, such as sea level changes (Figure 5.22),[28] may be more convincing indicators of the global threat we face. The so-called **anthropogenic** impact is associated with what goes on in the part of the planet occupied by humankind, known as the technosphere or the anthrosphere. The age of man, particularly that associated with the period when *homo sapiens* began to have a significant impact on the planet, has been given its own name, the Anthropocene epoch.[29] There are those who object to attempts to see the technosphere as separate, suggesting that it represents an artificial distinction, indeed a dangerous one, as it suggests that such 'spheres' are merely transactional (*i.e.* involving simple exchanges of material one with another and little else) rather than interacting, dynamically and mutually, in complex ways. The latter more holistic view, in which the Earth is seen as a single entity, can be described using the 'Gaia' concept—associated most with the independent scientist, James Lovelock.[30] We have already been introduced to some of Lovelock's rather trenchant thoughts about climate change.

5.11 GEOPHYSIOLOGY OR EARTH SYSTEMS SCIENCE

Lovelock's hypothesis is that the Earth is a self-regulating living system. This is not a new idea. The idea of the Earth as a 'super-organism' was first proposed by James Hutton (1726–1797), one of the fathers of the modern discipline of geology. However, Lovelock's ideas grew out of his consideration of the atmosphere of Mars, when he was working for NASA and was assessing what evidence there might be for life on Mars and how additional evidence might be

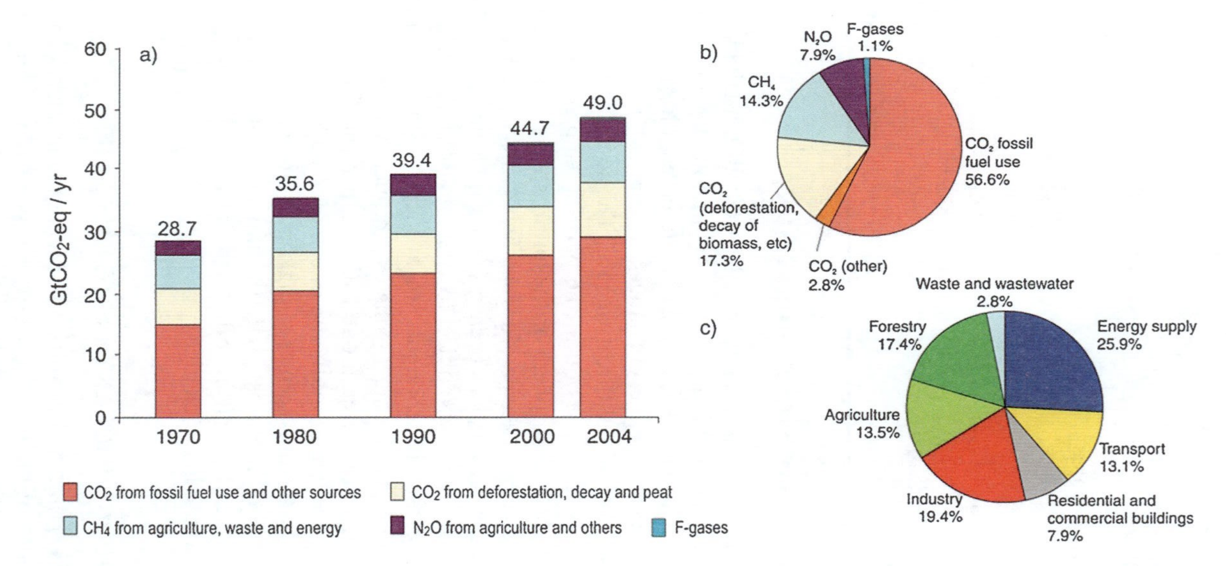

Figure 5.21 (a) Changes in global annual emissions of anthropogenic greenhouse gases in Gt(CO_2e) y^{-1} from 1970–2004; (b) the relative contributions (%) of important anthropogenic greenhouse gases in 2004; and (c) the relative contributions (%) from various anthropogenic greenhouse gas sources in 2004. Source: IPCC 4th Assessment Report (2007),[27] (Web 16). Original caption edited with the permission of the IPCC.

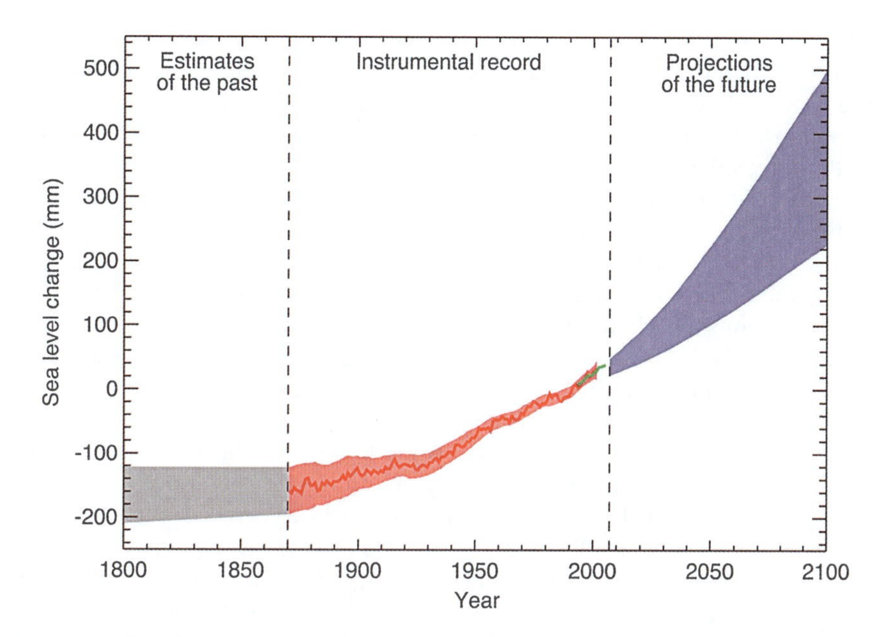

Figure 5.22 Time series of global mean sea level (with respect to 1980–1999 mean). There are no global measurements of sea level for the period before 1870. The grey shading shows the uncertainty in the estimated long-term rate of sea level change. Projections up to 2100 are calculated from observations. Up until 1990 measurements arise from tide gauges. From 1990, levels were observed from satellite altimetry. Source: IPCC 4th Assessment Report (2007)[28] (Web 8). Original caption edited with the permission of the IPCC.

Table 5.7 Composition of the atmospheres of Venus, Mars and Earth and the estimated composition of the atmosphere of a lifeless Earth. Data from ref. 30.

	Venus	*Mars*	*Earth without life*	*Earth with life*
CO_2 (%)	98	95	98	0.03
N_2 (%)	1.9	2.7	1.9	78
O_2 (%)	trace	0.13	trace	21
Ar (%)	0.1	2	0.1	1
Surface Temp (°C)	477	−53	290±50	13
Total pressure (bar)	90	0.0064	60	1.0

obtained. Lovelock was struck by the differences between the atmospheres of Venus and Mars, which do not appear to support life, and that of the Earth.

Venus and Mars are characterised (Table 5.7) by extremes of temperature and pressure, with atmospheres very low in oxygen and rich in carbon dioxide—characteristics significantly different from the equivalent concentrations for Earth. Lovelock calculated the values that might be expected if there was no life on Earth and came up with some estimates that were closer to those of Mars and Venus. He came to the view that the Earth's atmosphere and its pressure and temperature (and the fact that the concentrations of oxygen and methane in the atmosphere were not those predicted from simple considerations of chemical equilibrium) was a consequence of the existence of life forms. Bearing in mind changes in the Sun's irradiance over geological time, Lovelock was led to ask how the surface temperature of the Earth had remained so constant. Why is the Earth's atmosphere the composition it is and how has this remained essentially constant? Particularly, why is the concentration of dioxygen in the atmosphere 21%? Lovelock noted that spontaneous combustion occurs at >25% and human life requires more than about 15%. Also, why has ocean salinity remained essentially constant?

Lovelock's ideas were controversial when first propounded and were criticised for being 'teleological'. That is, critics suggested that Lovelock's ideas were based on a belief that all things have a pre-determined purpose. Such criticisms were not helped by the espousal of Gaia (the Earth goddess in Greek mythology) by some as a neo-Pagan New Age religion. Gaia can probably better be seen as a metaphor, though evidence obtained since Lovelock first put forward his hypothesis suggests that it now should be seen as a valid theoretical model of the Earth's global processes. It has given rise to new scientific disciplines (*e.g.* geophysiology—or even biogeophysiology) and is encompassed by existing disciplines (*e.g.* Earth systems science).

The basic idea is that the Earth functions 'homeostatically'. Homeostasis (first proposed by W. B. Cannon in 1932) is defined as the property of an open system to regulate its internal environment so as to maintain a stable constant condition by means of dynamic equilibrium adjustments *via* interrelated regulation mechanisms. It is fundamental to living beings to maintain their internal environment within tolerable limits. It is reliant on sets of feedback mechanisms that do not require conscious control and, therefore, are not 'teleogical'. Feedback is a phenomenon introduced in our discussion of the role of water vapour in the atmosphere. It can be illustrated (Figure 5.23) by the function of a mechanical 'governor' that, through changes in the flow of fuel, automatically controls shaft rotation speed by linkages that move up or down as a consequence of centrifugal forces.

The critical point that Lovelock saw was that the Earth exists in a quasi-equilibrium state and that this has intermittently undergone rapid changes (for which geological evidence exists). He believes that just such a rapid change (on the geological timescale) could arise as a consequence of greenhouse gas emissions. The Earth should therefore be seen as a complex entity involving the biosphere, atmosphere, oceans and soil; the totality constituting a feedback or cybernetic system which seeks an optimal physical and chemical environment for life on this planet. Precisely which of these various effects may dominate, and how each will evolve with time, is very difficult to model. This is one of several reasons why there is so much uncertainty in future predictions of the average temperature of the Earth.

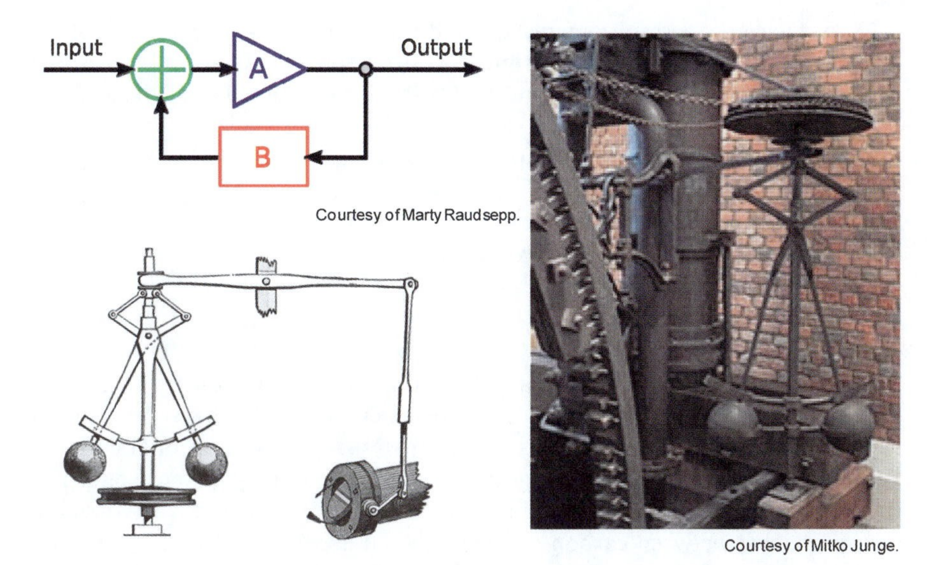

Figure 5.23 Various representations of feedback processes.

5.12 GEOENGINEERING

Bearing in mind the global scale of the effects that human activity is having (or is going to have) on the planet, then it is not surprising that consideration is being given—particularly by those who judge the situation to be critical—to projects that might have global impact. Geoengineering (*i.e.* the intentional alteration of the Earth's climate) is the discipline that has grown up to address this. Critics suggest that the cure may turn out to be worse than the disease and may bring dire unforeseen consequences. Others, including the Royal Society (Web 17), suggest that we should at least explore the possibilities to dismiss those with little hope of effecting the desired outcome and to assess those that may have more promise.

These proposals, summarised below, but not discussed in detail, include:[31,32]

- **Marine cloud albedo enhancement:** by mechanical generation of sea-salt spray to form cloud condensation nuclei (CCN); by generating CCN from dimethyl sulfide emissions by iron fertilisation of part of the Southern Ocean
- **Increase surface albedo:** by increasing reflectivity of deserts, roofs of buildings or other surfaces (though urban land is only 0.05% of the Earth's surface); increase reflectivity of vegetation using bioengineering
- **Space solar sunshade:** to offset the effect of the increase in atmospheric CO_2 by decreasing incoming solar radiation through the intervention of a 'sunshade' (total area *ca.* $4\times10^6\,km^2$) between the Earth and the Sun
- **Solar power generation**

- **Stratospheric aerosols:** increasing reflection of short-wave radiation by injection of sulfate aerosols into the stratosphere, the quantity needed (dependent on particle size and where the material is injected) being 1.5–5 $\mathrm{Tg\,S\,y^{-1}}$
- **Ocean fertilisation:** with phosphorus, nitrogen, iron
- **Ocean carbonate addition**
- **Air capture** (Section 12.2.2)
- **Enhanced ocean upwelling/downwelling**
- **Afforestation/reafforestation**
- **Biochar production** (Section 11.6.6).

These suggestions divide, essentially, between those that reduce short-wave radiation entering the Earth and those that increase the long-wave radiation leaving it. Lenton and Vaughan's initial analysis[32] (ignoring economic constraints and deployment costs) suggests that the former may be more effective than the latter; some are likely to have little benefit. The Royal Society divides the approaches between carbon dioxide removal and solar radiation management; the former have relatively low uncertainty and risks but will be slow to work, whereas the latter can act quickly, may create additional problems and do not correct the increased atmospheric concentrations of carbon dioxide and related effects such as ocean acidification.

These geoengineering ideas will remain speculative until they have been properly evaluated.

We need, therefore, to return to the more conventional view of the various 'spheres' of the Earth, dividing the Earth notionally into two:

- the **ecosphere**, which encompasses all living things and physical resources, other than human beings and the objects and systems created by human beings
- the **technosphere**, which comprises human beings and the objects and systems created by human beings.

This allows some useful ideas to be developed that can be put into mathematical form (a model or simplified representation). The resulting exergetic analysis (we will discuss exergy in Chapter 6) can then provide data that allow comparisons to be made between technologies and technological options. The conclusions from such analyses might then guide future decision making.

REFERENCES

1. R. M. Harrison, *Understanding our Environment*, Royal Society of Chemistry, Cambridge, 2nd edn, 1992.
2. T. O'Riordan, *Environmental Science for Environmental Management*, Prentice Hall, London, 2nd edn, 2000, pp. 13, 53.

3. S. Solomon, D. Qin, M. Manning, Z. Chen, M. Marquis, K. B. Averyt, M. Tignor and H. L. Miller, *Climate Change 2007: The Physical Science Basis. Contribution of Working Group I to the Fourth Assessment Report of the Intergovernmental Panel on Climate Change*, Cambridge University Press, Cambridge, 2007 (Web 1): (a) Frequently Asked Question 1.2, Figure 1, p. 96; and (b) Frequently Asked Question 1.1, Figure 1, p. 94; (c) J. Kiehl and K. Trenberth, *Bull. Am. Meteorol. Sci.*, 1997, **78**, 197.
4. J. E. Andrews, P. Brimblecombe, T. D. Jickells, P. S. Liss and B. J. Reid, *An Introduction to Environmental Chemistry*, Blackwell Publishing, Oxford, 2nd edn, 2004.
5. S. E. Manahan, *Environmental Chemistry*, CRC Press, Boca Raton, FL, 6th edn, 1994.
6. P. Rekacewicz, in *Our Changing Planet: The US Climate Change Science Program for Fiscal Year 2006*, US Global Change Research Information Office, Washington DC, 2005, Figure 1, p. 23 (see Web 3).
7. (a) S. Hong, J.-P. Candelone, C. C. Patterson and C. F. Boutron, *Science*, 1996, **272**, 246; (b) J. O. Nriagu, *Science*, 1996, **272**, 223.
8. D. J. Jacob, *Introduction to Atmospheric Chemistry*, Princeton University Press, Princeton, NJ, 1999.
9. K. L. Denman, G. Brasseur, A. Chidthaisong, P. Ciais, P. M. Cox, R. E. Dickinson, D. Hauglustaine, C. Heinze, E. Holland, D. Jacob, U. Lohmann, S. Ramachandran, P. L. da Silva Dias, S. C. Wofsy and X. Zhang, Couplings between changes in the climate system and biogeochemistry, in *Climate Change 2007: The Physical Science Basis. Contribution of Working Group I to the Fourth Assessment Report of the Intergovernmental Panel on Climate Change*, ed. S. Solomon, D. Qin, M. Manning, Z. Chen, M. Marquis, K. B. Averyt, M. Tignor and H. L. Miller, Cambridge University Press, Cambridge, 2007, ch. 7 (Web 4): (a) Figure 7.3, p. 515; (b) Table 7.1, p. 516; (c) Table 7.6, p. 542; (d) Table 7.7, p. 546.
10. J. H. Seinfeld and S. N. Pandis, *Atmospheric Chemistry and Physics: From Air Pollution to Climate Change*, John Wiley & Sons, Chichester, 1998.
11. I. C. Prentice, G. D. Farquhar, M. J. R. Fasham, M. L .Goulden, M. Heimann, V. J. Jaramillo, H. S. Kheshgi, C. Le Quéré, R. J. Scholes and D. W. R. Wallace, The carbon cycle and atmospheric carbon dioxide, in *Climate Change 2001: The Scientific Basis. Contribution of Working Group I to the Third Assessment Report of the Intergovernmental Panel on Climate Change*, ed. J. T. Houghton, Y. Ding, D. J. Griggs, M. Noguer, P. J. van der Linden, X. Dai, K. Maskell and C. A. Johnson, Cambridge University Press, Cambridge, 2001, ch. 3, Section 3.1, Figure 3.1, p. 188 (Web 5).
12. D. O. Hall and K. K. Rao, *Photosynthesis*, Cambridge University Press, Cambridge, 6th edn, 1999.
13. (a) R. A. Houghton, F. Hall and S. J. Goetz, *J. Geophys. Res., [Biogeosci.]*, 2009, **114**, G00E03; (b) K.-H. Erb, F. Krausmann, V. Gaube, S. Gingrich, A. Bondeau, M. Fischer-Kowalski and H. Haberl, *Ecol. Econ.*, 2009, **69**, 250.

14. J. T. Houghton, L. G. Meiro Filho, B. A. Callander, N. Harris, A. Kaltenberg and K. Maskell, eds, *Climate Change 1995: The Science of Climate Change: Contribution of Working Group I to the Second Assessment Report of the Intergovernmental Panel on Climate Change*, Cambridge University Press, Cambridge, 1996, Technical Summary, Fig. 1 (a), p. 16.
15. P. Forster, V. Ramaswamy, P. Artaxo, T. Bernsten, R. Betts, D. W. Fahey, J. Harwood, J. Lean, D. C. Lowe, G. Myhre, J. Nganga, R. Prinn, G. Raga, M. Schulz and R. Van Dorland, Changes in atmospheric constituents and radiative forcing, in *Climate Change 2007: The Physical Science Basis. Contribution of Working Group I to the Fourth Assessment Report of the Intergovernmental Panel on Climate Change,* ed. S. Solomon, D. Qin, M. Manning, Z. Chen, M. Marquis, K. B. Averyt, M. Tignor and H. L. Miller, Cambridge University Press, Cambridge, 2007 ch. 2 (Web 6) (a) Figure 2.3, p. 138; (b) Figure 2.16, p. 189; and (c) Table 2.14, p. 212; (d) R. F. Keeling, S. C. Piper, A. F. Bollenbacher and J. S. Walker, *Atmospheric Carbon Dioxide Record from Mauna Loa, New Zealand*, Carbon Dioxide Information Analysis Center (CDIAC), Oak Ridge National Laboratory, Oak Ridge, TN, 2009 (Web 7a); (e) R. F. Keeling, S. C. Piper, A. F. Bollenbacher and J. S. Walker, *Atmospheric Carbon Dioxide Record from Baring Head, New Zealand*, Carbon Dioxide Information Analysis Center (CDIAC), Oak Ridge National Laboratory, Oak Ridge, TN, 2008 (Web 7b); (f) R. C. Willson and A. V. Mordvinov, *Geophys. Res. Lett.*, 2003, **35**, 3; (g) C. Fröhlich and J. Lean, *Astron. Astrophys. Rev.*, 2004, **12**, 273.
16. K. Caldeira and M. E. Wickett, *Nature*, 2003, **425**, 365 (see Web 8).
17. K. E. Trenberth, P. D. Jones, P. Ambenje, R. Bojariu, D. Easterling, A. Klein Tank, D. Parker, F. Rahimzadeh, J. A. Renwick, M. Rusticucci, B. Soden and P. Zhai, Observations: surface and atmospheric climate change, in *Climate Change 2007: The Physical Science Basis. Contribution of Working Group I to the Fourth Assessment Report of the Intergovernmental Panel on Climate Change*, ed. S. Solomon, D. Qin, M. Manning, Z. Chen, M. Marquis, K. B. Averyt, M. Tignor and H. L. Miller, Cambridge University Press, Cambridge, 2007, ch. 3, (a) Frequently Asked Question 3.1, Figure 1 (top), p. 253 (Web 9); (b) Section 3.2, pp. 244–253; (c) Appendix 3.A, p. 336; (d) Appendex 3.B, p. 336.
18. J. R. Petit, J. Jouzel, D. Raynaud, N. I. Barkov, J.-M. Barnola, I. Basile, M. Bender, J. Chappellaz, M. Davis, G. Delaygue, M. Delmotte, V. M. Kotlyakov, M. Legrand, V. Y. Lipenkov, C. Lorius, L. Pépin, C. Ritz, E. Saltzman and M. Stievenard, *Nature*, 1999, **399**, 429.
19. E. Jansen J. Overpeck, K. R. Briffa, J.-C. Duplessy, F. Joos, V. Masson-Delmotte, D. Olago, B. Otto-Bliesner, W. R. Peltier, S. Rahmstorf, R. Ramesh, D. Raynaud, D. Rind, O. Solomina, R. Villalba and D. Zhang, Palaeoclimate, in *Climate Change 2007: The Physical Science Basis. Contribution of Working Group I to the Fourth Assessment Report of the Intergovernmental Panel on Climate Change*, ed. S. Solomon, D. Qin, M. Manning, Z. Chen, M. Marquis, K. B. Averyt, M. Tignor and H. L. Miller, Cambridge University Press, Cambridge, 2007, ch. 6, Figure 6.13, p. 477 (Web 11).

20. (a) J. V. Holder, *Handbook of Green Chemistry and Technology*, ed. J. Clark and D. Macquarrie, Blackwell Science, Oxford, 2002, Ch. 3, p. 28, Figure 3.2; (b) J. S. Sawyer, *Nature*, 1972, 239, Figure 3, p. 25; (c) D. J. Spedding, *Air Pollution*, Oxford University Press, Oxford, 1974, Figure 3.1, p. 21; (d) R. P. Turco, *Earth Under Seige: From Air Pollution to Climate Change*, Oxford University Press, Oxford, 1997, Figure 11.12, p. 338.
21. P. V. Hobbs, *Introduction to Atmospheric Chemistry*, Cambridge University Press, Cambridge, UK, 2000, p. 55.
22. J. Hansen, M. Sato, P. Kharecha, D. Beerling, R. Berner, V. Masson-Delmotte, M. Pagani, M. Raymo, D. L. Royer and J. C. Zachos, *Open Atmos. Sci. J.*, 2008, **2**, 217.
23. (a) F. Keppler, J. T. G. Hamilton, M. Braß and T. Röckmann, *Nature*, 2006, **439**, 187; (b) D. C. Lowe, *Nature*, 2006, **439**, 148.
24. D. L. Albritton, L. G. Meira Filho, U. Cubasch, X. Dai, D. J. Griggs, B. Hewitson, J. T. Houghton, I. Isaksen, T. Karl, M. McFarland, V. P. Meleshko, J. F. B. Mitchell, M. Noguer, B. S. Nyenzi, M. Oppenheimer, J. E. Penner, S. Pollonais, T. Stocker and K. E. Trenberth, Technical summary, in *Climate Change 2001: The Scientific Basis. Contribution of Working Group I to the Third Assessment Report of the Intergovernmental Panel on Climate Change*, ed. J. T. Houghton, Y. Ding, D. J. Griggs, M. Noguer, P. J. van der Linden, X. Dai, K. Maskell and C. A. Johnson, Cambridge University Press, Cambridge, 2001, Figure 11, p. 41 (Web 15).
25. F. Lefèvre and F. Forget, *Nature*, 2009, **460**, 720.
26. (a) D. J. Wuebbles, *Science*, 2009, **326**, 56; (b) A. R. Ravishankara, J. S. Daniel and R. W. Portmann, *Science*, 2009, **326**, 123; (c) P. J. Crutzen, A. R. Mosier, K. A. Smith and W. Winiwarter, *Atmos. Chem. Phys.*, 2008, **8**, 389; (d) J. W. Ersiman, H. van Grinsven, A. Leip, A. Mosier and A. Bleeker, *Nutr. Cycling Agroecosyst.*, 2010, **86**, 211.
27. IPCC Core Writing Team, R. K. Pachauri and A. Reisinger, *Climate Change 2007: Synthesis Report,* IPCC, Geneva, Figure 2.1, p. 36 (Web 16).
28. N. L. Bindoff, J. Willebrand, V. Artale, A. Cazenave, J. M. Gregory, S. Gulev, K. Hanawa, C. Le Quéré, S. Levitus, Y. Nojiri, C. K. Shum, L. D. Talley and A. S. Unnikrishnan, Observations oceanic climate change and sea level, in *Climate Change 2007: The Physical Science Basis. Contribution of Working Group I to the Fourth Assessment Report of the Intergovernmental Panel on Climate Change*, ed. S. Solomon, D. Qin, M. Manning, Z. Chen, M. Marquis, K. B. Averyt, M. Tignor and H. L. Miller, Cambridge University Press, Cambridge, 2007, ch. 5, Frequently Asked Question 5.1, Figure 1, p. 409 (Web 8).
29. (a) P. J. Crutzen, *Nature*, 2002, **415**, 23; (b) Editorial, *Nature*, 2003, **424**, 709; (c) J. Zalasiewicz, M. Williams, W. Steffen and P. Crutzen, *Environ. Sci. Technol.*, 2010, **44**, 2228.
30. J. Lovelock, *Gaia: A New Look at Life on Earth*, Oxford University Press, Oxford, 1995, p. 36.

31. S. H. Schneider, *Phil. Trans. R. Soc., A*, 2008, **366**, 3843 [taken from a special issue that also includes a contribution from J. Lovelock, p. 3883].
32. T. M. Lenton and N. E. Vaughan, *Atmos. Chem. Phys.*, 2009, **9**, 5539.

BIBLIOGRAPHY[x]

S. E. Manahan, *Environmental Chemistry*, Lewis Publishers, Boca Raton, FL, 6th edn, 1994.

R. M. Harrison, *Understanding Our Environment: an Introduction to Environmental Chemistry and Pollution*, Royal Society of Chemistry, Cambridge, 2nd edn, 1992.

J. R. Mann, R. S. Davidson, J. B. Hobbs, D. V. Banthorpe and J. B. Harborne, *Natural Products: Their Chemistry and Biological Significance*, Longmans Scientific and Technical, J. Wiley and Sons, New York, NY, 1994.

H. V. Thurman and A. P. Trujillo, *Essentials of Oceanography*, Prentice-Hall, Upper Saddle River, NJ, 7th edn, 2001.

G. Faure, *Principles and Applications of Geochemistry*, Prentice-Hall, Upper Saddle River, NJ, 2nd edn, 1998.

B. J. Skinner, S. C. Porter and J. Park, *Dynamic Earth: an Introduction to Physical Geology*, J. Wiley and Sons, Chichester, 5th edn, 2004.

M. C. Jacobson, R. J. Charlson, H. Rodhe and G. H. Orians, *Earth Systems Science: From Biogeochemical Cycles to Global Change*, Academic Press, London, 2000.

C. N. Banwell and E. M. McCash, *Fundamentals of Molecular Spectroscopy*, McGraw-Hill, Maidenhead, 4th edn, 1994.

J. M. Hollas, *Modern Spectroscopy*, J. Wiley and Sons, Chichester, 4th edn, 2004.

WEBLIOGRAPHY

1. www.ipcc.ch/publications_and_data/ar4/wgl/en/contents.html
2. www.usgcrp.gov/usgcrp/Library/ocp2006/ocp2006.pdf
3. http://ga.water.usgs.gov/edu/watercyclesummary.html
4. www.ipcc-wg1.unibe.ch/publications/wg1-ar4/ar4-wg1-chapter7.pdf
5. www.grida.no/climate/ipcc_tar/wg1/pdf/TAR-03.pdf
6. www.ipcc-wg1.unibe.ch/publications/wg1-ar4/ar4-wg1-chapter2.pdf
7. (a) http://cdiac.ornl.gov/trends/co2/sio-mlo.html
   (b) http://cdiac.ornl.gov/trends/co2/sio-nzd.html
8. www.ipcc-wg1.unibe.ch/publications/wg1-ar4/ar4-wg1-chapter5.pdf
9. www.ipcc-wg1.unibe.ch/publications/wg1-ar4/ar4-wg1-chapter3.pdf
10. www.grida.no/publications/vg/climate/page/3057.aspx
11. www.ipcc-wg1.unibe.ch/publications/wg1-ar4/ar4-wg1-chapter6.pdf
12. http://unfccc.int/kyoto_protocol/items/2830.php

[x]These texts supplement these already cited as references.

13. http://unfccc.int/2860.php
14. www.guardian.co.uk/science/2006/jan/12/environment.climatechange
15. www.grida.no/climate/ipcc_tar/wg1/pdf/WG1_TAR-FRONT.pdf
16. www.ipcc.ch/pdf/assessment-report/ar4/syr/ar4_syr.pdf
17. http://royalsociety.org/stop-emitting-co2-or-geoengineering-could-be-our-only-hope

All the web pages listed in this Webliography were accessed in May 2010.

CHAPTER 6

Waste, Pollution and the Second Law of Thermodynamics

A ramp has been built into probability
The universe cannot re-ascend.

'Ode to Entropy', John Updike

The twin strategies of dematerialisation and transmaterialisation (processing material more efficiently and using more renewables) lie at the heart of the transition to a more sustainable society. Central to both is the requirement to reduce waste. To reduce it (and its impact) requires an understanding of its origins and how it arises, recognition of past efforts to control it, and the challenges in bringing about further reductions.

This chapter surveys and introduces waste and its origins sufficient for an appreciation of what is needed to address questions of technological sustainability. Additional technical and techno-commercial matters are covered in later chapters.

6.1 WHAT IS WASTE?

'Waste', 'emissions', 'by-product' and 'pollutant' are terms often used interchangeably. However, their use in this book requires more precise definitions.

There are legal definitions of the term 'waste' such as that produced by the European Union. Under the Waste Framework Directive (European Directive 75/442/EC as amended), 'waste' is an object the holder discards, intends to discard or is required to discard.

Chemistry for Sustainable Technologies: A Foundation
By Neil Winterton

Published by the Royal Society of Chemistry, www.rsc.org

There are different types of waste that require different approaches to their minimisation (and monitoring) depending on whether they are:

- solid, liquid or gaseous
- concentrated or dilute
- point source or diffuse source
- hazardous or non-hazardous.

In the UK, we produce about 335 million tonnes (Mt) of waste each year. More than 60% of this arises from construction, quarrying and mining, with 12–13% each from industry and commerce, and 9% from domestic sources (Web 1).

Many developed countries have a waste processing infrastructure which collects municipal (usually solid) waste (this includes domestic waste), with the cost of collection and disposal (by incineration, recycling or landfill) being borne by the tax payer or householder. Likewise, water-borne waste is treated in extensive sewerage systems (the efficient operation of which has been, and remains, a major contributor to public health).

Industry is held responsible for the containment and treatment of the waste it generates in making the materials and products that we, directly or indirectly, buy or make use of. The disposal of industrial waste, particularly hazardous chemical waste, is now very tightly regulated in most developed countries and is an increasingly costly aspect of the operation of many businesses or technologies. There is, as a consequence, a significant motivation to reduce the quantity of waste itself. In the chemicals sector this has—historically through a combination of self-interest and regulation—led to progress in emission reduction some associated with improvements in energy usage efficiency. In addition, and this will increasingly be a factor in moving to more sustainable means of production, industry is more and more held responsible for its products when they are at the end of their useful life and are to be disposed of.

This chapter therefore introduces the following definition: **waste is a product without value whose disposal incurs cost**. It is what we have no use for and throw away. In some circumstances, it is useful to distinguish between such a material and one that escapes or is lost during a process or operation. The latter can be better described as an '**emission**'.

It is sometimes possible to turn waste into something whose disposal is less costly or, better still, to something of value that would generate income instead of incurring a cost. The potential benefits arising from cost reduction tend to drive efforts to minimise waste generation and/or the search for alternative uses for waste. In some cases, the discovery of uses for chemical waste has been the basis of whole new industries. The following two examples illustrate this point.

6.1.1 Coal-tar

Coke (a form of carbon) is generated from the carbonisation of coal and is used in iron and steel production. Coal was also gasified to generate town gas, initially used for street lighting. These processes, which were developed in the 19th century, created large quantities of coal tar—a waste made up of a complex mixture of organic compounds, rich in aromatics and heterocyclics. Seeking to use coal tar as a source of pure compounds was the origin of much of today's synthetic dyestuffs and pharmaceuticals industries.

6.1.2 Caustic Soda

Caustic soda (sodium hydroxide[i]) is manufactured for use in the glass and detergent industries. One of the most important processes for its manufacture is the electrolysis of brine, initially *via* the mercury cell process, shown in simplified form in eqn (6.1) to eqn (6.3), or *via* the membrane process (simplified, eqn 6.2 and 6.4):

$$\text{Cathode}: \quad 2\,\text{Na}^{+} + 2\,\text{e}^{-} + \text{n}\,\text{Hg} \rightarrow 2\,\text{Na(Hg)} \tag{6.1}$$

$$\text{Anode}: \quad 2\,\text{Cl}^{-} \rightarrow \text{Cl}_2 + 2\,\text{e}^{-} \tag{6.2}$$

$$2\,\text{Na(Hg)} + 2\,\text{H}_2\text{O} \rightarrow 2\,\text{NaOH} + \text{H}_2 + \text{n}\,\text{Hg} \tag{6.3}$$

$$\text{Cathode}: \quad 2\,\text{Na}^{+} + 2\,\text{H}_2\text{O} + 2\text{e}^{-} \rightarrow 2\,\text{NaOH} + \text{H}_2 \tag{6.4}$$

$$\text{Anode}: \quad 2\,\text{Cl}^{-} \rightarrow \text{Cl}_2 + 2\,\text{e}^{-} \tag{6.2}$$

One mole of dichlorine is the unavoidable **co-product**[ii] of the electrolysis for every two moles of caustic soda produced. Initially, there were few applications for dichlorine. However, as we mentioned above, it came to be widely used in water disinfection in the early part of the 20th century and, later, for the production of a large range of chloro-organic products—the most important of which today are materials based on the polymer, poly(vinyl chloride) (PVC).

6.2 WHEN WASTE BECOMES POLLUTION

The materials that make up waste and emissions would not be problems if they were wholly contained and safely processed. The problem arises when such materials are discharged to the environment or are accidentally lost through poor containment, whereupon they becomes **pollution** and can have a harmful impact on individuals or the environment. The harm may arise in a number ways, giving rise to a number of questions: Harm to what? How? Over what timescale?

---

[i] See footnote, xi, p. 18.

[ii] I introduce the term 'co-product' to specify a product formed stoichiometrically (thus inevitably) with the desired product. The term '**by-product**' is used to define any other material produced in a process in addition to the desired product. 'By-product' and 'co-product' are often used interchangeably though, in this book, I limit the use of 'co-product' to the meaning just given. A by-product is thus anything formed in addition to the desired product and any co-products.

BOX 6.1 LEBLANC PROCESS

The environmental performance of the chemical industry has certainly improved since the late 19th century when the Leblanc process for making soda ash (sodium carbonate) was operated in Widnes (Figure 2.8), before it was rendered obsolete by the more efficient Solvay ammonia–soda process. The Leblanc process (eqn 6.5–6.7) converted common salt (sodium chloride) first into sodium sulfate and then to soda ash:

$$2\,NaCl + H_2SO_4 \rightarrow Na_2SO_4 + 2\,HCl \tag{6.5}$$

$$Na_2SO_4 + 3\,C + CaCO_3 \rightarrow Na_2CO_3 + CaS + 2\,CO + CO_2 \tag{6.6}$$

Or, overall:

$$\begin{aligned} &2\,NaCl + H_2SO_4 + 3\,C + CaCO_3 \\ &\quad \rightarrow Na_2CO_3 + 2\,HCl + CaS + 2\,CO + CO_2 \end{aligned} \tag{6.7}$$

This process emitted hydrogen chloride and carbon oxides to the atmosphere and produced a foul-smelling sulfide-containing solid. Little effort was made to contain these emissions other than to disperse the gaseous effluent by sending it a hundred feet up into the atmosphere in a series of chimneys. The solid waste was simply dumped on land nearby, where it slowly weathered, emitting hydrogen sulfide in the process. The impact of these wastes and attempts to control them gave rise in 1863 to the Alkali Act, an early form of emission control legislation.

The wide distribution of plastic waste in the remote oceans and the detection of pollutants in the polar regions (see Box 5.1) suggest there can be global as well as local, regional and national impacts. Pollution can, therefore, be described as waste (though, in reality, any discarded or emitted product) in the wrong place. Initially, concerns were raised when such pollution was known to be doing harm. The passing of the Alkali Acts in the 19th century was in response to the damage being done by pollution arising from emissions of waste from processes such as the Leblanc process (Box 6.1).

However, it was not long before there were demands to implement preventative measures for control of materials that were known to be hazardous, even though it may not have been evident that they were emitted at concentrations sufficient to cause harm. As the population became more sensitised to the presence of pollution and the possible risks that exposure to it might represent, controls tightened so that only those emissions were tolerated that would lead to concentrations in the environment orders of magnitude lower than those that were thought to be safe. Such is the mistrust of institutions and companies (and those that work for them) that an expectation of the prevention of **all** emissions is now often evident in the pronouncements of those who might be affected.

While this may be an understandable response to the so-called 'risk society'[1] (an interesting and relevant topic that can only be mentioned in passing, with the curious reader wishing to know more referred to some useful sources listed in the Bibliography), it is impossible to know whether excessive restrictions may have had the unanticipated consequence of preventing the development of some useful material or product in the possibly spurious belief that public safety might be at risk by the emission of pollutants at many orders of magnitude lower concentrations at which harm is likely. Such a worrisome consideration has contemporary relevance, bearing in mind the timescale needed to develop and make widely available more sustainable technologies. To what extent may unnecessary regulation (driven by an over-zealous application of the precautionary principle[iii]) constrain future developments by delaying the point at which their operation is permitted? This is just one instance where pragmatic negotiation is needed to balance properly a variety of conflicting interests and principles.

6.3 CHEMICAL WASTE: SHELDON'S E-FACTOR

It is not just science and chemistry that are necessary for the transformation of society to one that is more sustainable. As we have seen (Chapter 3) many other disciplines, interests, perspectives and perceptions must be reconciled, or attempted to be reconciled, to bring this about. Perceptions about chemistry and chemical technology are mixed, with a significant part of the public suspicious (if not actively hostile) towards the discipline and its application—for which phenomena the terms 'chemophobia' and 'technophobia' have been coined. However many improvements there have been in the treatment of chemical emissions and however extensive continue to be the benefits to society and individuals of chemical technology, it must be said negative attitudes are hardly surprising bearing in mind some past industrial practice. More recent incidents such as that in Bhopal in India in 1984 allow such suspicions and hostility to perpetuate.

So, being subject to greater public pressure, legislation and regulation, as well as competitive pressure, the chemical industry has become increasingly concerned with waste and what to do with it. It is important, therefore, to know how much waste the chemical industry produces when making products for sale. One of the best known and most widely-quoted estimates of such waste, the E-factor, was proposed by Sheldon in 1992.[2] The E-factor represents the total waste associated with the manufacture of one unit of final useful product (eqn 6.8).

$$\text{E-factor} = \frac{\text{Mass of waste}}{\text{Mass of product}} \tag{6.8}$$

[iii] The precautionary principle suggests that nothing should be done if harm may result. While, in one interpretation, this is simply widely practiced common sense, it can also be used to suggest that nothing should be done until it can be scientifically proven that no harm of whatever sort will result to anything for the indefinite future. Solving the trilemma of sustainability requires the common sense interpretation. It is always wise to establish precisely what is meant when this term is employed. More discussion of the precautionary principle can be found in the text listed in the Bibliography.

In this estimate, therefore, waste is normalised to a unit of production—thereby providing a relative, rather than an absolute, measure of efficiency. On its own, the E-factor gives no indication of how damaging the waste might be if emitted to the environment untreated. Ideally, E should tend to the lowest value possible. To address the question of potential impact, Sheldon introduced a multiplier, Q, the Environmental Quotient. He proposed arbitrary values that would be low for those with limited impact such as sodium chloride, and high for those with serious impact such as heavy metals. The product, $E \times Q$, provides an estimate of the potential impact of a production process.

Importantly, Sheldon estimated the E-factor for four different sectors of the chemicals industry, *i.e.* refining, bulk chemicals, fine chemicals and pharmaceuticals. This showed that significant differences in efficiency and the total amount of waste produced existed between these sectors. (I deal with this in more detail in Sections 6.4–6.6). Table 6.1 lists the chemical industry sectors with corresponding orders-of-magnitude estimates of the volumes of production and their E-factors.[iv] Production by the oil refining sector is huge, 10^6–10^8 tons per year, though it has a low E-factor of 0.1; *i.e.* the sector produces just 0.1 unit of by-product or waste per unit of production, suggesting overall that the sector is very efficient in its processing. It is worth point out here that these figures are Sheldon's rough estimates; they are not based on actual measured figures—though values of the E factors for individual processes are now beginning to appear (*e.g.* ref. 4). This, however, does not lessen the usefulness of this metric in providing a focus on waste and by-product generation in the chemicals industry, the differences between sectors, and the origins (and significance for waste reduction) of these differences. This is why we will come back to this table several times in later chapters. At the other end of Table 6.1, it can be seen that the pharmaceuticals sector produces much less than the other sectors overall but it is also the least efficient as a sector, with its E-factor very much larger, perhaps surprisingly so, at between 25 and >100. Pharmaceuticals production can, therefore, lead to *ca.* 100 units of waste for each unit of useful active ingredient it manufactures. We will consider why this may be so in Section 6.5.

Table 6.1 Estimates of waste as a proportion of production (E-factor) and of total waste for different sectors of the chemical industry.[2a]

Industry sector	*Product tonnage*	*E-factor kg by-products per kg product*	*Waste (tonnes)*
Oil refining	10^6–10^8	~ 0.1	10^5–10^7
Bulk chemicals	10^4–10^6	<1–5	10^4 to 5×10^6
Fine chemicals	10^2–10^4	5–50	5×10^2 to 5×10^5
Pharmaceuticals	10^1–10^3	25–100+	250 to $>10^5$

[iv] An interesting additional perspective on the development of nanotechnology is provided by Zimmerman *et al.*[3] who suggest that the E-factor for this industry is likely to be in the range 100–100 000 kg waste per kg product. However, these estimates are based largely on laboratory synthesis rather than production-optimised data.

In the right hand column in Table 6.1 is found the (mathematical) product of the volume and E-factor values for each sector, providing rough estimates of the absolute amount of waste each produces. The range of values of annual tonnages of waste or by-product is now much narrower, though of course this takes no account of the potential impact of such waste. How this aspect is addressed is an important topic covered later in the book (Chapter 9).

6.4 APPROACHES TO CHEMICAL WASTE MINIMISATION

A range of technical responses exist that a business or operation can adopt when faced with evidence that a waste is being produced to an unacceptable degree or is having an unacceptable impact. These include:

- remediation
- end-of-pipe treatment
- retrofit
- intrinsically waste-minimising technology.

6.4.1 Remediation

If the waste has already been emitted and is now recognised to be an unacceptable pollutant, then if the impact is localised, the problem can be remediated, *i.e.* it can be dealt with by an after-the-event clean-up. This is clearly the least acceptable option in that the waste should not have been emitted in the first place. However, some historic waste (*e.g.* from the production of town gas in the late 19th and early 20th centuries) caused serious pollution at the sites where this operation took place. Subsequent use of the land may require massive investment for remediation. For example, some £20 million was needed for the treatment of the land prior to the construction of London's Millennium Dome (now known as the O_2 Arena).

6.4.2 End-of-Pipe

The simplest way of dealing with a waste from a process already in operation is to build a waste treatment plant on the end of it—an approach known as 'end-of-pipe' treatment. The use of a catalytic converter on automobiles is a literal and well-known example of such an approach. End-of-pipe methods are usually quite expensive, both in terms of the capital cost of plant construction and in ongoing operational costs. Nor do they reduce waste formation in the first place, which clearly would be the preferred method.

6.4.3 Retrofitting

Best, of course, is not to make the particular waste at all as 'what you don't make can't pollute'. (We cannot prevent all waste production for reasons covered in Section 6.10.) This can be achieved by modifying an existing process by so-called 'retrofitting' important technical changes. This might be done by separating the waste and recycling it (or some of it) back into the process, or by a more radical change to reduce the amount of waste formed.

6.4.4 Intrinsic Waste Minimisation

Alternatively, if a new investment is to take place and a new plant is to be constructed, the whole process can be redesigned, building waste reduction into the technology. Such 'intrinsic waste minimisation' purposefully develops 'cleaner' chemical processes.

In some areas of chemical technology, waste streams from one process are used as feedstock (starting material) for another. Such integration is part of the reason why the petrochemicals sector has been so successful in reducing its E-factor. The development of even more extensively integrated industrial systems forms the basis of the approach known as 'industrial ecology' or 'industrial symbiosis' which we will meet in Section 9.12.

6.5 WASTE MINIMISATION HIERARCHY

It would be wrong to conclude that, prior to Sheldon's publication on the E-factor in 1992, the chemical industry had been wholly uninterested in the issue of wastes and emissions. Avoiding the formation of a by-product that could not be sold produced from a raw material that had to be paid for (and that had incurred additional costs in its processing) has been, and continues to be, an important consideration in seeking to make processes more efficient and less costly to operate. The Waste Minimisation Hierarchy (Web 2) was enunciated to guide businesses, institutions and operations in approaches to reducing waste.

The most effective approach, however, may not turn out to be the cheapest, easiest or most practical. The preference will be governed by one's perspective. The owner of a chemical plant may well take a different view from a resident living downwind of a plant emitting malodorous waste, who could take yet a different view from a public health official who may see the problem as one of nuisance as opposed to one that entails a threat to health or life.

The hierarchy is as follows, with the most preferred (in an ideal world) first:

- **Eliminate source:** the most effective action would be one that prevented the formation of the waste in the first place.
- **Reduce source:** less waste is better than more waste, if the waste cannot be eliminated altogether.
- **Recycle in process:** if a process waste contains some starting material then it makes sense to feed it back into the process rather than throw it away. This may not always be straightforward, in which case the following option could be employed.

- **Reuse, recover or remanufacture:** it may well be that an additional process step is required to remove another by-product that would otherwise be deleterious in the process before it is possible to feed back a starting material or another by-product. This would incur additional costs and consume additional energy and resources.
- **Convert or treat to less harmful materials:** if a hazardous waste cannot be reused in any way, then it needs to be converted to something that poses a reduced (preferably zero) threat to the world beyond the production unit. A good example of this is the treatment of automobile emissions to reduce the concentration of carbon monoxide, hydrocarbons and other gaseous components using autocatalysts in the exhaust streams of cars.
- **Dispose safely:** if there is no other feasible option and the waste poses the minimum of threats to the environment, then it may be acceptable to release it back to the environment. An example would be the pumping into the marine environment of depleted brine (aqueous sodium chloride) from the electrolytic production of sodium hydroxide and chlorine. However, we should not forget that it was once imagined that emissions of carbon dioxide from combustion processes would be of limited environmental and human impact.

Efforts to minimise waste and emissions can focus on simple and straightforward factors such as operational and 'housekeeping' matters such as avoiding spills and only using materials and utilities as required (*i.e.* turning off supplies when not needed). While such actions are simple, they have a relatively limited (but still important) benefit.

To bring about bigger benefits can be more complicated, requiring expenditure and effort, with plans and budgets to manage them. In a chemical process, as we will see, it is possible to address changes in the chemistry to avoid the use of a hazardous raw material or one whose own production (or origin) may be problematic. However, to do this requires a process chemist and process engineer to examine the chemical reactions and establish how the reactor and overall process can be re-designed and re-engineered. This may well require modifications to the process or changes to the feedstock, meeting the costs of which will require justification. Considerations of process and reactor engineering relevant to more sustainable chemicals processing are dealt with in Chapter 9.

The approaches taken by businesses and operations within the four chemical sectors are likely to be different because of the nature, scale and requirements of the products each make and the processes used to make them.

6.5.1 Bulk or Commodity Chemicals Sector

The bulk or commodity chemicals sector is characterised by:

- **large tonnages** (tens of thousands to millions of tons) of a small number of individual products such as sulfuric acid, ammonia, ethene, benzene or poly(ethylene terephthalate), *e.g.* some 140 Mt of sulfuric acid and ammonia are produced annually

- products that are **articles of commerce** and have long been the basis of a wide variety of other technologies (not just chemical) and will continue to be in use for the foreseeable future. Indeed, the amount produced usually provides one measure of the general state of the economy
- the product from one supplier is usually **differentiated** from that of another primarily **by price**
- **continuous manufacturing processes** (as distinct from batch) integrated with other processes (thereby making difficult radical changes to one part of the system that do not impinge on another) and whose use is not generally constrained by patents
- reliance on the **use of catalysts** in their manufacture (of which more in Chapter 10).

High tonnage production plants are very expensive to build, utilising tens to hundreds of millions of pounds in capital. Consequently, with such money at risk, there must be a high probability that the technology being operated will work as expected. This means that it is very unlikely that those committing such large sums would routinely seek to introduce radical or untested technologies. So instead the working life of such processes is extended by evolutionary changes introduced by retrofitting rather than by step-change technology. This cautious attitude is a factor that will affect the pace with which new large-scale technology for more sustainable production will be taken up.

6.5.2 Fine Chemicals and Pharmaceuticals Sectors

The characteristics of the commodity sector can be contrasted with those of the fine chemicals or pharmaceuticals sectors:

- Fine chemicals tend to be manufactured on a **much smaller scale** (tens to thousands of tons).
- Production takes place in small volume, **multi-step, often batch, processes**, using stoichiometric rather than catalytic chemistry—sometimes even simply scaled-up laboratory transformations.
- The materials produced are, by and large, so-called **'effect' chemicals.** An effect chemical can be a component (surfactant, anti-oxidant, reactive monomer) in a complex formulation (*e.g.* a coating, adhesive or lubricant) that delivers some function of value to the user. The product is sold because of its function not because of its composition (which may be proprietary). Its composition and the effect may be protected by **patents**.
- Products have relatively **short lives in the marketplace.** An improved effect may be produced by some other compound or formulation, so new products in this sector are developed regularly with others being, just as regularly, rendered obsolete.

6.6 CHEMICAL WASTE: HISTORICAL TRENDS AND CHANGES

We have noted that reducing the Earth's environmental burden in the future can only realistically be achieved through a significant reduction in the general burden per unit of economic activity, which itself can only be brought about through technological advances. The current focus is on the emissions of greenhouse gases, predominantly from energy generation. We consider the transition from our current dependence on fossil fuels to more sustainable energy production technologies in Chapter 12, addressing some of the chemistry-related questions arising from their development. This shift of raw materials for energy production (wood to coal to oil to natural gas) led to parallel changes in feedstock for chemicals production. We examine the move to chemical technologies based on renewable feedstocks in Chapter 11.

Some changes, hitherto primarily driven by economic factors, have already been responsible for immense benefits in emissions reduction. For example, the replacement of coal as a chemical feedstock for production of acetylene[v] (ethyne)—from calcium carbide itself produced in an electric arc furnace from coal and lime (eqn 6.9 and 6.10)—by oil (for ethene) and later natural gas (for ethane) led to the displacement of one technology by another.

$$CaO + 3\,C \rightarrow CaC_2 + CO \qquad (6.9)$$

$$CaC_2 + 2\,H_2O \rightarrow HC{\equiv}CH + Ca(OH)_2 \qquad (6.10)$$

Bringing about waste minimisation at the level of an entire industry requires some external threat that is likely to force companies to set aside their natural competitive instincts and work with one another [as was the case, for instance, following the agreement of the Montreal Protocol (Web 3), the adoption of which led to the phasing out of chlorofluorocarbons (used as refrigerants and aerosol propellants) responsible for depletion of stratospheric ozone].

Changes to encompass an entire country or political unit become even more difficult to implement, both to agree on what should be done as well as to measure how effective the changes might be. Achieving change at the global level such as may be necessary to bring about sustainable development, while bringing enormous benefits if successful, nevertheless would be well-characterised as a 'wicked' problem as discussed in Section 2.9.

Other historical technological changes can be seen to have enabled humankind to avoid what were, at the time, perceived to be intractable and life-threatening global problems. For example, more than a century ago and as Malthus predicted (see Section 2.3), concern was being expressed that the demand for key raw materials was already outstripping supply. Of specific concern was the

[v] Even as late as 1965, the author of a book on acetylene manufacture and use,[5] published under the auspices of ICI, felt able to say: '*It seems probable that it will be many years, if ever, before the carbide route to acetylene is entirely superseded by processes using hydrocarbon raw materials*'. With hindsight, and considering how quickly oil-based feedstocks actually displaced coal-derived, this looks like breathtaking complacency. It may, however, simply be a reflection of how difficult it is to foretell the future.

continued availability of so-called 'fixed' nitrogen for use as fertiliser and as a source of nitrates and nitric acid for use in explosives production. At that time, the prime source of fixed nitrogen was nitre (a mineral form of potassium nitrate) or Chile saltpetre (a mineral form of sodium nitrate) available in South America. Sir William Crookes, then President of the British Association for the Advancement of Science, warned, in his presidential address in 1898, that '*All . . . nations stand in deadly peril of not having enough to eat*' as the then only source of fixed nitrogen available in sufficient quantity for use as a fertiliser, nitrate of soda, would be exhausted in 20–30 years. Without a suitable replacement there would be '*a catastrophe little short of starvation*'. Crookes believed that '*starvation may be averted through the laboratory*' and that '*the fixation of atmospheric nitrogen is one of the great discoveries awaiting the ingenuity of chemists*'. It was evident even at the end of the 19th century that humankind needed to 'live by the grace of invention' (see Section 2.7) but there were, at that time, no artificial means of economically producing such 'fixed' nitrogen. This became a very active area of research. (For fascinating discussions of this and associated topics surrounding the development of 'fixed' nitrogen, see the books by Smil and by Partington and Parker in the Bibliography.) It was not long after Crookes' apocalyptic address that Fritz Haber providentially discovered the catalyst for the process which now bears his name in which elemental nitrogen (readily separated from the atmosphere) is converted catalytically to ammonia by reaction with hydrogen (eqn 6.11). The centenary of this development was celebrated in 2009.

$$N_2 + 3\,H_2 = 2\,NH_3; \Delta H_{298} = -46\,\mathrm{kJ(mol\,of\,NH_3)^{-1}};\; K_P = \frac{p^2_{NH_3}}{p_{N_2}\cdot p^3_{H_2}} \qquad (6.11)$$

The discovery was developed into a technology by Bosch, Haber and Mittasch for the major German chemical company, BASF; 40% of all nitrogen fixed on Earth each year now results from the operation of the Haber process, a demonstration of the degree to which we now depend on technology.[6]

Table 6.2 shows the percentages of ammonia formed at equilibrium established at different temperatures and pressures. Le Chatelier's principle states that an exothermic reaction is favoured by low temperatures and a gas-phase

Table 6.2 Percentage ammonia at equilibrium with nitrogen and hydrogen (eqn 6.11) for various temperatures and pressures.

	% NH_3 at equilibrium			
	Process pressure			
Process temperature	*1 atm*	*10 atm*	*100 atm*	*1000 atm*
200 °C	15.3	50.7	81.5	98.3
400 °C	0.4	3.8	25.1	79.8
700 °C	0.0	0.0	2.1	12.9

Table 6.3 Changes in waste produced from, and resource requirements for, the industrial production of 300 000 t y^{-1} ammonia by the Haber process resulting from the use of different primary raw materials and technological development.[8]

Factor	*Unit*	*1960*	*1970*	*1980*	*1990*
Feedstock		coal	oil	oil	gas
SO_x	t y^{-1}	400 000	1.6	1.3	1
NO_x	t y^{-1}	high	160	140	12
Solids	t y^{-1}	5000	0	0	0
Liquids	t y^{-1}	high	low	low	very low
Land area	acres	60	6	5	3
Energy	therm y^{-1}	800	400	300	300
Staff		3000	75	60	30
Capital	£ million (1990)	400	120	120	80

reaction in which the number of moles of product is less than the number of moles of reactant benefits from the application of high pressure. Because of the substantially greater costs of plant able to operate at very high pressures, the precise conditions (a typical process is operated at 100–200 atm and 450 °C) represent a compromise: the reaction is carried out at temperatures at which, while conversion of reactants to products is not 100%, equilibrium can be reached relatively quickly. Unreacted nitrogen and hydrogen must be separated from the product and recycled. The catalyst employed today (iron oxide containing small amounts of chromium and cerium) is very similar to the one optimised by Mittasch in the early stages of technical development.[vi] Interestingly, most of the more important changes to the process—making it more efficient, producing less waste and costing less to operate—have come through changes to the production of the hydrogen used as feedstock. This is evident in Table 6.3, which lists for an industrial-scale plant making 300 000 tonnes of ammonia per year, the quantities of waste generated as well as other costs (*e.g.* capital used for the plant's construction, the number of personnel needed to operate it and energy consumed in its operation). During the 1950s and 1960s when coal was the source of hydrogen, very large amounts of oxides of sulfur (SO_x) and oxides of nitrogen (NO_x), as well as other solid and liquid wastes, were produced. The capital costs for such plant and the number of people needed to run it were also large. This changed significantly in the late 1960s and 1970s when the prime feedstock for the manufacture of hydrogen shifted from coal to oil, and again during the 1990s when natural gas was used in the process with a further reduction in capital costs. The land area needed for the process has also reduced significantly over this period.

These changes in efficiency were driven largely by competitive pressure and the seeking of competitive advantage, and are among the factors that enabled the planet to feed itself, not only to avoid Crookes' dire prediction but also the more recent one from Paul Erhlich mentioned in Chapter 2.

[vi] For Mittasch's own perspective on the associated catalyst development, see his review[7] published in 1950.

6.7 INEVITABILITY OF WASTE (BUT NOT NECESSARILY OF POLLUTION)

With the focus on industry both to accelerate the development of ever more efficient processes and to do so while shifting the basis of its technologies to more renewable feedstocks, we should perhaps confront a fairly obvious question: Why make any waste at all? Why cannot industry produce no waste?

We show below that waste, including chemical waste, is inevitable and is a direct consequence of the operation of the Second Law of Thermodynamics. In the words of Tim Jackson *et al.*[9] '*the concept of "non-waste" technologies is a chimera*'. The relevance of this can be appreciated from Figure 6.1, where process efficiency is plotted along the *x*-axis and waste produced along the *y*-axis. If P represents the amount of useful product manufactured in a process that consumes an amount, M, of raw material, then efficiency (the inverse of its E-factor) is simply P/M (a relative measure) and the associated waste produced is M-P (an absolute measure). P/M varies between 1 (for a perfectly efficient process where P = M) and 0 (where no product at all is formed), with M-P varying between 0 and M. The stages of technological development over time can be represented by vertical lines, with E_2 representing current technology and E_1 representing some foreseeable practical limit. Developments in technology resulting from greater experience and understanding, or the use of different processes or feedstocks, should move the vertical line to the right towards E_1, decreasing the area represented by waste. We will see in Section 6.10 that it is not possible, even in principle, to proceed beyond a limit, represented by E_0, in reducing waste. The line E_0 represents the absolute limit set by thermodynamics.

Historic trends have seen the line represented by E_2 move to the right as efficiency improves and the waste produced declines. The aim of technological

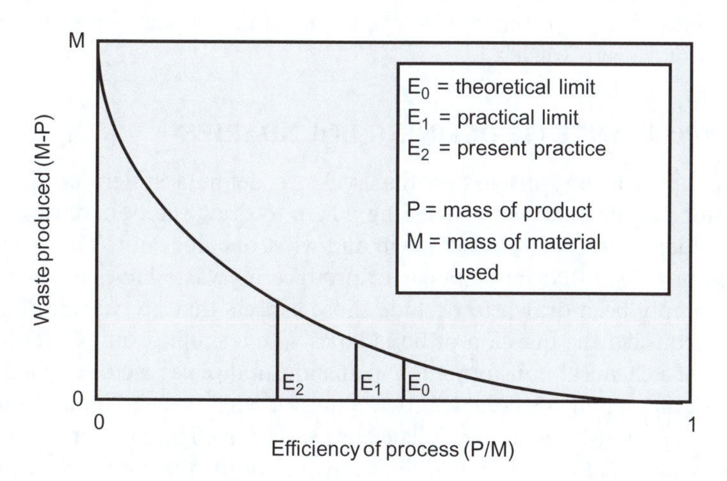

Figure 6.1 Process efficiency as a function of waste produced, highlighting theoretical and practical limits. Adapted from ref. 9.

development is to identify new products and processes that further shift the vertical line to the right. However, accepting that thermodynamic limitations mean that waste will always be produced, ways must be found either to transform the waste into harmless emissions that can be safely released (so-called **abatement**) or to collect the inevitable waste and place it somewhere where it can do no harm (so-called **sequestration**). We use the case of CO_2 abatement (Box 6.2) to illustrate this and other relevant questions, to some of which we will return in later chapters.

However, it is important to recognise that abatement and sequestration both have energy and material costs that must be met and an associated set of emissions that must be dealt with. The current debate surrounding the possibility that the carbon dioxide produced from energy generation *via* coal combustion may be captured and stored is an example of a sequestration technology that must be analysed in these terms (and is considered again in Chapter 12).

After these important diversions, let us move back to a consideration of the efficiency with which chemical technology processes raw materials into useful products. We saw that it was possible to express such efficiency in relative terms, though earlier we entered a caveat to suggest that this may simply result in increased consumption so that, while relative levels of waste might decrease, the absolute levels might not. So, the more difficult question needs to be asked: what if improvements in relative efficiency are not enough to achieve the Factor 4 or Factor 10 improvements believed to be necessary (see Section 3.2, particularly p. 45). A reasonable question for the general public to ask is: 'Is it possible to move towards chemical technology that produces no waste at all?' Unfortunately, as we will see in Section 6.10, processes that produce no waste (and are not associated directly or indirectly in any way with the production of waste) are impossible. The more practically focused question, therefore, is this: how do we extract all usable function or energy from any material or from any process using sufficient of it to aim for zero environmental impact, knowing that we cannot achieve zero waste?

6.8 IMPORTANCE OF DEFINING BOUNDARIES

A critical part of any discussion that seeks to define a system so that waste generation can be analysed is deciding where to draw the boundaries of what one considers to be part of the system and what one does not. The boundaries of those processes that may appear to produce no waste have, by accident or design, simply been drawn to exclude those aspects that are waste-producing.

Let us consider the question of boundaries with a simple example. The understanding of a chemical transformation in an industrial-scale reactor requires, at the very least, measurements designed to account for what goes in (inputs) and what comes out (outputs) (*i.e.* a mass balance) as shown in Figure 6.3. In Figure 6.3b, the waste treatment operation is included as part of the plant and, Figure 6.3a, it isn't. In both cases, any waste that may have been generated in the production of the input materials has been completely ignored.

BOX 6.2 ABATEMENT AND SEQUESTRATION OF CARBON DIOXIDE.

We now realise that, even though CO_2 is a low exergy material (Section 6.13), it still has a deleterious effect on the environment. In the transition from the use of coal to oil to natural gas for energy generation, the emissions of CO_2 and other greenhouse gases emitted per unit of electricity generation have gone down[10a,b] (Table 6.4). During the further transition to more sustainable sources of energy (nuclear energy, energy from renewable sources—dealt with in Chapter 12), fossil energy sources will still be consumed. To limit CO_2 emissions, abatement technologies will be required in which the CO_2 generated is prevented from being emitted to the atmosphere by being 'sequestered' and put somewhere where it cannot do harm.

Table 6.4 Production of greenhouse gases in g(CO_2e) kWh^{-1} from various energy sources.[10a]

Energy source	*g(CO_2e) kWh^{-1}*
Coal	960–1050
Oil	778
Natural gas	443
Nuclear	66

But before it can be sequestered, it must first be captured. Several approaches are under consideration. This book examines two 'post-combustion' processes:

- capture from a concentrated source (e.g. as power station emissions), see below
- capture from a diffuse source (the atmosphere) (Section 12.2.2).

The furthest developed **carbon capture and storage** (CCS) technology involves removal of CO_2 from the waste gases from power stations that would otherwise be emitted to atmosphere.[11a] This can be achieved by its reversible reaction with ethanolamine (eqn 6.12 and 6.13). Such a process has been demonstrated and is ready to be applied to the removal of CO_2 from flue gases of a fully operational power station (ref. 11b, Web 6). CO_2 must then be 'stripped' from the ethanolamine solution, dried and compressed (see Figure 6.2) in readiness for sequestration.[11c–e] All these are energy-consuming processes with associated CO_2 emissions in their own right.

$$HOCH_2CH_2NH_2 + CO_2 \rightleftharpoons HOCH_2CH_2NH_2.CO_2 \tag{6.12}$$

$$HOCH_2CH_2NH_2.CO_2 \rightleftharpoons HOCH_2CH_2NHC(O)OH \tag{6.13}$$

Some 3.5 Mt y^{-1} of recovered CO_2 are already used in the USA to enhance the recovery of oil from wells close to depletion, with the CO_2 thereby being stored in underground geological strata. This application creates a value for CO_2 of *ca.* US$25 per tonne that provides some incentive for its use. An attractive related use is enhanced coal-bed methane production[13] in which two moles of CO_2 displace one mole of methane. The stoichiometry of this process is attractive because more CO_2 is sequestered than is produced directly by the burning of the displaced methane. The capacity for such storage is enormous, being 3 GtC in the Netherlands alone.

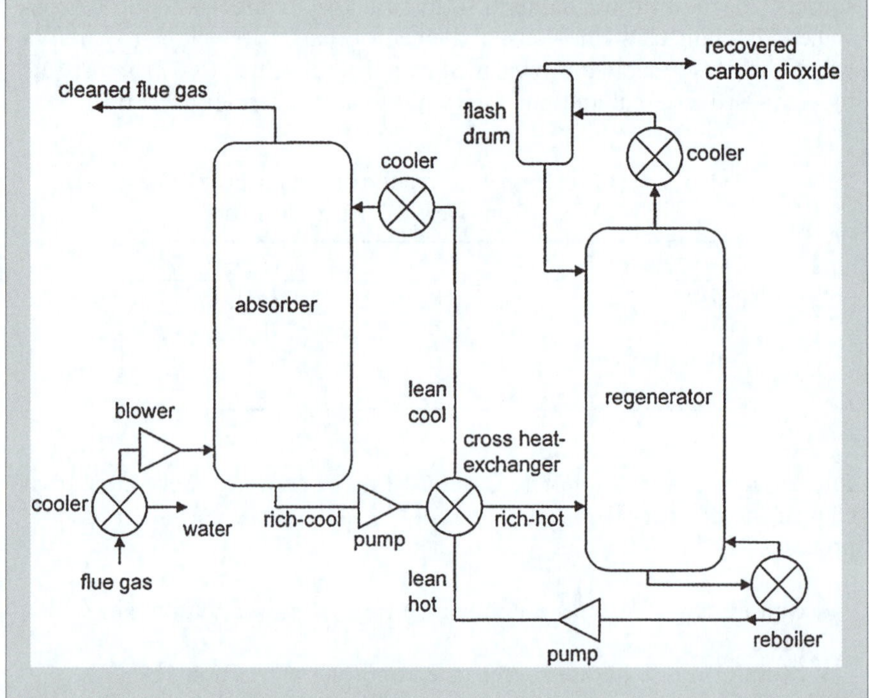

Figure 6.2 Flowsheet for process to strip and recover carbon dioxide from flue gas (ref. 12). (Reproduced with permission of T. U. Delft.)

Other anthropogenic 'sinks' for CO_2 could involve deep ocean disposal where, at below 1000 m, solid $CO_2.5.75H_2O$ clathrates are stable (though the rates at which such clathrates may be formed and the net energy balance of the overall process still needs fully to be assessed). Proposals have also been made for CO_2 disposal in deep wells and in saline aquifers. A serious concern associated with each of these methods is the degree to which the CO_2 is permanently sequestered, as any leakage would negate the benefit—perhaps catastrophically if it occurred suddenly and on a large scale. An attractive alternative is to convert the CO_2 *via* the growth of biomass into charcoal (or 'biochar') and to bury this material. The involatility and insolubility in water of biochar would make such an end point particularly attractive. (This matter is considered further in Chapter 11.)

An alternative that a chemist might suggest when faced with a large source of a low-cost waste such as carbon dioxide would be to use it as a product or as a chemical feedstock.[14a] However, in considering its more extensive use as a source of carbon in the production of organic chemicals, we must remember that CO_2 is of low exergy and energy would need to be expended to convert it into most useful products, making it economically unattractive in such cases. In addition, the use of CO_2 as a feedstock for the direct chemical production of hydrocarbons and oxohydrocarbons will be dependent on a source of dihydrogen whose production also avoids the use of fossil feedstocks. Were such a cheap and renewable source of dihydrogen to be found (see Chapters 11 and 12), then it might be possible to carry out a wider series of (catalysed) hydrogenations leading, for instance, to the important feedstocks, methanol, formaldehyde and carbon monoxide (eqn 6.14–6.17):

$$CO_2 + H_2 \rightarrow CO + H_2O \quad (6.14)$$

$$CO_2 + H_2 \rightarrow HC(O)OH \quad (6.15)$$

$$CO_2 + 2\,H_2 \rightarrow HCHO + H_2O \quad (6.16)$$

$$CO_2 + 3\,H_2 \rightarrow CH_3OH + H_2O \quad (6.17)$$

Methane, the ultimate product of such reductions, could be used to form synthesis gas in reaction (eqn 6.18) with more CO_2:

$$CH_4 + CO_2 \rightarrow 2\,CO + 2\,H_2 \quad (6.18)$$

$$C_6H_5C_2H_5 + CO_2 \rightarrow C_6H_5CH{=}CH_2 + CO + H_2O \quad (6.19)$$

CO_2 could be used in dehydrogenations such as in the catalysed formation of styrene from ethylbenzene (eqn 6.19).[14b] Fine chemicals such as organic carbonates, carbamates, isocyanates and amides may also be accessible (eqn 6.20 and 6.21). The formation of isocyanates (widely used in the production of urethanes and polyurethanes) in this way is currently the subject of intense effort[15] for another reason: the successful development of a process using CO_2 would avoid the use of phosgene, the current raw material of choice, though one that is hazardous to use.

$$CO_2 + 2\,RNH_2 \rightarrow 2\,RNH_2.CO_2 \quad (6.20)$$

$$2\,RNH_2.CO_2 \rightarrow [RNH_3][RNHCO_2] \rightarrow RNCO + H_2O + RNH_2 \quad (6.21)$$

While these processes may convert CO_2 into a form where it cannot be emitted to the atmosphere, after such material has come to the end of its working life it may be incinerated whereupon the carbon will be emitted. So, such uses could only delay for varying lengths of time, rather than prevent, the eventual emission of CO_2 to the atmosphere. Furthermore, such uses would only ever be modest consumers of CO_2 compared with the *ca.* 5 GtC as CO_2 emitted from energy generation. Even if CO_2 were to replace all fossil-sourced carbon used as feedstock in chemical processing, the impact on anthropogenic CO_2 emissions would still be minimal.

Even in this simple case, we can see that care is needed when comparing one process with another. Comparing full-scale technologies is even more complex. This can be generalised for any system and its surroundings as shown in Figure 6.4, something we do as a matter of course when considering a thermodynamic analysis. We have a bounded system comprising a series of components separated from an environment. The system receives inputs in the form of energy and materials from the environment and delivers a desired product, wastes and other emissions to it. All material and energy must be accounted for and, from the requirements of the First Law of Thermodynamics, can be accounted for.

6.8.1 Batteries vs. the Internal Combustion Engine

A topical example that illustrates the importance of defining system boundaries is provided by efforts to decide whether an electric car powered by batteries is more or less polluting than a car powered by an internal combustion engine.

Much of the media attention surrounding electric cars has suggested that they are 'pollution free' when being driven. From a consideration of the car alone, the simple answer might be that the internal combustion engine is more polluting as a power source as it emits fuel combustion products in use whereas the electric car does not. However, this fails to give a complete picture. A petrol (gasoline) fuelled car needs to fill up with fuel from a petrol station. The petrol is delivered by tanker from a refinery that has received its raw material, crude oil, by sea or pipeline after extraction from an oil well. In the same way, the batteries that power an electric car must be charged and recharged (a process that, at present, takes much longer than pumping petrol). This is done using electricity that has been generated in power stations driven by coal, oil or natural gas, or by nuclear fission, all of which involved the extraction and processing of the feedstock from a natural source. Only electricity from hydroelectric or other renewable sources is free from this association. A proper analysis of the material and energy flows associated with these two technologies is unlikely to conclude, therefore, that one is 'zero emitting' except in a narrow and misleading sense that has resulted from an ill-judged drawing of the system boundary used in the comparison. (It is worth pointing out that automobiles powered by hydrogen fuel cells generate only water as a waste. However, even here, hydrogen is currently almost entirely sourced from fossil feedstocks. Only when hydrogen can be obtained from ultimately renewable sources will it make a contribution to cleaner technology. Likewise, the use of photovoltaics to convert sunlight directly to electricity may appear attractive, but the efficiencies

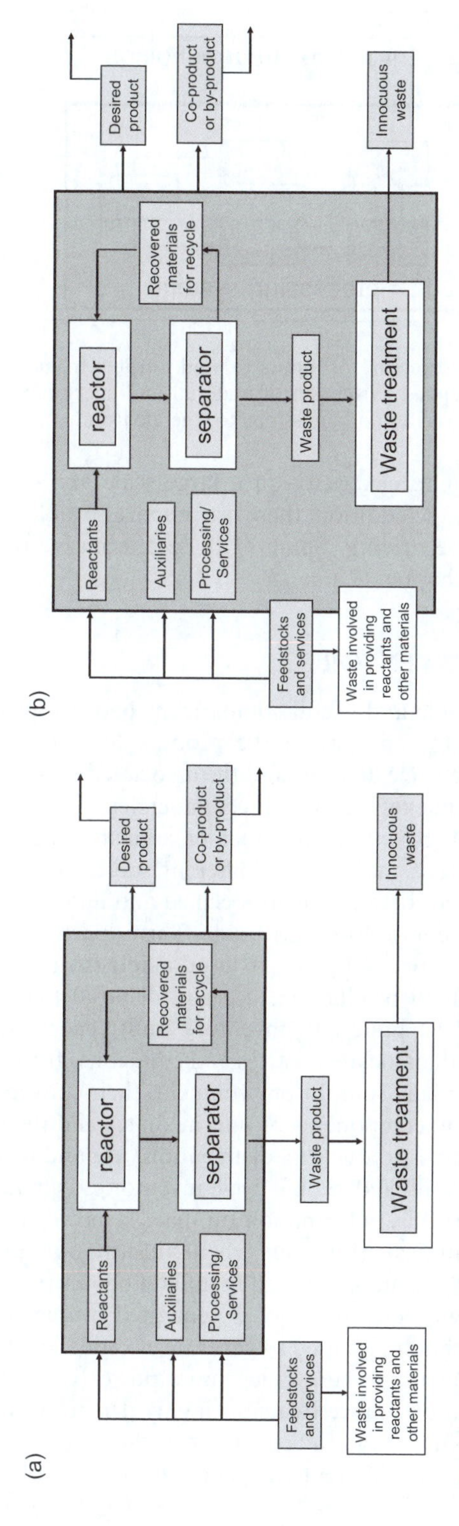

Figure 6.3 Elements of a process for chemical production showing some unit operations and key boundaries, particularly highlighting the different accounting boundaries (b) with and (a) without waste treatment.

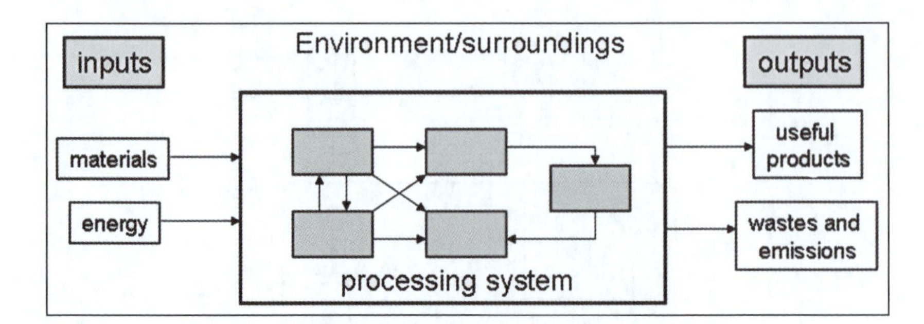

Figure 6.4 Generalised picture of inputs to, and outputs from, a processing system with associated extraction of material and energy from, and delivery of products, wastes and emissions to, the environment.

of conversion of light to electricity on a large scale are as yet insufficient—though are improving. In addition, there are practical problems, such as energy storage, for night-time driving. Some of these alternative power sources are explored further in Chapter 12.)

6.9 LIFE-CYCLE INVENTORY

Seeking to make reliable and useful comparisons between products, processes or systems leads to the concept of the product life-cycle. The best-known techniques rely on life-cycle analysis and the associated life-cycle inventory and life-cycle impact approaches, for which recommended methodologies have been published (Ref. 16, Web 7). Such methods have their recognised weaknesses.

A product life-cycle is made up of different stages (Figure 6.5), with any interfaces or boundaries between them specified and understood to ensure that meaningful and valid comparisons are made. This is particularly the case when the scope of the inventory is limited to production (so-called '**cradle-to-gate**' life-cycle) rather than to production, use and disposal ('**cradle-to-grave**' life-cycle) (Figure 6.6). The preferred 'inventory' must encompass material and energy consumed (with associated wastes and emissions) from the extraction of the primary raw materials from the environment, their conversion first into the materials that make up components or intermediates and their assembly in the manufacture of the final product, its distribution, use and disposal.

Producing reliable and credible life-cycle inventories is made difficult by the lack of reliable and precise information on many aspects needed for comparison. Major uncertainties arise even where information exists. Differences between processes or products operated in different countries can, for example, be dependent on how electrical power is generated in one country compared with another. The technologies used to generate electrical power are different in the UK and, say, Canada (where much more power generation comes from hydroelectric sources) or France (which is highly dependent on nuclear power generation). It is possible that choices between the relative efficiencies of different technologies are different, depending on the country where the

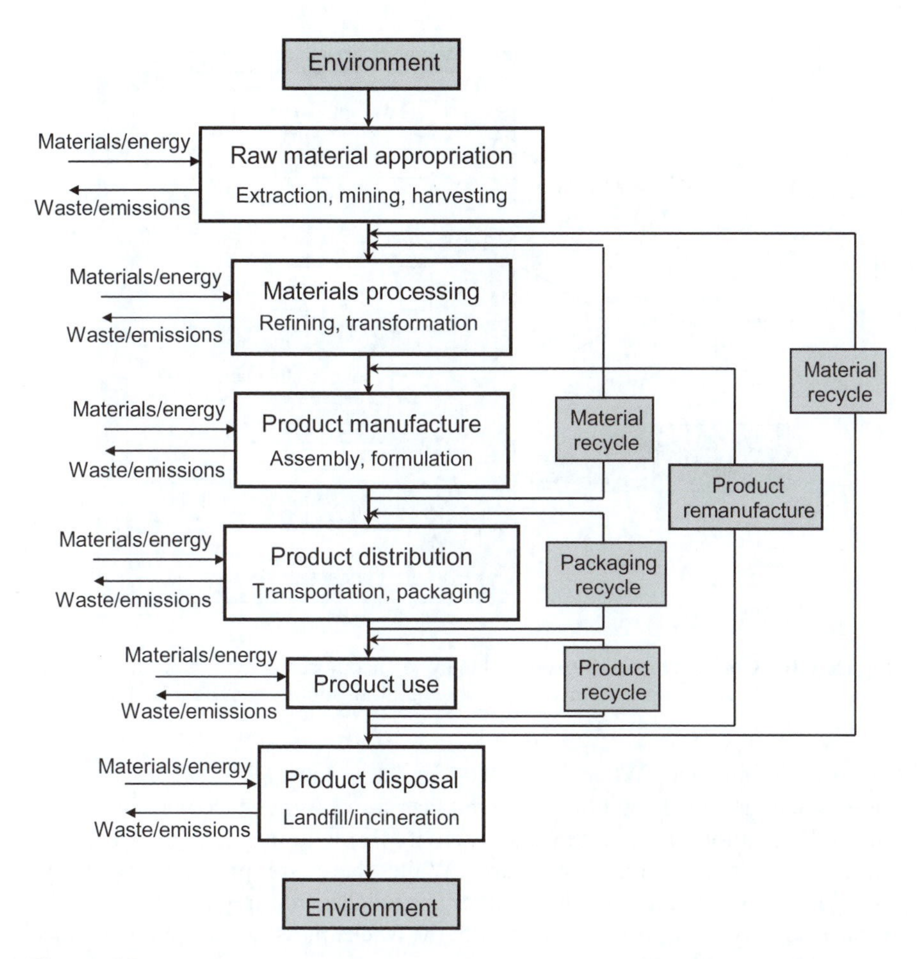

Figure 6.5 Generalised representation of the components of a product life-cycle.

technology is to be operated for reasons of the so-called 'energy mix'. An example of such an analysis, which illustrates some of the ideas and difficulties we are about to address, based on a comparison of the environmental impact of disposable *versus* reusable cups, is provided in Box 6.3.

When comparing the impact of technologies, difficulties also arise because of the 'incommensurable' character of some of the factors. By this we mean that one impact of one technological approach might be a contribution to ozone depletion, while the corresponding impact of another technological approach with which the first is being compared is on species biodiversity or encroachment into wilderness. Clearly, it is impossible to compare these two impacts directly. However, economists make indirect comparisons by converting such factors into a monetary equivalent by assessing the value that people might put on what may be at risk to determine, notionally, how much they might be willing to pay to prevent the impact or to be paid to

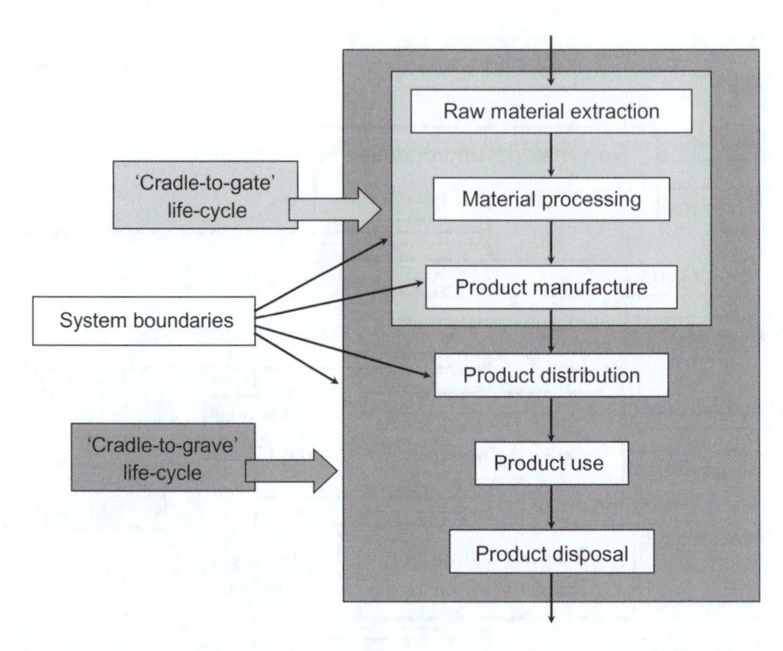

Figure 6.6 Components of a product life-cycle encompassed by 'cradle-to-gate' and 'cradle-to-grave' inventories, highlighting some key system boundaries.

tolerate the impact. While the need for 'internalising'[viii] such factors is increasingly accepted, building such impacts into business accounting is difficult to bring about unless all business activity throughout the world is assessed on a common accounting standard. While there are problems with this approach, it is difficult to envisage other approaches that might successfully be applied. In any event, other than to note the 'wickedness' of such comparisons, these matters (other than the discussion of process and project economics in Chapter 9) go beyond the scope of this book.

As noted earlier, cradle-to-gate inventories confine the analysis up to the point where a product leaves the producer's factory, ceases to be under the control of the manufacturer and becomes the responsibility of someone else. Here the inventory addresses usages from primary raw material extraction to manufacture. A full cradle-to-grave life-cycle is preferred as a further range of impacts (and sometimes bigger impacts) arising from the use and disposal of the product manufactured are taken into account. Such '**gate-to-grave**' impacts can be substantial, particularly for long-lived products such as refrigerators. Even here, concern has tended to be focussed on the 'direct' impact of the working fluid or refrigerant (associated with its global warming potential; see

[viii] 'Internalising' is a means by which it is possible financially to account for the costs of environmental and other 'external' impacts associated with the activities of organisations, companies or businesses. The disciplines of environmental and ecological economics and the technique of environmental accounting address these contentious matters but, as important as they are, they are not discussed here in detail. Some relevant texts are included in the Bibliography.

BOX 6.3 WHICH ARE MORE ENVIRONMENTALLY ACCEPTABLE: REUSABLE OR DISPOSABLE CUPS?

An everyday example comparing four different types of cup—two disposable, two reusable—highlights some central issues. The analysis is based on a paper, published by Hocking in 1994,[17a] though the analysis and conclusions derived from it remain valid.[17b,vii] Hocking set out to answer the question: 'Which is more environmentally acceptable: a ceramic, glass, paper or plastic cup?' This qualitative question was re-formulated into a different, more quantitative, question: 'How many times, *x*, must a reusable cup (glass, ceramic) be reused before the energy consumption per use equates to that required for a single use of a disposable cup (paper, plastic)?'

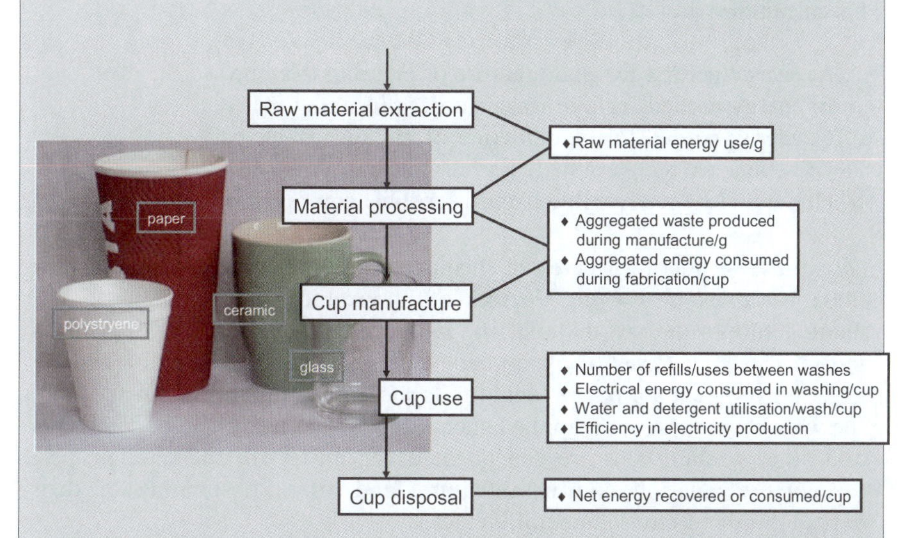

Figure 6.7 Key parts of an inventory of material and energy usage during various stages of the manufacture, use and disposal of cups made from four different materials.[17a]

To make the comparison valid we need to make an inventory of the material and energy usage in the various stages of manufacture, use and disposal for each of the cup types.

The five stages (the analysis does not explicitly include distribution) in the cradle-to-grave cycle are shown in Figure 6.7. This also shows the associated metrics that are used to define material and energy usage in a form that

[vii] A more recent life-cycle assessment/eco-efficiency analysis[17c] was unable to come to a conclusion regarding which was better for the environment: a reusable poly(carbonate) cup or disposable poly(propylene), poly(ethylene)-coated or poly(lactide) cups.

enables a valid comparison to be made between quite disparate technologies and materials. The relevant metrics include:

- the energy expended per gram to extract and process the material from which the cup is made
- the aggregated energy used in the fabrication of the cup (per cup)
- for those cups that will be reused, the nature of the use process (whether washed after a single use or after multiple uses), the energy and water (and washing aid) utilisation of the washing process (per cup) and the efficiency of the conversion of primary energy source into electricity
- the net values of waste produced and energy recovered following disposal of the different types of cup.

The variables (all expressed in a single unit, the Joule) that Hocking uses in his calculation are:

A: energy needed for manufacture of one reusable cup
B: energy needed for one wash (to desired standard)
C: energy needed for manufacture of one disposable cup
G: net energy recoverable from reusable cup on disposal
H: net energy recoverable from disposable cup on disposal.

(For those wondering where the chemistry comes in, it lies behind each of these variables to varying degrees of importance and remoteness. The manufacture of the raw material and its processing into the cups depend on the relative efficiencies of a number of chemical technologies for making glass, poly(styrene) or paper—and for its subsequent processing. In addition, the washing process is an applied chemical problem in which the detergents and other washing aids are dependent on chemistry for their design, performance evaluation and manufacture, and an understanding of their degradation and environmental impact.)

Various scenarios can be envisaged. We will use the simplest of Hocking's scenarios to exemplify the method: a single use of the reusable cup before washing without taking account of energy and material recovery on disposal. In this case, the energy associated with the single use of the disposal cup is:

$$\mathrm{E}_D = \mathrm{C} \tag{6.22}$$

The energy associated with the single use and wash of the reusable cup is:

$$\mathrm{E}_R = \mathrm{A} + \mathrm{B} \tag{6.23}$$

For x reuses with washing, this becomes:

$$\mathrm{E}_R(x) = \mathrm{A} + \mathrm{B}x \tag{6.24}$$

For the use of x disposable cups, then:

$$E_D(x) = Cx \tag{6.25}$$

We wish to establish whether, at some point, the overall energy usage becomes the same (*i.e.* the point at which the waste associated with production of a particular number of disposable cups becomes the same as the production of a single reusable cup and the same number of reuses). At this point:

$$E_R(x) = E_D(x) \tag{6.26}$$

$$A + Bx = Cx \tag{6.27}$$

The equivalence point is thus represented by:

$$x = \frac{A}{(C - B)} \tag{6.28}$$

A second scenario might take account of energy recovery on disposal, leading by a similar argument to:

$$x' = \frac{(A - G)}{[(B - H) - C]} \tag{6.29}$$

Other scenarios might take account of repeat uses before washing.

The basic data for the four materials are given in Table 6.5, expressed in kJ per cup. The energy requirement per gram of material from which the cup is fabricated is greatest for polystyrene and lowest for glass. However, we need to compare like with like, so we have to consider cups with approximately the same volume, estimated to be 8–9 fluid ounces (US) (236.6–266.2 ml). To produce cups with this volume requires different weights of material, with polystyrene being the least and glass being the most. The energy requirement per cup therefore goes in the order polystyrene < paper < glass < ceramic.

Table 6.5 Energy requirement (kJ cup^{-1}) for the production of a typical cup from four different materials.[17a]

	Energy requirement (kJ g^{-1})	*Weight of 8–9 fluid ounce cup (g)*	*Energy requirement (kJ cup^{-1})*
Expanded polystyrene	104.3	1.9	200
Paper	66.2	8.3	550
Ceramic	48.2	292.3	14 100
Glass	27.7	198.6	5500

In addition, we need to know the electrical energy required for washing the cup. This is estimated to be 105.5 kJ cup^{-1}. The primary energy required for washing per cup will be dependent on the so-called 'energy

mix'. This reflects the thermodynamic efficiency of the different methods employed to generate electricity (different from method to method, with the 'mix' different from country to country). The primary energy required for washing per cup is, therefore, 184 kJ cup^{-1} in Canada (where the overall efficiency is 57.3% because of the significant use of hydroelectricity generation) and 278 kJ cup^{-1} in the USA (where the overall efficiency is 38%).
Various comparisons can now be made between a reusable cup and a disposable cup to establish what the break-even point is (Table 6.6). It is the lowest for the glass/paper comparison. In this case, 15 reuses of the glass cup equates (for this scenario only in which no account is taken of energy recovery on disposal) to 15 paper cups. It is highest for the ceramic/polystyrene comparison in which over a 1000 reuses are needed for equivalence with 1000 polystyrene cups used once. Note that this value is highly dependent on the difference between the energy associated with the manufacture of the disposable cup and the energy associated with the washing process.

Table 6.6 Break-even point, x, for two scenarios at which the overall energy used for the production and x reuses of a glass or ceramic cup becomes equal to that used in the production of x disposable cups made from paper or polystyrene.[17a]

Pairs compared	*x, one use before wash or disposal*	*x, two uses before wash or disposal*
Glass/paper	15	30
Glass/polystyrene	393	786
Ceramic/paper	39	78
Ceramic/polystyrene	1006	2013

We can draw some other conclusions in what might, initially, have been thought to be a trivial and simple example.

Results are highly dependent on variables included or excluded (*e.g.* the energy recovered on disposal) and on local circumstances (*e.g.* the energy 'mix') and the nature of the washing process.

The foregoing takes no account of a range of other important factors that might govern the choices made as to whether or not to use a disposable instead of a reusable cup. Other use attributes include:

- reusability or disposability
- convenience such as the ease of holding
- circumstance of use, whether domestic, catering or institutional
- heat retention
- aesthetics
- safety including insulation, resistance to breakage or chipping

- resistance to leakage
- hygiene relating to use and reuse and the efficacy of washing process.

All these might result in usage that is non-optimal for the environment. However, because the benefits are so circumstance-specific there is no single scenario that could guide some sort of global decision to require disposables or reusables to be used. If such is the case for simple products such as drinking utensils, think how much more difficult such judgements are in the context of more complicated products.

Section 5.7) introduced during manufacture, as this may leak during use or be lost to atmosphere when the refrigerator is disposed of (if steps to recover refrigerant are not taken). However, the issue is more complex, as by far the greater impact of refrigeration (by a factor of 2.6–6.2[18]) arises from the greenhouse gases associated with the energy consumed in operating the refrigerator.

The cradle-to-grave assessment has the valuable consequence of encompassing final disposal. Ideally, such analyses challenge the producer to ensure that as much of the product as possible can be recycled. Such an outcome can be assisted by the purposeful design of the product in the first place. Much progress is being made, for instance, to engineer cars, photocopiers or mobile phones to make the process of disassembly easier so that recycle and remanufacture can be maximised. Some of this can be assisted by sensible materials design and materials chemistry is playing a role in this regard. However, the processes of disassembly of chemical products themselves are not so straightforward, as the example in Box 6.4 illustrates. It is generally the case that different functions in a composite material or product are provided by different chemical components. Will more energy and materials be consumed in bringing about the separation of the individual chemical components than are brought back into use from the components so recovered? If this is the case, there is no net benefit; indeed, there is an extra cost and additional waste production or energy usage.

6.10 THE CENTRAL IMPORTANCE OF THERMODYNAMICS

While it is not the intention of this book to present a comprehensive treatment of thermodynamics (there are many sources to which the reader may be directed—see Bibliography), nevertheless an awareness is necessary of certain fundamental concepts and terms such as free energy, enthalpy and entropy, reversibility and irreversibility, and the relationship between them. These underpin the discussion of the relevance of the laws of thermodynamics (summarised in Box 6.5) when we come to consider practical systems (and introduce the concept of exergy in Section 6.13) as well as questions of waste generation, its minimisation and amelioration relevant to the manufacture and processing of chemicals.

BOX 6.4 COMPLEXITY AND COMPOSITE MATERIALS: THE UBIQUITOUS CRISP PACKET.

A triumph of chemical and material processing technology has been the ability to produce composite products able to meet, at an acceptable monetary cost, a range of demanding performance criteria. Such enhanced efficacy is often the result of complex processing using a number of purposefully designed chemical components. Such sophistication can be seen in everyday, apparently simple, objects, such as the crisp (or, potato chip) packet (an example taken from ref. 19a).

To be acceptable to a consumer, a crisp packet must deliver the following functions:

- prevent spoilage (by providing a barrier to oxygen preventing oxidation of the lipids from the cooking oil)
- keep the crisps crisp (by providing a barrier to moisture)
- allow easy transport and handling (by being puncture and tear resistant)
- protect the contents (by ensuring seal integrity and being tamper-proof)
- convey information (by permitting ink adhesion).

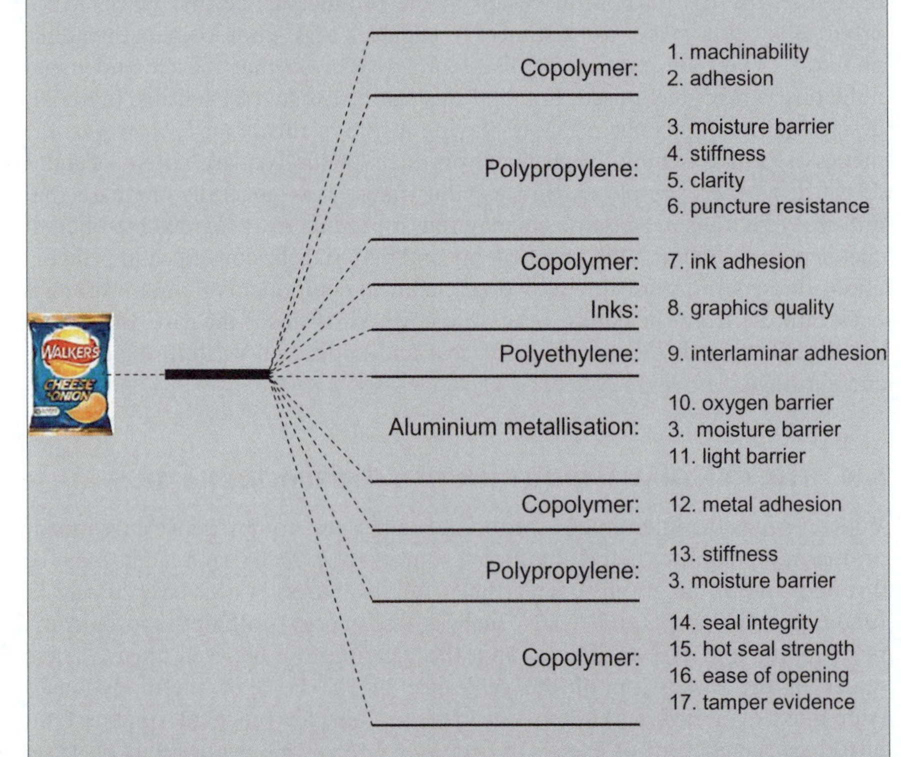

Figure 6.8 Multi-layer composite film used, for instance, in packaging of crisps providing 17 separate functions. Adapted from ref. 19a.

Optimising these functions leads to great technical complexity in the assembly of a composite material, 25–40 μm thick, made up of nine layers of five different materials, laid down in the sequence shown in Figure 6.8. Compared with aluminium foil, metallisation of a polymer to give an aluminium layer *ca.* 0.01 μm thick reduces the overall weight by *ca.* 40% and aluminium usage 320-fold,[19b] with consequent reductions in associated material and energy usage.
However, try to imagine disassembling the nine-layer composite into its component chemical parts. This would probably require even more complexity in processing than was involved in the original manufacture and would, in all probability, deliver impure material not fit for reuse. In this case, it is probably better simply to burn the crisp packet and recover its energy value. While currently the final stage of the life-cycle of too many crisp packets is the municipal landfill site, the optimum disposal method should preferably be burning to produce energy in a properly operated municipal solid waste incinerator. The problem with the alternative of making the crisp packet simpler is that the crisps inside would be less well-protected with the potential for greater spoilage and waste.

BOX 6.5 THE LAWS OF THERMODYNAMICS.

First Law: While heat and work are interconvertible, energy cannot be created or destroyed.
or: *You can't get something for nothing, ever*.

Second Law: Heat cannot be converted completely into work. Spontaneous processes are accompanied by an increase in entropy.
or: *You can't even break even (except at absolute zero)*.

Third Law: The entropy of a pure, perfect crystal is zero at 0 K.
or: *Catch 22: you can never reach absolute zero*.

Thermodynamics is concerned with the relationships between heat, work and energy as expressed in relationships such as eqn (6.30) and (6.31):

$$\boldsymbol{E} = U + pV - TS - \Sigma\mu \tag{6.30}$$

$$\Delta U = q + w \tag{6.31}$$

where U represents internal energy, p, pressure, V volume, T absolute temperature, S entropy and μ, chemical potential. Equation 6.31 represents the conservation of energy where q = heat and w = work.

We are also aware of other state functions[ix] (and the conditions under which they apply):

[ix] State functions have values that depend only on the 'state' of the system as governed by its pressure (or volume) and temperature, with changes in value determined only by the initial and final conditions irrespective of the path taken to go from one to the other.

- Gibbs free energy, G, and changes in free energy, ΔG
- entropy, S, and changes in entropy, ΔS.

Gibbs free energy, entropy and enthalpy, H, are related as follows: $G = H - TS$.

ΔG is the maximum amount of work available in a reversible process at constant T and P (in addition to work, $p\Delta V$, done in volume change), according to eqn (6.32):

$$\Delta G = \Delta H - \mathrm{T}\Delta S \tag{6.32}$$

For a general chemical process involving a series of reactants and products, for which individually, $G = G^{\circ} + RT \ln \mathrm{a}$ (where a is the activity), the free energy of a reaction is given by eqn (6.33) or eqn (6.34) (where Q is the reaction quotient):

$$\Delta G = G_{\mathrm{products}} - G_{\mathrm{reactants}} \tag{6.33}$$

$$\Delta G = \Delta G^{\circ} + RT \ln \mathrm{Q} \tag{6.34}$$

Or, at equilibrium (eqn 6.35):

$$0 = \Delta G^{\circ} + RT \ln K_{\mathrm{eq}} \tag{6.35}$$

The reaction isotherm, therefore, relates free energy change and K_{eq}, the equilibrium constant at temperature, T, according to eqn (6.36):

$$\Delta G^{\circ} = -RT \ln K_{\mathrm{eq}} \tag{6.36}$$

6.10.1 Heat of Reaction

Chemists' first practical contact with these ideas arises in the laboratory when studying thermochemistry. This is concerned with enthalpic changes, ΔH, of chemical and physical processes, endothermic ($+\Delta H$) or exothermic ($-\Delta H$). A chemical compound is characterised by its standard enthalpy of formation, $\Delta_f H^{\circ}$, from its constituent elements (for which, by definition, $\Delta_f H^{\circ} = 0$) referenced to a universal standard state of pressure of 1 bar (1.01325×10^5 Pa) and temperature, 298.15 K.

This enables questions of chemical stability to be considered, allowing equilibrium constants to be determined and reaction profiles to be constructed that enable how far a reaction will proceed to be determined, and how much energy is released or consumed in the process.

The exchange of heat between a reacting chemical system and its surroundings can be represented as in Figures 6.9 and 6.10, in which the conventional signs of the enthalpies of processes that are exothermic or endothermic are shown and for work done by or on a system as a consequence of volume changes.

Considerations of enthalpy change, ΔH, alone are insufficient in determining whether a particular reaction proceeds spontaneously[x] or not. To underline this

---

[x] Note that the word 'spontaneously', when used in this context, does not mean 'instantaneously'; rather, it means 'unforced' or 'unprompted'.

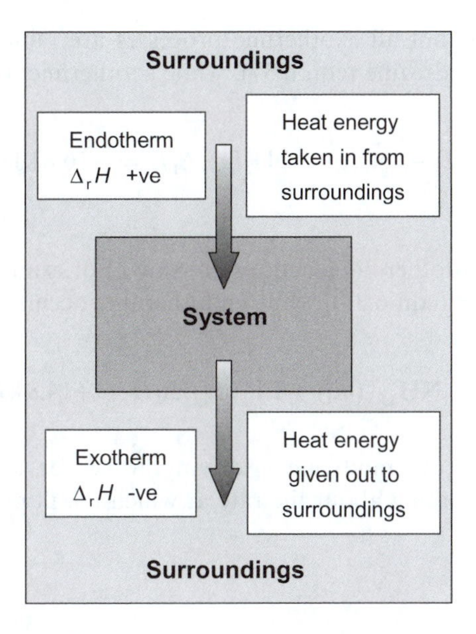

Figure 6.9 Relationship between a chemically reacting system and its surroundings for endothermic and exothermic processes.

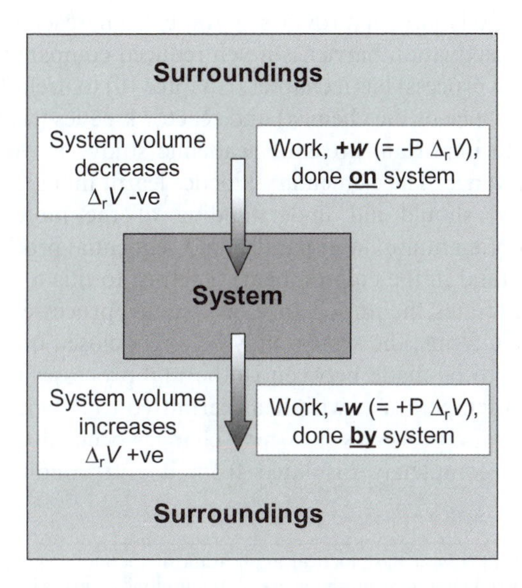

Figure 6.10 Relationship between a chemically reacting system and its surroundings for processes at constant pressure that result in either increases or decreases in volume.

point, we note that not all exothermic processes are rapid. For example the decomposition of hydrazine (eqn 6.37), while exothermic, is slow:

$$N_2H_4(g) \rightarrow N_2(g) + 2\,H_2(g); \Delta_r H = -50.6\,kJ\,mol^{-1} \tag{6.37}$$

Similarly, not all endothermic reactions are slow. For example, the dissolution of $[NH_4]Cl$ in water (eqn 6.38), while endothermic, occurs spontaneously.

$$[NH_4]Cl(s) \rightarrow [NH_4]^+(aq) + Cl^-(aq); \Delta_r H = +14.8\,kJ\,mol^{-1} \tag{6.38}$$

The existence of a thermodynamic driving force for a chemical process, therefore, tells us nothing about the rate at which it will proceed.

6.10.2 Kinetics

We also need to understand the importance of kinetics, which can tell us how fast a reaction might go under particular circumstances and leads us to a consideration of the reaction profile,[xi] the transition state (or activated complex) and activation energy, E_a.

If a reaction has a very high activation barrier it will be very slow, irrespective of how large the driving force for the reaction (as expressed in the free energy change) might be. Because of its relevance to cleaner technology and waste minimisation catalysis (which provides a route from reactants to product in which the highest activation barrier is much reduced compared with the barrier for the uncatalysed process) has a chapter (Chapter 10) to itself. Reaction kinetics (primarily the province of the chemist) and reactor kinetics (that of the chemical engineer) are both important areas of academic study, as well as of immense technological relevance. The elementary kinetics learnt in undergraduate courses (see Bibliography) should aid understanding of the more complex multicomponent (and often multiphase) parallel and sequential processes occurring in a chemical plant (and in the environment). I return to this topic in Chapter 9.

Figure 6.11 reiterates the importance of systems, processes and boundaries. In thermodynamic terms, the system may be open, closed or isolated. There is also a distinction to be made between isothermal processes (in which changes occur at constant temperature, with heat permitted to flow into and out of the system) and adiabatic processes (in which changes occur in an isolated system that is thermally completely insulated from its surroundings; or for which $\Delta q = 0$).

[xi] While we are concerned with the practical application of thermodynamics, in the interests of rigour it would be wrong not to mention the many assumptions we make in drawing such reaction profiles. We need to be aware that, in these cases, we are considering the path for the reaction of a single molecule (or small assembly of molecules) and not for the bulk material on the large scale. See ref. 20 for a discussion of this particular point.

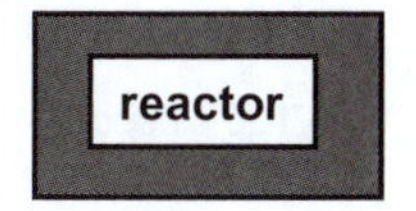

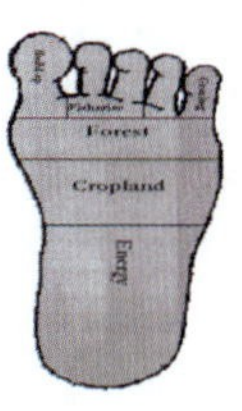

- Systems can be:

  open
  closed
  or isolated

- Processes can (ideally) be:

  isothermal (change occurring at constant temperature allowing heat flow into and out of system)

  or adiabatic (change occurring in isolated system completely insulated from surroundings)

Figure 6.11 Systems, processes and boundaries. Photo copyright Mark Smith 2010. Used under license from Shutterstock.com.

6.11 ENTROPY AND WASTE

This section provides a thermodynamic account of the origin of waste and how this is linked with the concept of entropy.

It may be helpful to start with a descriptive representation of entropy[xii] as a means of quantifying the degree of disorder or order in a system and changes thereto. Figure 6.12a shows two vessels, one containing gas 1, the other gas 2, both at the same temperature and non-reacting, with the tap between them closed. If the tap is opened (Figure 6.12b) and we wait long enough, then a situation represented by Figure 6.12c will arise in which there is an intimate mixture of the two gases. The mixing of the two gases has occurred spontaneously (if slowly). We might then ask why the reverse process, in which the gas mixture separates into its pure components, does not also occur spontaneously.

Intuitively, we can see that the degree of order has changed, with the mixed system being more disordered than before the two containers were connected. It is generally found that spontaneous processes always lead to decreases in order (*i.e.* an increase in entropy). Two real-life examples illustrate the point: breaking off at snooker and ink dispersing in water. We can ask how likely it is that, sometime during the game, the snooker balls would spontaneously revert to the positions they occupied immediately before breaking off. How long would you need to wait for the dispersed ink solution to reform the initial blob that was present immediately after it had been dropped into the water?

All these examples are associated with the state function, *S*, entropy, and its change (Figure 6.13). Entropy is a measure of disorder and increases for spontaneous processes. Heat flow also leads to changes in the degree of order.

[xii] The concept of entropy, first introduced by Rudolf Clausius, can be difficult to grasp and efforts have been made to present it in different ways. A full coverage of this is beyond the scope of this book, other than to refer the interested reader to texts on the subject (see Bibliography).

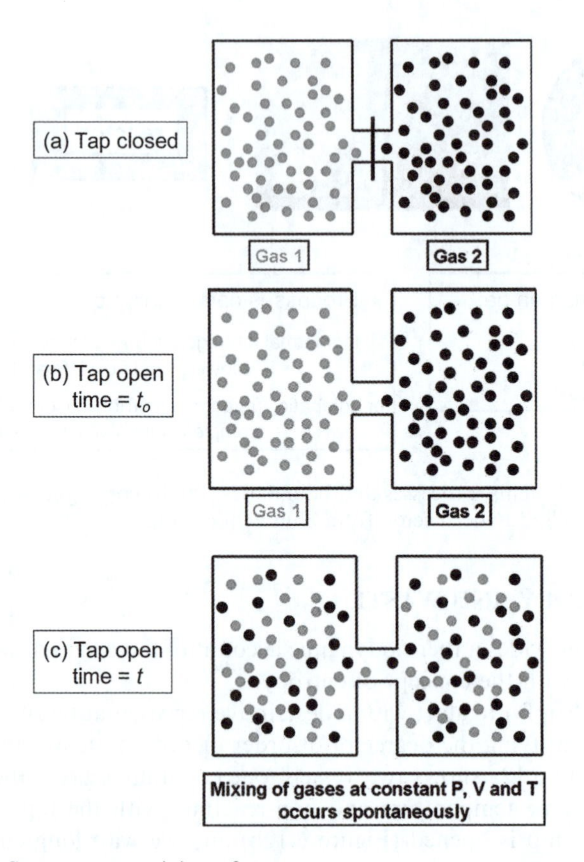

Figure 6.12 Spontaneous mixing of gases.

For example, in the part of the system that loses heat (the source) entropy decreases and in the part of the system that receives the heat (the sink) entropy increases. Importantly, work must be done to counter the increase in disorder that occurs in spontaneous processes.

One statement of the Second Law of Thermodynamics states that the entropy of a system and surroundings increases during a spontaneous irreversible change.

Entropy is precisely defined (eqn 6.39) as the product of the Boltzmann constant, k_b, multiplied by the natural logarithm of the number of microstates in which the system may exist. W is the number of microstates.

$$S = k_b \ln W; \quad k_b \propto T^{-1} \tag{6.39}$$

Importantly, the starting state (whether this is the position of snooker balls on a table, molecules of a volume of ink immediately after adding to water, or the distribution of gas molecules in two containers in which all the molecules in one are gas 1 and in the other gas 2) is simply one of many microstates that the

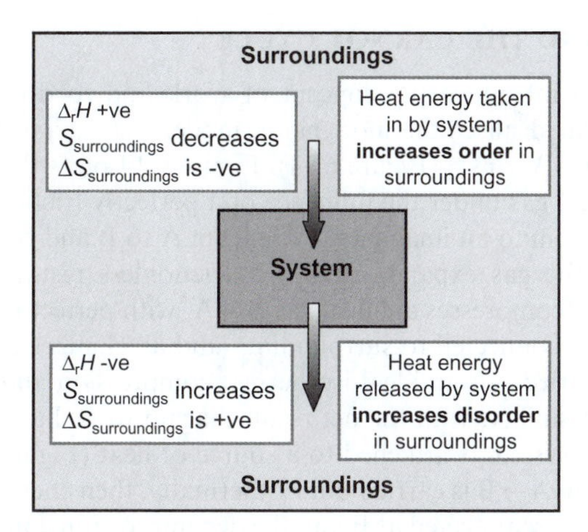

Figure 6.13 Changes in the entropy of the surroundings arising from endothermic and exothermic processes. Entropy changes associated with the chemical change itself are not included.

system can take up. For a chemical system, these may be energetic, positional, vibrational or rotational microstates.

The solution to the Rubik cube places all the elements of the six colours so each colour occupies a single face, this being one of 43 252 003 274 489 856 000 possible states (Web 8). K. J. Laidler's book, *Energy and the Unexpected*[21] provides some other examples. Were we asked to separate the 52 playing cards into suits and to put them into numerical sequence, 2 to ace, we could compute that this sequence is just one of 52! (factorial 52) or *ca.* 10^{67} possible sequences. Each one of the possible sequences represents a microstate that the system can adopt. Mixing a pack of cards, at one shuffle per second, a chance formation of this ordered sequence would take *ca.* 4 × the age of the universe.

Let us consider a more chemically relevant system (also taken from Laidler): what are the chances that a 38 g mixture of one mole of H_2O and one mole of D_2O would spontaneously separate to give the unmixed materials? A rough estimate[21] would be *ca.* 1 in 10 to the power of 10^{23}, *i.e.* 10 with 10^{23} zeros. If each 0 was 1 mm wide, the figure would stretch *ca.* 10^{17} km, or $10^{11}\times$ the distance from the Earth to the moon and back.

A starting state moving spontaneously to one of a huge number of equivalent states is therefore a highly favoured process, but occurs without the input of heat energy. Were one to attempt to reverse the process to go back to the starting state (*e.g.* separating protiated water from deuteriated water), then a large amount of energy would be needed (*e.g.* to drive a fractional distillation to separate them on the basis of the small differences in boiling point between the two isotopic forms).

6.12 WORK AND THE CARNOT CYCLE

We now need to introduce the concept of work and to do this we need to examine a classical idealised thermodynamic model system, a form of the Carnot cycle suggested by Atkins[22] (Figure 6.14). Figure 6.14 plots the pressure *versus* volume of an ideal gas under the influence of a perfectly frictionless piston. Let us subject the system to an imaginary cycle from A to B and back again carried out as follows: the gas expands against a frictionless piston A→B and the frictionless piston compresses an ideal gas B→A, with perfect reversibility of all changes, perfectly connected to surroundings and all changes being isothermal. We have thus carried out an ideal isothermal compression and decompression with perfect reversibility, with the net result that no work has been done.

If, instead, the system is attached to a source of heat (Figure 6.15) at A and the decompression A→B is carried out isothermally, then energy is absorbed. If the heat source is disconnected at B and the decompression continued to C, this change will occur adiabatically. If at C the system is connected to a cold reservoir, and the compression to D carried out isothermally, then energy is dumped, *i.e.* is transferred from the gas in the piston to the cold reservoir. The piston is then removed from the cold reservoir and compression continued adiabatically to return to the state A. The area enclosed by the curve is now a measure of the work done.

The crucial point is that work can be done only if some heat is discarded (*i.e.* wasted) into a cold sink. The latter has low entropy and absorbs the energy received into the system from the hot reservoir (which has high entropy), with enough entropy being generated in the cold sink to overcome the decrease in

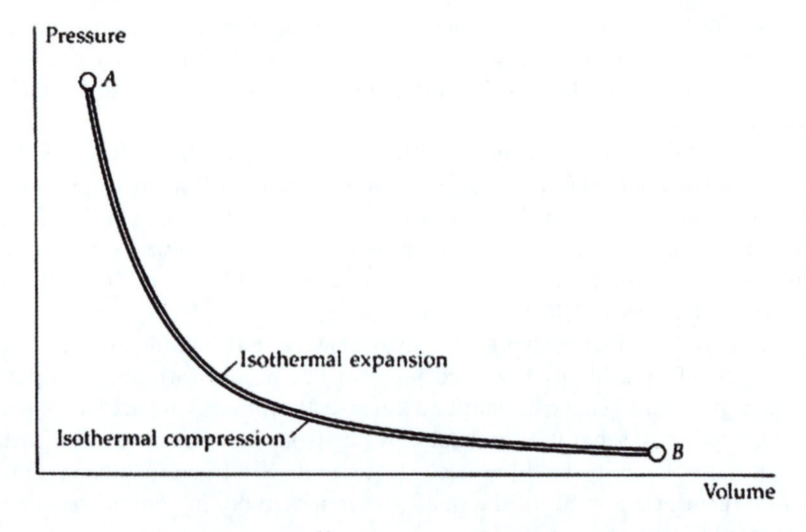

Figure 6.14 The 'Atkins' cycle:[22] perfectly reversible isothermal compression and decompression of an ideal gas in a perfectly frictionless piston. Copyright P. W. Atkins, reproduced with permission.

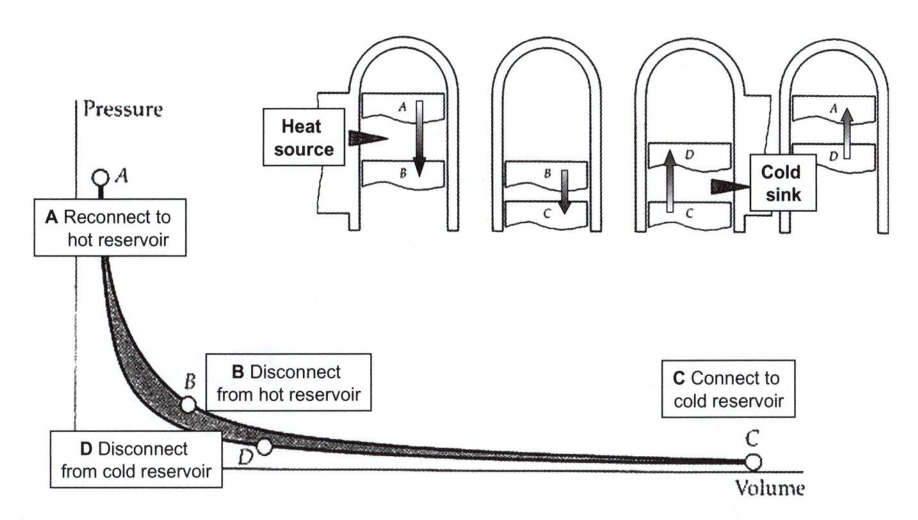

Figure 6.15 The Carnot cycle (based on ref. 22; Copyright P. W. Atkins, reproduced with permission).

entropy on the hot reservoir. The latter condition must exist as, if there is no overall increase in entropy, the process cannot occur spontaneously.

Another formulation of the Second Law of Thermodynamics states that the energy output of work is always less than the energy transformed to accomplish it. The difference is associated with entropic changes and represents lost work. In summary: energy dumped is no longer available to do work.

Entropy is high when T is high (when it is available to do work) and low when T is low (when it is not). If the temperature of the hot reservoir is T_1 and that of the cold reservoir is T_2, then the maximum efficiency of the system, the so-called **Carnot efficiency**, is represented by $1 - T_2/T_1$. Efficiency is maximised when the temperature difference is as high as possible (within clearly practical limitations) though can never be unity.

As we cannot create or destroy energy, we have not changed the quantity of energy, only its 'quality'. Entropy can be seen as a measure of the 'quality' of energy. The natural direction of change causes quality of energy to decline. Why is this relevant? Most energy that powers our society is derived from a fossil fuel such as oil, coal or natural gas. This is essentially stored solar energy captured initially by the process of photosynthesis (eqn 6.40):

$$6\,CO_2 + 6\,H_2O + h\nu \rightarrow C_6H_{12}O_6 + 6\,O_2 \qquad (6.40)$$

It is then subjected over millions of years to anaerobic transformations at high T and P. The extraction of energy from the combustion of these fuels (eqn 6.41) is subject to the Carnot efficiency mentioned above, but only in the context of power generation.

$$C_6H_{14} + 9.5\,O_2 \rightarrow 6\,CO_2 + 7\,H_2O \qquad (6.41)$$

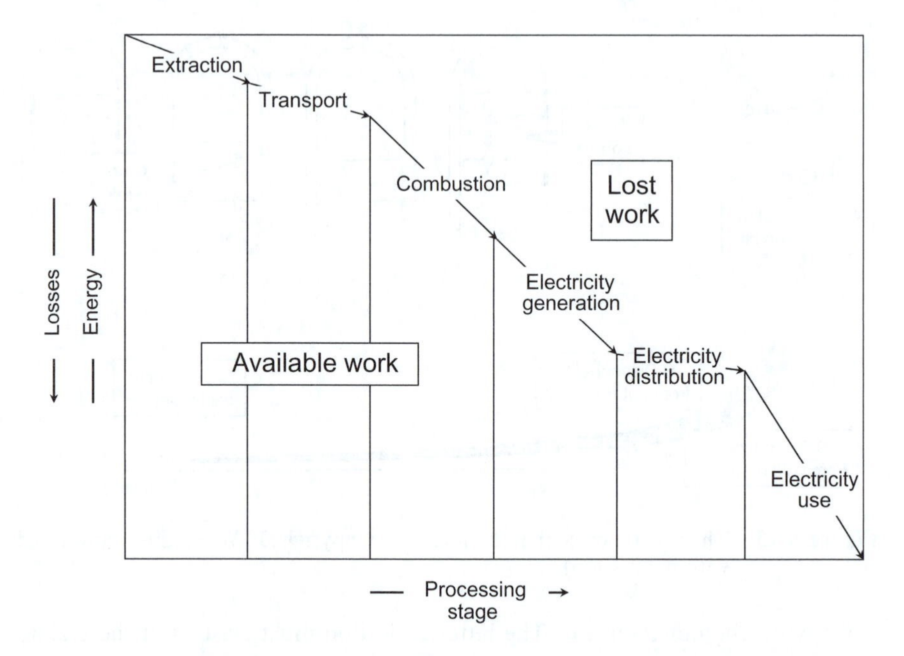

Figure 6.16 Losses in the availability of work for a primary raw material (*e.g.* oil or natural gas) used in electricity production associated with its extraction, transport, combustion and the generation and distribution of electricity.

Losses occur at all other stages in the process (Figure 6.16) from primary extraction from the environment, to transportation and processing, and the distribution of electricity generated. These can be expressed in losses of the energy available from each unit of raw material. When we turn on a light or boil a kettle, and finally come to use the electricity, all the earlier losses will already have taken place.

6.13 REAL PROCESSES: EXERGY

As all processes are governed by the laws of thermodynamics, any claims that products or processes are not waste generating are usually found to have failed to take account either of all the material and energy flows (First Law) or how the appropriate boundaries were defined, or both. We can conclude from the Second Law (the energy output of work is always less than the energy transformed to accomplish it) that waste is inevitable, resulting from the need in any spontaneous process for an increase in entropy to be brought about; this can only occur in a system from which work is extracted by transferring entropy from a high temperature source to a low temperature sink.

The starting state of a system undergoing spontaneous change is one amongst many, with spontaneous change bringing an increase in disorder, with energy distributed between many states. One cannot go back spontaneously to the starting state; in fact, energy would need to be expended to go back to this more ordered starting state. Following the spontaneous change, while the quantity of energy will not have changed (operation of the First Law), the 'quality' of energy will have been reduced.

How can we apply these ideas to chemical reactions, particularly to complex transformations operated industrially on the large scale? A physical change such as the movement of a (frictionless) piston in the pressurising and decompression of a gas (ideal) produces a maximum amount of work with 100% efficiency, but only under circumstances of perfect reversibility. Circumstances are somewhat more complicated if, instead of (or in addition to) a physical process, the system involves a chemical change. However, in principle, the issue is the same. Were we able to carry out the chemical transformation with perfect reversibility, then it could likewise be achieved with 100% efficiency. For the process to be reversible, all driving forces for change, be they physical (temperature, pressure, concentration) or chemical (changes in chemical potential) must be zero, and must be so throughout the entire process. As Leites *et al.* point out,[23] a completely reversible process, operating infinitesimally, would require an infinitely large plant and take infinite time to complete. The requirement, therefore, for practical systems to work is that they must operate with finite driving forces, losing some thermodynamic efficiency as a consequence. It is the concern of the process engineer to optimise these driving forces so as to maximise efficiency and limit losses. For example, having a driving force that is sub-optimal may result in a process operation requiring too large a design capacity (thereby adding to capital costs) or requiring too long a residence time (thereby reducing throughput or productivity).

As it is more practically focussed, the term **exergy** will now be used to take account of losses in the availability of work associated with entropy changes. We will also use a different standard state from the more rigorous ones used in physical chemistry [pressure of 1 bar (1.01325×10^5 Pa), temperature: 298.15 K] and reference states often used in academic study (*e.g.* infinite dilution). Exergetic analysis seeks instead to relate a process and its components to the condition in the environment (which clearly will not be as precisely defined). We introduce the term exergy (see ref. 24 for a brief introduction and the Bibliography for some additional useful sources) to define the maximum amount of work that can (ideally) be obtained from a quantity of energy or matter when it is brought (adiabatically $T \rightarrow T_o$, isothermally $p \rightarrow p_o$) from a defined set of conditions (T, p) to those of the natural ambient environment (the 'dead' state, *i.e.* at thermo-mechanical and chemical equilibrium, T_o, p_o). All materials, energy carriers and operations can be assessed on a common scale, the Joule. This allows more ready comparisons to be made between different processes, something we will illustrate below for three processes for ethanol production.

The energy change of a system from one state to another can be represented by eqn (6.42) in which the net change is the sum of changes in the internal energy, work done as a consequence of volume changes at constant pressure, the effect of entropic changes and the changes in chemical potential of all those materials undergoing some transformation. We have already noted the standard state to which these changes are usually referred.

$$\Delta E = \Delta U + p\Delta V - T\Delta S - \Sigma\mu\Delta n \quad (6.42)$$

Exergy can therefore be expressed as follows (eqn 6.43):

$$\text{Exergy} = U - U_{eq} + p_o(V - V_{eq}) - T_o(S - S_{eq}) - \Sigma\mu_o(n^i - n^i_{eq}) \quad (6.43)$$

By definition, a material in equilibrium with the environment and at a concentration and in a form in which it exists in the environment will have zero exergy.

We can express this in a form (eqn 6.44) that has more relevance when considering a technological transformation involving the release of waste:

$$E_{tot} = E_{phys} + E_{mix} + E_{chem} \quad (6.44)$$

Thus exergy is divided into its constituent physical, mixing and chemical components:

- $\boldsymbol{E_{phys}}$ represents the sum of the enthalpy and entropy changes for the combination of individual pure components (in vapour and liquid phases) as the pressure and temperature change from p, T to p_o, T_o (*i.e.* to the temperature and pressure of the ambient environment).
- $\boldsymbol{E_{mix}}$ represents the loss of exergy when the pure components are mixed together. This makes intuitive sense when we recall the mixing of two pure gases (see Figure 6.12) and the need for work to be done to separate the mixed gases into pure components.
- $\boldsymbol{E_{chem}}$ is the work done when a pure component, i, which also exists in the reference environment, is diffused from its initial concentration to the reference concentration (*i.e.* its concentration in the undisturbed environment). (A moment's thought at this point highlights a difficulty of defining such a concentration—as well as the notion of an undisturbed environment. The reader can explore this further in the sources included in the Bibliography and form an independent view of the degree to which this loss of rigour is justifiable on grounds of practical applicability.) If the component i does not exist in the reference environment, as will be the case for many synthetic chemical products, i must be chemically converted to substances that are present within the reference environment. In these circumstances, E_{chem} is then the free energy of reaction for this conversion plus the work done by diffusing the decomposition products formed in this conversion to the reference concentration in the environment.

An exergy balance can be produced for each process component, in-going and out-going, to calculate exergy losses that occur in the process. Noting as before, the importance of identifying and understanding the significance of system boundaries, internal exergy losses are those occurring within the process boundaries and external exergy losses consist of exergy flowing to surroundings.

Fortunately, compilations of exergy values (see Bibliography) of common materials and energy sources have been published which allow exergy changes to be calculated. Raw materials and auxiliaries that have been extracted from natural sources and subjected to a series of additional processes to convert them into useful materials will have relatively high exergy values. It is perhaps not surprising that water (at atmospheric pressure and ambient temperatures) and CO_2 (at its concentration in the ambient environment) have near zero exergy values. Even though this is the case, this tells us nothing about the environmental impact of these low exergy materials. Superheated water as used in power generation or compressed carbon dioxide captured for storage will have an exergy content represented by the difference between their energy in these forms and that in the environment, and which may be high enough to be a resource from which useful exergy may be extracted.

6.14 EXERGETIC ANALYSIS

Exergetic analysis as we have seen:

- takes account of the irreversibility of each technological process and the inherent degradation of the quality of energy (associated with entropy increase) that is related to a loss of exergy; and
- permits the quantification of a measure of the sustainability of technologies by expressing material and energy use in a single unit, the Joule.

Dewulf and colleagues[24] have developed expressions to characterise material and energy contributions after identifying all energy carriers in a production process traced back to the primary resources needed to produce them [using a (simplified) life-cycle inventory].

We will describe the approach developed by Dewulf and colleagues that enabled them to compare three methods for producing the important commodity, ethanol. Although the approaches of Dewulf and others have become more sophisticated, we will illustrate the methodology in its basic form rather than present a more detailed up-to-date analysis. The cited texts and those in the Bibliography are suggested as starting points for those wanting to pursue this area further.

Any anthropogenic process can be represented (imperfectly) as a series of flows of energy and material between the ecosphere and the technosphere. Resources, energy and matter are removed from the ecosphere and enter the technosphere where they are processed (such processed materials, as we have seen, will have relatively high exergetic content) and then used. Waste materials (of lower exergy) and energy are returned to the ecosphere. This cyclic system can be compared with the historical linear one (dig up, use, throw away)

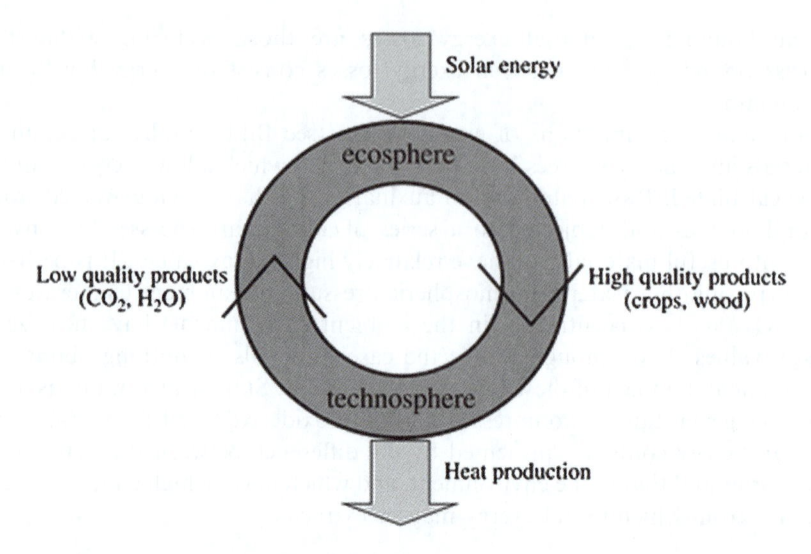

Figure 6.17 A simple model for the exchanges between the technosphere and ecosphere.[24a] Reproduced by permission of the Royal Society of Chemistry.

that we saw in the picture of the smoke stacks over Widnes (Figure 2.8). The extraction, processing, use and disposal of raw materials can be represented in terms of interactions between the ecosphere and the technosphere, as illustrated in Figure 6.17. In this simple model, sunlight drives the production of crops and biomass in the ecosphere. These are harvested, transformed and used by humankind in the technosphere, with low exergy waste such as water and carbon dioxide being returned to the ecosphere. Ideally, to be sustainable, this system should be in balance.

However, as we have already seen, the rate of extraction and consumption of materials and energy from the environment is currently greater than the replacement rate. In addition, the wastes returned to the ecosphere, whether direct and indirect, are having deleterious impacts.

From this simple analysis, two conditions necessary for sustainability can be specified:

- resources required for producing goods should not run out; and
- emissions must not endanger ecological systems.

The first can be achieved by minimising resource consumption through greater efficiency and beneficial recycling (dematerialisation) and through the use of renewable resources, ultimately sourced from solar energy (transmaterialisation). The second can be achieved by reductions in emissions through improved process efficiency, extended product life and by the treatment of unavoidable emissions. Emissions can either be abated by converting them to harmless wastes or sequestered or isolated—recognising always, the associated additional energy costs.

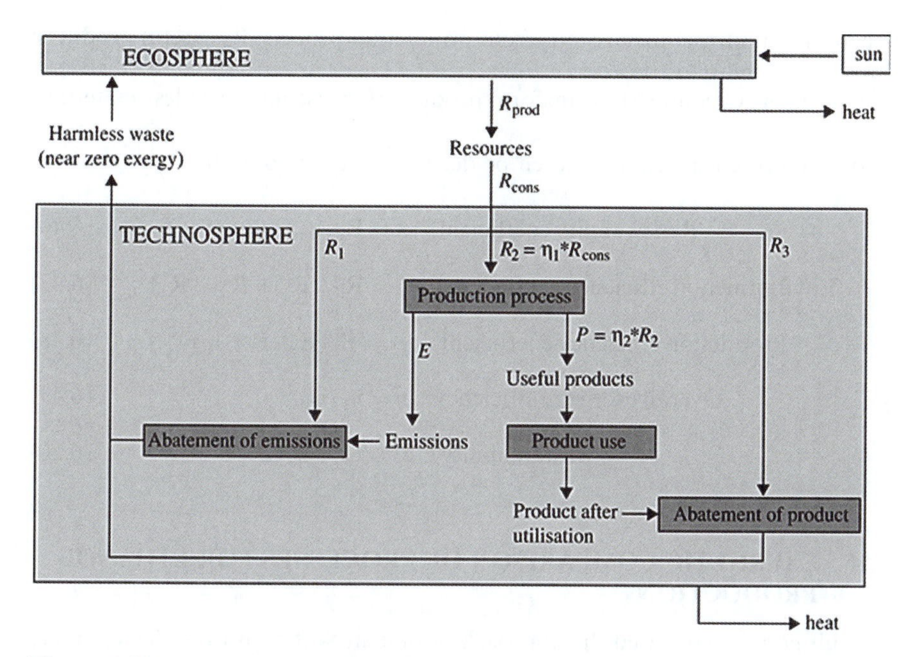

Figure 6.18 Important material and energy flows between the technosphere and ecosphere, including (a) those associated with the process of production and (b) those needed to treat process emissions and the product on disposal to return only harmless waste to the ecosphere.[24a] Reproduced by permission of the Royal Society of Chemistry.

Figure 6.18 identifies the various material and energy flows that need to be accounted for in a system whose sustainability we want to assess. Resources, R, are extracted from the ecosphere and R_{cons} are those consumed in the technosphere. A proportion of R_{cons}, R_2, is used in the production process itself. Some (R_1) is used to abate emissions and some (R_3) to treat the product at the end of its useful life. Ideally, the fraction of R_{cons} used in abatement should be as small as possible. R_2 is itself made up of three components. E is the proportion that ends up as waste to be abated and P the proportion that ends up in useful product, the latter fraction being maximised, preferably. Some additional exergy, I_P, is lost because of the irreversibility of the production process and the compromise the design engineer needs to make (in terms of reaction time and reactor size) to maximise efficiency within the requirements of practicality.

A series of relationships can be defined (eqn 6.45):

$$R_{cons} = R_1 + R_2 + R_3 \qquad (6.45)$$

where:

R_1 = Exergy required for abatement of emissions (GJ y^{-1})
R_2 = Exergy required for production process; $R_2 = E + P + I_P$

where: E = non-useful emissions from production process; P = useful products; I_P = irreversibility of process;
R_3 = Exergy required to transform product after use into harmless materials.

A series of coefficients can then be derived[24] (eqn 6.46–6.50):

$$\text{Renewability coefficient} \quad \alpha = R_{cons,renewable}/R_{cons} \tag{6.46}$$

$$\text{Environmental efficiency coefficient} \quad \eta_1 = R_2/(R_1 + R_2 + R_3) \tag{6.47}$$

$$\text{Production efficiency coefficient} \quad \eta_2 = P/(E + P + I_P) \tag{6.48}$$

$$\text{Overall exergetic efficiency} \quad \eta = \eta_1/\eta_2 \tag{6.49}$$

$$\text{Sustainability} \quad \Sigma = (\alpha + \eta)/2 \tag{6.50}$$

6.15 EXERGETIC COMPARISON OF PROCESSES FOR ETHANOL PRODUCTION

Dewulf *et al.*[24] have used the approach outline in Section 6.14 to derive values of these coefficients for three different technologies for the production of ethanol in a study that exemplifies the potential of the approach. These examples also illustrate some the factors this book is designed to highlight.

The three routes permit a comparison to be made between:

- the classical fossil-sourced chemical route to ethanol (from naphtha-cracking followed by catalysed ethene hydration)
- a viable biotechnological process using fermentation of an agricultural biomass feedstock
- a more speculative transformation (H_2 from solar-driven photoelectrolysis used in the catalysed reduction of CO_2 captured from a flue gas) that, if reduced to large-scale practice, could claim to be ultimately sustainable.

Before proceeding to the analysis, a question must be addressed arising from the possible implication that materials of low or zero exergy which are close to their form, concentration and condition in the dead state of the ambient environment are, as a consequence, of low environmental impact. This question arises in particular in the context of emissions of carbon dioxide, a material of low exergy. Being present in the atmosphere at low concentrations (or mixing ratios), the increase from 250 ppm to 3–400 ppm might be thought to have little effect (though we now know differently). The small amount of exergy associated with CO_2 at a concentration of 350 ppm compared with that at a pre-industrial level, therefore, takes on a major environmental significance.

Account must also be taken of the exergy expended in the collection and sequestration of the (fossil-derived) carbon dioxide generated in the different processes.

Carbon dioxide from a process stream is first absorbed into aqueous ethanolamine and then 'stripped' from this solution, dried and compressed to 80 atmospheres. It can then, as we have seen, be stored underground in depleted oil and gas wells or other geological strata. The overall process requires the expenditure of exergy of 5.9 MJ (kg CO_2)$^{-1}$. To avoid adding more CO_2 to the atmosphere from this expenditure of exergy than is diverted to underground storage by the process of CCS, then CCS must overall be better than carbon 'neutral'. It seems obvious, now we know to define proper boundaries for the processes we consider, that to be of any use there must be a net diversion of CO_2 to CCS. Even so, the CCS process will be costly, increasing (according to estimates published in 2006[11c]) the cost of electricity generation from coal in the UK from 2.6 to 3.7 p per kWh.

The comparison of the three processes for ethanol production, simplified as they are, nevertheless exposes the challenges surrounding the implementation of the transition from chemical technologies wholly based on fossil resources to those that use what are believed to be more sustainable feedstocks, particularly renewable sources of material and energy.

6.15.1 Conventional Process for Ethanol Manufacture

The first method involves the use of fossil-sourced raw materials in a conventional widely practised chemical process involving naphtha cracking to deliver ethene followed by catalysed ethene hydration (eqn 6.51 and 6.52). Dewulf and colleagues[24] analysed all the inputs and outputs associated with the various stages of ethene and ethanol production. Details are taken from ref. 12. All the inputs and outputs are expressed in terms of MJ (kg ethene)$^{-1}$.

$$\text{Oil} \rightarrow \text{Naphtha} \rightarrow \text{Ethene} \tag{6.51}$$

$$\text{Ethene hydration over zeolite (60 atm, 300 °C)} \rightarrow \text{Ethanol} \tag{6.52}$$

Inputs and outputs are:

Ethene production:	Inputs:	feedstock: oil, gas energy: coal, oil, gas, hydro, nuclear
	Outputs:	ethene
Ethene hydration:	Inputs:	ethene, water, power
	Outputs:	ethanol

This approach involves a significant input of energy in the process of ethene production. This needs to take account of different electricity generation technologies and the fact that individual countries have different energy 'mixes'. The 'mix' is the proportion of each energy generation technology used. The importance of this is that each energy generation technology has a different exergetic efficiency, which will affect the value of the parameter used in the

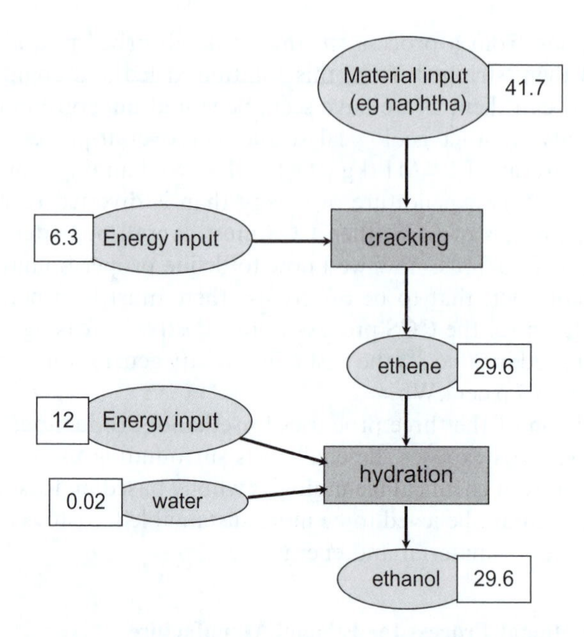

Figure 6.19 Flow diagram for the exergetic analysis of the production of ethanol from fossil-derived feedstocks. All energy and material inputs expressed in MJ (kg ethene)$^{-1}$. Based on data from ref. 12.

calculation. The consequence of this is that the values to be plugged into the calculations change from country to country, so the results of the calculation will differ depending on where the process is to be operated and what local energy prices will apply. We will use the figure used by Mulder.[12] However, this will have little impact on the comparison of technologies we are seeking to make.

The data are given in the flow diagram shown in Figure 6.19. Putting the relevant values into the individual equations, we obtain the production efficiency for this technology as follows:

Efficiency of cracking step = 0.616
Efficiency of hydration step = 0.706
Overall production efficiency = 0.49

It is clear that the production efficiency of a well-established chemical process is relatively high, making it difficult to displace on economic grounds. (We will explore some additional reasons for this in Chapter 9).

6.15.2 Biotechnological Process for Ethanol Manufacture

The second method involves the use of 'biomass' from an agricultural feedstock (*e.g.* corn or straw) and its fermentation to produce ethanol (eqn 6.53). The vegetable material is produced by photosynthesis, in which solar energy converts CO_2 from the atmosphere into biomass.

$$\text{Light} + CO_2 + H_2O \rightarrow \text{Biomass} \rightarrow \text{Ethanol} \quad (6.53)$$

Dewulf *et al.*[24] analysed all the inputs and outputs associated with the various stages of planting, growing and harvesting of the vegetable matter, the fermentation process and the recovery of product ethanol in a saleable form. Additional details are taken from ref. 12.

Inputs and outputs include:

Wheat production:	Inputs:	seed, cereal growth, light, CO_2, H_2O pesticides, nutrients, harvesting
	Outputs:	straw, wheat grain
Fermentation:	Inputs:	water, hydrolysis, yeast, decantation
	Outputs:	'cake' (fibre), gluten (protein)
Ethanol recovery:	Inputs:	fermentation, distillation, rectification dehydration, heat
	Outputs:	ethanol (96%), ethanol (100%).

While sunlight is available at no cost, a proper comparison requires an estimate of the exergy input represented by solar energy, expressed in units of flux per unit area per unit time (GJ ha^{-1} y^{-1}). Clearly, there are wide variations in sunlight depending on geography and local topography, time of day (diurnal variations), seasonal and longer timescale variations.

The flow diagram, with numerical values of each input and output, also expressed in GJ ha^{-1} y^{-1}, is shown in Figure 6.20. These data can be used to calculate the various coefficients shown above for each stage and then overall, permitting a comparison between the technologies.

$$\text{Production efficiency coefficient, } \eta_2 = \frac{\text{Sum of outputs}}{\text{Sum of inputs}}$$

$$\text{Efficiency of grain production} = \frac{(91 + 147)}{(33480 + 2.4 + 2.2 + 11 + 5.6)}$$

$$= 238/33501 = 0.0071$$

$$\text{Efficiency of ethanol production} = \frac{(92 + 6 + 44)}{(147 + 1.6 + 20 + 2.1)} = 142/171 = 0.83$$

$$\text{Overall efficiency} = \frac{(91 + 92 + 6 + 44)}{(33501 + 171)} = 233/33672 = 0.0075$$

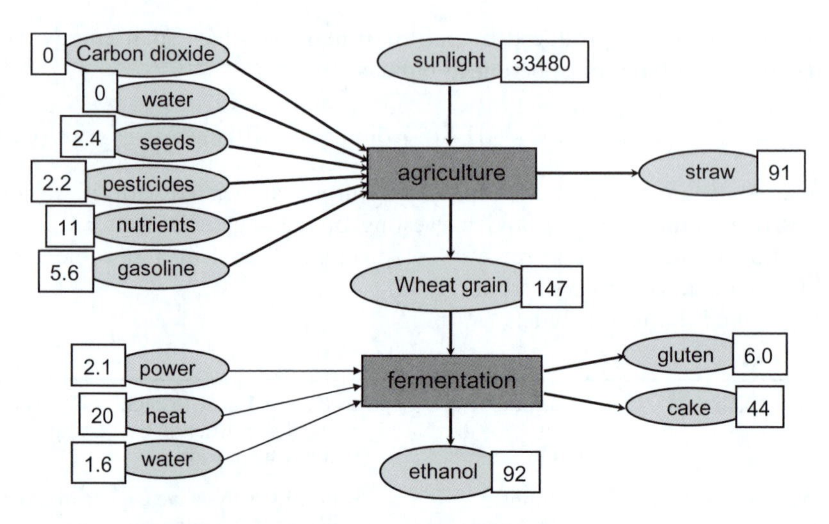

Figure 6.20 Flow diagram for the exergetic analysis of the production of ethanol from biomass by a biotechnological process. All energy and material inputs expressed in GJ ha^{-1} y^{-1}. Based on data from ref. 12.

Recently, the use of food crops such as wheat and corn for conversion into transportation fuels has become a matter of controversy, with concerns being raised about the impact of the demand for such raw materials on food prices[xiii] and particularly the impact that this might have on the poor and those in the developing world. This is a good illustration of the points made in Chapter 2 that social and political factors, viewed from a range of perspectives, will have as much, if not more, influence on what is considered to be an acceptably sustainable technology as technological judgements themselves.

Figure 6.21 illustrates a comparison between a series of technologies for the production of transportation fuel (including ethanol from a series of sources). In Figure 6.21, total environmental impact is plotted against greenhouse gas emissions, with the data all normalised such that gasoline or petrol scores 100 on each axis. Clearly, technologies scoring lower than 100 on either count are better than gasoline, with the best technologies having the lowest value of the product of the two factors. This topic is revisited in Chapters 11 and 12. However, it is pertinent to point out that, according to this analysis, ethanol from corn is associated with almost the same greenhouse emissions as gasoline, and a greater total environmental impact. On the other hand, it should be noted that ethanol from corn is in its early stages of technological development and is therefore not optimised, whereas gasoline use is well-established and this has some bearing on their relative technical performance (as opposed to social acceptability).

[xiii] It should be pointed out that others take a different view,[25] suggesting that '*index-based investments in agricultural futures markets*' were a more significant influence.

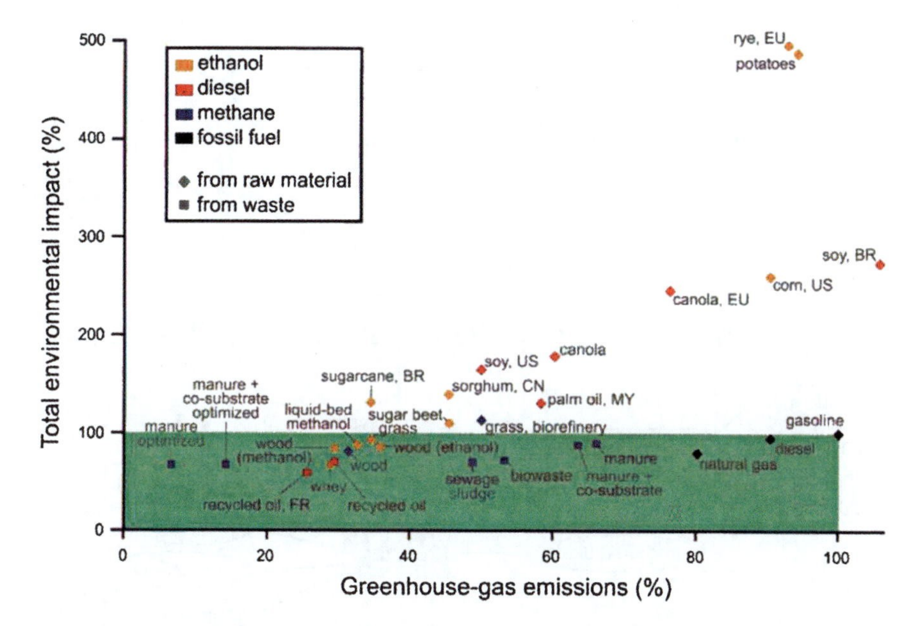

Figure 6.21 Total environmental impact and greenhouse gas emissions [relative to petrol (gasoline) = 100] for a series of technologies for the production of transportation fuel.[26b] Reproduced with permission of the copyright owners Zah *et al.*, Empa[26a] and AAAS.

6.15.3 Idealised Sustainable Process for Ethanol Manufacture

The third approach is not yet an operable technology, though it has features that might make it attractive. First, we might generate hydrogen by water electrolysis (eqn 6.54) using electricity produced in a photovoltaic device able to convert solar energy into electricity. We might use as the carbon source CO_2 from the atmosphere and, having discovered a catalyst, convert these two feedstocks into ethanol (eqn 6.55).

$$\text{Light} + \text{Water} \rightarrow H_2 + 0.5\,O_2 \tag{6.54}$$

$$6\,H_2 + 2\,CO_2 \rightarrow C_2H_5OH + 3\,H_2O \tag{6.55}$$

Inputs and outputs are made up as follows:

Solar cell electricity:	Inputs:	sun (conversion efficiency) power, materials (PV cells)
	Outputs:	electricity
Water electrolysis:	Inputs:	water, electricity
	Outputs:	hydrogen, oxygen
CO_2 from flue gas:	Inputs:	flue gas, power, heat
	Outputs:	carbon dioxide
Ethanol synthesis:	Inputs:	H_2, CO_2, power
	Outputs:	EtOH, MeOH, CH_4, CO, water Steam

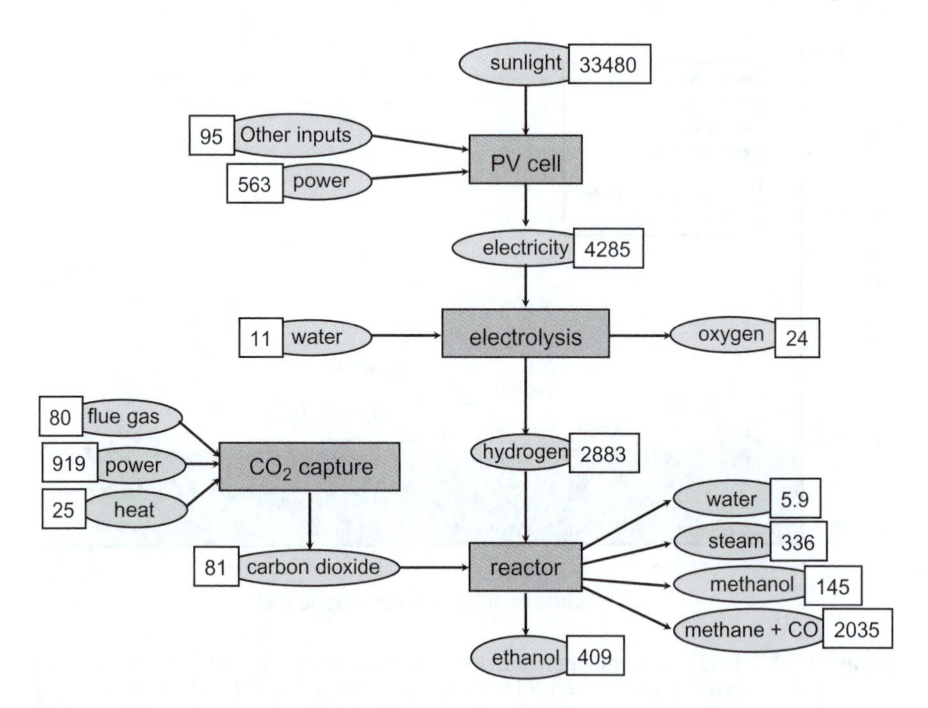

Figure 6.22 Flow diagram for the exergetic analysis of the possible production of ethanol using hydrogen derived ultimately from solar energy and captured carbon dioxide as feedstock. All energy and material inputs expressed in GJ ha^{-1} y^{-1}. Based on data from ref. 12.

Using the flow diagram and numerical values of the inputs and outputs shown in Figure 6.22, production efficiencies of the individual steps are obtained as follows:

Efficiency of solar conversion = 0.126
Efficiency of hydrogen production = 0.68
Efficiency of CO_2 capture = 0.79
Efficiency of ethanol synthesis = 0.75
Overall production efficiency = 0.082

6.15.4 Comparison of the Processes

Table 6.7 collates the values of the production efficiency, η_2, calculated by Dewulf *et al.* and those of the related coefficients, α (the renewability coefficient), η_1 (environmental efficiency coefficient) and η (overall exergetic efficiency).

This table highlights what we probably know intuitively, namely, that the renewability coefficients of fossil-sourced technologies are very low and their production efficiency is high, whereas the reverse is true for fermentation technology and the more speculative hydrogen-based approach.

Table 6.7 Comparison of various parameters for three processes for the large-scale manufacture of ethanol. Data from ref. 24a.

Ethanol production technology	*α renewability coefficient*	*η_1 environmental efficiency coefficient*	*η_2 production efficiency coefficient*	*η overall exergetic efficiency*	*Σ sustainability coefficient*
Fossil	0.0002	0.744	0.490	0.365	0.18
Fermentation	0.998	0.9995	0.007	0.00694	0.50
CO_2/solar H_2	0.909	0.978	0.082	0.0792	0.48

Clearly, much can be done to make biomass-based technologies more efficient and to discover more efficient methods for capturing solar energy than *via* photosynthesis. However, while such developments occur (we discuss the timescales on which this might be reasonably expected in Chapter 9), there is every reason during the process of transition to more sustainable technologies to continue to exploit the better examples of conventional technology.

REFERENCES

1. U. Beck, *Risk Society: Towards a New Modernity*, Sage, London, 1992.
2. (a) R. A. Sheldon, *Chem. Ind.*, 1992, 903; (b) R. A. Sheldon, *Green Chem.*, 2007, **9**, 1273; (c) R. A. Sheldon, *Chem. Commun.*, 2008, 3352.
3. M. J. Eckelman, J. B. Zimmerman and P. T. Anastas, *J. Ind. Ecol.*, 2008, **12**, 316.
4. P. J. Dunn, S. Galvin and K. Hettenbach, *Green Chem.*, 2004, **6**, 43.
5. D. W. F. Hardie, *Acetylene Manufacture and Use*, Oxford University Press, Oxford, 1965, p. ix.
6. J. W. Erisman, M. A. Sutton, J. Galloway, Z. Klimont and W. Winiwarter, *Nat. Geochem.*, 2009, **1**, 639.
7. A. Mittasch, *Adv. Catal.*, 1950, **2**, 81.
8. *Industry and the Environment – A Strategic Overview*, Centre for Exploitation of Science and Technology, 1991, London, p. 48.
9. T. Jackson, R. Costanza, M. Overcash and W. Rees, in *Clean Production Strategies: Developing Preventive Environmental Management in the Industrial Economy*, ed. T. Jackson, Lewis Publishers, London, 1993, p. 17.
10. (a) B. K. Sovacool, *Energy Policy*, 2008, **36**, 2950; (b) Parliamentary Office of Science and Technology, *Carbon Footprint of Electricity Generation*, POST, London, 2006, Postnote No. 268 (Web 4).
11. (a) IPCC, *Special Report on Carbon Dioxide Capture and Storage*, ed. B. Metz, O. Davidson, H. C. De Coninck, M. Loos and L. A. Meyer, Cambridge University Press, Cambridge, 2005 (Web 5); (b) G. T. Rochelle, *Science*, 2009, **325**, 1652; (c) Report of the House of Commons Science and Technology Committee, *Meeting UK Energy and Climate*

Needs: The Role of Carbon Capture and Storage: Government Response to the Committee's First Report of Session 2005–06, The Stationery Office, London, 2006, HC 1036, Third Special Report of Session 2005–06 (Web 9); (d) R. S. Haszeldine, *Science*, 2009, **325**, 1647; (e) J. Gibbons and H. Chalmers, *Energy Policy*, 2008, **36**, 4317.
12. J. M. Mulder, *Alternative Routes to Ethanol Production from Renewable Resources: A Thermodynamic Approach*, TU Delft, Delft, The Netherlands, 2000.
13. T. J. Tambach, J. P. Mathews and F. Van Bergen, *Energy Fuels*, 2009, **23**, 4845.
14. (a) M. Mikkelsen, M. Jorgensen and F. C. Krebs, *Energy Environ. Sci.*, 2010, **3**, 43; (b) J. C. S. de Araujo, C. B. A. Sousa, A. C. Oliveira, F. N. A. Freire, A. P. Ayala and A. C. Oliveira, *Appl. Catal. A*, 2010, **377**, 55.
15. R. Juárez, A. Corma and H. García, *Top. Catal.*, 2009, **52**, 1688.
16. (a) A. Azapagic, *Chem. Eng. J.*, 1999, **73**, 1; (b) A. Azapagic, in *Handbook of Green Chemistry and Technology*, e. J. H. Clark and D. Macquarrie, Wiley-Blackwell, 2002, ch. 5, p. 62; (c) G. Rebitzer, T. Ekvall, R. Frischknecht, D. Hunkeler, G. Norris, T. Rydberg, W.-P. Schmidt, S. Suh, B. P. Weidema and D. W. Pennington, *Environ. Int.*, 2004, **30**, 701; (d) D. W. Pennington, J. Potting, G. Finnveden, E. Lindeijer, O. Jolliet, T. Rydberg and G. Rebitzer, *Environ. Int.*, 2004, **30**, 721; (e) G. Finnveden, M. Z. Hauschild, T. Ekvall, J. Guinée, R. Heijungs, S. Hellweg, A. Koehler, D. Pennington and S. Suh, *J. Environ. Manage.*, 2009, **91**, 1.
17. (a) M. B. Hocking, *Environ. Manage.*, 1994, **18**, 889; (b) M. B. Hocking, personal communication; (c) A. Vercalsteren, C. Spirinckx and T. Geerken, *Int. J. Life Cycle Assess.*, 2010, **15**, 221.
18. P. Blowers and J. M. Lownsbury, *Environ. Sci. Technol.*, 2010, **44**, 1526.
19. (a) D. J. Richards and R. A. Frosch, in *The Industrial Green Game*, ed. D. J. Richards, National Academy of Engineering, Washington DC, 1997, p. 12; (b) M. A. Del Nobile, G. Mensitieri, A. Aldi and L. Nicolais, *Packag. Technol. Sci.*, 1999, **12**, 261.
20. F. R. Cruickshank, A. J. Hyde and D. Pugh, *J. Chem. Educ.*, 1977, **54**, 288.
21. K. J. Laidler, *Energy and the Unexpected*, Oxford University Press, Oxford, 2002.
22. P. W. Atkins, *The Second Law*, W. H. Freeman & Co., New York, 1984.
23. I. L. Leites, D. A. Sama and N. Lior, *Energy*, 2003, **28**, 55.
24. (a) J. Dewulf, H. Van Langenhove, J. Mulder, M. M. D. van den Berg, H. J. van der Kooi and J. de Swaan Arons, *Green Chem.*, 2000, **2**, 108; (b) J. Dewulf, H. Van Langenhove, B. Muys, S. Bruers, B. Bakshi, G. F. Grubb, D. M. Paulus and E. Sciubba, *Environ. Sci. Technol.*, 2008, **42**, 2221.
25. C. L. Gilbert, *J. Agric. Econ.*, 2010, **61**, 398.
26. (a) R. Zah, H. Böni, M. Gauch, R. Hischier, M. Lehmann and P. Wäger, *Ökobilanz von Energieproduckten: Ökologische Bewertung von Biotreibstoffen*, Empa, St. Gallen, Switzerland, 2007; (b) J. P. W. Scharlemann and W. F. Laurance, *Science*, 2008, **319**, 43 and supporting online material.

BIBLIOGRAPHY[xiv]

F. Furedi, *Culture of Fear: Risk-Taking and the Morality of Low Expectations*, Cassell, London, 1997.

P. Slovic, *Science*, 1987, **236**, 280.

P. Harremoës, D. Gee, M. MacGarvin, A. Stirling, J. Keys, B. Wynne and S. Guedes Vaz, *The Precautionary Principle in the 20th Century: Late Lessons from Early Warnings*, Earthscan, London, 2002.

V. Smil, *Enriching the Earth*, MIT Press, Cambridge, MA, 2001.

J. R. Partington and L. H. Parker, *The Nitrogen Industry*, Constable, London, 1922.

J. W. Moore and R. G. Pearson, *Kinetics and Mechanism: A Study of Homogeneous Chemical Reactions*, John Wiley & Sons, New York, 3rd edn., 1981.

P. W. Atkins, *The Second Law*, Scientific American Books, New York, 1984.

G. Price, *Thermodynamics of Chemical Processes*, Oxford University Press, Oxford, 1998.

E. B. Smith, *Basic Chemical Thermodynamics*, Imperial College Press, London, 5th edn, 2004.

J. B. Ott and J. Baerio-Goates, *Chemical Thermodynamics: Principles and Applications*, Academic Press, London, 2000.

M. A. Rosen, I. Dincer and M. Kanoglu, *Energy Pol.*, 2008, **36**, 128.

I. Dincer and M. A. Rosen, *Exergy: Energy, Environment and Sustainable Development*, Elsevier Science, Oxford, 2009.

WEBLIOGRAPHY

1. www.defra.gov.uk/evidence/statistics/environment/waste/kf/wrkf02.htm
2. http://en.wikipedia.org/wiki/Waste_hierarchy
3. (a) http://en.wikipedia.org/wiki/Montreal_protocol
   (b) http://ozone.unep.org/Publications/MP_Handbook/Section_1.1_The_Montreal_Protocol/
4. www.parliament.uk/documents/upload/postpn268.pdf
5. www.ipcc.ch/pdf/special-reports/srccs/srccs_wholereport.pdf
6. www.rsc.org/chemistryworld/News/2008/September/15090801.asp
7. (a) www.iso.org/iso/catalogue_detail?csnumber = 37456 (b) www.iso.org/iso/iso_catalogue/catalogue_tc/catalogue_detail.htm?csnumber = 38498
8. http://en.wikipedia.org/wiki/Rubik%27s_cube
9. www.publications.parliament.uk/pa/cm200506/cmselect/cmsctech/1036/1036.pdf

All the web pages listed in this Webliography were accessed in May 2010.

[xiv] These texts supplement those already cited as references.

CHAPTER 7

Measurement

'Everything that can be counted does not necessarily count; everything that counts cannot necessarily be counted.'

Albert Einstein

We highlighted in Chapter 5 the important part that chemistry plays in identifying and understanding the components of the natural environment (both pristine and perturbed) particularly in detecting, characterising and quantifying materials that may also be pollutants. It is crucial to observe and monitor what is going on in the environment to be able to assess the impact that human activity is having on it, including that arising from the production and use of materials manufactured and processed by the chemical and associated industries.

As wastes and emissions arising from the production (and our use) of chemical products are of particular concern, we need to be certain of the strengths and weaknesses of the methods used to assess the effectiveness of approaches that minimise waste formation from chemical reactions, particularly on the manufacturing scale. To be of use, the better sort of process 'metrics', whether for the laboratory or the large scale, must be:

- informative
- relevant (to waste minimisation)
- universal or widely used
- easy to measure and understand
- quantitative
- use generally accepted units (*e.g.* $Mol\,L^{-1}$, $kg\,m^{-3}$)
- allow comparisons to be made (between plant, processes, companies and over time)
- realistic.

Chemistry for Sustainable Technologies: A Foundation
By Neil Winterton

Published by the Royal Society of Chemistry, www.rsc.org

We are already familiar with the E-factor and the environmental quotient, Q, and will, later in this chapter, be introduced to the metrics **atom efficiency** (also known as atom economy or atom utilisation) (Section 7.5), **reaction mass efficiency** (Section 7.8) and **balance yield** (Section 7.7).

The study of reaction efficiency is an area of much activity: Constable and colleagues[1] have summarised much in a short, accessible article. We return in Sections 7.8 and 7.9 to the important work of Andraos[2] who has developed some of these metrics and unified them into powerful and useful tools for analysing reaction efficiency in ways relevant to fine chemical processing. See the Bibliography for texts that review the most recent work including the wider (and more complex) topics of metrics for sustainable development.

7.1 REACTION YIELD

A question for chemists to ask (and to answer) is whether the metric long used in assessing the effectiveness or otherwise of a chemical transformation, the **reaction yield**, is useful in assessing its efficiency and the extent of associated waste produced. Process chemists or chemical engineers may also be familiar with some related terms such as **mass balance** (Section 7.2), **conversion** (Section 7.3) and **selectivity** (Section 7.4), the importance and relevance of which will become evident in Chapter 9.

Let us first have a look at reaction yield for estimating reaction efficiency in the context of sustainable technologies. We will consider a very simple transformation (eqn 7.1). While this chemistry is apparently simple, it is also extremely important technologically, being operated on a very large scale. The conversion of methanol to chloromethane provides raw materials for the manufacture of silicones, amongst other products.

$$CH_3OH + HCl \rightarrow CH_3Cl + H_2O \tag{7.1}$$

Say, we begin with 32 g (1 mole) of methanol and react it with sufficient hydrogen chloride, producing 20 g chloromethane. Water is a **co-product**.[i] The yield is given in eqn (7.2):

$$\text{Yield} = \frac{M_1}{M_2} \times 100 \tag{7.2}$$

where M_1 is the mass of a specified product (usually after isolation and purification; in this case 20 g of chloromethane) and M_2 is the maximum mass predicted from 100% conversion of the specified precursor in a reaction with an assumed stoichiometry (*i.e.* the so-called 'theoretical' amount of chloromethane).

[i] See footnote ii, p. 118.

The 'theoretical' yield is given by eqn (7.3):

$$M_2 = \frac{\text{MWt of product} \times \text{Mass of precursor}}{\text{MWt of precursor}} \tag{7.3}$$

In the case of chloromethane formed from methanol according to eqn (7.1), the theoretical yield is:

$$\frac{50.5 \times 32}{32} = 50.5\,\text{g}$$

The yield is therefore

$$\frac{20 \times 100}{50.5} = 39.6\%$$

This figure fails to tell us anything about why the yield of this reaction is as low as it is. Nor does it give any indication of the extent of the waste that might be formed in the reaction. It is possible, of course, that the reaction has reached equilibrium and this is all the chloromethane that is formed under the particular set of reaction conditions. However, the reaction may also have gone to completion (*i.e.* has consumed all the limiting reactant), so there may be additional products formed that we have not so far taken into account. It is possible, even, that the initial product reacts further to give something other than chloromethane and the stoichiometry is not as initially thought. A second reaction does indeed occur, leading to the formation of dimethyl ether, CH_3OCH_3 (eqn 7.4). This is a **by-product**, the result of a different (parallel) or subsequent (series) reaction from that giving the desired product, chloromethane, and which may or may not have saleable value. By-product formation is something that, in general, is an inconvenience in laboratory chemistry but assumes major significance in industrial processes.

$$CH_3OH \rightleftharpoons 0.5\,CH_3OCH_3 + 0.5\,H_2O \tag{7.4}$$

In our example, we also see the formation of, say, 10 g of dimethyl ether from the 32 g of methanol we started with. Using the relationship for yield and assuming the stoichiometry of eqn (7.4), we can easily establish that the maximum amount of dimethyl ether that can be obtained from 32 g of methanol is $0.5 \times 46\,\text{g} = 23\,\text{g}$ and that the yield of dimethyl ether is therefore $(10/23) \times 100 = 43.5\%$.

The combined yield of 83.1% leaves us with some material unaccounted for. Is there another unsuspected product? Or have we not allowed the reaction go to completion? Or has the reaction gone to completion to give an equilibrium mixture? If the latter, some unconverted methanol should still be present in the system. Or have we been sloppy in how we set up the experiment and there have been more physical losses than we might have first suspected. It is clear, even in this apparently simple example, how many questions there are relevant to assessing reaction efficiency and accounting for losses which, from the consideration of yield as the reaction metric, we cannot fully answer.

7.2 MASS BALANCE

We need to recall a very fundamental chemical law, namely the Law of Conservation of Mass, whose expression in its modern form is generally associated with Antoine Lavoisier (1743–1794). The Law states that mass cannot be created or destroyed. The modern statement of the First Law of Thermodynamics would talk, more strictly, in terms of energy instead of mass; however, chemical energies are small enough that the distinction is academic.

If we consider a batch reactor (Figure 7.1), the Law of Conservation of Mass requires that if we start with an empty reactor into which we put our reactants and, after a specified time, we remove the entire contents of the reactor, then the mass of material removed must be the same as the mass of the material put in, even if reaction has taken place. This can be checked by establishing a **mass balance**, a crucial tool in the understanding of waste generation and its minimisation.

If we carry out a complete chemical analysis of the contents of our reactor we find that it contains only the following:

- the desired product chloromethane
- the by-product dimethyl ether
- the co-product water
- unused reactants, hydrogen chloride and methanol.

The mass balance would then allow us to test the equality (eqn 7.5) and establish whether the sum on the right-hand side is the same as, or less than, that on the left, and, if the latter, by how much.

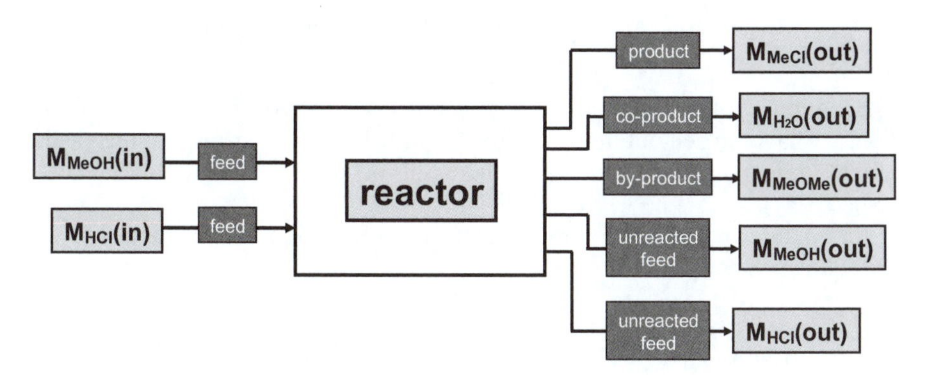

Figure 7.1 Inputs to and outputs from a batch reactor in which methanol is reacted with hydrogen chloride.

$$M_{\text{MeOH}}(\text{in}) + M_{\text{HCl}}(\text{in}) = M_{\text{MeOH}}(\text{out}) + M_{\text{HCl}}(\text{out}) + M_{\text{MeCl}}(\text{out}) + M_{\text{MeOMe}}(\text{out}) + M_{H_2O}(\text{out}) \quad (7.5)$$

We can now answer the following important questions: Have we accounted for all specified inputs? If not, why not? The latter may be because there are further unidentified products—still possible, if our analytical methods are not sensitive enough or cannot detect particular types of product. There may also be physical losses associated with the manipulation of reactants and products. This may be thought of as being relatively trivial in a laboratory context, as reactions are carried out in a fume hood. However, such losses are a critical question in an industrial context where they may have either environmental or occupational health consequences (in addition to the obvious economic consequences arising from the loss of useful material).

Because it is often necessary, particularly during the early (laboratory) stages in the development of a chemical process, to anticipate the foregoing questions, certain additional metrics are usually applied that bear upon questions of reaction efficiency and the investigation of the sources of waste and their minimisation. These include reaction **conversion** and reaction **selectivity**.

7.3 CONVERSION

Conversion examines how much of a reactant is transformed in the reactor. Clearly, if a reactant is used in excess, then there must always be unreacted material at the completion of reaction. If the reactant is present in deficiency, then the expectation is that all will be converted to the desired product in the reactor. The focus in our case will be on the methanol introduced and how much of it is converted. To establish this we need to know how much unreacted methanol is to be found in the contents of the reactor at the end of the reaction (eqn 7.6):

$$2\,CH_3OH + HCl \rightleftharpoons CH_3Cl + 0.5\,CH_3OCH_3 + 1.5\,H_2O \quad (7.6)$$

Equation (7.6) now represents the overall stoichiometry of the transformations we have carried out in the reactor, and for which the corresponding yields of chloromethane and dimethyl ether will be different (79.2 and 87.0%, respectively).

An organic analysis of the reactor contents at the end of the reaction showed the following:

$M_{\text{MeOH}}(\text{in}) = 32\text{ g}$
$M_{\text{MeCl}}(\text{out}) = 20\text{ g}$
$M_{\text{MeOMe}}(\text{out}) = 10\text{ g}$
$M_{\text{MeOH}}(\text{out}) = 5.4\text{ g}$

Thus 5.4 g of unreacted methanol is found. The conversion of methanol is given by eqn (7.7):

$$\text{Conversion} = \frac{\text{Mass of MeOH consumed}}{\text{Initial mass of MeOH}} \times 100 \qquad (7.7)$$

Or, if we plug in the relevant values:

$$\frac{[M_{MeOH}(in) - M_{MeOH}(out)]}{M_{MeOH}(in)} \times 100 = \frac{26.6}{32} \times 100 = 83.1\%$$

So, only 83.1% of the methanol introduced into the reactor is converted into products. Note it is immaterial whether the calculation is by weight or by mole.

7.4 SELECTIVITY

Selectivity[ii] tells us something different, measuring the fraction of the products formed that is the one desired. Here we have two products: chloromethane and dimethyl ether. We set out to make chloromethane, so we should be concerned if any of the methanol we have paid for ends up being converted to a material we do not want. The selectivity metric tells us how good or bad we have been in producing the thing we do want. We would like this to be as close to 100% as possible. Using the analytical data for the reactor contents shown above, the selectivity is given by eqn (7.8a) or (7.8b):

$$\text{Selectivity} = \frac{\text{Mass of required product}}{\text{Total mass of all products from limiting feed}} \times 100 \qquad (7.8a)$$

$$\text{Selectivity} = \frac{M_{MeCl}(out)}{M_{MeCl}(out) + M_{MeOMe}(out)} \times 100 = \frac{20}{30} \times 100 = 66.7\%$$

Selectivity can also be estimated using the mole as the basis rather than weight. Does this make any difference?

$$\text{Selectivity} = \frac{\text{No. of moles of required product}}{\text{Total no. of moles of all products from limiting feed}} \times 100 \qquad (7.8b)$$

$$\text{Selectivity} = \frac{\text{Moles}_{MeCl}(out) \times 100}{\text{Moles}_{MeCl}(out) + \text{Moles}_{MeOMe}(out)}$$
$$= \frac{20/50.5 \times 100}{(20/50.5) + (10/46)} = 64.6\%$$

As the molecular masses of methyl chloride and dimethyl ether are quite similar, the difference in this case is small. It is more usual to estimate selectivity on a mole basis.

[ii] The different types of chemical selectivity are dealt with in Section 10.3.4.

If the reaction has gone to completion and there is no interconversion between the products, then it is a simple calculation to show that the selectivity is the ratio of the rates of formation of the two products.[iii] Much effort is expended by chemists to try to maximise the amount of the desired product and to minimise the amount of undesired product being formed (*e.g.* by seeking to catalyse the reaction of choice). Sometimes this is not possible, so other strategies are used such as feeding the by-product back into the reactor or seeking to convert it to the required material in a different process.

So, the process chemist would investigate the extent to which dimethyl ether might be converted to chloromethane in the presence of hydrogen chloride (eqn 7.9 and 7.10):

$$CH_3OCH_3 + HCl \rightleftharpoons CH_3Cl + CH_3OH \quad (7.9)$$

$$CH_3OCH_3 + 2\,HCl \rightleftharpoons 2\,CH_3Cl + H_2O \quad (7.10)$$

If dimethyl ether is converted to chloromethane (or *vice versa*) in the process, then the selectivity calculation based on reaction rates is somewhat different. This is left to the reader to carry out as an exercise.

7.5 ATOM EFFICIENCY

A strategy adopted by chemists, which was formally enunciated during the early 1990s, seeks to render chemical synthesis (and also it was hoped chemical production) more efficient by searching purposefully for reactions leading to products that incorporated reactants to the greatest degree. This was variously called atom economy, atom efficiency (AE) or atom utilisation; Trost[4] is primarily associated with its development. The attraction of this approach is that it automatically focuses attention on the **elimination of co-products** and the maximising of the incorporation of the elemental components of the reactants into a single product.

Ideally, and in the limit, atom efficiency will be 100%, in which all atoms of the reactants are incorporated into a single product. An example of a transformation with 100% AE (well-known before the AE concept was propounded) is cycloaddition, *e.g.* the dimerisation of cyclopentadiene (eqn 7.11)

[iii] It is rare for the process chemist to be so fortunate as to have a well-behaved system such as the one we have described. Often there are both parallel and sequential reactions involving reactants and products. In addition, equilibria can be set up involving a reactant and two possible products, P_1 and P_2. If the forward rate leading to P_1 is much greater than that to P_2 (even if the equilibrium constant favours P_2) and the reverse process is slow, then P_1 builds up and the process is said to be under 'kinetic control'. If all reactions (forward and back) are rapid, then any P_1 reverts to starting material and will be converted to the thermodynamically favoured P_2. The reaction is then said to be under 'thermodynamic control'. These factors have a significant bearing on optimum processing conditions to deliver a targeted compound and to control the amount of waste generated. Some examples (and a more detailed discussion of the kinetics) are given in ref. 3.

or the addition of cyclopentadiene to cyclohexenone[5] (eqn 7.12) to give tricyclo[6.2.1.0^{2,7}]undec-9-en-3-one.

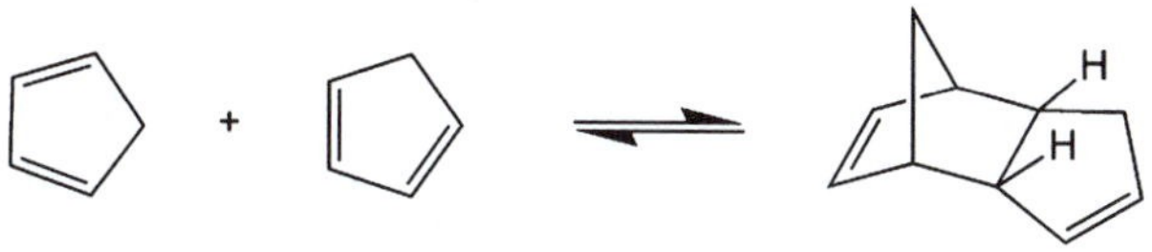

(7.11)

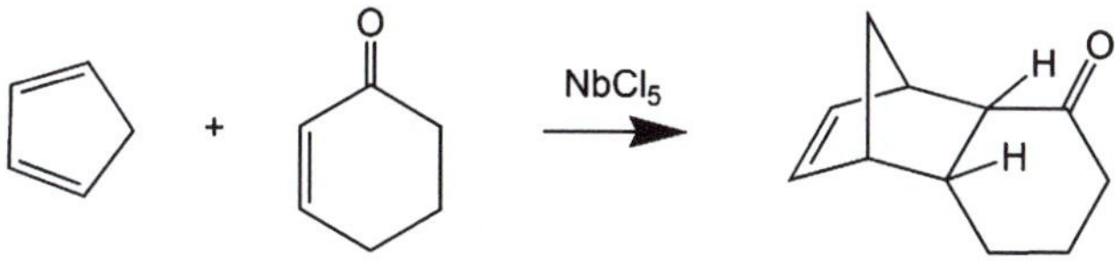

NbCl$_5$

(7.12)

The hydroformylation of ethene (as in eqn 7.13) with carbon monoxide and dihydrogen leads to propionaldehyde, a reaction of industrial importance and one that displays, in principle, 100% atom efficiency.

$$CH_2{=}CH_2 + CO + H_2 \rightarrow CH_3CH_2C(O)H \quad (7.13)$$

However, whether hydroformylations with substituted ethenes such as in eqn (7.14) can strictly be said to display 100% atom efficiency very much depends whether the demand for each product is equal to the quantity produced (otherwise some part of the production of one or other would have to be seen as a waste).

$$2\,CH_2{=}CHR + 2\,CO + 2\,H_2 \rightarrow CH_3CHRC(O)H + RCH_2CH_2C(O)H \quad (7.14)$$

Similarly, the formation of tricyclo[6.2.1.0^{2,7}]undec-9-en-3-one (eqn 7.12) requires the presence of a catalyst and is carried out at very low temperatures to avoid polymerisation.[5]

Most useful chemicals production is *via* processes that do not display 100% atom efficiency. Nevertheless, the development of this concept has stimulated new chemistry[iv] and focused attention on atom efficiency in chemical technology, particularly in the pharmaceuticals and fine chemicals sector which, as Sheldon's E-factor (Section 6.3) estimates suggest, are relatively waste producing.

For the general reaction eqn (7.15) in which the desired product is C, the atom efficiency is given by eqn (7.16), *i.e.* the ratio (usually expressed as a percentage) of the molecular weight (MWt) of the desired product to the sum of molecular weight of the stoichiometric reactants.

$$A + B \rightarrow C + D \quad (7.15)$$

[iv] Notable examples of improved atom efficiency include the total synthesis of the natural product, bryostatin 16, described in Section 8.3.

$$\text{Atom efficiency} = \frac{\text{MWt of C}}{\text{MWt of A} + \text{MWt of B}} \times 100 \tag{7.16}$$

If no D is formed, then atom efficiency is 100%.

More generally, for a sequence of reactions (eqn 7.17–7.19):

$$A + B \rightarrow C + D \tag{7.17}$$

$$C + E \rightarrow F + G \tag{7.18}$$

$$F + H \rightarrow I + J \tag{7.19}$$

$$A + B + E + H \rightarrow D + G + I + J \tag{7.20}$$

or overall for the reaction sequence (eqn 7.20) leading from A + B to the desired product, J, atom efficiency is given by eqn (7.21):

$$\text{Atom efficiency} = \frac{\text{MWt of J}}{\sum \text{MWts of A, B, E and H}} \times 100 \tag{7.21}$$

It is evident that atom efficiencies for multi-step processes, each using additional reactants, are likely to be lower than for single-step processes.

The importance of the AE concept becomes evident when we compare traditional stoichiometric reactions (*e.g.* the classical laboratory reduction using sodium borohydride) with the equivalent catalytic reaction operated widely and on a huge scale in the chemical industry. The reduction of acetophenone to 1-phenylethanol may be represented in eqn (7.22) by the simplified overall stoichiometry. (As an exercise, think why this is a *simplified* stoichiometry.)

$$PhC(O)Me + 0.25\,NaBH_4 + H_2O \rightarrow PhCH(OH)Me + 0.25\,NaB(OH)_4 \tag{7.22}$$

C_8H_8O	$+ 0.25 \times BH_4Na$	$+ H_2O$	$\rightarrow$	$C_8H_{10}O$
$(96 + 8 + 16)$	$+ (11 + 4 + 23)/4$	$+ (2 + 16)$	$\rightarrow$	$(96 + 10 + 16)$
120	+ 9.5	+ 18	$\rightarrow$	122

Atom efficiency is then calculated from the molecular weights of the desired product and the individual reactants in their stoichiometric proportions:

$$\begin{aligned} &\text{AE} \\ &= \frac{\text{MWt of 1-phenylethanol}}{\text{MWt of acetophenone} + 0.25 \times \text{MWt of sodium borohydride} + \text{MWt of water}} \times 100 \\ &= [122/(120 + 9.5 + 18)] \times 100 = 83\% \end{aligned}$$

This may be compared with a catalytic reduction of acetophenone (eqn 7.23) leading to the same product:

$$\begin{array}{llll} PhC(O)Me & + H_2 & \rightarrow & PhCH(OH)Me \\ C_8H_8O & + H_2 & \rightarrow & C_8H_{10}O \\ 120 & + 2 & \rightarrow & 122 \end{array} \tag{7.23}$$

$$AE = [122/(20 + 2)] \times 100 = 100\%$$

A similar comparison can be made between the stoichiometric and catalytic oxidations of 1-phenylethanol to acetophenone (which can be done as an exercise), giving 49% AE for the former using potassium permanganate/sulfuric acid, compared with 87% AE for the latter using dioxygen.

Some industrial processes that have been operated on a large scale up to quite recent times have employed a series of stoichiometric reactions to obtain the desired product. For instance, the 'chlorhydrin' process for the production of the industrial intermediate, ethylene oxide (also known as oxirane or epoxyethane, and important among other things for the manufacture of synthetic detergents and surfactants) involves the sequence shown in eqn (7.24) and eqn (7.25). It can immediately be seen that none of the chlorine used ends up in the desired product.

$$CH_2{=}CH_2 + Cl_2 + H_2O \rightarrow ClCH_2CH_2OH + HCl \tag{7.24}$$

$$ClCH_2CH_2OH + Ca(OH)_2 + HCl \longrightarrow \text{(epoxyethane)} + CaCl_2 + 2\,H_2O \tag{7.25}$$

The overall reaction is shown in eqn (7.26):

$$CH_2{=}CH_2 + Cl_2 + Ca(OH)_2 \longrightarrow \text{(epoxyethane)} + CaCl_2 + H_2O \tag{7.26}$$

$$\begin{array}{lllll} C_2H_4 & + Cl_2 & + CaH_2O_2 & \rightarrow & C_2H_4O \\ 28 & + 71 & + 74 & \rightarrow & 44 \end{array}$$

$$AE = (44/173) \times 100 = 25\%$$

An industrial process for epoxyethane production that avoids the use of chlorine has been developed which uses a heterogeneous silver catalyst and has the following (idealised) stoichiometry (eqn 7.27):

$$CH_2{=}CH_2 + 0.5\,O_2 \longrightarrow \text{(epoxyethane)} \tag{7.27}$$

$$\begin{array}{llll} C_2H_4 & + 0.5\,O_2 & \rightarrow & C_2H_4O \\ 28 & + 16 & \rightarrow & 44 \end{array}$$

$$AE = (44/44) \times 100 = 100\%$$

Unfortunately this is not the complete story as there is a 'side' reaction, one that is not desired and cannot be completely suppressed, and which consumes both reactant and product, generating unwanted by-products, in this case carbon dioxide and water. This so-called 'burning' reaction, associated with the use of dioxygen as a feedstock and the formation of acetaldehyde as a reactive intermediate, lowers selectivity to the desired product epoxyethane.

Taking account of burning leads to a revised stoichiometry (eqn 7.28):

$$1.1\ CH_2{=}CH_2 + 0.8\ O_2 \longrightarrow \text{(epoxyethane)} + 0.2\ CO_2 + 0.2\ H_2O \qquad (7.28)$$

From which a more accurate atom efficiency can be determined:

$$AE = 100 \times 44/(1.1 \times 28 + 0.8 \times 32) = 78\%$$

Even taking account of this by-product formation, the atom efficiency is significantly better than for the older process. This goes some way to explain why the catalysed process has displaced the chlorhydrin process.

The related chlorhydrin process for the production of propylene oxide (1,2-epoxypropane) shown in eqn (7.29) has an atom efficiency of 31%.

$$CH_2{=}CH(CH_3) + Cl_2 + Ca(OH)_2 \longrightarrow \text{(propylene oxide)} + CaCl_2 + H_2O \qquad (7.29)$$

$$\begin{array}{llll} C_3H_6 & + Cl_2 & + CaH_2O_2 & \rightarrow & C_3H_6O \\ 42 & + 34 & + 74 & \rightarrow & 58 \end{array}$$

This traditional process is slowly being displaced by less waste-producing technologies, involving catalysed oxidations with hydrogen peroxide with the formal stoichiometry shown in eqn (7.30) and having the significantly higher atom efficiency of 76%.

$$CH_2{=}CH(CH_3) + H_2O_2 \longrightarrow \text{(propylene oxide)} + H_2O \qquad (7.30)$$

$$\begin{array}{llll} C_3H_6 & + H_2O_2 & \rightarrow & C_3H_6O \\ 42 & + 34 & \rightarrow & 58 \end{array}$$

The story does not end here, however, as hydrogen peroxide is an expensive oxidant compared with dichlorine and one which is also hazardous to handle and to transport. Hydrogen peroxide is currently manufactured in a two-step liquid phase process in which a quinone is first reduced to a hydroquinone, which is then oxidised with dioxygen to form hydrogen peroxide and to regenerate the quinone. In reality, the process is more complicated (Scheme 7.1[6]) because additional hydrogenated and oxygenated intermediates are formed in the reaction cycle that themselves engage in hydrogenations and oxidations.

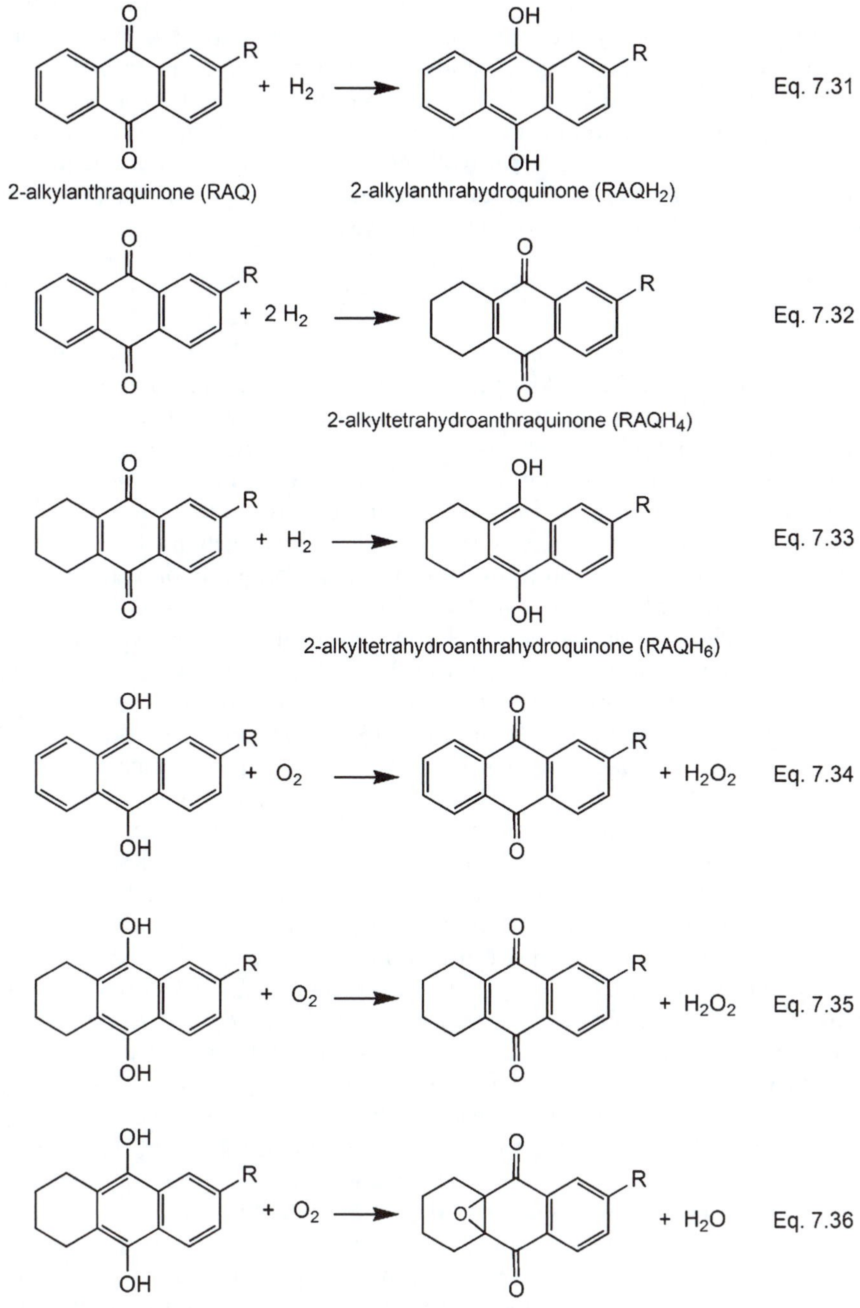

2-alkylanthraquinone (RAQ) $+ H_2 \longrightarrow$ 2-alkylanthrahydroquinone ($RAQH_2$) Eq. 7.31

RAQ $+ 2\,H_2 \longrightarrow$ 2-alkyltetrahydroanthraquinone ($RAQH_4$) Eq. 7.32

$RAQH_4 + H_2 \longrightarrow$ 2-alkyltetrahydroanthrahydroquinone ($RAQH_6$) Eq. 7.33

$RAQH_2 + O_2 \longrightarrow RAQ + H_2O_2$ Eq. 7.34

$RAQH_6 + O_2 \longrightarrow RAQH_4 + H_2O_2$ Eq. 7.35

$RAQH_6 + O_2 \longrightarrow$ (epoxide) $+ H_2O$ Eq. 7.36

Scheme 7.1 Key reactions in the manufacture of hydrogen peroxide using 2-alkylanthraquinone as reaction mediator.

2-Alkylanthraquinone (RAQ; R is usually ethyl) is first reduced with dihydrogen (eqn 7.31) to 2-alkylanthrahydroquinone ($RAQH_2$, or 2-alkylanthracene-9,10-diol) using a modified nickel catalyst. Hydrogenation of an aromatic ring (eqn 7.32 and eqn 7.33) also occurs, giving a tetrahydroanthrahydroquinone (6-alkyl-1,2,3,4-tetrahydroanthracene-9,10-diol). Both diols appear able to react with oxygen to form hydrogen peroxide (eqn 7.34 and eqn 7.35) in free radical chain reactions. The tetrahydroanthrahydroquinone also gives an epoxide as a by-product (eqn 7.36) in its reaction with oxygen.

$$H_2 + O_2 \rightarrow H_2O_2 \quad (7.37)$$

While the net reaction, eqn (7.37), is 100% atom efficient, no account is taken of process-related inefficiencies associated with the degradation of the anthraquinone, the nature of the catalyst and questions of the recovery of RAQ, catalyst and the product (usually as a 35–40% aqueous solution).

It is not surprising that much research[6b] is devoted to a so-called 'direct' process in which dihydrogen and dioxygen are reacted together to form hydrogen peroxide. The difficulty, of course, is to make such a reaction selective for the desired product and to avoid (or at least limit) the formation of water (and to avoid explosions while doing so). The development of a palladium–platinum catalyst to effect this change earned Headwaters Technology Innovation a Presidential Green Chemistry Challenge Award in 2007 (Web 1). More importantly, a demonstration plant has operated this NxCat™ technology on the semi-technical scale, although product is formed in a more dilute aqueous solution than in the anthraquinone process. A joint venture called EvonikHeadwaters 'aims to invest' (March 2009, Web 2) in a full-scale production unit, though this will need to compete with product from the operation of two new very large-scale operations using the conventional process. I discuss in Section 9.1 some of the reasons why such new and apparently less waste-producing technologies sometimes find it difficult to displace conventional processes.

Atom efficiency has some merit in focusing on waste associated with a particular transformation and in suggesting means by which it might be reduced. However, William H. Glaze whilst editor of the journal, *Environmental Science and Technology*, urged that green chemistry should focus explicitly on the quantifiable minimisation and elimination of problems at the process scale.[7] In other words the 'greenness' of a transformation can only properly be assessed in the context of its scale-up, its application and its practice. There is, thus, a need to take a much wider view of chemical technology and the means by which its efficiency can be measured and improved than simply by focusing on the chemical transformation itself. Atom efficiency has some serious weaknesses in this regard that are relevant to its utility to chemicals processing on the large scale. For instance, it assumes that the stoichiometry of the reaction leading to the desired product is precisely known, *i.e.* the reactant is converted to the specified product with perfect selectivity. As we have seen in the case of the catalysed oxidation of ethene to ethylene oxide, this need not always be the case and requires additional investigation to check.

In its simplest form, the method tends to ignore the atom efficiency of processes leading to feedstocks and starting materials (currently being addressed although the arithmetic becomes complex[8]). There is little point in carrying out a very atom-efficient reaction using a feedstock whose own preparation is characterised by very poor atom utilisation. Neither does atom efficiency take account of the presence of molar excesses of reactants, the use of solvents or of catalysts. Catalysts used industrially can often be very expensive. Efficient catalyst recovery is thus essential, but losses arising from less than perfect catalyst recovery are not taken into account by atom efficiency.

Atom efficiency also ignores:

- the issue of recycle (*e.g.* as arises when conversion is not 100% in an equilibrium-limited reaction, or when, as in the case of chloromethane production, unreacted methanol or by-product dimethyl ether can productively be recycled)
- the consumption of energy and utilities (*e.g.* water, electricity, gases) used in the process and in the important aspects of product separation, isolation and purification—all of which contribute to resource consumption that will affect process economics.

The question of more meaningful metrics for assessing efficiencies of chemical processes must take into account factors that academic chemists have, by and large, ignored. They have been thought to lie within the province of the chemical or process engineer. But no longer.

7.6 PROCESS CHEMISTRY

The highly idealised process shown in Figure 7.1 takes no account of the auxiliaries that might be employed in the reactor such as solvents, catalysts, or blanketing or inerting gases. Nor is account taken of the separation of the desired product from everything else and its purification. This is often carried out in a separate piece of equipment such as a still for distillation or a crystalliser after the contents have been removed from the reactor. (Engineers call these different process stages 'unit operations'.) In both cases, such operations will consume material such as extraction solvents or filtration aids which so far we have taken no account of. It is likely that the reactor will need to be heated, cooled and stirred; the still heated and provided with water or other coolant for condensation; and liquids and slurries will need to be pumped about. All require the appropriate utilities to be provided.

The discussion has so far focused on what is going on in and downstream from the reactor. For a full assessment of the waste generated in manufacturing a chemical product, we also need to understand the contribution to waste generation arising from the production of the feedstocks, as well as from the manufacture of solvents and catalysts used in the process.

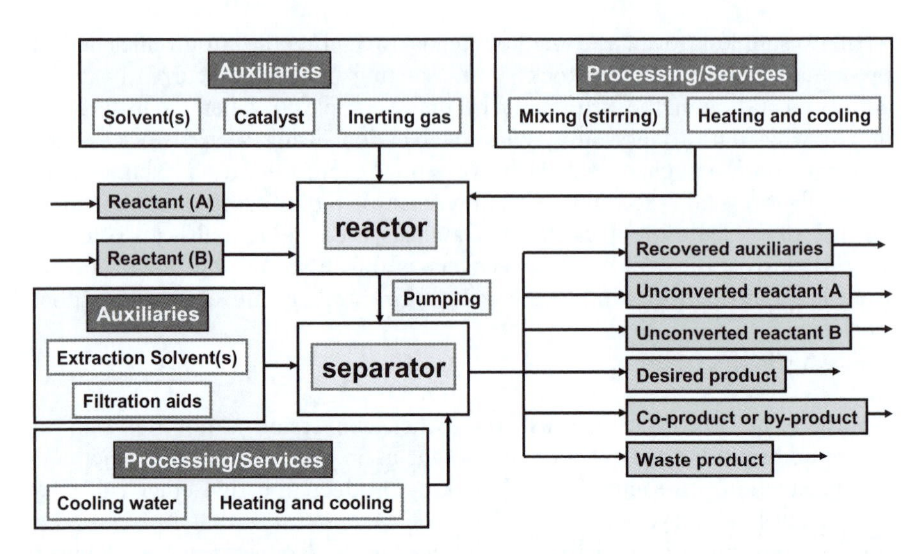

Figure 7.2 Components, functions, inputs and outputs of an idealised chemical production unit.

We have now described, in very simple form, the model for a chemical plant shown in Figure 7.2. Where does the waste from the process appear? In the case of the old Leblanc process, the waste appeared outside of the process by its emission to the environment. However, this is no longer acceptable and it is now the expectation that such toxic (or simply unpleasant) waste is treated by the producer to convert it into something that is less impacting.

Different types of waste can often be collected together and then treated in a single process. Alternatively, hazardous waste treatment can be incorporated into the chemical plant itself. It is thus critical in addressing how efficient processes are (and how well they treat waste) to know which type of waste is under consideration. This highlights again the importance of an awareness of system boundaries in any discussion of reaction efficiency, as the use of different boundaries can lead to different conclusions. The discussion of Sheldon's E-factor glossed over the distinction between 'waste' and 'by-product' which is resolvable only when we are more precise in our definitions. Sheldon's original approach, useful as it is, did not, in fact, focus on individual processes or products.

7.7 BALANCE YIELD

Some of the process-related factors we have introduced can be exemplified with an examination of the process for making the important fine chemical intermediate, *p*-anisaldehyde (4-methoxybenzaldehyde) on an industrial scale. The data (based on a real process) are taken from a paper on productivity functions, by Steinbach and Winkenbach.[9] Anisole is converted (Scheme 7.2) to *p*-anisaldehyde by hydrolysis of an intermediate formed from the

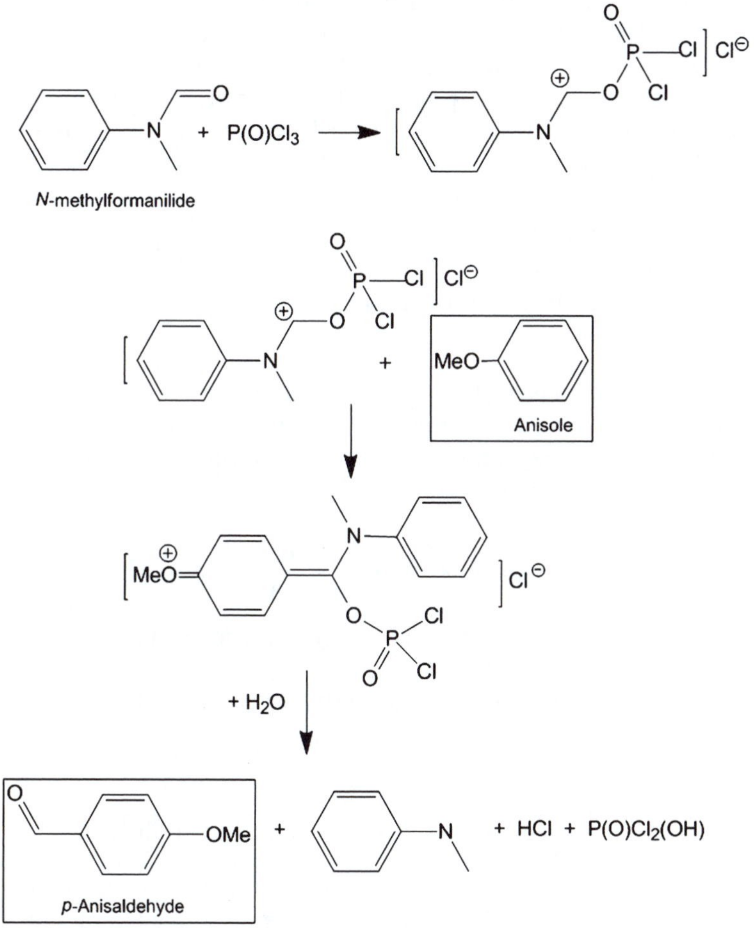

Scheme 7.2 Chemical transformation of anisole to *p*-anisaldehyde in an industrial-scale process.[9]

electrophilic attack at the *para* position of the starting material by an intermediate formed from the phosphorylation of *N*-methylformanilide.

The overall process can be represented simply by eqn (7.38), with the atom efficiency derived as before:

$$\begin{array}{llll} \mathrm{PhMeNC(O)H} & +\,\mathrm{POCl_3} & +\,\text{4-}\mathrm{MeOC_6H_5} & +\,\mathrm{H_2O} \\ \mathrm{C_8H_9NO} & +\,\mathrm{Cl_3OP} & +\,\mathrm{C_7H_8O} & +\,\mathrm{H_2O} \\ 135 & +\,153.5 & +\,108 & +\,18 \end{array} \tag{7.38}$$

$$\begin{array}{l} \rightarrow \text{4-}\mathrm{MeOC_6H_4C(O)H} + \mathrm{PhMeNH} + \mathrm{HCl} + \mathrm{HOP(O)Cl_2} \\ \rightarrow \mathrm{C_8H_8O_2} \\ \rightarrow 136 \end{array}$$

The atom efficiency of a reaction with this stoichiometry is $(136/414.5) \times 100 = 33\%$. This serves to highlight how much chemical material is needed simply (if formally) to insert CO into the *para* C–H bond of anisole.

A better understanding of process efficiency can be obtained by using some different metrics such as theoretical balance yield (eqn 7.39) and balance yield (7.40):

$$\textbf{Theoretical balance yield} = \frac{\text{Maximum mass of product (in 100\% yield from assumed stoichiometry)}}{\text{Total mass of materials used as reactants}} \times 100 \quad (7.39)$$

Using the stoichiometry in eqn (7.38), we can estimate the number of kilograms of input reactants needed to produce a metric ton (1000 kg) of *p*-anisaldehyde and the quantity of co-products formed. These are shown in Table 7.1.

$$\text{Theoretical balance yield} = \frac{1000}{3046} \times 100 = 33\%$$

This value is the same as atom efficiency. Remember, this assumes 100% yield in an idealised stoichiometry.

The balance yield takes account of the total mass of materials used in the process:

$$\textbf{Balance yield} = \frac{\text{Mass of desired product} \times 100}{\text{Total mass of materials used to produce it}} \quad (7.40)$$

This provides a more realistic measure of the amount of material utilised in the chemical plant. Table 7.2 shows the main inputs and main outputs of the process (as operated).

We immediately see from Table 7.2 that 29 000 kg of material (some used in downstream processing, including the extraction of the product with cyclohexane and the neutralisation of excess acid for subsequent disposal) are needed and not the 3046 kg required by the stoichiometry (Table 7.1).

Table 7.1 Calculated inputs and outputs for the process for the conversion of anisole to 1000 kg of *p*-anisaldehyde according to the stoichiometry of eqn (7.38). Data from ref. 9.

Inputs	*(kg)*	*Outputs*	*(kg)*
N-methylformanilide	993	Anisaldehyde	1000
Phosphorus oxychloride	1127	*N*-methylaniline	787
Anisole	794	Dichlorophosphoric acid	991
Water	132	Hydrochloric acid	268
TOTAL	3046	TOTAL	3046

Table 7.2 Summary of main inputs and outputs for the industrial process for the conversion of anisole to 1000 kg of *p*-anisaldehyde. Data from ref. 9.

Main inputs		*(kg)*	*Main outputs*		*(kg)*
Primary raw materials	anisole	2656	Main product	anisaldehyde	1000
	N-methylformanilide	5000			
	others	4344	Residues for incineration	*N*-methylaniline	200
				N-methylformanilide	3500
Secondary raw materials	cyclohexane	6000		cyclohexane	5900
	sodium hydroxide	7600			
	others	3400	Residues to wastewater	sodium chloride	6400
				Na_2HPO_4	5300
				others	3100
			Residues to vent	cyclohexane	100
				others	3500
	TOTAL	29 000		TOTAL	29 000

These materials are accounted for using a **mass balance** approach, as shown, that includes identifying and accounting for materials that end up in wastewater, are vented to atmosphere, or are waste residues sent for incineration.

$$\text{In this case, the } \textbf{product yield} = \frac{1000 \times 100}{2656 \times (136/108)} = 29.9\%$$

However, the 1000 kg of product results from the use of 29 000 kg of material.

$$\text{The } \textbf{balance yield} = \frac{1000 \times 100}{29\,000} = 3.5\%$$

From this, we get the **specific balance yield**:

$$= \frac{\text{Balance yield} \times 100}{\text{Theoretical balance yield}} = \frac{3.5 \times 100}{33} = 11\%$$

Importantly, *ca.* 17 000 kg of auxiliaries are used in the preparation, representing 59% of the materials used.

This example goes someway to explain why the fine chemical and pharmaceutical sectors are so waste-producing. It is worth remembering also that our analysis does not include the services or the utilities used (nor even the costs of labour, *etc.*).

7.8 REACTION MASS EFFICIENCY

It is clear that each of the metrics described so far has its use in particular and possibly limited circumstances. For the more complex processes operated on the industrial scale (particularly, as we have seen, in the relatively more waste-producing fine chemical and pharmaceuticals sector), more sophisticated and more generally applicable methods are needed. Composite metrics such as reaction mass efficiency (RME) have been introduced by Curzons *et al.*[10] and developed by Andraos.[2]

Their value is that they bring together reaction efficiency (formation of co- and by-products, excess reactants) and process efficiency (reaction and processing auxiliaries). RME combines atom efficiency, yield and stoichiometry according to eqn (7.41)[2] and can be applied to sequential or convergent reaction series, including reactions leading to by-products and isomeric products:

$$\text{RME} = \varepsilon \times \text{AE} \times (1/\text{SF}) \times (\text{MRP}) \tag{7.41}$$

where

ε = yield of target product (as in eqn 7.2)

AE = atom efficiency
SF = a stoichiometric factor that takes account of excess reagents so that:

$$SF = 1 + \frac{\sum \text{excess masses of all reagents}}{\text{stoichiometric masses of all reagents}} \quad (7.42)$$

MRP = a material recovery parameter that takes account of the use of auxiliary materials. It is determined as follows:

$$MRP = 1/[1 + (\varepsilon)(AE)(c + s + w)/m_p(SF)] \quad (7.43)$$

where: m_p = mass of collected target product; c = mass of catalyst; s = mass of reaction solvent; w = mass of all other auxiliary materials used in work up and purification.

Andraos' metrics are now of demonstrable value on the large scale, having been used to analyse all published routes to the anti-viral, Tamiflu[2d] (Section 8.5).

7.9 OTHER METRICS

Andraos[2] has introduced a series of additional metrics (particularly useful in comparing complex multi-step syntheses); these are discussed below in Sections 7.9.1–7.9.4. In addition, process and product metrics designed to take a broader view of chemicals processing using exergetic analysis,[11] the principles of industrial ecology (Section 9.12)[12] or introducing other factors that are more techno-commercially (energy and water use) and environmentally relevant (toxics formation and CO_2 emissions) have been proposed by Shonnard, Sikdar and others (Section 7.9.5).[13,14] Interestingly, a recent application of a series of life-cycle impact assessment methods[15] to the industrial production of a pharmaceutical showed that 65–85% of impacts were caused by energy production and use, with process emissions (other than those resulting from energy generation) minor contributors to environmental impact.

7.9.1 μ_1

$$\mu_1 = \frac{\sum_j [MW(\text{intermediate product, j}) - MW(\text{target product})]}{(N + 1)} \quad (7.44)$$

where N is number of reaction stages.

μ_1 as determined by eqn (7.44) is a parameter that tracks the difference in molecular weight between any given synthesis intermediate and the target product over the course of synthesis. Good syntheses are characterised by large negative values since they use low molecular weight materials which are then used to produce intermediate products that in turn progressively increase in molecular weight with minimal overshoots.

7.9.2 f(sac)

$$\text{f(sac)} = 1 - \frac{\sum \text{MW(reagents in whole or part that end up in target product)}}{\sum \text{MW (all reagents)}} \tag{7.45}$$

$$= 1 - \text{AE(overall)} \frac{\sum \text{MW(reagents in whole or part that end up in target product)}}{\text{MW(target product)}} \tag{7.46}$$

f(sac) as determined by eqn (7.45) and eqn (7.46) traces the origin of each atom back to a corresponding starting material to provide a measure of reagents, protecting or chiral auxiliary groups whose atoms are not incorporated into the final target structure.

7.9.3 B/M

B/M is the number of target bonds made per reaction, the higher the better. M is the number of reaction steps, a step starting with one isolated intermediate and ending with the next isolated intermediate.

7.9.4 HI

HI is hypsicity[v] or oxidation level index. This highlights reactions in which the redox state of an atom is changed [*e.g.* the conversion of a >S group to >S(O) or >$S(O)_2$], as such reactions are usually of low atom utilisation. The preferred value of HI is zero.

All such considerations are clearly relevant to the design of synthetic strategies and are considered again in Chapter 9.

7.9.5 'Global' Metrics

In an ideal world, 'global' metrics (that take account of all relevant material and energy consumed and all significant environmental impact) would be developed that allow all-encompassing, universally accepted comparisons to be made between product or process options. Unfortunately, not only are there disagreements about which factors need to be included and how they should be weighted, but also what form the aggregation should take and how the information should be presented.

BASF has developed an Eco-Efficiency Analysis that uses life-cycle methods with economic data to compare different approaches to the delivery of a defined customer benefit.[13] Figure 7.3 shows an environmental fingerprint for five

[v] An isohypsic process is a synthetic scheme that involves no redox steps. When a reaction step uses a chemical oxidant or reductant, it is inherently atom inefficient, leading to the related idea of 'redox economy'. For a recent review, see ref. 16.

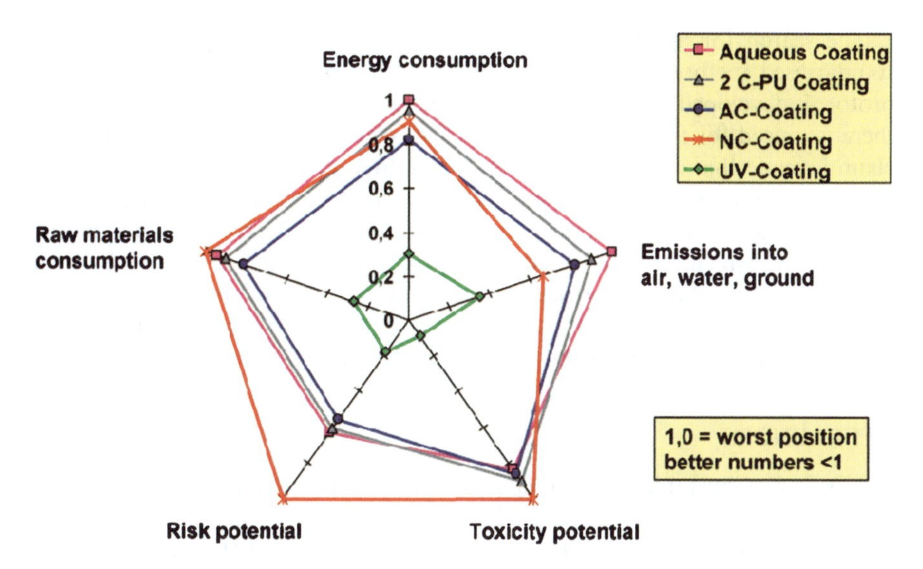

Figure 7.3 Environmental fingerprint for five processes for curing a coating. Reprinted with permission from ref. 13. Copyright 2003 American Chemical Society.

alternative processes for curing a coating in the form of a 'spider' diagram in which energy consumption, emissions, toxicity potential, risk potential and raw materials consumption (each of which, itself, represents an aggregate) are presented on five axes having been normalised and weighted to provide numerical values between 0 (most favourable) and 1 (least favourable). It is clear that one curing process is by far superior to the other four.

Sikdar[14] has generalised the notion of aggregation of the different metrics obtained from a multifactorial comparison of the relative sustainability of process options of the sort evaluated in the BASF study. The advantage presented by Sikdar's method is that the values of a metric for two options being compared are 'ratio-ed' to give a dimensionless number. These ratios can be weighted and a geometric mean of these ratios obtained (eqn 7.47):

$$D = \left(\prod_{i}^{n} [c_i/(y_i/x_i)] \right)^{1/n} \tag{7.47}$$

where x_i and y_i are values of n different metrics (m_1 to m_n) that describe the behaviour of a system in two different states, S_1 and S_2. The difference between the states, D, is the geometric mean of the ratio of the values of these metrics, each weighted by a factor, c_i, an estimate (inevitably subjective but hopefully informed) of relative importance.

Sikdar's analysis, applied to Shonnard's data,[13] reproduces the conclusion from the fingerprint method. He also applied the method to the comparison of three processes for making dichlorine and two for making 1,6-diaminohexane.

The search for additional metrics continues, particularly for some agreed standardised methodology for assessing reaction efficiency along the lines of protocols developed for the assembling of life-cycle inventories. While this is so, there is a need for full and proper disclosure of underpinning data whenever a claim is made that a reaction, process of product is 'environmentally friendly'. The editors of journals in which such claims are made have a duty to require that any suggested environmental or process efficiency benefit arising from a new or modified transformation is backed by supporting experimental data. As such information is made more generally and more widely available, then independent comparisons such as those made by Andraos will become possible—a necessary pre-requisite for the development of cleaner chemicals processing.

REFERENCES

1. D. J. C. Constable, A. D. Curzons and V. L. Cunningham, *Green Chem.*, 2002, **4**, 521.
2. (a) J. Andraos, *Org. Process. Res. Dev.*, 2005, **9**, 149; (b) J. Andraos, *Org. Process. Res. Dev.*, 2005, **9**, 404; (c) J. Andraos, *Org. Process Res. Dev.*, 2006, **10**, 212; (d) J. Andraos, *Org. Process Res. Dev.*, 2009, **13**, 161.
3. (a) M. E. Brown, K. J. Buchanan and A. Goosen, *J. Chem. Educ.*, 1985, **62**, 575; (b) R. B. Snadden, *J. Chem. Educ.*, 1985, **62**, 653.
4. B. M. Trost, *Science*, 1991, **254**, 1471.
5. M. G. Constantino, V. L. Júnior and G. V. José da Silva, *Molecules*, 2002, **7**, 456.
6. (a) F. Sandelin, P. Oinas, T. Salmi, J. Paloniemi and H. Haario, *Ind. Eng. Chem. Res.*, 2006, **45**, 986; (b) J. M. Campos-Martin, G. Blanco-Brieva and J. L. G. Fierro, *Angew. Chem., Int. Ed.*, 2006, **45**, 6962.
7. W. H. Glaze, *Environ. Sci. Technol.*, 2000, **34**, 449A.
8. M. Eissen, R. Mazur, H.-G. Quebbemann and K.-H. Pennemann, *Helv. Chim. Acta*, 2004, **87**, 524.
9. A. Steinbach and R. Winkenbach, *Chem. Eng.*, 2000, **107**(4), 94.
10. (a) A. D. Curzons, D. J. C. Constable, D. N. Mortimer and V. L. Cunningham, *Green Chem.*, 2001, **3**, 1; (b) D. J. C. Constable, A. D. Curzons, L. M. Freitas dos Santos, G. R. Geen, R. E. Hannah, J. D. Hayler, J. Kitteringham, M. A. McGuire, J. E. Richardson, P. Smith, R. L. Webb and M. Yu, *Green Chem.*, 2001, **3**, 7.
11. H.-S. Yi, J. L. Hau, N. U. Ukidwe and B. R. Bakshi, *Environ. Prog.*, 2004, **23**, 302.
12. J. Dewulf and H. Van Langenhove, *Resour., Conserv. Recycl.*, 2005, **43**, 419.
13. D. R. Shonnard, A. Kicherer and P. Saling, *Environ. Sci. Technol.*, 2003, **37**, 5340.
14. S. K. Sikdar, *Clean Technol. Environ. Policy*, 2009, **11**, 157.
15. G. Wernet, S. Conradt, H. P. Isenring, C. Jiménez-González and K. Hungerbühler, *Int. J. Life Cycle Assess.*, 2010, **15**, 294.
16. N. Z. Burns, P. S. Baran and R. W. Hoffmann, *Angew. Chem. Int. Ed.*, 2009, **48**, 2854.

BIBLIOGRAPHY[vi]

A. Lapkin and D. J. C. Constable, *Green Chemistry Metrics: Measuring and Monitoring Sustainable Processes*, Wiley, Chichester, 2008.
A. E. Marteel, J. A. Davies, W. W. Olson and M. A. Abraham, *Annu. Rev. Environ. Resour.*, 2003, **28**, 401.
T. M. Parris and R. W. Kates, *Annu. Rev. Environ. Resour.*, 2003, **28**, 559.
J. Dewulf and H. Van Langenhove, *Resour., Conserv. Recycl.*, 2005, **43**, 419.
S. K. Sikdar, *AIChE J.*, 2003, **49**, 1928.
M. Eissen and J. O. Metzger, *Chem.–Eur. J.*, 2002, **8**, 3580.
F. G. Calvo-Flores, *Chem. Sus. Chem.*, 2009, **2**, 905.

WEBLIOGRAPHY

1. www.epa.gov/greenchemistry/pubs/docs/award_entries_and_recipients2007.pdf
2. www.headwaters.com/elements/pdfs/InvestmentProfile.pdf

All the web pages listed in this Webliography were accessed in May 2010.

[vi] These texts supplement those already cited as references.

CHAPTER 8

Chemistry: Necessary but not Sufficient

'We live by the grace of invention.'

Norbert Wiener, 1953

So far, we have learnt something of the chemistry of the environment, the development (and limitations) of new concepts to bring into sharper focus the efficiency of chemical reactions, and the development and use of new reaction metrics (and their rootedness in the laws of thermodynamics). Chapter 9 examines the critical role of the reactor and the process in developing better chemical technology and surveys the use of novel stimuli and contacting methods to effect more selective and less energy-consuming transformations. Chapter 11 examines the investigation and the development of processes using renewable feedstocks, the use of which is thought to be more sustainable.

Before we move on, we will look at the richness of chemistry and its relevance. This chapter first takes a brief snapshot[i] of some areas of chemical enquiry chosen specifically because they do not have any explicit sustainability motivation. I include these in the belief that potential solutions to key technical, scientific or transdisciplinary problems may, depending on their scope or nature, arise (or be sparked) from quite unrelated work. It thus pays to take the widest view of the subject. The selection illustrates the discipline's vibrancy, breadth and relevance. Examination of the chemical literature including leading scientific journals such as *Nature* and *Science* shows that this continues to be so. Indeed a new journal, *Nature Chemistry* (Web 1), began publication recently

[i]The bare essentials of each topic (particularly the relevance to the purpose of this book) will be introduced, referencing one or more recent papers. Obtaining further information on each topic will require an examination of these original papers and of important earlier publications cited by them, relevant later papers that cite them or of additional material obtained from wider searches of the literature. Appendix 1 introduces some of ways in which this might be done.

Chemistry for Sustainable Technologies: A Foundation
By Neil Winterton

Published by the Royal Society of Chemistry, www.rsc.org

with a number of leading chemists provide their own perspectives of the many ways the subject will advance.[1]

Secondly, the chapter highlights a small selection of the many areas not yet covered in which the fundamental insights and practical methodology of chemistry assist and enable other scientific and technical disciplines to make contributions towards more sustainable technologies. These include:

- engineering—in the development of processes employing novel reaction media
- device development—novel materials for solar cells
- biotechnology—in developing organisms capable of efficiently breaking down lignocelluloses
- geochemistry—in determining the capacity of geological strata to retain captured carbon dioxide.

It does not include, for instance, important and exciting areas such as the chemistry of materials, molecular machines or nanoparticles, or systems chemistry and chemistry associated with genomics. However, by keeping an eye on the scientific press it is not difficult to find examples of current developments in these and other fields.

This book is concerned with efforts (placed in a broad technical and societal perspective) to minimise the environmental impact of chemical technology and to find ways to use or transform materials, increasingly derived from renewable sources, more efficiently. In setting out the way more sustainable technologies are being developed, I seek to illustrate the ways in which chemistry and chemical knowledge contribute to this process. Much of this particular work (even that which is exploratory) has had a technological and industrial focus. Some of it aligns itself with a green chemistry motivation (of which more in Section 8.10); some of it had no such professed motivation. All of it is relevant.

8.1 PREBIOTIC CHEMISTRY (ORGANIC, INORGANIC AND PHYSICAL)

It is appropriate to begin at the beginning. The classic experiments undertaken by Miller and Urey[2] in the 1950s were devoted to the most profound of questions: 'how did life-forms arise?' Miller and Urey showed that amino acids, the building blocks of proteins, could be formed from simple compounds such as methane, ammonia, water and hydrogen (though, one has to stress, these may not necessarily all have been present in the early atmosphere). Orgel,[3] building on the work of Oro (who showed that adenine, one of the four nucleobases, could be obtained from refluxing aqueous ammonium cyanide) has also suggested, counter-intuitively, that the formation of a proposed precursor, a tetramer of HCN, and its subsequent reaction to give adenine (Scheme 8.1), would proceed more favourably in cold conditions as a

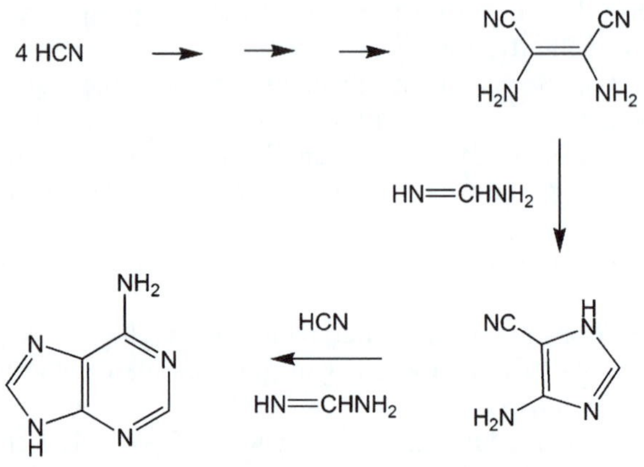

4 HCN → → → NC, CN, H_2N, NH_2

$HN{=}CHNH_2$

NH_2 N N N N H

HCN

$HN{=}CHNH_2$

NC, H N, N, H_2N

Scheme 8.1 Formation of adenine, one of the four nucleobases, from aqueous ammonium cyanide.

consequence of the concentration that can occur in solutions from which water is crystallising (so-called 'eutectic concentration').

These studies lead on to many additional questions: how do we define life?[ii] What stages can be identified that link an abiotic ('without life') planet to one in which forms that we might recognise as living exist? When and where did this happen? What were conditions like then? What was the temperature, the pH, was it oxidising or reducing, what were the key chemical components?

While it is widely believed that small molecules were formed in the 'primordial soup', not all agree, the argument centring on what chemical driving force would be needed to produce early genetic material. Would the initial spark be enough? It is possible to envisage that the initially formed small molecules subsequently aggregated and polymerised, with some having the capacity to self-organise and self-replicate. At what point these precursors produced primitive or prototype cells which we would recognise as the basis of all living systems is difficult to say. Indeed, how would all the complex sets of chemical processes (not all of which are thermodynamically favoured), necessary for sustained replication (far from chemical equilibrium) to evolve, be coupled together? Cells replicate through processes involving proteins (chains of amino acids) and nucleic acids (chains of monomeric nucleotides).[iii] Did these arise over the same period or did one arise before the other? If so, which

[ii] Leslie Orgel, a pioneer in prebiotic chemistry (see Bibliography), coined the term 'specified complexity' to define (in the context of Darwinian evolution) what distinguishes living things from non-living things. However, this term has been appropriated by those who believe in (the non-scientific) idea of 'intelligent design'. Bear this in mind if you come across it.

[iii] The formation of polypeptides in cells takes place in the ribosome, a complex of RNA and protein. The contributions of Ramakrishnan, Steitz and Yonath to the understanding of the structure and function of ribosomes led to their receipt of the 2009 Nobel Prize for Chemistry.

one? Why do natural proteins consist of only L-amino acids, not their D form? Has this arisen from an extraterrestrial cause or from a random process associated with so-called 'spontaneous resolution' of enantiomers from a homochiral environment.[4]

While this book was in preparation, a paper[5] appeared describing chemical studies in which a ribonucleotide, a monomeric component of RNA (ribonucleic acid, central to protein synthesis), was synthesised from 'plausible prebiotic feedstock molecules'[iv] including glycolaldehyde, glyceraldehyde, cyanamide, cyanoacetylene and inorganic phosphate. The key point of the paper was to suggest that it may not have been necessary to pre-form the sugar (ribose) or the nucleobase (cytosine) separately before the nucleotide could be made. Interestingly, phosphate plays a key role in the proposed mechanism (Scheme 8.2) acting as both catalyst and buffer. Just as experiment can aid our thinking about what might have happened prebiotically, our understanding of chemical kinetics can also point to proposals or ideas that may be more or less chemically plausible. Blackmond[6] has recently analysed chemical processes that are distinguished by being either autoinductive (in which a reaction product accelerates a reaction step though without producing more of itself) and autocatalytic (in which a reaction product acts as the catalyst for its own formation). In various simulations, she concludes that autoinductive processes are unlikely to have been important in abiogenesis.

An alternative view[7a] addresses explicitly the question of the continuous energy source needed to drive prebiotic chemistry, *i.e.* to ask what conditions on the early Earth might permit both the formation of compounds with high energy bonds (necessary as energy carriers) and their primitive organisation to take advantage of a replenishing source of energy. Martin and colleagues[7a] proposed, in an essay that builds on arguments made by others, that certain alkaline (non-volcanic) hydrothermal vents provide a credible and continuous source of both chemicals and energy. The key process is that of 'serpentinisation' (eqn 8.1). This involves the highly exothermic reaction of seawater with the mineral, olivine [a magnesium iron silicate, $(Mg,Fe)_2SiO_4$], to produce an alkaline (pH 9–11) vent[v] rich in dihydrogen (formally represented by eqn 8.2) and capable of forming [along with serpentine, $(Mg,Fe)_3Si_2O_5(OH)_4$] methane from carbon dioxide from sea water (eqn 8.3). The vents are also associated with porous superstructures in which higher hydrocarbons and oxygenated compounds such as acetate are known to form abiotically. The continuous source of energy necessary to allow biosynthetic chemistry to develop comes from primordial 'chemiosmosis'[vi] (this results from a proton

[iv] We cannot of course know what the compositions of the atmosphere or the oceans were at this time (Web 2).

[v] On pages 204 and 414 we consider the possibility that the generation of similar alkaline geological environments could be exploited for the long-term sequestration of carbon dioxide, identifying geological strata that would render such confinement essentially permanent.

[vi] The chemiosmotic mechanism of biological energy transfer was proposed first by P. Mitchell in 1951.[7b] He was awarded the Nobel Prize for Chemistry in 1978.

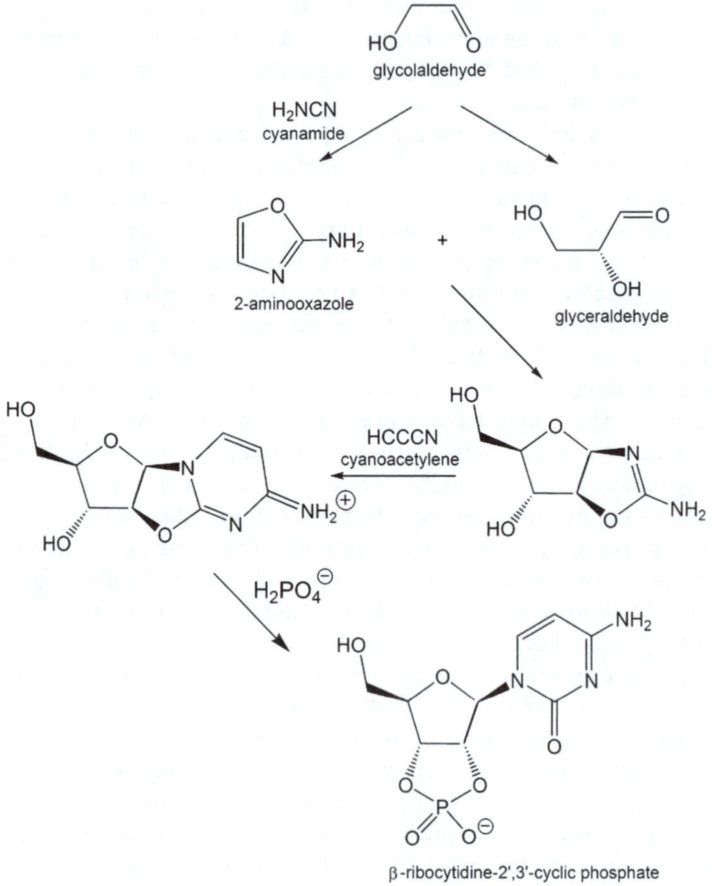

Scheme 8.2 Proposed route to the formation of the ribonucleotide, β-ribocytidine-2′,3′-cyclic phosphate, from plausible prebiotic precursors. Based on ref. 5a.

concentration gradient across a membrane which can be seen today in all living systems), resulting from the pH difference between the vent and the (then) acidic oceans.

$$3\,Mg_2SiO_4 + SiO_2 + 4\,H_2O \rightarrow 2\,Mg_3Si_2O_5(OH)_4 \qquad (8.1)$$

$$3\,Fe_2SiO_4 + 2\,H_2O \rightarrow 2\,Fe_3O_4 + 3\,SiO_2 + 2\,H_2 \qquad (8.2)$$

$$(Mg, Fe)_2SiO_4 + n\,H_2O + CO_2 \rightarrow (Mg, Fe)_3Si_2O_5(OH)_4 + Fe_3O_4 + CH_4 \qquad (8.3)$$

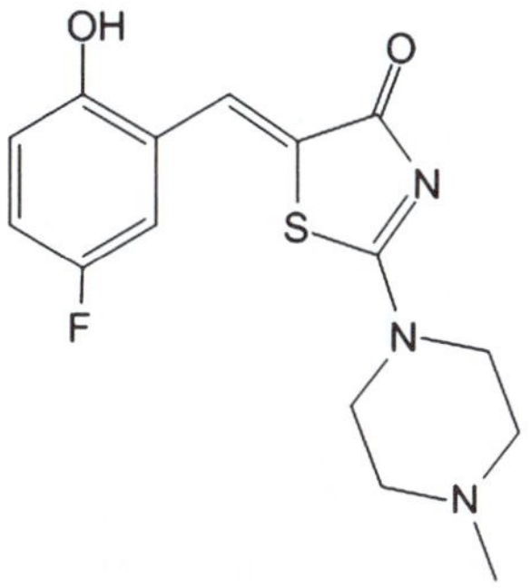

Figure 8.1 (5^{Z})-5-[(5-fluoro-2-hydroxyphenyl)methylene]-2-(4-methyl-1-piperazinyl)-4(5^{H})-thiazolone, the 50 millionth chemical substance registered by the Chemical Abstracts Service.[8]

8.2 THE 50 MILLIONTH CHEMICAL SUBSTANCE

A recent editorial in *Chemical and Engineering News*[8] reported that (5^{Z})-5-[(5-fluoro-2-hydroxyphenyl)methylene]-2-(4-methyl-1-piperazinyl)-4(5^{H})-thiazolone (Figure 8.1) was registered by the Chemical Abstracts Service (CAS) on 7 September 2009 as the 50 millionth chemical substance. The CAS registry is a vital source of information for chemists and others, providing unique numbers (Registry Numbers) for new compounds (in this case 1181081-51-5) that enable critical data to be found easily in the Chemical Abstracts database. The fact that CAS registers about 25 unique new substances per minute suggests that the productivity and ingenuity of chemists continues to be prodigious and the rich variety of chemical compounds being discovered never-ending.

8.3 CAS REGISTRY NUMBER 173075-49-5

Evolution of living things has followed many paths including that leading to the bryozoans, a phylum[vii] of sea animals. These and similar marine organisms are the source of an immense range of natural products. Bryostatins, isolated from the bryozoan, *Bugula neritina*, are a class of such materials. They have biological activity that may lead, ultimately, to the identification of compounds for clinical use.

Unfortunately, bryostatins are not very abundant. Only 18 g of the most abundant, bryostatin 1, could be isolated from 14 t of the sea animal. Pharmacological and clinical investigation and development cannot proceed with such small quantities of material. Bryostatin 16 (Figure 8.2, CAS RN = 173075-49-5) is one of this series and, because it has a structure from which (in principle, at least) all other bryostatins (and analogues) can be derived, it has become a target for total synthesis. Various strategies may be

[vii] Phylum is one of the eight major taxonomic groups of living things, below 'kingdom' and above 'class'.

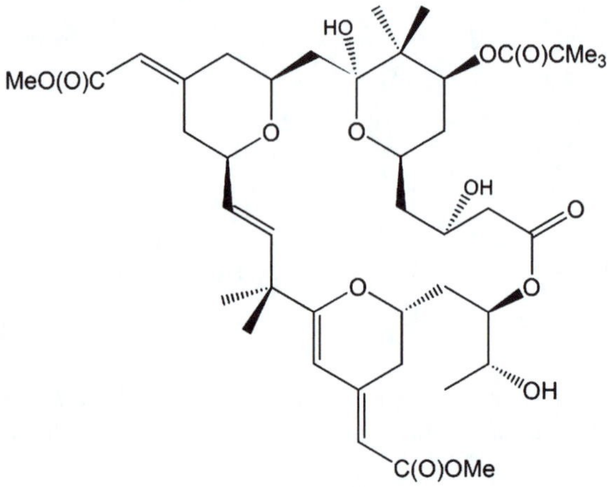

Figure 8.2 Bryostatin 16.

adopted that all usually begin with some form of retrosynthetic analysis (Bibliography). Trost and Dong[9a] have recently made a substantial advance by shortening, from >40 steps to 26, the longest linear sequence in the synthesis from the readily available precursor, 2,2-dimethylpropan-1,3-diol. The synthetic efficiency is associated (Scheme 8.3) with three catalytic steps:

- a ruthenium-catalysed alkene/alkyne coupling of **3** and **4** to give **5** (in tandem with a Michael addition)
- a palladium-catalysed intramolecular alkyne–alkyne coupling in **8** leading to **9** with a 22 atom ring; and, finally
- a gold-catalysed cyclisation of **9** to give bryostatin 16.

These particular steps are highly chemoselective and proceed with 100% atom efficiency (AE). Attention is now being devoted to other, less atom efficient, steps.

8.4 THE SIGNIFICANCE OF SMALL THINGS

It is worth contemplating for a moment the world before the first bryostatin was isolated (1971) and characterised (1982).[10] How did discovery of the bryostatins first come about, particularly when they could only be isolated from a highly complex mixture in $<10^{-4}$ % yield? Someone had been curious, persistent[10b] and observant. A different type of curiosity arises when someone is sceptical, as mentioned in the footnote x on p. 59. Such scepticism may also focus on an apparently unimportant, but ultimately crucial, detail. For example, a concern over the precise nature of a catalyst[11] used in iron-catalysed coupling reactions (Scheme 8.4, Table 8.1) revealed that the catalytic effect arose from the copper impurity in the technical grade ferric chloride initially

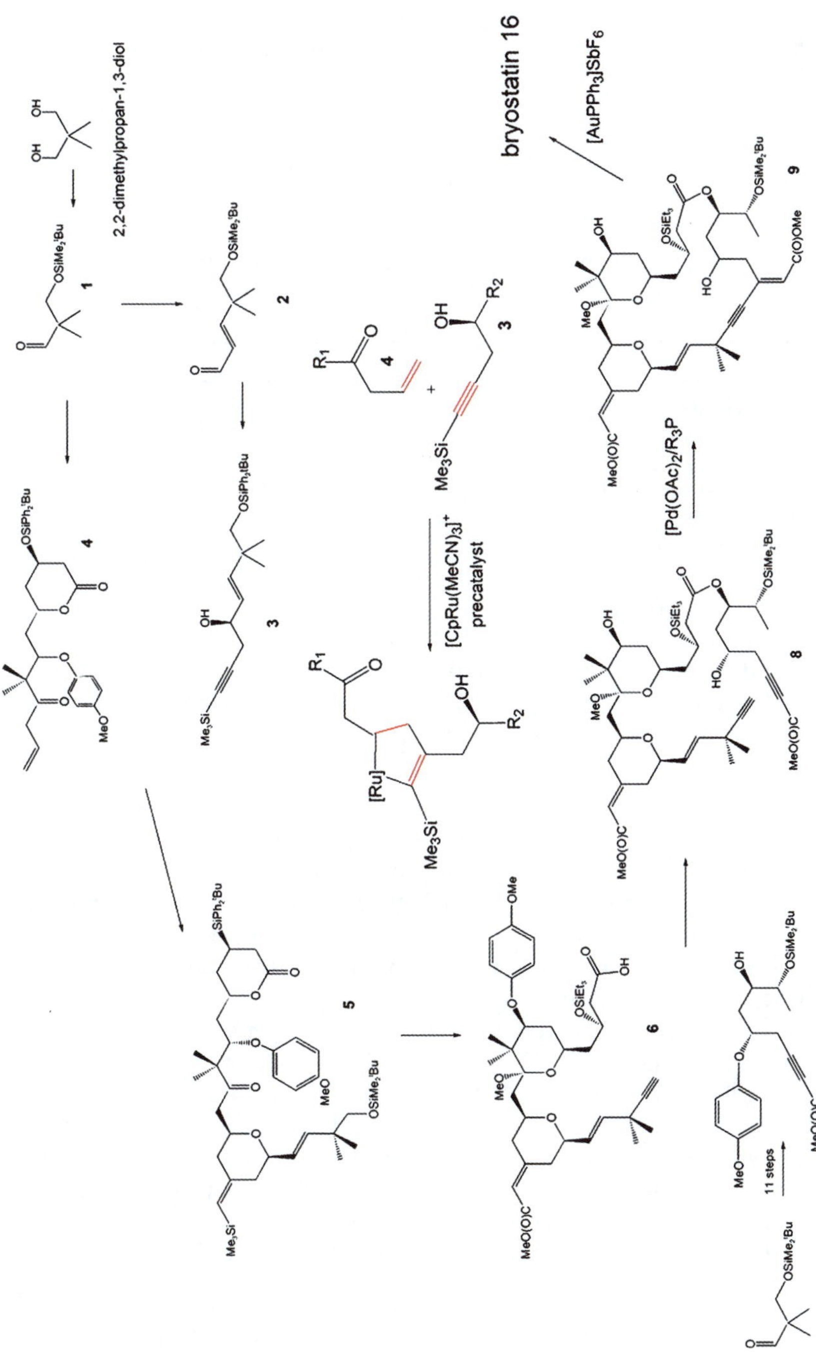

Scheme 8.3 Three key catalysed steps, 3 + 4 → 5, 8 → 9 and 9 → final product, in the total synthesis of bryostatin 16.[9]

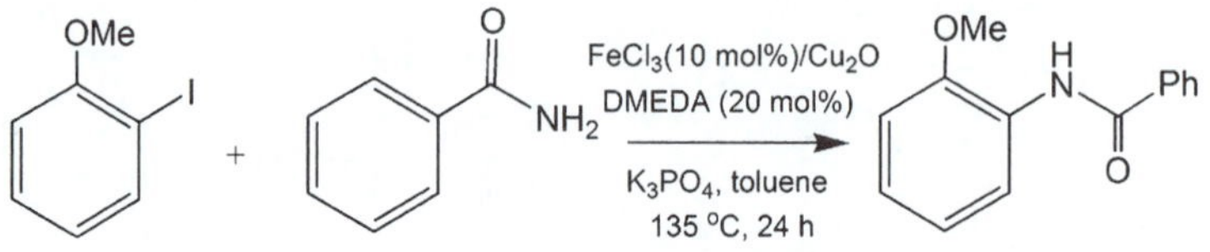

Scheme 8.4 Catalysed *N*-arylation of benzamide in the presence of ferric chloride.[11] DMEDA = *N*,*N*′-dimethylethylenediamine.

Table 8.1 The effect of changes in catalyst purity and the purposeful addition of copper(I) oxide on the yield of the *N*-arylation of benzamide (Scheme 8.4).[11]

Catalyst	*Yield (%)*
$FeCl_3$ (>98%) (Merck)	79
$FeCl_3$ (>98%) (Aldrich)	16
$FeCl_3$ (>99.99%) (Aldrich)	trace
$FeCl_3$ (>99.99%) + 5 ppm Cu_2O	98
$FeCl_3$ (>99.99%) + 10 ppm Cu_2O	99
DMEDA[a] + 5 ppm Cu_2O	97
5 ppm Cu_2O	34

[a] *N*,*N*′-dimethylethylenediamine

used. Using increasingly pure catalyst reduced the yield; the yield increased when copper was added to the pure ferric chloride. The effect was confirmed when the reaction proceeded in the complete absence of iron.

8.5 TAMIFLU

The UK and other governments have stockpiled and made available millions of doses of oseltamivir phosphate ('Tamiflu') as the main treatment and prophylactic for swine flu. Because of the threat of a pandemic that might arise from a human variant of avian flu, improved routes to Tamiflu have been investigated. Andraos,[12] using his reaction mass efficiency approach, has analysed 15 synthesis plans (six from industry and nine from academia) that use a range of starting materials (Scheme 8.5) with the mass of waste (per kg of product) varying between 3.1 and 150.3. The least waste-producing process,[viii] using shikimic acid (Scheme 8.5) as the starting material, is that used in commercial production (a good example of the alignment of economic or commercial objectives with the objectives of developing cleaner chemical processes). Shikimic acid is another natural product, though in this instance, is readily available from the Chinese star anise (a shrub). It may also be obtained *via* a

[viii] At the end of this notable paper, Andraos makes a series of statements about the importance of complete and accurate experimental data in producing reliable estimates of metrics. Such metrics are only of credible value if they permit comparisons to be made and if they both identify particular aspects of the syntheses that are points of weakness and provide a quantitative basis for defining a target for measuring any improvement.

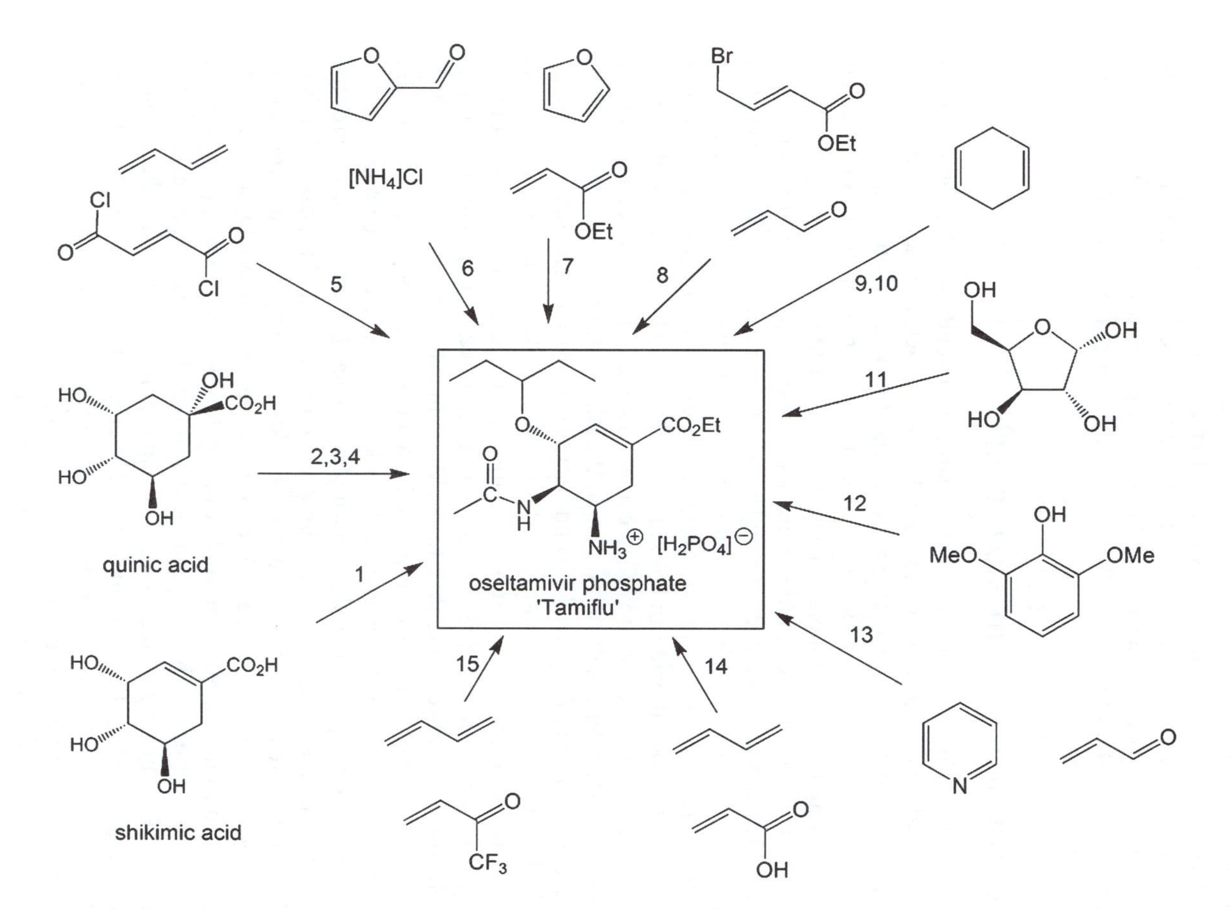

Scheme 8.5 Starting points for 15 synthesis plans to oseltamivir phosphate ('Tamiflu') analysed by Andraos.[12]

biotechnological route[13] from glucose using a genetically engineered strain of *Escherichia coli* (discussed further in Scheme 11.14). Such developments rely on fundamental investigations of the biosynthesis of shikimic acid from glucose that go back to the 1950s.[14]

An eight-step patented synthesis of oseltamivir phosphate has recently been reported by Hoffmann-La-Roche, its manufacturers,[15] that converts shikimic acid to the tris(*O*-methanesulfonate) derivative of its ethyl ester, followed by a series of steps shown in Scheme 8.6, requiring fewer purifications and no group protection steps.

8.6 CHEMISTRY IN THE REAL WORLD

As we saw in Chapter 5 and will see in Chapter 9, an understanding of what goes on in a laboratory flask is a necessary but not sufficient condition for understanding what goes on in a large-scale chemical process, whether this occurs in the environment or in an industrial reactor. The conditions (temperature and pressure) and scale (temporal and spatial) of natural processes give rise to unexpected circumstances and events that appear, at first sight, to be unusual but which can be explained and understood (and possibly ameliorated) by a consideration of basic physical and chemical principles. Such a circumstance exists at Lake Kivu in the Democratic Republic of Congo which contains 300 km^3 of carbon dioxide and 60 km^3 of methane.[16] The movement of tectonic plates along the Great Rift Valley exposes molten rock that releases carbon dioxide into the dense salt-rich waters at the lake bottom. At its greatest depth, the lake is anoxic (dioxygen-free) and bacteria slowly convert some of the carbon dioxide to methane. Fresh water at the surface is of lower density and this difference prevents vertical mixing, trapping the gases at the lake bottom. Currently, the waters of Lake Kivu are not saturated and are unlikely to emit their gases spontaneously (unlike Lakes Monoun and Nyos in Cameroon which erupted in 1984 and 1986, respectively, killing many people). At least an awareness of the threat may enable the risk to human lives to be minimised.

The very same physical and chemical phenomena that are responsible for the condition of these lakes also lie behind some ideas for the disposal of carbon dioxide, captured from the combustion of fossil fuels in energy generation, in the deep ocean, aquifers or geological formations (which might also include depleted oil and gas wells or coal seams). However, here the challenge is to maximise the rate at which carbon dioxide is dissolved and to prevent a future reverse of the process, doing so without using so much energy that the net effect is to emit more CO_2 than is sequestered. The chemical problem is essentially one of mass transfer (gas to liquid or gas to solid) and reaction kinetics, albeit under somewhat unusual circumstances. One paper (of very many)[17] highlights some of the chemical and associated questions relating to the storage of carbon dioxide in deep sea basalt formations. These appear to have the capacity for large-scale storage, though the CO_2 must be transported to the sea bed and dissolve into seawater, following which it must migrate into the pores of the basalt (an igneous mineral comprising mainly the silicates of

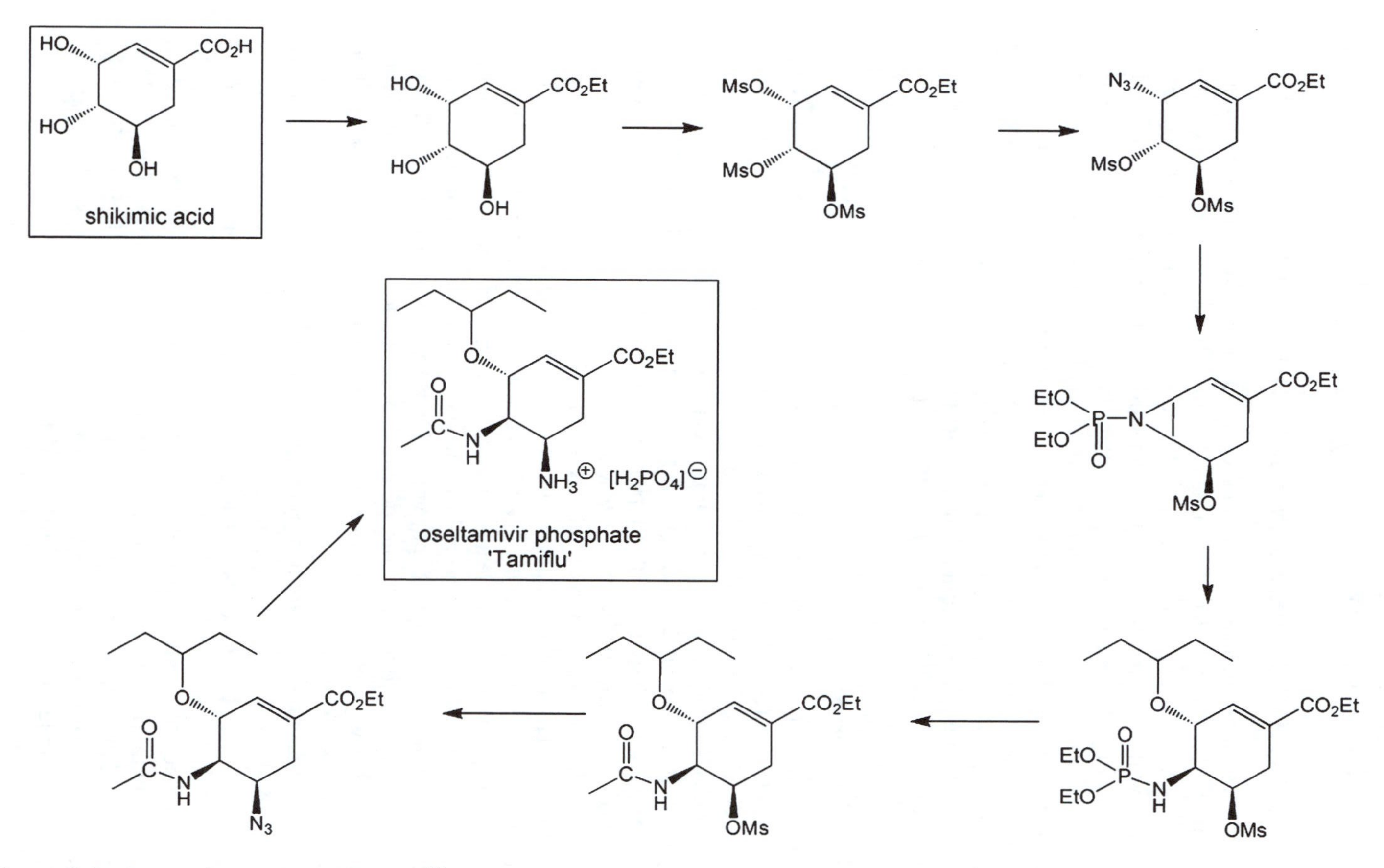

Scheme 8.6 Patented manufacturing route[15] to oseltamivir phosphate ('Tamiflu') from shikimic acid.

magnesium, calcium and iron) and there react to give essentially inert carbonate-containing minerals.

8.7 METATHESIS, FULLERENES AND THE NOBEL PRIZE

The successes of chemists are recognised by awards, the most prestigious of which are the Nobel Prizes awarded by the Royal Swedish Academy of Sciences (Web 3). The history of the main developments in chemistry over the last 100 years is encompassed within the work of its recipients from Jacobus van't Hoff in 1901 ('*in recognition of the extraordinary services he has rendered by the discovery of the laws of chemical dynamics and osmotic pressure in solutions*') to Venkatraman Ramakrishnan, Thomas A. Steitz and Ada E. Yonath in 2009 '*for studies of the structure and function of the ribosome*'. Many of the developments recognised are relevant to the contribution chemistry makes to the quality of human life and to minimising our environmental impact referred to in this book; the prizes for 2007, 2005, 2001 and 1995 most directly so.

8.7.1 Metathesis

The 2005 prize (Web 4) was awarded to Yves Chauvin, Robert H. Grubbs and Richard R. Schrock '*for the development of the metathesis method in organic synthesis*'. While olefin metathesis had long been practised industrially, the catalytic mechanism of the process was not understood. The discoveries from industry-relevant fundamental research made by the recipients over the last 30–40 years have enabled catalysts to be developed on which improved (that is less waste-producing) industrial processes are now based. The 2005 Nobel Prize is seen[18] as an award for green chemistry (and indeed, the Nobel Committee's press release (Web 4) (though not their citation) suggests that this is so). However, it is more true to say, bearing in mind the period over which the work for which the prize was awarded was done, that cleaner chemical processing is a likely outcome rather than green chemistry having been the prime motivation. None of the recipients, in their Nobel lectures,[19] suggest that green chemistry was a particular motivation, though Grubbs[19c] does note the potential of metathesis in providing greater process efficiency in the transformation of renewable raw materials such as vegetable oils.

Metathesis can be exemplified by the cross-metathesis (CM) reaction shown in eqn (8.4).

$$2\,CH_3CH{=}CH_2 \rightarrow CH_2{=}CH_2 + CH_3CH{=}CHCH_3 \tag{8.4}$$

Two families of catalysts were discovered by Grubbs and Schrock based on molybdenum and ruthenium, both shown to involve the formation of alkylidene–metal intermediates (Figure 8.3) from the precatalysts. The identity of the likely nature of the key intermediate arose from mechanistic studies, particularly those undertaken by Grubbs[19d] and Katz.[19e]

The two most likely mechanisms (Scheme 8.7) for the metathesis reaction involve either (a) the initially favoured concerted interchange, in which both

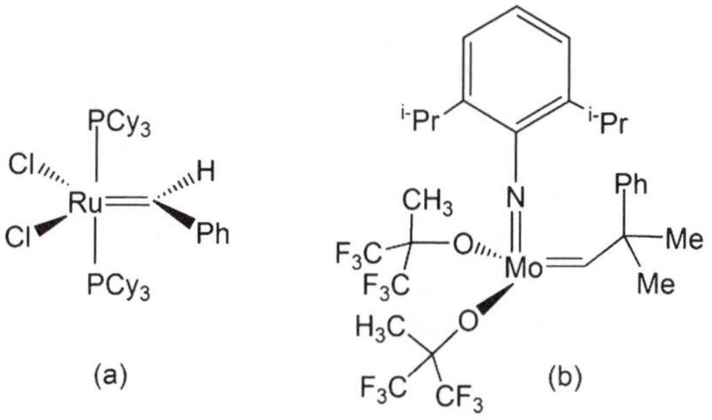

Figure 8.3 Examples of ruthenium and molybdenum alkylidene complexes, key intermediates in metathesis catalysis.

alkene entities were bound simultaneously to the same metal centre or (b) first proposed by Chauvin,[19f] a chain process involving a sequence of intermediates in which cleavage of the carbon–carbon double bond had occurred. These two possibilities were elegantly tested by the use of isotopically labelled alkenes. The results from Katz[19e] on a ring-closing metathesis (RCM) (shown in Scheme 8.8) illustrate the argument: if the reaction proceeded *via* mechanism (a), then the bonding of both ends of the diene to a single metal centre at the same time would require any interchange of the alkylidene units to be concerted, and $CH_2{=}CH_2$ and $CD_2{=}CD_2$ would be formed in a 1 : 1 ratio. None of the third isotopomer, $CD_2{=}CH_2$, would be expected. (Importantly, Katz was also able to establish that none of the d_2-divinylbiphenyl, from which $CD_2{=}CH_2$ could be formed, was produced during the course of the reaction.) On the other hand, if the process occurred sequentially *via* mechanism (b), then a statistical distribution of the d_4-, d_2- and d_0- isotopomers of 1 : 2 : 1 would be predicted. Katz showed that the average percentage of d_4-, d_2- and d_0-ethene formed was 24.3 (SD 1.5), 48.7 (2.3) and 25.9 (1.8), respectively. Related studies by Grubbs[19d] using a mixture of 1,1,8,8-d_4-octa-1,7-diene, $CD_2{=}CHCH_2CH_2CH_2CH_2CH{=}CD_2$, and its fully protiated isotopomer, $CH_2{=}CHCH_2CH_2CH_2CH_2CH{=}CH_2$, also support the carbene mechanism. This understanding then permitted a further evolution of catalyst design and the extension of the chemistry to other olefinic substrates (and also to alkane substrates[19h]).

Metathesis catalysis has now been extended to include polymerisations, ROMP (ring-opening metathesis polymerisation) (Scheme 8.9) and ADMET (acyclic diene metathesis), with the former the basis of a range of industrial products. Both ruthenium[19g] and molybdenum[19b] metathesis catalysis has been used in pharmaceuticals production. Schrock, for instance, describes the shortening of the synthetic route to tipranavir, an HIV protease inhibitor, using asymmetric metathesis (Scheme 8.10) to synthesise an enantiomerically pure tertiary ether intermediate.[19b] The intermediate produced in the ring-closing metathesis shown in Scheme 8.11 has been produced on the 400 kg scale.[19g]

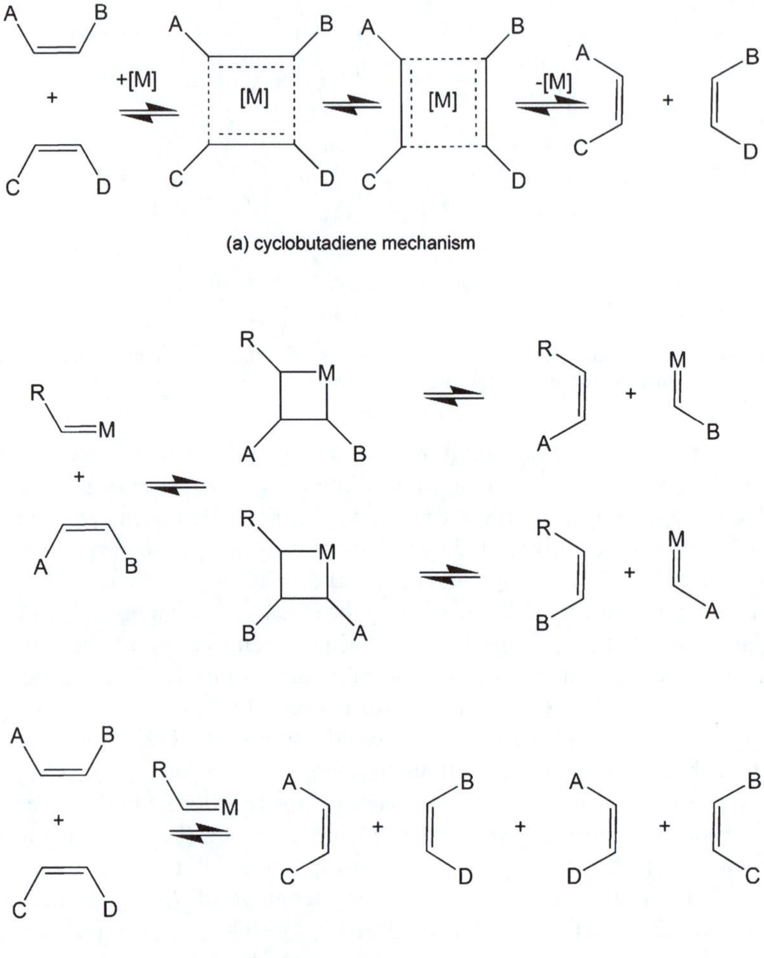

(a) cyclobutadiene mechanism

(b) carbene mechanism

Scheme 8.7 Two possible mechanisms for metathesis catalysis: (a) the interchange or cyclobutadiene mechanism; or (b) the carbene mechanism.

8.7.2 C_{60} or [60]fullerene

The work that led to the award of the 1996 Nobel Prize for Chemistry (Web 4) to R. F. Curl, H. W. Kroto and R. E. Smalley '*for their discovery of fullerenes*',[ix] exemplifies why both science and the study of scientists doing science are fascinating in themselves. The fullerene story also illustrates the fortuitous connections, intuitive leaps, frustrations, personal conflicts, triumphs and

[ix] In 2001 to mark the 100th anniversary of the Nobel Prize, the UK Post Office issued a series of commemorative stamps, one for each prize. The prize for chemistry was represented by an image of C_{60}.

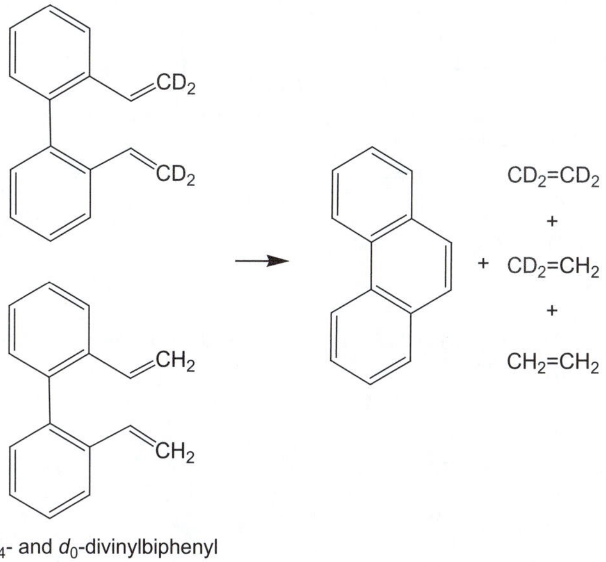

CD_2

CD_2

$CD_2=CD_2$

+

+ $CD_2=CH_2$

+

$CH_2=CH_2$

CH_2

CH_2

d_4- and d_0-divinylbiphenyl

Scheme 8.8 Mechanistic test[19e] to distinguish between two mechanisms of the ring-closing metathesis reaction in which the isotope distribution in the product, ethene, from isotopically enriched reactants was predicted to be different.

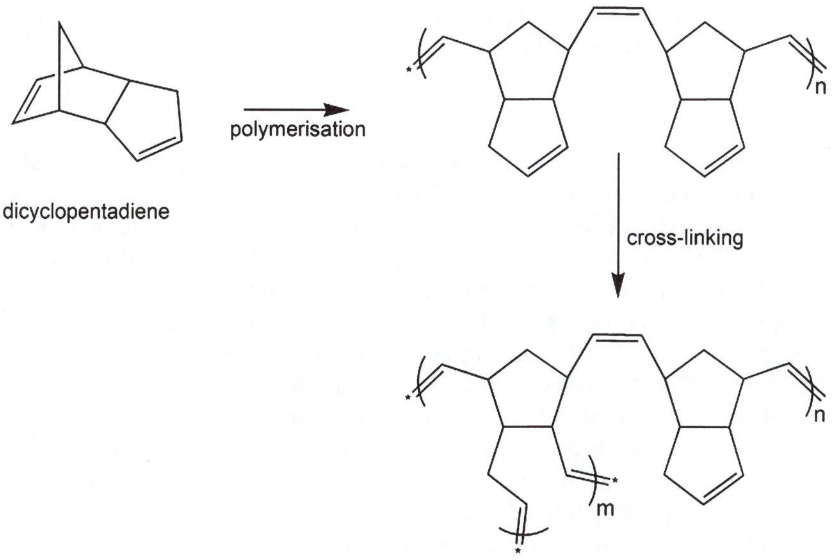

polymerisation

n

dicyclopentadiene

cross-linking

n

m

Scheme 8.9 Ring-opening metathesis polymerisation (ROMP) of dicyclopentadiene.

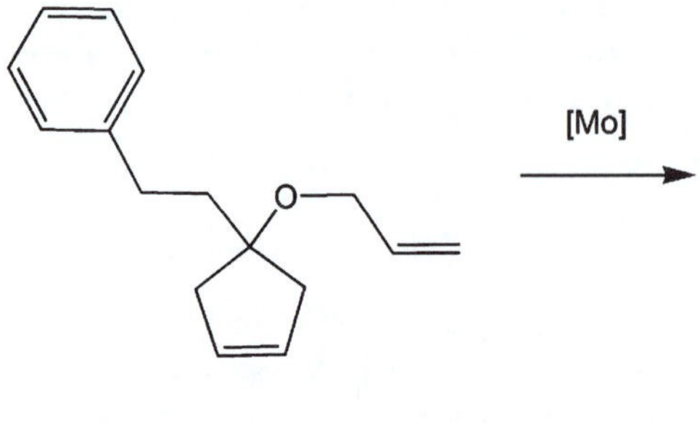

[Mo]

95% ee; 95% yield

Scheme 8.10 Asymmetric metathesis for the production of a chiral intermediate to tipranavir.[19b]

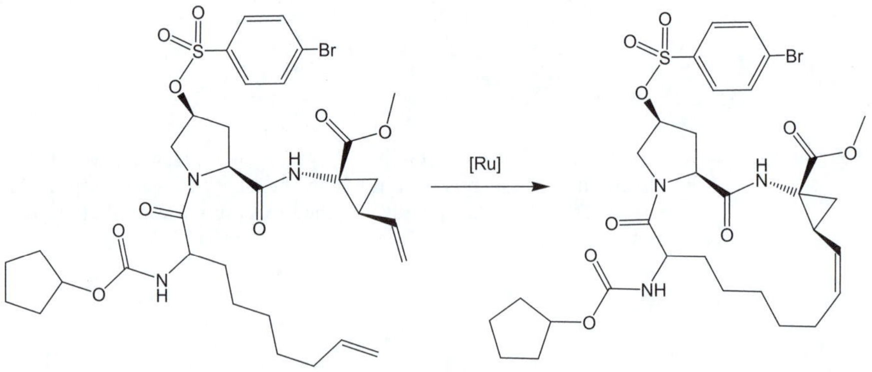

[Ru]

Scheme 8.11 Ring-closing metathesis operated industrially for the production of a pharmaceutical intermediate.[19g]

failures that are associated with scientific discovery. Books have been written on the subject (see, for instance, ref. 20a).

Fullerenes are new forms of the element, carbon, different from graphite and diamond. The prototypical fullerene, C_{60}, initially called buckminsterfullerene by its discovers,[20b] was detected in surprising abundance in vapour generated from laser irradiation of graphite. It is often the case that major scientific discoveries do not arise through an anticipated progression of steps, but quite fortuitously. The discovery of C_{60} (and that of all the related materials, including carbon nanotubes, that followed) came about because of Harry Kroto's interest (with his colleague David Walton) in the nature of long-chain carbon molecules (polyynes and cyanopolyynes) found in interstellar space, particularly some unexplained spectral features of interstellar matter (that, ironically, this research did not resolve and that remain unexplained). This led Kroto to spend time in Smalley's laboratory, which had the necessary equipment to investigate carbon vaporisation and carbon vapours.

Once convinced of their observations that C_{60} was remarkably stable, the question relating to its structure and the reasons for its stability followed, with initial considerations ruling out structures based on the known graphite and diamond. It was an interest in, and awareness of, another intellectual domain, architecture, particularly of Buckminster Fuller's geodesic domes, that led to the consideration of how to link carbon hexagons together (and the importance of pentagons to achieve curvature[20c]). The proposal that C_{60} was a truncated icosahedron (a soccer ball has such a structure) (Figure 8.4a) then followed. Not long before ref. 20b appeared, workers at Exxon had described[20d] the presence of a series of even-numbered carbon clusters, C_{30}–C_{100} in carbon vapour. While their work showed somewhat higher abundance of C_{60} compared with clusters of similar carbon number, they were more concerned with the reasons for the abundance of even-numbered over odd-numbered clusters. They were sceptical of the claims (and the experiments) of Kroto, Smalley and their colleagues and the two research groups fell out. What was needed was the isolation of sufficient C_{60} fully to characterise it. However, this proved frustrating, not least because Kroto, having returned to the University of Sussex, found it difficult to secure research funding to pursue investigations into the chemistry of C_{60}. Interestingly, the race was not won by a synthetic chemist (there were obvious attempts to design rational routes to C_{60}) nor by the discoverers of C_{60} (though they came close and had the distinction of isolating another fullerene, C_{70} for the first time[20e]), but (five years after the initial report) by a collaboration of physicists[20f] who separated C_{60} from soot. Further work confirmed[20g] the suggestion that crystalline C_{60} consists of 60 sp^2-hybridised carbon atoms linked together to form a truncated icosahedron. Shortly thereafter, even naturally occurring C_{60} and C_{70} were detected in a mineral, shungite.[20h]

The availability of C_{60} in amounts that would permit additional investigation of its bulk properties and its chemical reactivity led to further excitement, with (among many other developments) studies of superconductivity (of metal derivatives such as K_3C_{60}) and exploration of series of related carbon clusters, some of which were large enough to have cavities in which atoms could be confined (so-called 'endofullerenes', *e.g.* Figure 8.4b containing a C_2Sc_2 cluster) and which at the limit extended to long tubes of carbon (carbon nanotubes), opening up an array of opportunities for material and chemical investigation and discovery, engaging both academic and industrial scientists. The first products of chemical reaction of C_{60}, the formation of a series of methyl derivatives in radical reactions and characterised in solution,[20i] supported the idea that C_{60} was a special type of polyene. The formation of water-soluble derivatives (Figure 8.4c)[20j] opened up the possibilities of investigating the aqueous chemistry of fullerene derivatives, their possible biological function and their use as pharmaceuticals.[20k] A compound (Figure 8.4d) in which the natural product, paclitaxel (Section 8.8), is linked to C_{60} has been investigated as a slow-release anti-cancer agent.[20l] Finally, give a thought to Osawa, who in 1970 first proposed, in an article in Japanese, the idea that C_{60} might be a stable entity.[20m]

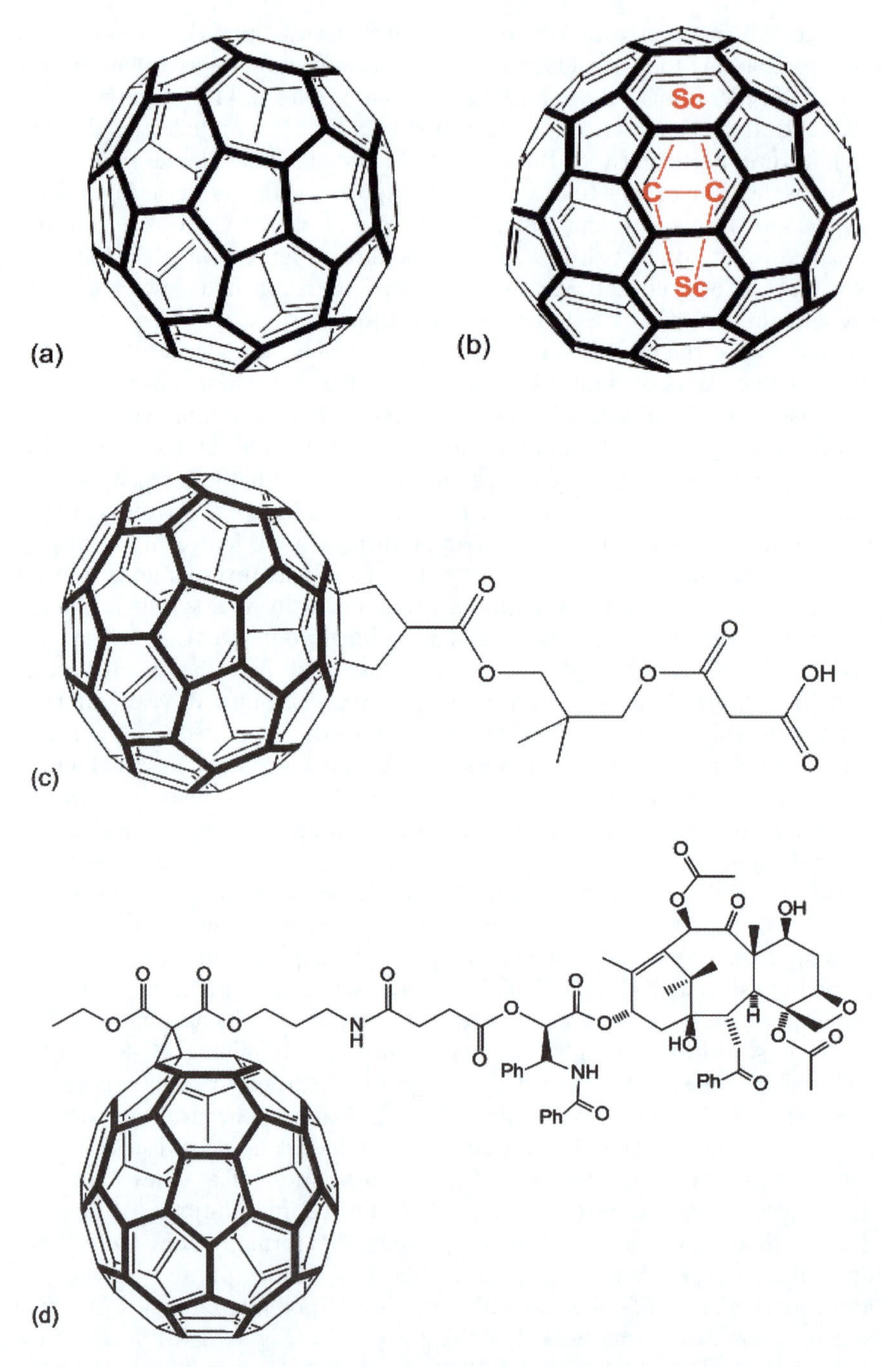

Figure 8.4 Fullerenes (a) prototypical [60]fullerene, C_{60}; (b) 'endohedral' fullerene with the cluster Sc_2C_2 encapsulated in [82]fullerene, C_{82}; (c) a water-soluble fullerene derivative;[20j] and (d) a fullerene derivative with C_{60} linked to paclitaxel investigated as a drug delivery compound.[20l]

8.8 PRESIDENTIAL GREEN CHEMISTRY CHALLENGE AWARDS

Many developments that either have the potential to contribute, or indeed, have already demonstrated a contribution, to more sustainable chemical technologies have since 1996 been recognised through the US Presidential Green Chemistry Challenge Awards. Some of these developments are discussed in this book (Table 8.2). All have been fully documented with details available from the website of the US Environmental Protection Agency (Web 5), including those for the winners in 2009.

A 2004 award was made to Bristol-Myers-Squibb for their part in the development of more sustainable ways of obtaining paclitaxel (Figure 8.5a), the active ingredient of Taxol®. Paclitaxel (Web 6) is another natural product, first isolated in the 1960s (this time from the bark of the Pacific yew, *Taxus brevifolia*) and shown to have anti-tumour activity. Unfortunately it is present in the yew in only very low concentrations (0.0004%). Stripping the bark from the

Table 8.2 Selected developments awarded the US Presidential Green Chemistry Challenge Award.

Year	*Winner*	*Topic*	*Section in this book*
1997	BHC	Ibuprofen	10.8.6
1998	Barry Trost	Atom economy/efficiency	7.5
1998	Karen Draths/John Frost	Adipic acid from glucose	10.8 and 11.6.8
1999	Biofine	Levulinic acid	10.9 and 11.6.10
2002	Cargill-Dow (now Nature Works)	Poly(lactic acid)	10.9
2003	DuPont	Sorona®	11.6.10
2004	Bristol-Myers-Squibb	Paclitaxel	8.8
2006	Galen Suppes	Propylene glycol from glycerol	10.9 and 11.6.10

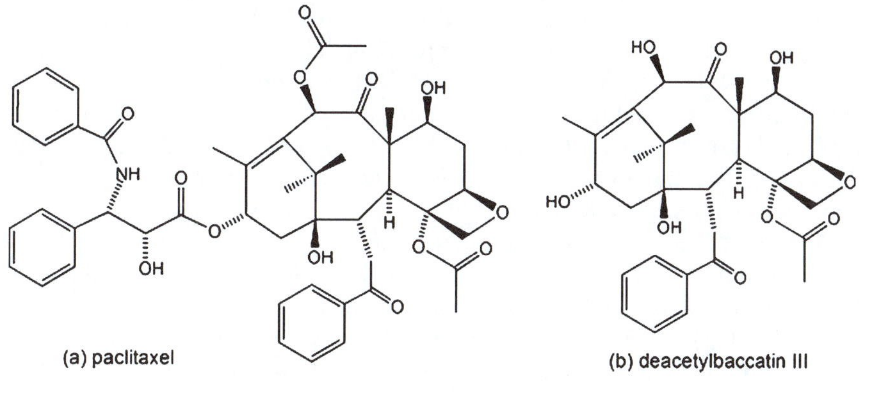

Figure 8.5 (a) Active ingredient of Taxol®, paclitaxel; and (b) natural product, deacetylbaccatin II, used in the semi-synthetic route to paclitaxel.

yew kills the trees, which take 200 years to mature. The bark from six trees would be needed to treat one patient. This source of supply, while natural and, in principle, renewable, is therefore quite unsustainable. To provide the treatment for a range of cancers for which it is found effective, synthetic routes from more readily available starting materials were required. Two total syntheses were published almost simultaneously.[x] At one stage, 30 research groups were involved, with six reporting a series of routes. The Holton synthesis (Web 6c) involved a sequence of about 40 steps. Even if each could be completed at 80% yield, the overall yield would be $(0.8)^{40}$ or 0.013%. Likewise, the overall atom efficiency would be exceptionally low, with any broader assessment of reaction efficiency, such as undertaken by Andraos, being even worse. A semi-synthetic route from the natural product, 10-deacetylbaccatin III (Figure 8.5b), more readily available by extraction from the European yew, was subsequently developed. Even so, this involved 11 steps (five of which required low temperatures). The manufacturer, Bristol-Myers Squibb, was awarded the Presidential Green Chemistry Challenge Award for the development of a biocatalytic process involving direct fermentation of cells isolated from the Chinese yew.

8.9 GREEN CHEMISTRY: A BRIEF HISTORY

Green chemistry has, in various guises, been an important influence over the last 15 years and we examine briefly how it has developed. It is most closely associated with the 12 Green Chemistry Principles enunciated by Anastas and Warner (Section 8.10). We analyse these and some related principles.

'Green' chemistry represents an approach to the discipline that consciously takes account of the potential for environmental impact (to an extent greater than its proponents believe chemists have exhibited in the past) and that seeks to prevent (or at least lessen) this impact through the application of its principles.

The adjective 'green' is one that, like others terms associated with sustainability, does not have a single, agreed or objective definition or meaning. Of the several definitions of 'green' to be found in The *Oxford English Reference Dictionary*,[21] the definition 'concerned with/supporting protection of the environment as a political principle' appears to associate an academic discipline, chemistry, with some political philosophy or objective. Recently, Paul Anastas, a founding figure and pioneer of green chemistry, has suggested[22] the definition 'young, fresh and new'.

Whether anticipated or not, a unique social movement has arisen;[23a,b] these papers also provide rather more detailed histories of green chemistry than I do here. Because of the growth of Green political parties (primarily in Europe), green chemistry was perceived negatively by some, not because of indifference to the objectives of improved environmental performance but unease with

[x] The papers were from groups led by R. A. Holton (first to be accepted for publication; Web 6c) and K. C. Nicolau (first to be published; Web 6d). Who should have precedence?

seeming to mix science with political discourse. It is possible (and of interest to historians of science) that this perceived association delayed the wider acceptance of its objectives amongst practitioners. On the other hand, it would be surprising if chemists (like any other citizens) did not want to play a role in determining responses to environmental concerns through participation in the democratic process. Indeed, there is an honourable tradition of such activism, though largely concerned with how the discoveries arising from research are applied rather than prescriptions about its academic practice. I explore these issues further in Chapter 13.

The Pollution Prevention Act enacted in the USA in 1990 provided a particular focus on waste minimisation and environmental protection, one that defines the motivating purpose of green chemistry. Whether or not others may have anticipated its precepts (see, for example, ref. 24), an additional question that historians will ponder, it would be wrong to understate the important stimulus this piece of legislation provided, not only among chemists but also among producers of chemicals.

Whether or not chemists agree with the presumption, green chemistry seeks to respond to public perceptions that chemistry, and its application through chemical technology, has been primarily responsible for many of the ills of the world manifest through the degradation of the environment. Its ideas have changed the discipline in important ways, as we shall see. As a consequence, all practising chemists (experienced or newly qualified) should know what is meant by green chemistry and (adherents and sceptics alike) should have an appreciation of its strengths and weaknesses.

The topic may also be called 'clean' or 'cleaner' chemistry or 'sustainable' chemistry. The term I prefer, 'chemistry for sustainable technologies', places the prime focus on the subject's technological application. During a period in which public pressure on industry to improve its environmental performance grew, green chemistry was one important influence. The chemical industry sought to respond by showing how its activities were (indeed, had been[25]) consistent with its principles. Green chemistry can now be seen as encompassing (or, depending on your perspective, being encompassed by) a variety of related interdisciplinary approaches, methodologies and metrics for waste minimisation and environmental protection including environmental chemistry, life-cycle analysis (LCA), industrial ecology, cleaner production, benign-by-design or green engineering, sustainability science and even environmental and ecological economics.

Concise and informative definitions of green chemistry have been provided by:

- Sheldon[26] (chemistry that '... *efficiently uses (preferably renewable) raw materials, eliminating waste and avoiding the use of toxic and/or hazardous reagents and solvents in the manufacture and application of chemical products*') and
- Marteel *et al.*[27] ('*the design of chemical products and processes that eliminates or reduces the generation of hazardous substances*').

Both underline the need for an appreciation of chemistry operated on the large scale (addressed in Chapter 9) and not just in the laboratory (the traditional focus of chemists).

Whatever we choose to call the topic, it has, nevertheless, brought about some real benefits. It has been a stimulus to new chemistry. Successes arising from its wider appreciation should begin to reduce some of the more unthinking chemophobia that exists and thereby restore public acceptance of chemistry and attract a new generation to its study. It should also provide a wake-up call to older professional chemists who, in the past, may not have concerned themselves sufficiently with the impact that their chemistry might have.

The advent of green chemistry is very much associated with the Green Chemistry Program initiated by Environmental Protection Agency and the National Science Foundation in the USA in 1991. Subsequently, the Green Chemistry Institute (Web 7) was set up in the USA and the Green Chemistry Network (Web 8) in the UK, with associated research and educational programmes. Other networks have followed. The topic was given a major boost when political leaders in the USA put their weight behind the idea of green chemistry with the founding, in 1996, of the US Presidential Green Chemistry Challenge Awards (Web 5).

Green chemistry now has its own journals (*e.g.* Web 9) and is the subject of many texts and monographs, though mostly concerned with research aspects. A selection can be found in the Bibliography. The book by Lancaster,[28] although now a little old (added in proof: A new edition has recently been published), is the most useful companion to the current text. No single text takes the wider view I am attempting though Paul Anastas is editing a 12-volume series entitled the *Handbook of Green Chemistry*.

An important milestone was the publication in 1998, authored by two key figures, Anastas and J. C. Warner, of *Green Chemistry: Theory and Practice*,[29] in which their famous 12 principles were enunciated. In response, others were propounded, focusing further on chemistry (ref. 30, reproduced in Appendix 3) and on chemical engineering[31]—with some handy mnemonics (Figure 8.6)

Condensed Principles of Green Chemistry

- P - Prevent wastes
- R - Renewable materials
- O - Omit derivatization steps
- D - Degradable chemical products
- U - Use safe synthetic methods
- C - Catalytic reagents
- T - Temperature, Pressure ambient
- I - In-Process Monitoring
- V - Very few auxiliary substances
- E - E-factor, maximise feed in product
- L - Low toxicity of chemical products
- Y - Yes, it is safe

Principles of Green Engineering

- I - Inherently non-hazardous and safe
- M - Minimize material diversity
- P - Prevention instead of treatment
- R - Renewable material and energy inputs
- O - Output-led design
- V - Very simple
- E - Efficient use of mass, energy, space & time
- M - Meet the need
- E - Easy to separate by design
- N - Networks for exchange of local mass & energy
- T - Test the life cycle of the design
- S - Sustainability throughout product life cycle

Figure 8.6 Green chemistry and engineering mnemonics: 'productively' and 'improvements'.[32] Reproduced by permission of the Royal Society of Chemistry.

being invented. The parallels between the three sets will be apparent. They should be seen as entirely complementary. There are now even ten commandments of sustainability,[33] which may imply a theological (or at least an ethical[34]) motivation to the topic as well as an ideological or philosophical[xi] one.

8.10 PRINCIPLES OF GREEN CHEMISTRY

Much fundamental chemical enquiry devoted to waste minimisation and pollution prevention and chemical technology developed with the minimisation of environmental impact in mind predates the formulation of the green chemistry principles. It is, therefore, important to recognise that much chemistry was (and continues to be) published that may not explicitly link itself to green chemistry principles but may well make a contribution to pollution prevention and sustainable development. Indeed, it is in the nature of research that the relevance of, and the benefits and opportunities arising from, new work may not be apparent until long after its publication.

Many chemical research publications now cite these principles as a motivation. While relatively few specify or quantify (except possibly in the most general of terms) environmental or efficiency benefit relevant to the large scale, the advent of improved metrics should see this number increase. Even fewer, though, provide any form of critique of the principles and their foundation. It is possible that these principles have, unintentionally, come to be seen as articles of faith with a belief that they may be simply and universally applied. Failure to accept this may be taken to indicate indifference, or worse. To subject them to critical analysis in a manner usual for any other scientific proposition may, in this case, lay the critic open to the accusation of being 'anti-green' (or 'brown'![23a]) and of not being sufficiently concerned about pollution and environmental degradation. However it is necessary, bearing in mind the extent to which the concepts of green chemistry are now being introduced into the school curriculum and university courses,[36] to scrutinise each in turn and examine its implications—particularly to reinforce the point that chemistry is governed, first and foremost, by basic physical and inviolable universal thermodynamic principles.

Anastas and Warner's principles are listed in Table 8.3. Twelve more were added subsequently (Table 8.4). Each is discussed in turn below.

8.10.1 Principle 1: It is better to Prevent Waste Formation than to Treat it after it is Formed

No-one should argue with a principle based on the notion that 'what you don't make can't pollute' or contest the implication that improvements in reaction and process efficiency should be the first priority. On the other hand,

[xi] The idea that green chemistry is a philosophy rather than preferred practice or a series of objectives has explicitly been suggested by Tucker.[35]

Table 8.3 The twelve green chemistry principles of Anastas and Warner.[29]

1. It is better to prevent waste formation than to treat it after it is formed
2. Design synthetic methods to maximise incorporation of all material used into final product
3. Synthetic methods should, where practicable, use or generate materials of low human toxicity and environmental impact
4. Chemical product design should preserve efficacy whilst reducing toxicity
5. Avoid auxiliary materials (eg solvents, extractants) if possible or otherwise make them innocuous
6. Energy requirements should be minimised: conduct syntheses at ambient temperature/pressure
7. A raw material should, where practicable, be renewable
8. Unnecessary derivatisation (such as protection/deprotection) should be avoided, where possible
9. Selectively catalysed processes are superior to stoichiometric processes
10. Chemical products should be designed to be degradable to innocuous products when disposed of and not be environmentally persistent
11. Monitor processes in real time to avoid excursions leading to the formation of hazardous materials
12. Materials used in a chemical process should be chosen to minimise hazard and risk.

Table 8.4 An additional twelve green chemistry principles (ref. 30, Appendix 3).

1. Identify and quantify by-products
2. Report conversions, selectivities and productivities
3. Establish full mass-balance for process
4. Measure catalyst and solvent losses in air and aqueous effluent
5. Investigate basic thermochemistry
6. Anticipate heat and mass transfer limitations
7. Consult a chemical or process engineer
8. Consider effect of overall process on choice of chemistry
9. Help develop and apply sustainability measures
10. Quantify and minimise use of utilities
11. Recognise where safety and waste minimisation are incompatible
12. Monitor, report and minimise laboratory waste emitted.

expectations should not be raised that zero waste[xii] processes are possible. The principle, as stated, does not make the important distinction between a waste and a pollutant.

Considerations of thermodynamics force the conclusion that even the 'greenest' chemical process will produce waste. The challenge is in its containment. Not all of the waste from a chemical process results from

[xii] A 'waste' should be distinguished from 'pollutant' (see Sections 6.1 and 6.2).

the chemical reaction, *per se*, nor even from the production of feedstocks. Contributions will come from other components and unit operations of the engineered chemical process and from the product in use. These considerations highlight the importance of full life-cycle methods for determining material and energy usage and emissions.

Few chemical processes convert a starting material completely to a single pure product that can be isolated easily. Forcing the completion of real processes characterised by the equilibria and kinetics of competing reactions would, except in a limited number of cases, be at the expense of overall reaction efficiency. For instance, if the desired product is formed under kinetic control[xiii] with a thermodynamically favoured alternative being accessible, then an extended reaction time may allow significant concentrations of an undesired second product to be formed, making the separation more complex (and wasteful). Similarly, if the product of the reaction is susceptible to further reaction, then an extended reaction time may allow sufficient of the second product to be formed—again to pose a separation problem. For a pseudo-first order reaction (one done in the presence of an excess of one reactant and presupposing the optimum process involves the use of a batch reactor), each 50% conversion of the remaining limiting reactant to product takes the same amount of time (the so-called half-life). It is a simple calculation (see Bibliography) to estimate that six half-lives will bring about 98.4% reaction. A further two half-lives will effect a further 1.2% conversion, leaving (assuming this to be the only reaction) 0.4% unconverted. The consideration whether or not to leave the reaction to react for longer will depend on a series of factors, specific for a particular set of circumstances:

- the level of purity required in a product and the ease with which the product can be separated from the reaction mixture
- the costs associated with operating the reactor for longer
- the possible effects of competing reactions.

The general historic trend in technological development has been to reduce waste because of the ever present economic or competitive stimulus. Where such stimuli are insufficient (or perceived to be insufficient), or where the impact of a waste as a pollutant requires it, regulation can be applied. The chemical industry has its advantage served by adopting waste minimisation practices, either through public or stakeholder pressure or from regulation.

8.10.2 Principle 2: Design Synthetic Methods to Maximise Incorporation of All Material Used Into Final Product

This principle is essentially a restatement of the atom efficiency or atom economy concept propounded by Trost in 1991. The shortcomings of atom

[xiii] See footnote, p. 176.

efficiency described in Section 7.5 also apply here. On the other hand, this principle has a wider scope in that it refers to 'all material', not just reagents.

The target of improved chemical reaction efficiency has motivated generations of synthetic chemists. Atom efficiency is one of a number of criteria (Chapter 7) by which the efficiency of chemical transformations can be evaluated. The application of unified metrics (Section 7.9) to evaluate individual reactions and more complex reaction schemes, and addressing factors relating to both chemical reaction and chemical process, highlights the value of taking into consideration a number of criteria and not focusing on one to the exclusion of another.

8.10.3 Principle 3: Synthetic Methods Should, where Practicable, Use or Generate Materials of Low Human Toxicity and Environmental Impact

It is true to say that this principle is unlikely to have been the prime motivation of many chemical practitioners when approaching some novel chemistry. On the other hand, few chemists would choose to use a hazardous material for the sake of it if an alternative was available, a point that Anastas and Warner recognise in the qualifying statement, 'where practicable'. However, while the research chemist may not have addressed this point, then the process chemist, process engineer or project manager involved in scale-up most certainly would, as additional containment and other requirements necessary for the handling of hazardous materials add to costs.

As there are no materials that are wholly free of toxic or environmental impact, there is also a need to counter the perception that entirely non-toxic materials can always be identified. An optimum combination of risk and benefit is the best that can be achieved and it can be wasteful of effort (that could be more fruitfully devoted elsewhere) to seek an unachievable perfection.

It is sobering to consider that, had avoidance of toxicity been a key factor in determining an approach to chemical synthesis, it is then possible that many important compounds or classic syntheses would not have been identified or developed. It is a key experimental skill that chemists develop that permits them to manipulate, generally in a safe manner, chemical materials and intermediates that are unstable, reactive, corrosive, toxic, explosive or otherwise hazardous. Assessment procedures such as COSHH (Control of Substances Hazardous to Health) in the UK (Web 10) have now been codified that guide experimenters and others in the assessment of the risks associated with compounds, materials, procedures and processes.[xiv]

A particular difficulty, when contemplating the synthesis of new chemical compounds, is that toxicological data may not be available or may be

[xiv] Care should be exercised when using the terms 'risk' and 'hazard': they should not be used interchangeably. The difference between them can be simply expressed as follows: a hazard can be represented by a six-shot revolver loaded with a single bullet; the risk (the combination of the seriousness of the hazard and the probability of exposure) is the chance that, if the trigger is pulled, the bullet would be fired.

incomplete. Predictions of toxicity are often unreliable. Do we mean chronic or acute toxicity? Are we concerned whether a material is a carcinogen, an endocrine disruptor, a neurotoxin or a teratogen? Should we take the toxicity of possible metabolites [breakdown products in the body (and whose body? human, primate, invertebrate?)] into account? A false negative (suggesting the absence of a hazard when there is one) from such predictions can be seriously problematic; a false positive (suggesting the presence of a hazard when there isn't one) can mean an unnecessary loss of opportunity.[37] (Ethical concerns regarding the current need to carry out toxicological testing on animals is driving the search for effective alternatives: their absence in important instances raises the question as to the appropriate balance between gaining critical information that truly enhances public safety and minimising suffering to animals on which necessary tests are carried out.)

All materials, even water (as a recent sad case demonstrates, see Web 11), can endanger life if taken in excessive amounts. Paracelsus (1493–1541), who pioneered the use of minerals and chemicals in medicine, put it succinctly: '*The dose is the poison*'. The estimation of what may be considered safe, therefore, depends on a number of factors that include:

- the intrinsic toxicity of a compound (however determined)
- the extent (a small amount, once only for a short time or for an entire working life) and nature of exposure (by inhalation, skin absorption)
- the susceptibility of the individual exposed (generally healthy adult, a child or someone with a medical condition).

Statutory exposure limits (Web 12) build in extra margins of safety, though they are often seen as indicators of harm even for brief one-off exposures.

Some materials are designed to be poisonous so attempts to synthesise them would fall foul of this principle, if taken too literally. Pesticides and pharmaceuticals are expected to be selectively toxic, killing a pest or an invading infective agent that causes illness, while leaving the infested plant or infected patient unharmed. Pharmacologists express this in terms of the therapeutic ratio, *i.e.* the ratio of the toxic dose (for 50% of the population) to the therapeutic dose (the minimum effective dose for 50% of the population). The value should be as large as possible.

Precipitately seeking to avoid using chemical compounds that are believed to represent an unacceptable toxic or environmental risk can bring about unforeseen consequences that may be worse than those avoided. The consequences to those at risk of malaria from the banning of DDT and of a hasty end to water chlorination in Peru have already been highlighted. In his entertaining and informative book, *Molecules at an Exhibition*,[38] John Emsley makes a convincing case that the appearance of bovine spongiform encephalopathy (BSE—known colloquially as 'mad cow' disease) was associated with the decision to phase out the use of dichloromethane in the extraction of fat from treated abattoir waste because of environmental and human toxicity concerns that were largely unfounded.

8.10.4 Principle 4: Chemical Product Design Should Aim to Preserve Efficacy Whilst Reducing Toxicity

Again, those involved in product development might well say this has been an ever present consideration. We are faced again with the question: what is meant by toxicity? Everything is toxic to something: a judgement must be made that balances benefit and risk.

8.10.5 Principle 5: Avoid Auxiliary Materials (e.g. Solvents, Extractants) if Possible or Otherwise Make Them Innocuous

This more concrete statement focuses on the specific objective of eliminating solvents and related materials or at least replacing those materials believed to be more harmful. We saw in Chapter 7 that the efficiency with which fine chemicals and pharmaceuticals are produced is relatively poor, largely associated with the use (in multi-step transformations) of solvents and other reaction and process auxiliaries. The possibility of reducing emissions and making savings in the costs of process operation has spurred much innovative chemistry, particularly seeking to minimise the use of solvents of concern. Approaches have included:

- **'solventless' processes** (*e.g.* mechano-chemical reactions[39,xv] may result from grinding two solid reactants together, but solvents are often used subsequently in isolation and purification)
- **'cascade' reactions**[41] (a sequence of transformations carried out in a single medium without isolation of intermediates). The production of pharmaceuticals often involves multi-step transformations each requiring a different solvent, necessitating separation at each stage. On the other hand, the process of isolation and re-dissolution often removes an impurity whose presence may be problematic at a later stage
- **the use of water** (Section 8.11.1): not especially attractive because of its solvency characteristics and its very high latent heat of evaporation (and low molecular mass), a critical factor in its removal and purification
- **novel reaction media** (Section 8.11)
- **novel reactors and reaction stimuli** (Chapter 9).

While there has been much research on these topics, there has been insufficient work on comparing processes for converting A into B that differ only in the medium used to effect the transformation.

[xv] The authors of ref. 39 distinguish between **solid phase synthesis** (where a compound in a fluid phase reacts with a solid substrate) and the truly solventless processes: **solvent-free synthesis** (where neat reagents react together in the absence of a solvent) and **solid state synthesis** (in which two macroscopic solids interact to form a third, solid, product without the involvement of a liquid phase). Recent studies[40] of the physical as well as the chemical changes occurring when *o*-vanillin and *p*-toluidine are ground together establish that a low-melting eutectic is formed and that reaction to give the corresponding azomethine occurs in the liquid phase.

8.10.6 Principle 6: Energy Requirements should be Minimised: Conduct Syntheses at Ambient Temperature/Pressure

The first part of the statement hardly sets the green chemistry approach apart as being either original or unique since, as energy is a cost, economic pressures already drive this objective through the application of, and developments in, exergetic analysis and process synthesis.

The second part of the statement is more contentious as it fails to take proper account of thermochemical factors. For reactions that produce heat (and ones that need to be operated at elevated temperatures to achieve necessary rates), recovery of the reaction exotherm, or the recovery of sensible heat at high temperatures, may enable steam to be raised. This can be used to do additional useful work (*e.g.* spinning a turbine to generate electricity). There seems to be no benefit in artificially cooling a reaction to bring it about at ambient temperature when this might mean throwing away potentially useful energy.

As we have seen, the Haber process for the manufacture of ammonia is operated at high temperatures and pressures to achieve practical conversions and economic rates of production. Biological nitrogen fixation, on the other hand, occurs at ambient temperatures and pressures (and in water) and may appear, superficially, green. However, this attraction is illusory (see Box 8.1) because the very large quantities of energy, needed to facilitate the conversion in nature, are available 'free' from the Sun. Until such time as solar energy can be captured on the industrial scale at low cost, the Haber process will remain attractive (and even then it is likely to be the preferred manufacturing route if the dihydrogen feedstock can be obtained cheaply from a renewable solar-energy-driven source).

8.10.7 Principle 7: A Raw Material should, where Practicable, be Renewable

This is a sensible idea, this time suitably qualified in terms of practicability. We deal with renewable raw materials and feedstocks (and the extent to which we can eliminate the use of fossil-sourced materials altogether) in Chapter 11. We also consider the environmental implications of satisfying the food, material and energy needs of global population from renewable resources.

8.10.8 Principle 8: Unnecessary Derivatisation (such as Protection/Deprotection) should be Avoided, where Possible

Implementation of this principle represents a major challenge to synthetic organic chemists[xvii] and is an active area of research.[43] Protection of one functional group while modifying another is often necessary because of the lack

[xvii] In a very readable personal perspective on synthesis, Cornforth[42] suggested in 1994 that the use of protecting groups should be regarded as an '*imperfection*' or an '*artistic failure*', likening some examples to '*a medieval knight buckling on a hundred-weight of armour before being hoisted onto a carthorse to go into battle*'. Cornforth was awarded a part share in the 1975 Nobel Prize for Chemistry (Web 4). His life and work feature in a video produced by the Vega Science Trust (Web 13).

BOX 8.1 BIOLOGICAL NITROGEN FIXATION

An argument that is often heard is that, as nature fixes nitrogen and does so under conditions of ambient temperature and pressure (in water and at neutral pH, for good measure), why cannot industrial technology do the same?

Nature fixes nitrogen using enzymes (biological catalysts); in this case, the nitrogenases (eqn 8.5).

$$N_2 + 8\,H_2O + 8\,e^- + (16\,MgATP + 16\,H_2O) \downarrow 2\,NH_3 + H_2 + 8\,OH^- + (16\,MgADP + 16[PO_4]^{3-} + 32\,H^+) \quad (8.5)$$

To proceed, this process relies on a co-reaction in which ATP (adenosine triphosphate, Figure 8.7) is hydrolysed to ADP (adenosine diphosphate) and one mole of phosphate (eqn 8.6). Crucially, this delivers[xvi] $30\,kJ\,mol^{-1}$ to drive the reduction of dinitrogen to ammonia.

$$ATP(aq) + H_2O\,(l) \rightarrow ADP(aq) + phosphate(aq) + 2\,H^+ + 30\,kJ\ (310\,K, pH\,7) \quad (8.6)$$

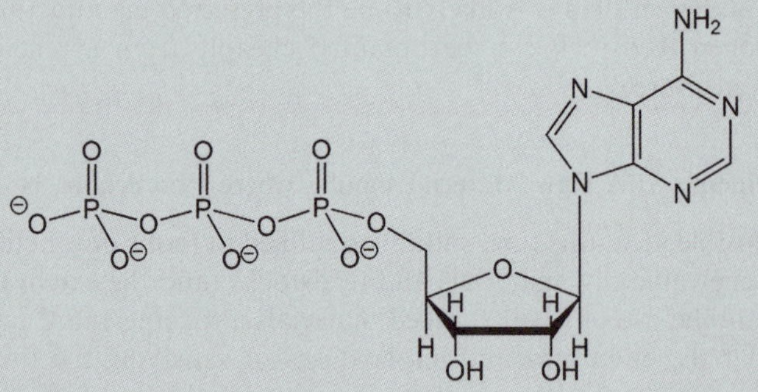

Figure 8.7 Adenosine triphosphate (ATP).

So, to achieve nitrogen reduction in water at ambient conditions of pressure, temperature and pH in an aqueous environment, 16 moles of ATP must be hydrolysed to provide the *ca.* 500 kJ required to produce two moles of NH_3.

If one wanted to operate the conversion of nitrogen to ammonia this way one would need to find a way of delivering this amount of energy to effect the conversion. The reason why this occurs in nature is that biological processes have found a way of capturing and utilising the energy of the Sun, which is

[xvi] The precise stage in the hydrolysis process at which the energy is released to drive biological processes (and how this is transferred) remains largely unknown.

essentially free and continually available. The technological challenge of mimicking nature, then, reduces to one of efficient and cost-effective solar energy capture and utilisation. This is in the process of being solved and is discussed in more detail in Chapter 12. However, even if one were to manufacture ammonia this way, there is an additional problem. As most of the ammonia used industrially is in the form of the anhydrous product, there would be a need to separate it from aqueous solution (quite a dilute solution at that), itself a particularly energy-consuming process. So, the current industrial process is much more energy efficient and produces an anhydrous product (but uses fossil energy sources).

of desired selectivity to the sought-after product in the absence of such protection. When it is done on the small scale for research purposes, protection/deprotection may be defensible. However, the problem arises when such reactions are done on the large scale, where reagents and auxiliaries will be used that do not find their way into the final product and simply end up as waste. However, a 2006 survey[44] of methods used in three major pharmaceutical companies showed that about one in five processes involved some form of protection/deprotection. Functional group protection leads to low overall atom efficiency and Andraos has particularly focused on this in his development of unified metrics and their application in a comparative analysis of many laboratory syntheses and routes to the antiviral, Tamiflu.[12]

While the associated extra steps involved in protection and deprotection may also be wasteful in solvents and other auxiliaries, their use may still be justified in a synthetic sequence that, overall, is less waste-producing than an alternative in which they are not used. An example is highlighted by Andraos[45] in his comparison of the Woodward–Rabe and Stork syntheses of quinine (Figure 8.8) (an alkaloid isolated from the bark of the cinchona tree and an historically important treatment for malaria). The former route leads to the racemic product in 19 steps, with an average yield/step of 65% with 12.5% of steps involving protection/deprotection. The latter route, on the other hand,

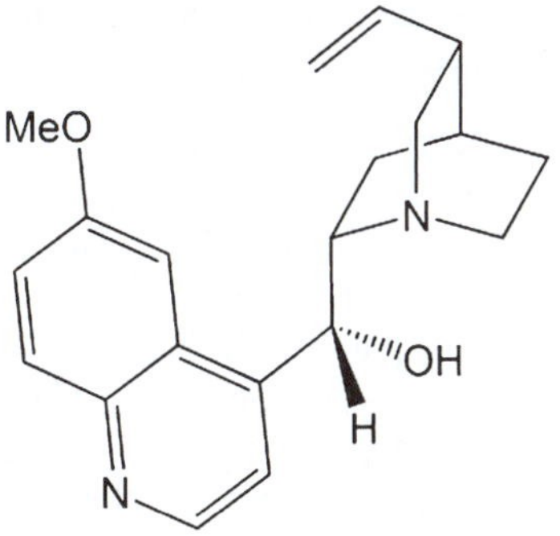

MeO
N
OH
H
N

Figure 8.8 The alkaloid, quinine, historically important in the treatment of malaria.

gives (–)-quinine in 17 steps, with an average yield/step of 86% with 24% of steps involving protection/deprotection.

A classic example (Ref. 25, p. 53–56) is given in Scheme 8.12, which shows two routes to 7-aminocephalosporanic acid (**5**) from cephalosporin C (**1**). The first is a classical method using organic solvents in which (**1**) is first converted to its zinc salt (**2**) before three functional groups capable of reacting with PCl_5 are converted to unreactive silyl esters (**3**) in the presence of base. These are subsequently removed from the chlorinated product (**4**), formed by treatment with PCl_5 at low temperatures, by hydrolysis giving the desired product (**5**). Calculating atom efficiency in the usual way (eqn 8.7) shows this to be 20%.

$$\begin{array}{llllll} C_{16}H_{20}N_3O_8S & + 0.5\,Zn(OH)_2 & + 3\,C_3H_9ClSi & + 3\,C_8H_{11}N & + Cl_5P & + H_2O \\ 414 & + 0.5 \times 99 & + 3 \times 108.5 & + 3 \times 121 & + 208.5 & + 18 \\ & & & & & \rightarrow C_{10}H_{12}N_2O_5S \\ & & & & & \rightarrow 272 \end{array} \quad (8.7)$$

$$AE = (272/1378.5) \times 100 = 20\%$$

The biotechnological route oxidatively deaminates (**1**) using the enzyme, D-α-amino acid oxidase, with an N–C bond then being cleaved by another enzyme, glutaryl amidase, to give (**5**) directly at temperatures slightly above ambient, without the need for derivatisation or organic solvents (at least as the reaction medium). The reaction is significantly more atom efficient (eqn 8.8).

$$\begin{array}{llll} C_{16}H_{20}N_3O_8S & + H_2O & \rightarrow & C_{10}H_{12}N_2O_5S \\ 414 & + 18 & \rightarrow & 272 \end{array} \quad (8.8)$$

$$AE = 272 \times 100/414 = 66\%$$

The benefits of the direct biotechnological route avoiding protection/deprotection are immediately apparent, although the volume of aqueous waste requiring treatment increases substantially.

8.10.9 Principle 9: Selectively Catalysed Processes are Superior to Stoichiometric Processes

The benefits of avoiding stoichiometric reagents through the use of catalysts were evident when exemplifying atom efficiency. The better relative performance of the petrochemical and commodity chemical sectors compared with the fine chemical and pharmaceutical sectors when using the E-factor for comparison can be ascribed, in part, to the widespread use of catalysts in the former and their more limited use in the latter. The application of catalysis represents historically one of the triumphs of chemistry and chemical technology. The greater use of catalysts for more effective waste minimisation in fine chemicals and pharmaceuticals production remains a major challenge.[44] This is elaborated on in Chapter 10.

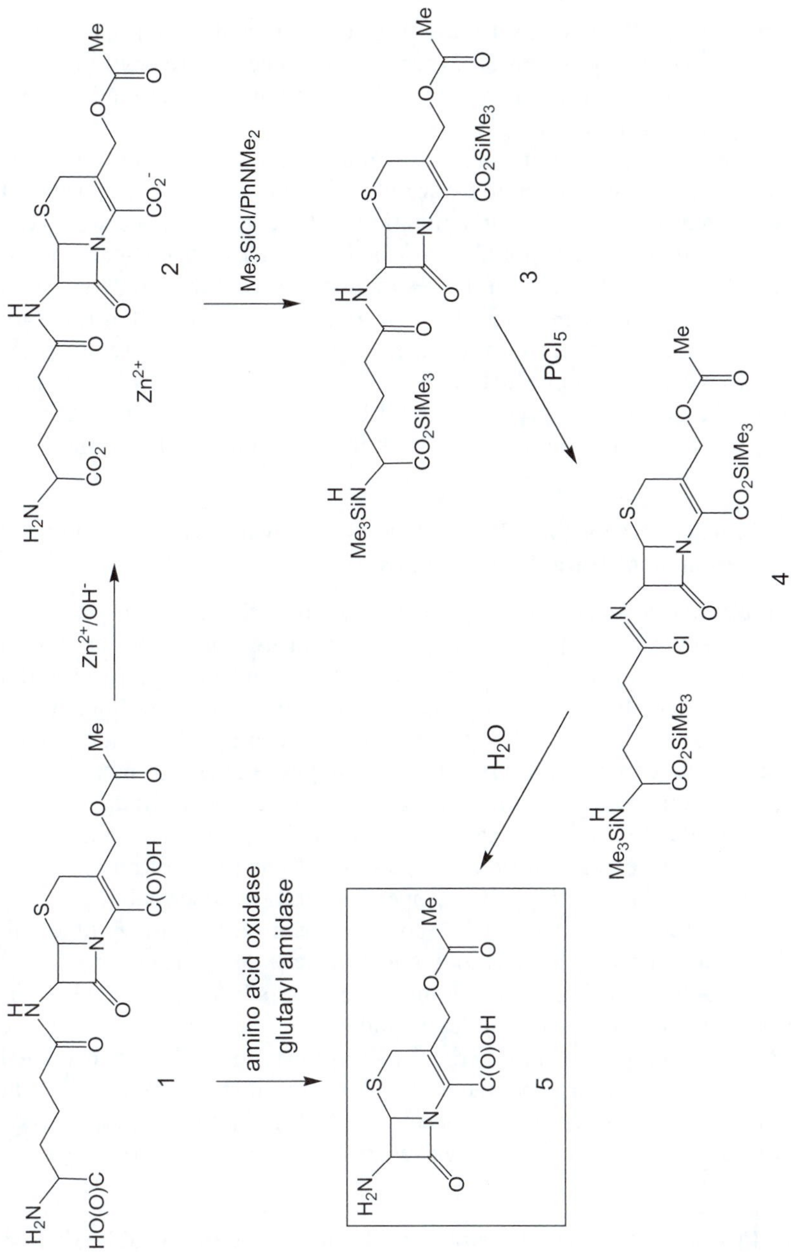

Scheme 8.12 Production of 7-aminocephalosporanic acid (**5**) from cephalosporin C (**1**) by a classical synthetic route and by a biotechnological route.

8.10.10 Principle 10: Chemical Products should be Designed to be Degradable to Innocuous Products when Disposed of and not be Environmentally Persistent

While persistent pollutants may have serious environmental impact and their presence continues to cause concern, there are instances where environmental persistence can be seen as a benefit. The principle as expounded does not properly recognise this.

For example, underground pipelines distribute gas, water and other important products. They are required to carry out their function for decades without degradation that might lead to contamination or leakage. Getting a balance is clearly important: a product should ideally be serviceable or fit-for-purpose for as long as it is required to perform the desired function and then undergo rapid degradation to harmless products (if it is not to be recovered for reuse). Building such characteristics into a material or a product represents major challenges to material and product design. One such idea would be to design a material so the degradative processes could be 'switched on' at the end of its useful life. This is an active area of research in the chemistry of materials.

8.10.11 Principle 11: Monitor Processes to Avoid Excursions Leading to the Formation of Hazardous Materials

This principle explicitly focuses on the operation of processes carried out on a large scale. Working outside the specified design limits of process operation can lead to poor selectivities, associated with the formation of more (and different) by-products or decomposition products that may be hazardous or deleterious to the quality of a product (and incurring costs to bring about their removal). Only by appropriate monitoring of the chemical process can it be established that the reaction and associated unit operations are functioning within design limits. Indications of excursions outside of these ranges can be detected and corrective action taken. In the extreme, of course, an excursion that gets out of control can lead to serious incidents that may threaten the safety of operators and the wider public.

Process monitoring is an important aspect of chemical technology (and will use well-established analytical methods known to chemists). There are special challenges to achieve such monitoring on a continuous basis. An added factor that relies on the skills of those that design and operate chemical processes is the avoidance of unnecessary shut-downs. As shut-down conditions (and related ones experienced during a subsequent start-up) are different from those that exist when a process is operating as specified, then the material being produced may not be suitable for use and may have to be discarded or reprocessed.

8.10.12 Principle 12: Materials Used in a Chemical Process should be Chosen to Minimise Hazard and Risk

I cannot think of any chemist or chemical technologist who would disagree with this statement as it stands. There may be some debate as to what represents a hazard or a risk (and what particular individuals mean when they use these terms—particularly how risk and hazard differ) and what may be deemed to be acceptable when set against the potential benefits that arise when incurring it (as happens every day on thousands of manufacturing sites around the world).

While these 12 principles cover some important matters others, necessary to reduce the impact on the environment of chemical technology (in its widest sense) and to make more efficient the processing of chemicals, are not explicitly dealt with. As we noted in Section 7.5, the 'greenness' of a process can only be assessed in the context of its scale-up, its application and its practice. This is why those chemists who have not already pondered on scale-up should consider the following, more process-orientated, principles (Table 8.4).

8.10.13 Principle 13: Identify and Quantify By-Products

Concentrating on the target product is all very well but reactions, as carried out practically and industrially, rarely follow the perfect stoichiometry represented by a chemical equation. This is also a limitation of the atom efficiency concept. There may be as much benefit in waste minimisation in understanding how by-products arise in conventional reactions as in seeking to identify new reactions with 100% AE.

Consider the process in Scheme 8.13: synthetic chemists generally are primarily interested in the target product (*e.g.* E). However, its large-scale manufacture (and often its use) requires the full accounting of all material used and a knowledge of other products (*e.g.* C, D or F) which may be present during the process. Sometimes, co-products are intermediates (*e.g.* C) leading to, or are formed in equilibria (*e.g.* D), with the desired product, E. In such cases, they can be separated and recycled. Sometimes, a by-product must be removed from a product to below a specified concentration. Such separations can be costly and even, in the limit, can make a process non-viable if the required concentration of the impurity cannot be achieved.

A practical example is given in the manufacture of ethanediol from ethene oxide.[46] Conventionally, this process involves a catalysed hydration of ethene oxide, a reaction (eqn 8.9) nominally having 100% AE. However, 10% oligomeric by-products are formed (eqn 8.10) even in the presence of excess water.

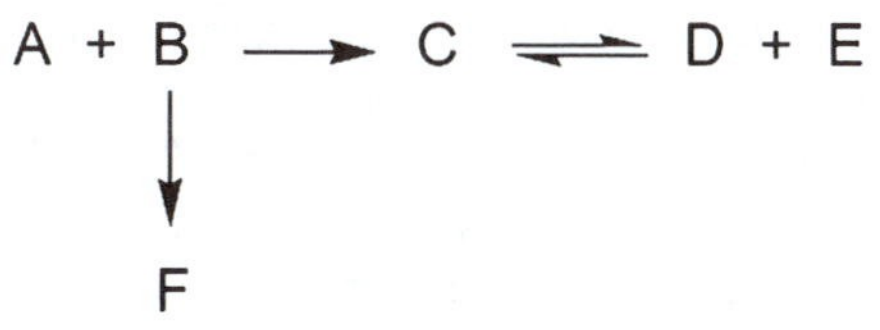

$$\begin{array}{ccccc} A + B & \longrightarrow & C & \rightleftharpoons & D + E \\ \downarrow & & & & \\ F & & & & \end{array}$$

Scheme 8.13 Origin of process by-products.

By dividing the reaction into two separate steps (eqn 8.11, 8.12) (and even though an additional reagent is included, reducing atom efficiency), $<1\%$ oligomers are formed, usage of water is reduced by 30% and of steam by 20%.

$$\text{ethylene oxide} + H_2O \longrightarrow HOCH_2CH_2OH \tag{8.9}$$

$$n\ \text{ethylene oxide} + H_2O \longrightarrow HOCH_2CH_2(OCH_2CH_2)_{n-1}OH \tag{8.10}$$

$$\text{ethylene oxide} + CO_2 \longrightarrow \text{ethylene carbonate} \tag{8.11}$$

$$\text{ethylene carbonate} + H_2O \longrightarrow HOCH_2CH_2OH + CO_2 \tag{8.12}$$

The technical importance of by-products can be exemplified with a topical example.[47] γ-Valerolactone (Scheme 8.14) meets the performance specification for use as a fuel additive. However, when prepared by hydrogenation from the renewable platform chemical, levulinic acid, a small amount of 2-methyltetrahydrofuran is formed as a by-product from over-hydrogenation. This is readily hydroperoxidised and γ-valerolactone containing it may fail fuel peroxidation stability tests.

8.10.14 Principle 14: Report Conversions, Selectivities and Productivities

Classic metrics such as yield are of little value in assessing reaction efficiency. Conversions, selectivities (Sections 7.3 and 7.4) and the unified metrics developed by Constable, Andraos and others are better measures.

Productivity is an important practical measure necessary for the design and operation of a chemical reactor or process. The space-time yield (STY) is defined as the amount of a specified material produced per unit volume of a reactor per unit time. It is an important factor in the techno-commercial assessment of a process with a slow reaction requiring a larger reactor to produce a defined quantity of product than that for a fast one (other things being equal). Estimates of productivity, therefore, influence either the size (and, therefore, capital cost) of a new reactor needed to produce the required quantity of product, or the time an existing reactor will need to be tied up (affecting costs of utilities, labour) to manufacture the required quantity.

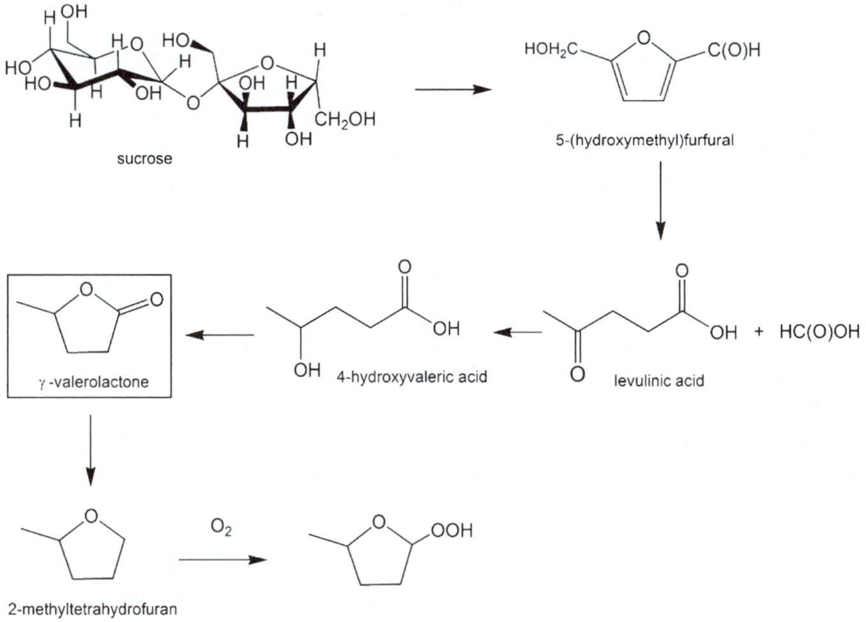

Scheme 8.14 Formation of γ-valerolactone using sucrose as a starting material and hydroperoxidation of 2-methyltetrahydrofuran formed on over-hydrogenation of γ-valerolactone.[47]

8.10.15 Principle 15: Establish full Mass-Balance for Process

This important point was made earlier (Section 7.2): reaction efficiency and avoidable waste generation can only be assessed if all materials used are fully accounted for. This will include, in addition to the reactants introduced into the process, solvents, catalysts and other reaction auxiliaries that have been used (including those used to extract, separate and purify the required product).

8.10.16 Principle 16: Measure Catalyst and Solvent Losses in Air and Aqueous Effluent

If a full mass balance reveals losses that cannot be accounted for, it is necessary to establish their fate (something on which industrial chemists can expend much effort). The analysis of waste streams (*e.g.* gaseous or aqueous effluents) may detect and quantify such losses. This can result in some unexpected observations. For example, the process for manufacturing the nylon intermediate, adipic acid, $HO_2CCH_2CH_2CH_2CH_2CO_2H$, by the oxidation of cyclohexanone oxime by nitric acid, was found to emit nitrous oxide, N_2O, a potent greenhouse gas.

8.10.17 Principle 17: Investigate Basic Thermochemistry

When carrying out syntheses on the small scale under laboratory conditions, the issue of thermochemistry is usually addressed in terms of experimental safety rather than any implications there may be for operating the reaction on the large scale. The concern is usually to find conditions under which the reaction can proceed at a reasonable rate. There are reactions in which an induction time occurs (*e.g.* in the formation of a Grignard reagent), which may provide a trap for the unwary. Some reactions are exothermic and, to avoid runaway conditions, require cooling when the reaction has begun.

These heating and cooling effects are achieved *via* **heat transfer** through the vessel walls. A key question is: will this be the same on the small and large scale? What would happen if a laboratory reaction was simply scaled up?

Let us consider a simple case on the lab scale using a round-bottomed flask, volume, V, of 1 L, with heat transfer occurring through glass of area, A, $= 5 \times 10^{-2}\,m^2$. The area to volume ratio is: $A/V = 0.05\,m^2\,L^{-1}$. A commercial reactor, with $V = 10\,m^3$, $(1\,m^3 = 10^3\,L)$ may have a heat transfer area, A, of $20\,m^2$. In this case the area to volume ratio is $A/V = 0.002\,m^2\,L^{-1}$. There is thus a 25-fold smaller ratio of surface area to volume for the larger reactor. The heat generated will be volume dependent; the heat transferred is area dependent. A smaller value of A/V has the consequence that heating up and cooling down is much slower on the larger scale, which may lead to altered conversions and selectivities, and will certainly affect productivities. Circumventing such constraints by process intensification, for example, by designing a reactor with much higher A/V ($>100\,m^2\,L^{-1}$), is discussed in Section 9.9.

8.10.18 Principle 18: Anticipate Heat and Mass Transfer Limitations

For a reactor or vessel, heat transferred is given by Q which is dependent on a number of variables (eqn 8.13):

$$Q = U \times A \times \Delta T \tag{8.13}$$

where:

U is the heat transfer coefficient [poor for glass, better for stainless steel (industrial reactors are often made from metal because of their better conductivity of heat, as well as for other reasons)];

A is the heat transfer area;

ΔT is the temperature difference between the reactor contents (where the heat may be generated) and the cooling medium (to which the heat of reaction is to be transferred).

In an industrial process, both U and A may change during the course of a reaction (A may increase as reactants are added or product is removed; U may change because of the deposition of a solid at the heat transfer surface). Providing a large ΔT can be expensive in energy terms.

The heat generated in a reaction is dependent on the heat of reaction (ΔH), the rate of reaction, r and the rate of addition of reactant.[xviii] The capacity for the removal of heat across the container wall must be greater than the heat generated (eqn 8.14); otherwise, a 'runaway' reaction may occur.

$$\mathrm{U} \times \mathrm{A} \times \Delta T \gg \Delta H \times \mathrm{r} \quad (8.14)$$

Mass transfer may occur across a range of interfaces:

- gas–liquid
- solid–liquid
- gas–solid
- liquid–liquid.

The consequences of poor mass transfer across liquid–liquid interfaces are also examined in the next chapter.

8.10.19 Principle 19: Consult a Chemical or Process Engineer

The chemical reaction, *per se*, is just one part of a multi-stage chemical process—though this is the primary and sometimes the only consideration of the university chemist. The industrial or process chemist will, on the other hand, be aware of the requirements to operate the process on the large scale within limits defined by process economics. Scale-up will be taken into account even in the exploratory stages of research and development. This may be very constraining to researchers because of the limited number of options that may be considered practical by those responsible for taking the process to the next stage. Therein lies the challenge to the chemist's ingenuity.

Synthetic chemists intending to address the practical issues of waste and energy minimisation and other related matters will not usually have been trained in process engineering principles such as the benefits or otherwise of using a 'batch' process or a 'continuous' process or the reasons for choosing different types of reactor. To fully engage with the principles of green chemistry, it is important to have an awareness of chemical and process engineering factors (*e.g.* the issues surrounding the scale-up of a laboratory reaction to a process operating on 10 000 tonnes per year). This is best done by consulting an expert (see the Bibliography to Chapter 9).

[xviii] The process chemist and process engineer will ask what might happen if there was a loss of cooling to a reactor (for which the cooling capacity was sufficient to remove the reaction exotherm). Which of the following two situations is likely to be of the greater concern: a process exhibiting fast kinetics in which there is no accumulation of unreacted feed, or one exhibiting slow kinetics in which there is a build-up of unreacted feed?

8.10.20 Principle 20. Consider Effect of Choice of Chemistry on Overall Process

It may well be that a new transformation has been found to produce a target chemical intermediate more efficiently, with better selectivity or better conversion, and with the consumption of less energy. However, if this is to be scaled up into a process the transformation must meet some additional criteria.

Working within the constraints laid down from a consideration of the overall process represents a major intellectual challenge to the industrial chemist. It may well have been the case that a novel transformation was demonstrated using an unusual laboratory reactor. How can the same chemistry be persuaded to occur under more conventional conditions as the time, cost and uncertainty of attempting to scale up a non-standard reactor to work on the industrial scale are usually prohibitive.

Has the laboratory reaction been undertaken using specially purified reactants? Will the laboratory-scale reaction proceed in the same way if technical grade materials for the large scale are used instead? Even if highly purified materials are acceptable, in terms of objective judgements about the benefits to waste minimisation, account would need to be taken of the fact that the waste process stream resulting from such purified products is generated at the works of the supplier. It won't have gone away.

8.10.21 Principle 21: Help Develop and Apply Sustainability Measures

We have seen the use made of the exergy concept in the comparison of three manufacturing routes to ethanol. Can such approaches (or the more sophisticated methods more recently developed) be applied for other transformations? The unified metrics developed by Andraos should be applied more widely.

8.10.22 Principle 22: Quantify and Minimise Use of Utilities

The usage of electricity, water and compressed gases is usually significant for processes operated on the industrial scale, though they are usually neglected in laboratory studies. It would be good practice to consider this even in the laboratory environment.

8.10.23 Principle 23: Recognise where Safety and Waste Minimisation are Incompatible

When a chemical process is modified in an attempt to improve energy and material utilisation or to reduce emissions, it is important always to consider the impact on other critical factors such as operator safety. As both are necessary conditions for an acceptable process, the issue then hinges on the relative costs associated with achieving these two outcomes. Deciding whether this makes economic sense can often be difficult.

In the discussion on atom efficiency, we saw that the use of oxygen in catalysed oxidations is generally superior to the use of stoichiometric oxidants such as $KMnO_4$. However, partial oxidations of hydrocarbons using dioxygen must be carried out in such a way as to avoid the formation of explosive mixtures. The prevention of conditions under which such mixtures might arise must be built in to the design and monitoring of the chemical process. Such considerations motivate research on the use of hydrogen peroxide as a liquid phase alternative to dioxygen in gas phase oxidations such as ethene oxidation to 1,2-epoxyethane (see p. 229 *et seq.*), in which explosive flammable mixtures oxygen and 1,2-epoxyethane may form. On the other hand, the use of hydrogen peroxide is not without its own hazards.

8.10.24 Principle 24: Monitor, Report and Minimise Laboratory Waste Emitted

The focus of waste production is usually on large-scale chemicals manufacture. Laboratories are sources too.

8.10.25 Some Final Thoughts on Green Chemistry and its Principles

Anastas and Warner are rightly fêted for bringing to the fore, in an accessible way, the importance of developing cleaner chemical processing. The principles (theirs and those of others) reiterate, concisely and usefully, the standards and objectives that should govern the way chemists work—even if many chemists, particularly in industry, believe this is how they have always worked. It is worthwhile, therefore, to keep the following in mind:

- The principles should be seen as aspirational ideals to be applied pragmatically and are not articles of faith.
- The principles are most usefully employed in making comparisons between options or choices. Comparisons require the proper definition of boundaries to ensure that like is always being compared with like. The principles should not lead to absolute statements about a process or a product.
- Any change based on the application of the principles needs to be examined critically for unforeseen consequences.[xix]
- The application of the principles may lead to cost savings. However, the 'greener' option cannot automatically be assumed to be less expensive than any other change being considered.
- Green chemistry may be necessary, but it is not always sufficient. The discussion leading up to this chapter has focused on many non-chemical factors that impinge on the acceptability or otherwise of a chemical process or product. The whole chain of supply may need to be checked (using life-cycle methods) to ensure that the problem has not simply been moved somewhere else without any net benefit to the environment.

[xix] This is the proper application of the Precautionary Principle; see footnote, p. 120.

- Whilst recognising the wider technological, social and economic issues relating to pollution prevention and the development of sustainable technologies, one should avoid suspending critical faculties because the ends may be worthy. Always remain sceptical of claims or assertions of an environmental benefit (from whatever quarter) made without available robust evidence. Find out the facts before coming to a judgement (and then remain open minded to new information and insight).
- Approach each problem with realism and rigour. All scientific laws still apply whether or not environmental issues are at stake. There are no 'green' versions of the laws of thermodynamics that avoid the inevitable formation of waste and there is no 'green' periodic table containing only non-toxic elements.
- Problems should be approached as far as possible using the scientific method which requires us to use (dispassionate) analysis, (fearless) objectivity, (merciless) self-criticism and (open-minded) scepticism.
- Be clear what you (and, just as importantly, others) mean when using words such as 'safe', 'environmentally benign' or 'non-toxic' as these beg many questions, including 'to what?'; 'under what circumstances?'; 'in what amounts?'; 'over what periods of time?'
- Understand what you and others mean when you use terms such as 'renewable', 'sustainable', 'waste' or 'clean'.
- Appreciate and anticipate the challenges associated with taking chemistry with potential for reducing risk, discovered in the laboratory, to its operation on the large-scale and thereby accelerate its introduction.

8.11 'GREEN' REACTION MEDIA

Any analysis of reaction efficiency of industrial chemical processes readily identifies solvent usage as a significant cost and solvent losses as a significant source of harmful emissions. The commercial production of sertraline hydrochloride (Scheme 10.5, p. 317) involves the formation of an imine intermediate (Scheme 8.15) *via* a dehydration using titanium tetrachloride. A modified process, in which the condensation between the precursor to the imine and methylamine, and its subsequent dehydration (without $TiCl_4$) is carried out in ethanol, leads to a marked reduction in solvent usage from 101 400 L (1000 kg sertraline hydrochloride)$^{-1}$ of five solvents to 24 000 L of two solvents.[48]

The search for processes that minimise solvent use or avoid it altogether has run alongside academic studies that have sought alternative reaction media that are more environmental benign.[49] This section, therefore, concentrates on three media that have been of considerable academic and technical interest: water, ionic liquids and supercritical fluids (for which Beckman received a Presidential Green Chemistry Challenge Award in 2002 and Eckert and Liotta in 2004). The technological benefits of this aspect of green chemistry, while very important, are difficult to assess fully as there are very few studies in which a proper comparison is made between 'conventional' and supposedly 'green' solvents. The reader is directed to

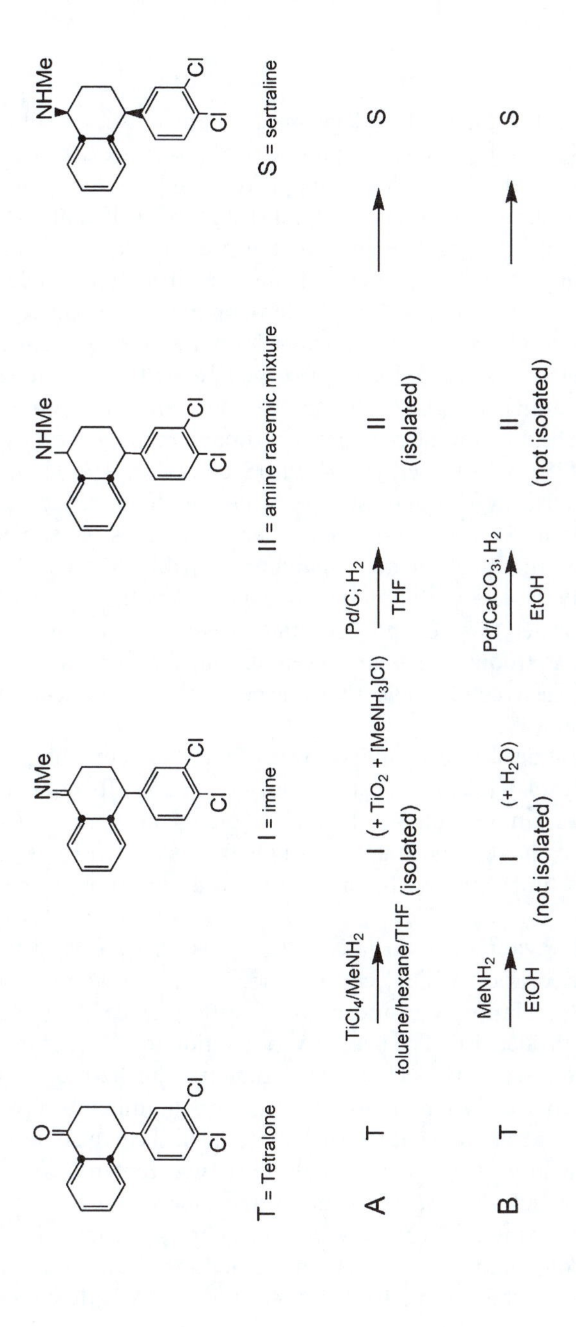

Scheme 8.15 Two routes to the pharmaceutical sertraline (S):[48] (A) operated commercially using a series of solvents; and (B) a modified process carried out in ethanol.

other texts (Bibliography) for details of the fundamental chemistry related to the topics discussed. The focus here is on the contribution or otherwise that the developments might make towards more sustainable chemical processing.

8.11.1 Water

When considering the physical and chemical characteristics of many hydrocarbon-based industrial products, it is perhaps not surprising that water is barely used as a reaction medium in conventional petrochemical processing. However, the nature of the components of biomass makes it more likely that water will be a more important medium, associated with the greater use of biotechnological chemical processing.[50a] The processing of biomass (including cellulose and lignocellulose and derived carbohydrates) by hot-compressed or supercritical water (in the absence of catalysts) has the advantage of not requiring dried feedstock as used in conventional processing. Its technological possibilities are under vigorous investigation. Treatment of glucose or fructose in compressed water at 200 °C under microwave heating (in the presence of a zirconium dioxide catalyst) leads to the formation of 5-(hydroxymethyl)furfural (Section 11.7.4) after five minutes reaction.[51] The solvency characteristics of water in the supercritical state are significantly different from those under ambient conditions of temperature and pressure, not least the much greater solubility of hydrocarbons. The ability to separate such material from solution by moving to below the critical temperature and pressure of water (374 °C and 218 atmospheres, respectively) is seen as a potential advantage (though there are severe technical problems associated with operating under such conditions). Other aspects of supercritical solvents are covered in Section 8.11.2.

Water has some decided advantages over organic solvents, being essentially non-toxic, readily available, low cost and non-flammable. Its poor solvency for organic substrates can be overcome by facilitating the delivery of water-soluble ionic reactants to an organic phase through the use, for example, of phase transfer catalysis (not explored further here, see Bibliography) or dispersants and emulsifiers.

Among the disadvantages associated with the use of water as a solvent is its high heat of evaporation which, coupled with a molecular mass of only 18 Daltons, makes distillation or evaporation of water very costly in energy terms per unit of weight processed. The heat of vaporisation for a kilogram of water is 2257 kJ compared with 413 $kJ\,kg^{-1}$ for toluene. Furthermore, wastewater can be difficult to treat to remove all organic contaminants (even though, paradoxically, it is a poor solvent for such materials). It can also, particularly in conjunction with ionic solutes, be highly corrosive towards metals used as materials of construction of reactors, tanks and pipework.

Its solvation characteristics, however, can bring about significant rate enhancements compared with equivalent reactions undertaken in organic media. For example, the data in Table 8.5 (from work by Rideout and Breslow in 1980[52a]) show that the Diels–Alder reaction (eqn 8.15) is 730-fold more rapid in water than in the anhydrous non-hydroxylic non-polar solvent, octane;

Table 8.5 The effect of reaction medium on the second-order rate constant and approximate relative rates of Diels–Alder reaction (eqn 8.15).[52a]

Medium	*Second order rate constant (10^{-5} M^{-1} s^{-1})*	*Approximate relative rate*
2,2,4-Trimethylpentane	5.9	1
Methanol	75.5	12.5
Water	4400±70	730
4.86 M aqueous LiCl	10 800	1400
10 mM aqueous β-cyclodextrin	10 900	1400

reaction in a solvent intermediate in polarity and hydrogen bonding, methanol, is 12.5-fold quicker.

cyclopentadiene + butenone → (8.15)

quadricyclane + dimethyl azodicarboxylate → (8.16)

More recently, Sharpless and his group reported[52b] unusual and marked accelerations when reactants that were insoluble in water were stirred together in aqueous suspension. For instance, a neat mixture of quadricyclane and dimethyl azodicarboxylate (eqn 8.16) gives no detectable reaction after 2 h at 0 °C. However, in the presence of water, yields of the addition product of 93% and 82% are formed after 1.5 h at 0 °C and 10 min at 23 °C, respectively. This reaction 'on' water rather than 'in' water has excited much discussion and is still not fully understood. (Added in proof: However, see ref. 52c.) Bearing in mind that natural waters have a thin surface organic layer (*e.g.* the sea surface microlayer[52d]), reactions of such systems may have wider relevance than simply chemical laboratory synthesis.

$$CH_3CH{=}CH_2 + H_2 + CO \rightarrow n\text{-}C_3H_7CHO(98\%) + CH_3CH(CH_3)CHO \qquad (8.17)$$

The homogeneously catalysed hydroformylation reaction (eqn 8.17) (also known as the oxo process) is very important technically. A process, developed by Ruhrchemie and Rhône-Poulenc, that uses water as reaction medium has been operated on the large industrial scale since 1984. Advantage is taken of the reaction exotherm to generate steam. The rhodium catalyst is based on $HRh(CO)L_3$ which, for conventional ligands, L [*e.g.* $P(C_6H_5)_3$], is insoluble in water. However, the use of the functionalised triphenylphosphine ligand, $L = P(3\text{-}C_6H_4SO_3Na)_3$, makes the complex water-soluble.[53] The product from the hydroformylation phase-separates from the aqueous medium and a negligible proportion of the rhodium catalyst is lost to the separated product. The process is operated on the large scale (600 000 t y^{-1}), representing about 10% world capacity.

8.11.2 Supercritical Fluids

A supercritical fluid (SCF)[50b,54] is a material above both its critical pressure (P_c) and critical temperature (T_c) (Figures 8.9 and 8.10). SCFs have densities nearer to those of liquids and viscosities nearer to those of gases. Because of the latter, they can display better diffusivity which can lead to better heat and mass transfer—both important technical characteristics.

An advantageous flexibility of a fluid in the supercritical state is that its properties (including solvency—with some marked differences between that of the SCF and the liquid at ambient conditions) can be varied in a controlled manner by changes to its temperature and pressure. In addition, it is possible to separate a solute dissolved in, or mixed with, a SCF by reducing T or P to below the critical values. This makes the separation and recycle of the reaction medium, in principle at least, easier than with classical solvents. However, this flexibility comes with a penalty and the evident benefits may be offset by the energy costs associated with fluid compression and recompression and the capital costs associated with the need for high pressure equipment.

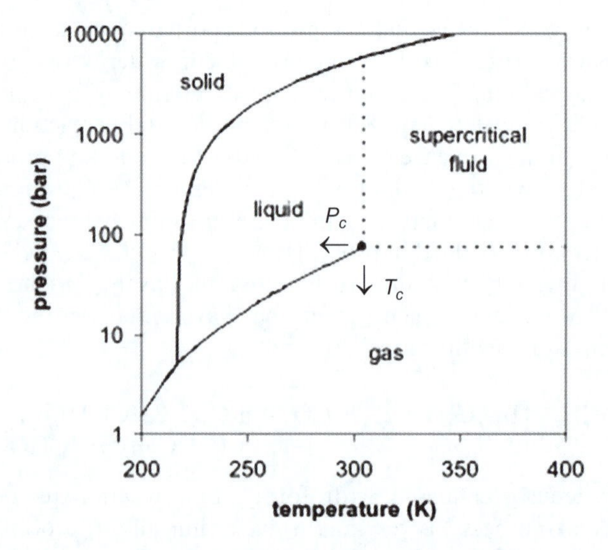

Figure 8.9 Pressure *vs.* temperature phase diagram of a fluid showing the supercritical region and its critical temperature, T_c, and critical pressure, P_c.

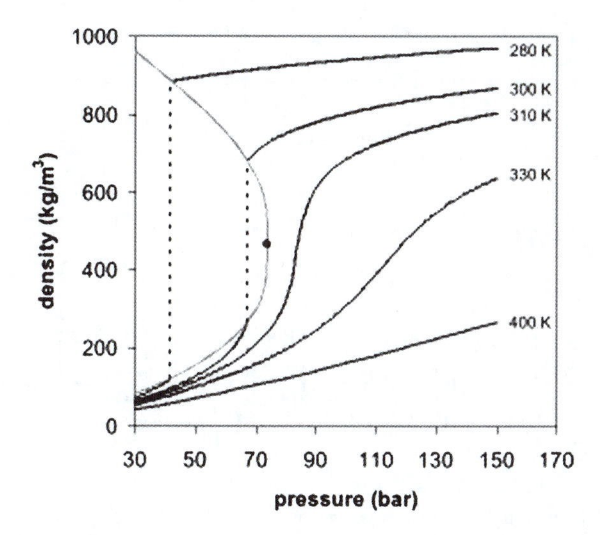

Figure 8.10 Density *vs.* pressure phase diagram of a fluid at a range of temperatures.

Particular interest has focused on supercritical carbon dioxide as a reaction medium. Supercritical CO_2 has a reasonably accessible critical temperature of 31.1 °C and critical pressure of 73.8 bar. It is non-toxic (though an asphyxiant), non-flammable and readily available. It is currently used (in conjunction with high-pressure water) to decaffeinate coffee for which, clearly, the relatively high capital cost of plant is no deterrent (with the added benefit that significant learning will have taken place on engineering large-scale processes that may be applied or adapted for use in other chemicals processing). These advantages have led to extensive investigations into the use of CO_2 as a reaction solvent.

Unfortunately, supercritical CO_2 is generally regarded, on its own, as a poor solvent; in fact, its 'antisolvent' characteristics (*i.e.* where it may be added to a medium to reduce substrate solubility) may equally be of interest. It is, however, reactive towards amines (and other basic materials) that may restrict its applicability to some substrates without functional group protection. Improvements in solvency can be brought about by the addition of solvent modifiers, though these may add operational complexity arising from their addition, removal, recovery and storage. The phase behaviour of complex mixtures involving supercritical fluids is an understudied area of research that may pose challenges to the development of particular processes for which special programmes of study will be needed fully to understand the chemistry occurring in a reactor or a process.

$$\text{isophorone} \xrightarrow[H_2]{\text{Pd/support}} \text{trimethylcyclohexanone} + \text{trimethylcyclohexanol} + \text{trimethylcyclohexane} \quad (8.18)$$

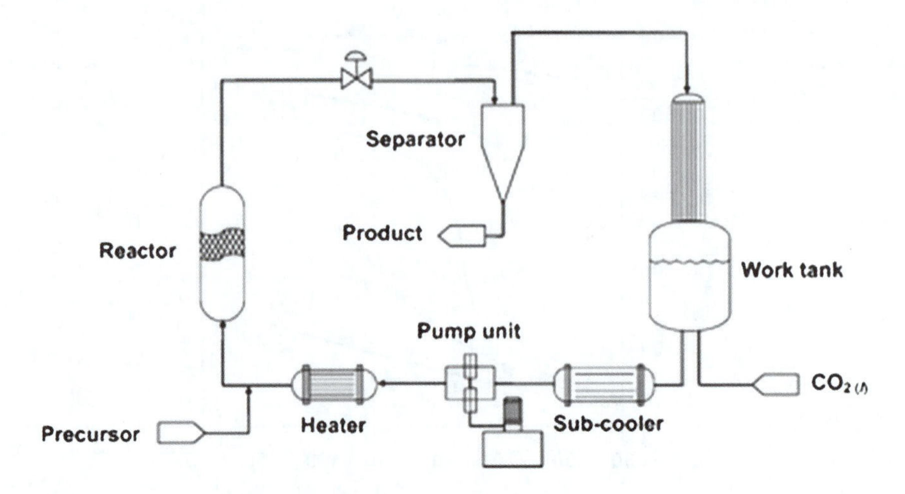

Figure 8.11 Flow diagram for a process for the catalysed hydrogenation of isophorone (eqn 8.18) using supercritical carbon dioxide as the process medium.[54a] Reproduced by permission of the Royal Society of Chemistry.

While this is so, some important developments have been described.[54] Working with the chemical company, Thomas Swann & Co., a research group at the University of Nottingham led by Martyn Poliakoff discovered and developed a novel process for the catalysed hydrogenation of isophorone (eqn 8.18).[54a] The process (Figures 8.11 and 8.12) involves the flow of substrate, carbon dioxide and dihydrogen (H_2 : substrate mole ratio; 1.7 : 1) over a 2% Pd/support catalyst in isothermal operation at 104–116 °C giving a product, trimethylcyclohexanone, of 99.4% purity and containing $<0.1\%$ isophorone and 0.3% trimethylcyclohexane with other by-products in amounts less than obtained from more conventional processes. The product from the reactor was sufficiently pure to use directly (other than to remove dissolved CO_2 by means of a water wash). Despite these additional complications, the opportunity was sufficient for Thomas Swann to build and operate a plant on the 1000 $t\,y^{-1}$ scale, with the additional benefit that they now have the experience gained from the development and operation of this process so that they may now be able to apply the technology in the production of other materials.

Another application that has developed beyond the research laboratory is the polymerisation of fluorinated monomers. DuPont has invested an initial $40 million in a production unit for the manufacture of the fluorinated polymer, FEP, (a co-polymer of tetrafluoroethylene and hexafluoropropylene) using $scCO_2$ as the continuous phase giving a product with improved properties. The process avoids by-products arising from chain transfer that occurs in the conventional medium, an aqueous surfactant mixture or the chlorofluorocarbon, 1,1,2-trichlorotrifluoroethane.[55]

Figure 8.12 Reactor at Thomas Swann & Co. plc for catalysed hydrogenations using supercritical carbon dioxide as process medium.[54a] Reproduced by permission of the Royal Society of Chemistry.

8.11.3 Ionic Liquids

The industrial use of molecular solvents poses a serious technical problem to minimise losses to the environment during the processes of separation and purification of a desired solute and during solvent recovery. Ionic liquids have attracted the interest of those looking for less waste-emitting processes. Ionic liquids are molten salts with, by convention, a melting point (mp) $<100\,°C$—very low in comparison with classical salts such as sodium chloride (mp 803 °C). Being salts, most ionic liquids have negligible vapour pressure and, as a consequence, volatile materials dissolved in them can be readily separated. They are therefore attractive when compared with conventional molecular solvents that, to varying extents, are co-volatile with products of reaction.

Ionic liquids also have an ability to dissolve a wide range of solutes and their use, in the laboratory at least, has been demonstrated in a very extensive series of preparative reactions.[56] Most ionic liquids are very expensive, costing

£1–20 g^{-1}, with a few available on the tonne scale at £20–30 kg^{-1}. This may be compared with typical solvents used industrially such as toluene, methyl ethyl ketone or dichloromethane, which cost in the range £0.60–1 kg^{-1}.

Because of their unique characteristics, ionic liquids are nevertheless under active study technologically and not just as reaction media. It is most likely that ionic liquids will find the greatest use in high-value applications (that can bear their high cost), especially those in which involatility is a particular advantage. Ionic liquids will find particular application in sealed devices such as batteries, capacitors and solar cells—especially where ionic liquid recovery is not necessary (except until ultimate disposal).[57]

Ionic liquids are made up of cations, $[Ct]^+$, and anions, $[An]^-$, that are in the main polyatomic with diffuse charges so that the lattice energies of the salts, [Ct][An], are low, contributing to their relatively low melting points (some as low as −50 °C). $[Ct]^+$ is usually quaternary ammonium, phosphonium and tertiary sulfonium—particularly quaternised heterocyclic amines such as pyridinium and imidazolium (see Figure 8.13). Typical anions $[An]^-$ are also shown.

There are, however, some examples in which an ionic liquid plays an important part in industrial chemicals processing. For example, BASF has manufactured dialkoxyphenylphosphines, $(RO)_2PPh$, as photoinitiators, according to eqn (8.19) (carried out in xylene):

$$PhPCl_2 + 2\,EtOH + 2\,Et_3N \rightarrow PhP(OEt)_2 + 2\,[Et_3NH]Cl \qquad (8.19)$$

$$PhPCl_2 + 2\,EtOH + 2\,mim \rightarrow PhP(OEt)_2 + 2\,[mimH]Cl \qquad (8.20)$$

The $[Et_3NH]Cl$ co-product, the result of scavenging HCl, is a solid and is difficult and slow to separate, making the process inefficient. By replacing the amine scavenger with 1-methylimidazole (mim)[58] (eqn 8.20), a liquid salt is formed (at the temperature of the reaction, *ca.* 70 °C). The [mimH]Cl separates as a second liquid phase enabling much easier separation and recovery. The salt can then be treated with base to regenerate the scavenger for reuse. An additional benefit is the catalytic role played by the scavenger which, with the use of a jet reactor, has increased process productivity by a factor of *ca.* 80 000. BASF has patented this process (the 'BASIL' Process[58]) and operate it on the large scale (Figure 8.14). In a separate large-scale application, ionic liquids based on choline chloride, $[Me_3NCH_2CH_2OH]Cl$,[59] which is relatively cheap, are (with other components) providing important technical benefits in the industrial electropolishing of stainless steel instead of the conventional mixed sulfuric/phosphoric acids.

It has been known since 1934 that ionic liquids will dissolve cellulose. As discussed in Chapter 11, cellulose is a component of lignin and both are difficult to process. A key stage in the processing of cellulose into useful chemicals (as opposed simply to generate energy by burning it) is its dissolution. Very few media are able to do this and ionic liquids appear to do so by breaking the multiple hydrogen bonds that hold the polysaccharide chains together.

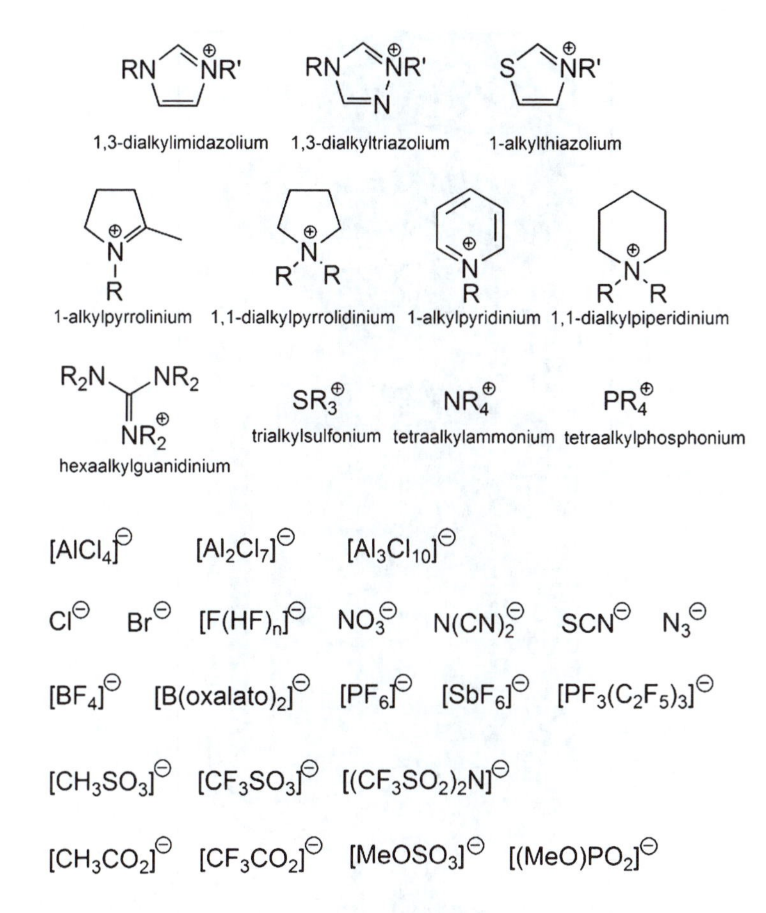

Figure 8.13 A selection of typical cations, $[Ct]^+$, and anions, $[An]^-$, that may be combined to form ionic liquids.

Attention has particularly focused on ionic liquids with hydrogen-bonding anions such as chloride, formate and dicyanamide. More recently, 5-(hydroxymethyl)furfural (5-HMF)—a biomass-derived 'platform' chemical (Section 11.6.10)—was reported[60a] to be formed in 55% yield from cellulose in the ionic liquid [emim]Cl (1,3-ethylmethylimidazolium chloride) in the presence of a mixed catalyst comprising $CuCl_2$ and $CrCl_2$. 5-HMF is also produced from untreated lignocellulose biomass in a single step in a related system.[60b] The challenge, of course, is to take such fascinating laboratory observations and establish whether a commercially viable process can be developed from them (or whether a compressed water-based process to such compounds might be more economic despite having to operate at very much higher temperatures).

There is a further question regarding the degree to which ionic liquids may be said to contribute towards more sustainable processing. They are complex compounds in their own right and, even if it were found possible cost-effectively to recover them after use and to recycle them, one is still forced to ask whether

Figure 8.14 Reactor employed by BASF in the BASIL process for the manufacture of dialkoxyphenylphosphines. Photograph courtesy of BASF.

from a life-cycle perspective (cradle-to-grave, using objectively comparable metrics) ionic liquids are more or less waste-producing and energy- or material-consuming than classical molecular solvents. This comparison has been carried out for a limited series of ionic liquids.[61a] A second comparison, using conventional molecular solvents, the ionic liquid [hmim]BF_4 (1,3-hexylmethylimidazolium tetrafluoroborate) and a citric acid/*N,N′*-dimethylurea melt as the medium for a Diels–Alder reaction came to similar conclusions.[61b] The results from these studies force the question whether or not the widespread use of ionic liquids, particularly those currently subject to the most intense scrutiny as reaction media, will bring any general environmental benefit. This is not to say that benefits may not arise from the use of particular ionic liquids in certain applications that outweigh other factors. The use of ionic liquids obtained from renewable precursors may well lead to such developments.

Ionic liquids remain a hot area of research that is capable of regularly producing excitement and surprises—some of which are relevant to the way they might contribute, indirectly, to more sustainable technologies. For example, the development of high-temperature superconducting materials would (among other things) reduce the significant ohmic losses associated with electricity transmission around the grid. A recent report[62a] describes the use of a conventional ionic liquid, *N*,*N*,-diethyl-*N*-(2-methoxyethyl)-*N*-methyl-ammonium bis(trifluoromethanesulfonyl)imide, as a gate dielectric to switch on the superconducting state of a thin atomically flat film of the layered insulator, zirconium nitride chloride, when subject to an electric field. An expert commentary[62b] suggests that this technical breakthrough opens up new directions in superconductivity research.

There should now be no doubt about the major sustainability challenges we face, their complexity, and the role of science and technology in meeting them. Chemistry will have a continuing role to play:

- to explore and understand fundamental phenomena
- to aid in the solution of current and future challenges as we seek to reconcile humankind's needs with the consequences to the environment.

In this, at least, we are not '*facing the limits of knowledge in the twilight of the scientific age*', as suggested by John Horgan.[63,xx]

REFERENCES

1. (a) R. Noyori, *Nat. Chem.*, 2009, **1**, 5; (b) H. B. Gray, *Nat. Chem.*, 2009, **1**, 7; (c) M. A. Johnson, *Nat. Chem.*, 2009, **1**, 8; (d) B. Imperiali, *Nat. Chem.*, 2009, **1**, 9; (e) G. M. Hieftje, *Nat. Chem.*, 2009, **1**, 10; (f) J. H. Clark, *Nat. Chem.*, 2009, **1**, 12; (g) A. Müller, *Nat. Chem.*, 2009, **1**, 13; (h) F. A. Stoddart, *Nat. Chem.*, 2009, **1**, 14.
2. (a) S. L. Miller, *Science*, 1953, **117**, 528; (b) S. L. Miller and H. C. Urey, *Science*, 1959, **130**, 245.
3. L. E. Orgel, *Orig. Life Evol. Biosph.*, 2004, **34**, 361.
4. C. F. Chyba, *Nature*, 1997, **389**, 234.
5. (a) M. W. Powner, B. Gerland and J. D. Sutherland, *Nature*, 2009, **459**, 239; (b) J. W. Szostak, *Nature*, 2009, **459**, 171.
6. D. G. Blackmond, *Angew. Chem., Int. Ed.*, 2009, **48**, 386.
7. (a) N. Lane, J. F. Allen and W. Martin, *BioEssays*, 2010, **32**, 271; (b) P. Mitchell, *Nature*, 1961, **191**, 144.
8. M. Toussant, *Chem. Eng. News*, 2009, 87, 14 Sept., p. 3.
9. (a) B. M. Trost and G. Dong, *Nature*, 2008, **456**, 485; (b) A. K. Miller, *Angew. Chem., Int. Ed.*, 2009, **48**, 3221.

[xx] For a rebuttal, read John Maddox's book, *What Remains to be Discovered* (see Bibliography).

10. (a) G. R. Pettit, C. L. Herald, D. L. Doubek and D. L. Herald, *J. Am. Chem. Soc.*, 1982, **104**, 6846; (b) G. R. Pettit, J. F. Day, J. L. Hartwell and H. B. Wood, *Nature*, 1970, **227**, 962.
11. S. L. Buchwald and C. Bolm, *Angew. Chem., Int. Ed.*, 2009, **48**, 5586.
12. J. Andraos, *Org. Process Res. Dev.*, 2009, **13**, 161.
13. S. S. Chandran, J. Yi, K. M. Draths, R. Von Daeniken, W. Weber and J. W. Frost, *Biotechnol. Prog.*, 2003, **19**, 808.
14. P. R. Srinivasan, H. T. Shigeura, M. Sprecher, D. R. Sprinson and B. R. Davis, *J. Biol. Chem.*, 1956, **220**, 477.
15. (a) M. Karpf and R. Trussardi, *Angew. Chem., Int. Ed.*, 2009, **48**, 5760; (b) R. Trussardi, *World Patent* WO2009037137, 26 March 2009.
16. A. Nayar, *Nature*, 2009, **460**, 321.
17. D. S. Goldberg, T. Takahashi and A. L. Slagle, *Proc. Natl. Acad. Sci., U. S. A.*, 2008, **105**, 9920.
18. P. G. Jessop, S. Trakhtenberg and J. Warner, in *Innovations in Industrial and Engineering Chemistry: A Century of Achievements and Prospects for the New Millennium*, ed. W. H. Flank, M. A. Abraham and M. A. Matthews, *ACS Symp. Ser.*, 2009, **vol. 1000**, p. 401.
19. (a) Y. Chauvin, *Angew. Chem., Int. Ed.*, 2006, **45**, 3740; (b) R. R. Schrock, *Angew. Chem., Int. Ed.*, 2006, **45**, 3748; (c) R. H. Grubbs, *Angew. Chem., Int. Ed.*, 2006, **45**, 3760; (d) R. H. Grubbs, D. D. Carr, C. Hoppin and P. L. Burk, *J. Am. Chem. Soc.*, 1976, **98**, 3478; (e) T. J. Katz and R. Rothchild, *J. Am. Chem. Soc.*, 1976, **98**, 2519; (f) J.-L. Hérisson and Y. Chauvin, *Makromol. Chem.*, 1970, **141**, 161; (g) T. Nicola, M. Brenner, K. Donsbach and P. Kreye, *Org. Process Res. Dev.*, 2005, **9**, 513; (h) J.-M. Basset, C. Copéret, D. Soulivong, M. Taoufik and J. T. Cazat, *Acc. Chem. Res.*, 2010, **43**, 323.
20. (a) H. Aldersey-Williams, *The Most Beautiful Molecule: An Adventure in Chemistry*, Aurum Press, London, 1995; (b) H. W. Kroto, J. R. Heath, S. C. O'Brien, R. F. Curl and R. E. Smalley, *Nature*, 1985, **318**, 162; (c) D. E. H. Jones, *New. Sci.*, 1966, **32**, 245; (d) E. A. Rohlfing, D. M. Cox and A. Kaldor, *J. Chem. Phys.*, 1984, **81**, 3322; (e) R. Taylor, J. P. Hare, A. K. Abdul-Sada and H. W. Kroto, *J. Chem. Soc., Chem. Commun.*, 1990, 1423; (f) W. Krätschmer, L. D. Lamb, K. Fostiropoulos and D. R. Huffman, *Nature*, 1990, **347**, 354; (g) J. M. Hawkins, A. Meyer, T. A. Lewis, S. Loren and F. J. Hollander, *Science*, 1991, **252**, 312; (h) P. R. Buseck, S. J. Tsipursky and R. Hettich, *Science*, 1992, **257**, 215; (i) P. J. Krusic, E. Wasserman, B. A. Parkinson, B. Mallone, E. R. Holler Jr., P. N. Keizer, J. R. Morton and K. F. Preston, *J. Amer. Chem., Soc.*, 1991, **113**, 6274; (j) E. Nakamura and H. Isobe, *Acc. Chem. Res.*, 2003, **36**, 807; (k) S. Marchesan, T. Da Ros, G. Spalluto, J. Balzarini and M. Prato, *Bioorg. Med. Chem. Lett.*, 2005, **15**, 3615; (l) T. Y. Zakharian, A. Seryshev, B. Sitharaman, B. E. Gilbert, V. Knight and L. J. Wilson, *J. Amer. Chem., Soc.*, 2005, **127**, 12508; (m) E. Osawa, *Kagaku*, 1970, **25**, 854; *Chem. Abstr.*, 1971, **74**, 75698v.
21. J. Pearsall and B. Trumble, *The Oxford English Reference Dictionary*, Oxford University Press, Oxford, 2nd edn, 1996, p. 614.

22. P. T. Anastas, *ChemSusChem*, 2009, **2**, 391.
23. (a) E. J. Woodhouse and S. Breyman, *Sci. Technol. Human Values*, 2005, **30**, 199; (b) J. A. Linthorst, *Found. Chem.*, 2010, **12**, 55.
24. A. Albini and M. Fagnoni, *ChemSusChem*, 2008, **1**, 63.
25. C. Christ, *Production-Integrated Environmental Protection and Waste Management in the Chemical Industry*, Wiley-VCH, Weinheim, 1999.
26. R. A. Sheldon, *C. R. Acad. Sci., Ser. IIC: Chim.*, 2000, **3**, 541.
27. A. E. Marteel, J. A. Davies, W. W. Olson and M. A. Abraham, *Annu. Rev. Environ. Resour.*, 2003, **28**, 401.
28. M. Lancaster, *Green Chemistry: An Introductory Text*, Royal Society of Chemistry, 2002.
29. P. T. Anastas and J. C. Warner, *Green Chemistry: Theory and Practice*, Oxford University Press, Oxford, 1998.
30. N. Winterton, *Green Chem.*, 2001, **3**, G73-5 [This article is reproduced in full in Appendix 3.]
31. W. McDonough, M. Braungart, P. T. Anastas and J. B. Zimmerman, *Environ. Sci. Technol.*, 2003, **37**, 434A.
32. (a) S. L. Y. Tang, R. L. Smith and M. Poliakoff, *Green Chem.*, 2005, **7**, 761; (b) S. Tang, R. Bourne, R. Smith and M. Poliakoff, *Green Chem.*, 2008, **10**, 268.
33. S. E. Manahan, *Green Chemistry and the Ten Commandments of Sustainability*, ChemChar Research Inc., Columbia, MO, 2nd edn, 2005.
34. G. D. Bennett, *Perspect. Sci. Christian Faith*, 2008, **60**, 16.
35. J. L. Tucker, *Org. Process. Res. Dev.*, 2010, **14**, 328.
36. (a) T. J. Collins, *J. Chem. Educ.*, 1995, **72**, 965; (b) P. T. Anastas, I. J. Levy and K. E. Parent, Eds., *ACS Symp. Ser.*, 1011 (*Green Chemistry Education: Changing the Course of Chemistry*), 2009.
37. (a) T. Hartung, *Nature*, 2009, **460**, 208; (b) N. Gilbert, *Nature*, 2009, **460**, 1065; (c) T. Hartung and C. Rovida, *Nature*, 2009, **460**, 1080.
38. J. Emsley, *Molecules at an Exhibition: Portraits of Intriguing Materials in Everyday Life*, Oxford University Press, Oxford, 1998, p. 165.
39. G. W. V. Cave, C. L. Raston and J. L. Scott, *Chem. Commun.*, 2001, 2159.
40. O. Dolotko, J. W. Wiench, K. W. Dennis, V. K. Pecharsky and V. P. Balema, *New J. Chem.*, 2010, **34**, 25.
41. K. C. Nicolaou, D. J. Edmonds and P. G. Bulger, *Angew. Chem., Int. Ed.*, 2006, **45**, 7134.
42. J. W. Cornforth, *Aldrichimica. Acta*, 1994, **27**, 71.
43. (a) I. S. Young and P. S. Baran, *Nat. Chem.*, 2009, **1**, 193; (b) P. S. Baran, T. J. Maimone and J. M. Richter, *Nature*, 2007, **446**, 404.
44. J. S. Carey, D. Laffan, C. Thomson and M. T. Williams, *Org. Biomol. Chem.*, 2006, **4**, 2337.
45. J. Andraos, *Org. Process Res. Dev.*, 2005, **9**, 149.
46. J.-P. Lange, *ChemSusChem*, 2009, **2**, 587.
47. V. Fábos, G. Koczó, H. Mehdi, L. Boda and I. T. Horváth, *Energy Environ. Sci.*, 2009, **2**, 767.

48. G. P. Taber, D. M. Pfistere and J. C. Colberg, *Org. Process Res. Dev.*, 2004, **8**, 385.
49. (a) R. A. Sheldon, *Green Chem.*, 2005, **7**, 267; (b) C. A. Eckert, C. L. Liotta, D. Bush, J. S. Brown and J. P. Hallett, *J. Phys. Chem. B*, 2004, **108**, 18108.
50. M. Lancaster, Organic solvents: environmentally benign solutions, in *Green Chemistry: An Introductory Text*, Royal Society of Chemistry, 2002, ch. 5. (a) p. 149; (b) p. 135.
51. X. Qi, M. Watanabe, T. M. Aida and R. L. Smith Jr., *Catal. Commun.*, 2008, **9**, 2244.
52. (a) D. C. Rideout and R. Breslow, *J. Am. Chem. Soc.*, 1980, **102**, 7816; (b) S. Narayan, J. Muldoon, M. G. Finn, V. V. Fokin, H. C. Kolb and K. B. Sharpless, *Angew. Chem., Int. Ed.*, 2005, **44**, 3275; (c) J. K. Beattie, C. S. P. McErlean and C. B. W. Phippen, *Chem. Eur. J.*, 2010, **16**, 8972; (d) R. Chester, *Marine Geochemistry*, Blackwell Science, Oxford, 2000, ch. 4.
53. C. W. Kohlpaintner, R. W. Fischer and B. Cornils, *Appl. Catal., A*, 2001, **221**, 219.
54. (a) P. Licence, J. Ke, M. Sokolova, S. K. Ross and M. Poliakoff, *Green Chem.*, 2003, **5**, 99; (b) J. M. DeSimone and W. Tumas, *Green Chemistry Using Liquid and Supercritical Carbon Dioxide*, Oxford University Press, New York, 2004; (c) P. N. Gooden, R. A. Bourne, A. J. Parrott, H. S. Bevinakatti, D. J. Irvine and M. Poliakoff, *Org. Process Res. Dev.*, 2010, **14**, 411.
55. C. D. Wood, J. C. Yarbrough, G. Roberts and J. M. DeSimone, in *Supercritical Carbon Dioxide in Polymer Reaction Engineering*, ed. M. F. Kemmere and T. Meyer, Wiley-VCH, Weinheim, 2005, ch. 9, p. 189.
56. M. Maase, in *Ionic Liquids in Synthesis*, ed. P. Wasserscheid and T. Welton, Wiley-VCH, Weinheim, 2nd edn, 2007, ch. 9, p. 663.
57. (a) Y. Bai, Y. Cao, J. Zhang, M. Wang, R. Li, P. Wang, S. M. Zakeeruddin and M. Grätzel, *Nat. Mat.*, 2008, **7**, 626; (b) S. M. Zakeeruddin and M. Grätzel, *Adv. Funct. Mater.*, 2009, **19**, 2187.
58. M. Maase and O. Huttenloch, BASF, *World Patent* 2005/061416; BASF, *US Patent* 2005/0020857, 27 Jan 2005.
59. A. P. Abbott, G. Capper, K. J. McKenzie, A. Glidle and K. S. Ryder, *Phys. Chem. Chem. Phys.*, 2006, **8**, 4214.
60. (a) Y. Su, H. M. Brown, X. Huang, X. Zhou, J. E. Amonette and Z. C. Zhang, *Appl. Catal., A*, 2009, **361**, 117; (b) J. B. Binder and R. T. Raines, *J. Am. Chem. Soc.*, 2009, **131**, 1979.
61. (a) Y. Zhang, B. R. Bakshi and E. S. Demessie, *Environ. Sci. Technol.*, 2008, **42**, 1724; (b) D. Reinhardt, F. Ilgen, D. Kralisch, B. König and G. Kriesel, *Green Chem.*, 2008, **10**, 1170.
62. (a) J. T. Ye, S. Inoue, K. Kobayashi, Y. Kasahara, H. T. Yuan, H. Shimotani and Y. Iwasa, *Nat. Mat.*, 2010, **9**, 125; (b) K. Prassides, *Nat. Mat.*, 2010, **9**, 96.
63. J. Horgan, *The End of Science: Facing the Limits of Knowledge in the Twilight of the Scientific Age*, Abacus, London, 1996.

BIBLIOGRAPHY[xxi]

J. Maddox, *What Remains to be Discovered: Mapping the Secrets of the Universe, the Origins of Life and the Future of the Human Race*, Macmillan Publishers, London, 1998.

L. E. Orgel, *The Origins of Life: Molecules and Natural Selection*, Chapman and Hall, London, 1973.

S. L. Miller and L. E. Orgel, *The Origins of Life on the Earth*, Prentice-Hall, Englewood Cliffs, NJ, 1974.

E. J. Corey and X.-M. Cheng, *The Logic of Chemical Synthesis*, John Wiley & Sons, New York, 1989.

V. Balzani, A. Credi and M. Venturi, *Molecular Devices and Machines: A Journey into the Nanoworld*, Wiley-VCH, Weinheim, 2008.

J. Clark and D. Macquarrie, *Handbook of Green Chemistry and Technology*, Blackwells Science, 2002.

A.S. Matlack, *Introduction to Green Chemistry*, Marcel Dekker, New York, 2001.

R. A. Sheldon, I. Arends and L. Handfeld, *Green Chemistry and Catalysis*, Wiley-VCH, Weinheim, 2007.

M. Lancaster, *Green Chemistry: An Introductory Text*, Royal Society of Chemistry, Cambridge, 2002.

P. T. Anastas and J. C. Warner, *Green Chemistry: Theory and Practice*, Oxford University Press, Oxford, 1998.

J. W. Moore and R. G. Pearson, *Kinetics and Mechanism: A Study of Homogeneous Chemical Reactions*, John Wiley & Sons, New York, 3rd edn, 1981.

F. M. Kerton, *Alternative Solvents for Green Chemistry*, Royal Society of Chemistry, Cambridge, 2009.

C. M. Starks and C. L. Liotta, *Phase Transfer Catalysis: Principles and Techniques*, Academic Press, New York, 1979.

P. Wasserscheid and T. Welton, *Ionic Liquids in Synthesis*, Wiley-VCH, Weinheim, 2nd edn, 2007.

WEBLIOGRAPHY

1. www.nature.com/nchem/index.html
2. http://en.wikipedia.org/wiki/Abiogenesis
3. http://nobelprize.org
4. http://nobelprize.org/nobel_prizes/chemistry/laureates/2005/info.html
5. (a) www.epa.gov/greenchemistry/pubs/docs/award_recipients_1996_2010.pdf
   (b) www.epa.gov/greenchemistry/pubs/pgcc/presgcc.html
6. (a) http://en.wikipedia.org/wiki/Paclitaxel
   (b) http://en.wikipedia.org/wiki/Paclitaxel_total_synthesis

[xxi] These texts supplement those already cited as references.

(c) http://en.wikipedia.org/wiki/Holton_Taxol_total_synthesis
(d) http://en.wikipedia.org/wiki/Nicolaou_Taxol_total_synthesis
7. www.epa.gov/greenchemistry/
8. www.greenchemistrynetwork.org
9. www.rsc.org/Publishing/Journals/gc/index.asp
10. www.hse.gov.uk/coshh/
11. http://news.bbc.co.uk/1/hi/england/bradford/7779079.stm
12. www.hse.gov.uk/press/2003/c03014.htm
13. www.vega.org.uk

All the web pages listed in this Webliography were accessed in May 2010.

CHAPTER 9

Chemicals Processing

'Commit your blunders on the small scale and make your profits[i] on the large scale: this should guide everyone who enters a new chemical enterprise, even if it taxes the patience of some who cannot conceive that one single, apparently minor, detail[ii] in a chemical process may upset all the good points and lead to ruin.'

L. H. Baekeland, Perkin Medal Address, 1916

Chemistry for more sustainable technologies must include an awareness of basic process chemistry and a familiarity with the key concepts and concerns of chemical and reactor engineering. In addition, an appreciation is needed of the degree of technological and infrastructure integration that currently exists in large-scale chemicals processing and the challenges that must be recognised and addressed in moving such systems to those based more on renewable feedstocks. To go deeper than the introductory treatment given here, you will need to find a friendly chemical engineer to consult along with some standard texts and articles (see Bibliography), and the 12 principles of green engineering[1] reproduced in Table 9.1. These have parallels with Anastas and Warner's original 12 principles, focusing on the prevention of waste, the use of renewables, energy and material efficiency (including recycling), and are generally considered alongside

[i] In the context of this book, the idea of profit should be taken to include all the benefits of a process or a technology, social, environmental as well as economic.

[ii] This notion arises from the very simple idea that, in a complex system in which a series of sequential steps must work with overall maximum efficiency, the combination of the efficiencies of individual steps is multiplicative and not additive. All it takes for low (or even zero) efficiency is for the value for one step to be low or zero, however good those for all the other steps are.

Chemistry for Sustainable Technologies: A Foundation
By Neil Winterton

Published by the Royal Society of Chemistry, www.rsc.org

Table 9.1 Twelve principles of green engineering.[1]

1. All material and energy inputs and outputs to be as inherently nonhazardous as possible
2. Better to prevent waste than to treat it or clean it up after it is formed
3. Design separation and purification operations to minimise energy consumption and materials use
4. Design products, processes and systems to maximise mass, energy, space and time efficiency
5. Products, processes and systems should be 'output pulled' rather than 'input pushed'
6. See embedded entropy and complexity as an investment when making design choices on recycle, reuse
7. Durability rather than immortality should be the targeted design goal
8. Unnecessary capacity or capability should be considered a design flaw
9. Minimise material diversity in multicomponent products to enable economic disassembly
10. Integrate and interconnect (close loop) design with available energy and material flows
11. Design products, processes and systems for performance in a commercial 'afterlife'
12. Use renewable material and energy inputs

the 12 green chemistry principles. The engineering principles introduce some additional ideas such as the role of design to ensure that the object or material is just fit-for-purpose (*i.e.* is not overdesigned) and can be readily disassembled (or perform some additional function) at the end of its (first) useful life.

We have already noted the critical role that engineering plays in bringing about waste minimisation. First, we saw how historic developments in the smelting of copper ore altered pollution emitted to atmosphere. We also have seen that changes to the primary raw materials used to make dihydrogen—a feedstock for the Haber process—brought about major reductions in waste, emissions and other costs that had little to do with the essential chemistry of the industrial process for nitrogen fixation.

We have also learned to look beyond the chemical transformation of reactant to product and to consider the entire sequence of process steps when seeking to assess the origins of waste and its minimisation. We have been introduced to the usefulness of lost work (exergy) analysis, exemplified by the stages of fossil fuel extraction, processing, transmission and use. Fossil fuel is, essentially, stored solar energy. Photosynthesis involves the following overall (highly simplified) biochemical process (eqn 9.1) (see ref. 2 for more details) to produce carbohydrate of the general formula $C_6H_{12}O_6$:

$$6\,CO_2 + 6\,H_2O + h\nu \rightarrow C_6H_{12}O_6 + 6\,O_2 \tag{9.1}$$

$$C_6H_{12}O_6 + 14\,\text{'H'} \rightarrow C_6H_{14} + 6\,H_2O \tag{9.2}$$

$$C_6H_{14} + 9.5\,O_2 \rightarrow 6\,CO_2 + 7\,H_2O + X\,\text{kJ} \tag{9.3}$$

Over geological time and, if the conditions are appropriate, biotic matter is transformed reductively and anaerobically under pressure to generate (eqn 9.2) the complex mixture of hydrocarbons (represented in idealised form by C_6H_{14}) that, with other organic compounds, make up the organic fractions of coal and associated gas. (Oil reserves arise from similar processes involving dead marine organic matter.) The energy content of the hydrocarbon can be released on combusting such products,[3] as represented by eqn (9.3). As we saw in Figure 6.16 (in the case for natural gas), in addition to exergetic losses arising from combustion and the conversion of the heat evolved into steam to drive turbines that generate electricity, losses of useful energy occur at each stage of extraction and transport of the primary raw material and the transmission of the electricity around the grid to industrial and domestic users. Such losses are incurred in any process that extracts raw materials from the environment, transports, transforms and then uses them. Efforts directed towards waste minimisation, therefore, need to recognise the importance of each of the additional steps in the life-cycle. A similar view needs to be taken of any chemical process.

In addition, scale-up from laboratory to manufacture is not straightforward and usually necessitates testing the technology out at intermediate stages (as Baekeland recommended). Such investigations use what are known as 'pilot' or 'semi-technical' plants to learn more about operating the process on a larger and more realistic but not yet the full scale (and also to produce the larger quantities of material needed for additional evaluation and test marketing). Much time and money are expended to engineer and to investigate the partial scale-up and, thereafter, even more is spent to build the full-scale plant. Such investment will only be committed if there is a reasonable expectation of an acceptable financial return. (What might be thought acceptable will depend on a range of circumstances and perspectives introduced in Section 9.3. The Bibliography lists texts for those wishing to pursue this further.) Any return will arise only after a significant period of time has elapsed, with the greater part of the investment already irretrievably committed. Income will only start to flow when product of the right specification is produced and made available in the required volumes to customers prepared to pay the price asked. During this time the money invested is at risk, particularly if a serious defect in the process is found such that completion of the project is delayed or cannot occur. Chemical and process engineering rather than chemistry—which nevertheless still plays an important supporting role—are primarily concerned with these phases of development.

The central objective for the identification and development of more sustainable chemical technologies is to find ways of reducing the material and energy usage per unit of useful output of chemical production by at least a factor of four and possibly a factor of 10 over a period of *ca.* 20 years. As noted earlier (Figure 3.6), developments in technology have, by and large (particularly in the developed economies) been following an improving trend, but neither to the required degree nor at the required pace. Improvements have come about in the past as a result of change driven by the way business and commerce compete in the marketplace and how we, as consumers, make our purchasing decisions rather than by a conscious wish to protect the environment. It is a measure of the challenge to develop better chemical technology that the additional improvements in materials and energy utilisation are both needed on a shortened timescale and need to be effected on novel (or, at least, adapted) processes employing more sustainable feedstocks that give products with reduced environmental impact (and without causing economic dislocation). The characteristics of the development cycle for more sustainable chemical processes are likely to be similar and it is helpful to understand them.

9.1 TECHNOLOGICAL DEVELOPMENT AND EXPERIENCE CURVES

First, most of us are familiar with the concept of economies of scale.[iii] If 50 000 cars per year are manufactured on an assembly line then the cost per car, other things being equal, is likely to be lower than that for 50 hand-built cars built per year. Which of these approaches is adopted will depend on the market being sold into, whether it is a mass market or a specialist (often called 'niche') market, and the capabilities of the operation to manufacture the product the market wants.

In addition, the concept of 'experience' or 'learning' curves is important since, as something is made repeatedly, ways will be found in which the process can be made more efficient. We will deal with this below and examine the basic product/process cycle from initial conception and initiation through growth, maturity and eventual decline—a process that may take 20 years or more,[5] with the need for significant levels of investment at different stages through this period. Such learning in the chemical industry can be seen, for instance, by comparing current technology for the manufacture of poly(propylene) with that first operated in the 1960s, when plants were much smaller (5000 t y^{-1}) and catalysts employed were of low selectivity and activity.[6]

The factors that affect the way processes or products develop with time are many and various, and I touch on only a few here. Critical are the activities of companies seeking to gain competitive advantage by offering something better or cheaper than a competitor and for which there are believed to be customers willing to buy. Underlying this is the process of discovery, invention and innovation (for a description of the processes of R&D see Bibliography) that goes on within a company (or in collaboration with universities and other institutions). These activities that formerly will have been undertaken outside public scrutiny now occur in a rather different social and regulatory environment. Increasingly, with the development of global media and the internet, we have seen (along with corporate advertising) the growth of industry-, product- or company-focused campaigns initiated by activists, customers and consumers,

[iii] The relationship between plant size and plant capital cost per unit of production can be expressed empirically in the so-called scaling law: cost per unit of production $\propto$ (plant size)$^{\alpha-1}$ (where $\alpha < 1$). See ref. 4 for a discussion of such scaling in the context of the development of a biorefinery, particularly addressing the extra costs ('dis-economies' of scale) associated with transporting biomass from the widely distributed sites where it is grown to the central point where it is processed.

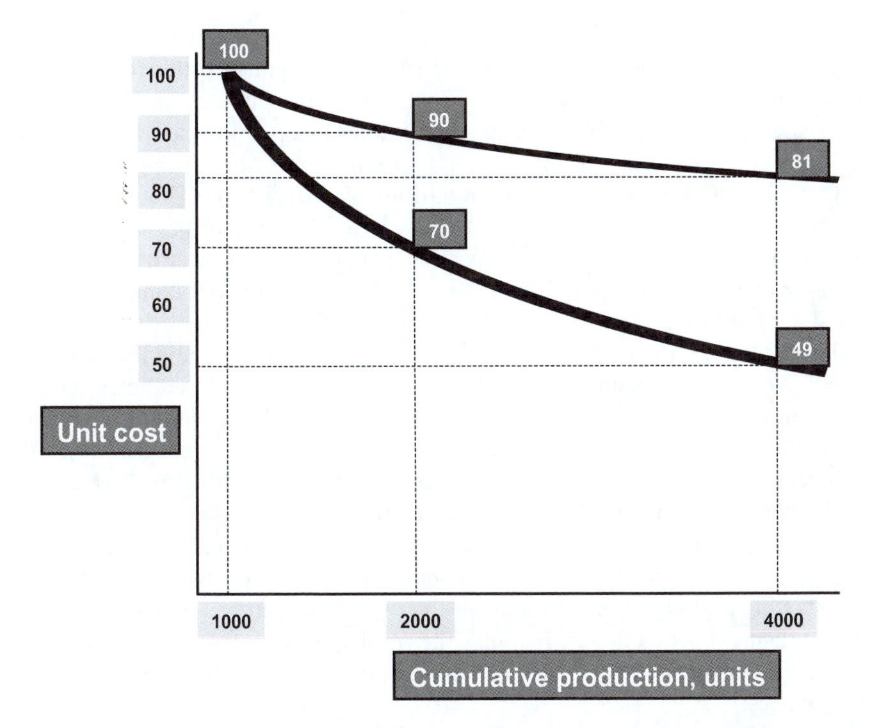

Figure 9.1 Idealised 'experience' curves showing two successive reductions in unit costs (100→90→81 and 100→70→49) as a function of successive doublings of production.

including the efforts of investor activists who take their concerns to meetings of shareholders and investors. We should be aware of these developments, particularly where the various protagonists use (or misuse) scientific and technical evidence in their efforts to influence what is developed and what is not.

From so-called experience curves such as those shown (in idealised form) in Figure 9.1, it can be seen that as cumulative experience is gained [as in the production of poly(propylene)], unit costs go down. This is primarily because the process for manufacture or production has become more efficient, but also because the large initial investment has been recouped. Efficiency gains may arise from the combination of incremental improvements such as the more efficient use of material (including the better terms for its acquisition because of a more assured long-term demand), better organisation that uses manpower more effectively, and investment in equipment to automate parts of the process so they can be operated round the clock.

If a manufacturer begins by making 1000 items of a product with a unit cost of 100 then, following the 90% experience curve, as output is doubled the unit cost for the second thousand is reduced from 100 to 90. For the fourth thousand (a further doubling), the unit cost would become 81. Similarly, following a 70% experience curve would see the unit cost, arising from the same doubling of output, reduced to 70 and 49, respectively. Clearly, if these represented two companies making the same or a similar product, then it is clear that the company following the 70% experience curve is likely to be the more successful, being able to reduce the selling price to increase the share of the market, to be more profitable, or to be in a position to invest more in further improvements (or a combination of all three). The process of 'learning' tends to make the job of a new entrant wishing to begin the manufacture of this product and who has not benefitted from this accumulation of experience much more difficult. This can have the related consequence of making traditional and well-established technologies difficult to displace, a matter of some importance in seeking to move to more sustainable technologies. We saw an example of this effect when we considered technologies for production of hydrogen peroxide (Section 7.5). This was also evident, in a different context, when we compared three technologies for the production of ethanol (Section 6.15). The conventional production technology used fossil petroleum feedstocks and the learning associated with operation of this technology over 50 or 60 years has led to improvements in process efficiency which the newer technologies—based on renewable feedstocks and starting from a different point on the experience curve—would inevitably find difficult to match. It is only the anticipation of future profit that encourages investment in continued development of a technology during its early stages.

For these reasons, there may well be merit in persisting with apparently non-optimum developing technologies based on renewables to allow the benefits from this early learning to be realised. This may well be preferable to the premature abandonment of one novel technology for another because the first one is less efficient than might be considered acceptable. There may be an argument, therefore, that some processes for making first generation biofuels (currently being criticised) should continue in development to achieve the associated learning rather than simply believing it possible to leap-frog to a second or third generation biofuel technology without the benefit of the earlier experience. Such difficult judgements can only be decided on a case-by-case basis and highlight the unintended consequences that may arise from less than fully informed public or media pressure. The development of biofuels is discussed further in Chapter 11.

On the other hand, the entrant (willing to accept the risk) who can make a step-change in unit cost by using innovative technology or some major advance in production (or even is able to deliver the benefit of an old product in a new way) can succeed in displacing an existing operation. The use of incandescent bulbs replaced gas lamps in the early 20th century and the development of compact fluorescent bulbs is now doing the same to conventional electric light bulbs—though more by regulation rather than through customer choice (Web 1). The advent of light-emitting diode technology may repeat the process. Clear-cut benefits, sufficient to offset higher prices, are not always apparent to consumers and this may well be the case in the transition to technologies based on renewable feedstocks.

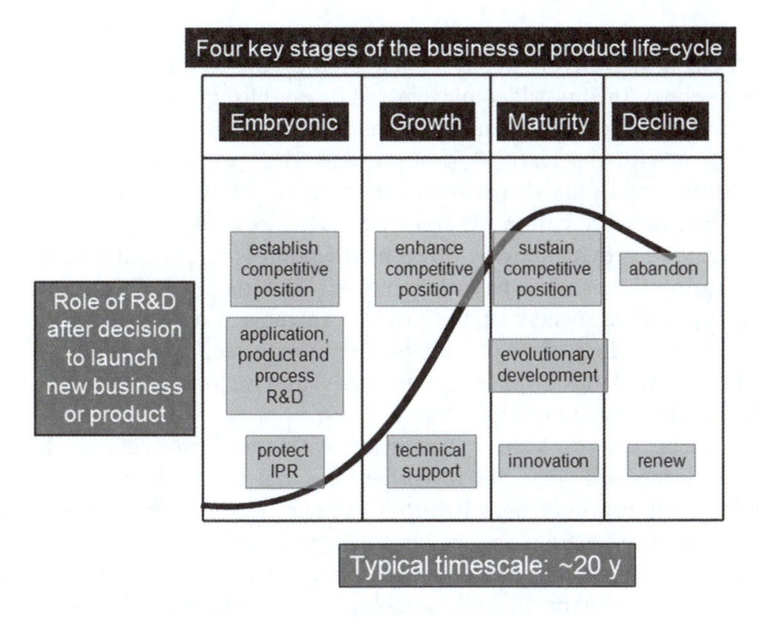

Figure 9.2 Changes in the role of research and development at key stages in the initiation, development and maturity of a technology-based industry.

9.2 STAGES OF TECHNOLOGICAL DEVELOPMENT

Figure 9.2 illustrates the different stages through which any technological development evolves from the point at which the decision is taken to market a new product. For a wholly new product this may take place only after a process of research and development that can take as long as 10–20 years from the original inception of an idea or from the scientific observations that prompted the possibility for exploitation.[iv] Much attention has been given to the management of research and development (see Bibliography) that seeks to reduce the time taken for new projects to move to larger-scale realisation. The chemical industry's efforts to dissolve discipline boundaries (*e.g.* between chemists and chemical engineers) and, where possible to carry out activities in parallel rather than in sequence, are motivated by such time pressures. While some researchers in academia are keen to see their discoveries exploited, examples are rare of a successful rapid translation of a basic scientific advance into its application driven primarily by the discoverer through their proactive integration of the skills and expertise needed. The clinical application of the drug, olaparib (CAS No. 763113-22-0), may be one example[7]—though, it has to be said, olaparib had already passed many of the regulatory stages for its use.

[iv] Ziegler and Natta's research on olefin polymerisation catalysis on which poly(propylene) manufacture is based was carried out in the 1940s and 1950s, well before the first plants were commissioned in the 1960s. See footnote p. 63 for another example.

Once the decision (or series of decisions) is made to offer a new product for sale, the next step involves the launch of the new technology (*e.g.* the first mobile phone or the first video player) and its establishment in the marketplace. If the launch is successful, then the product or the technology enters the growth phase, often with the learning process delivering price reductions (but generating enough surplus to permit further development) or improvements in performance or specification. At some point, the technology becomes 'mature', with the rate of growth slowing down as the point is reached at which most of those wanting the product have purchase it. The market is then sustained by efforts to persuade owners of the product to replace it with a newer model. Later, the technology may be superseded by a newer concept (*i.e.* becomes obsolete such as DVDs displacing videotapes) and the process of winding down the older technology may begin.

This process can be seen in operation with most products we buy. The timescale clearly will vary, but for something that is based on a new piece of technology, the process from birth to death can be 10–20 years and very little can be done to accelerate this beyond a certain point.[v] This is not simply a question of manpower (increasing the number of people working on a development to make it happen more quickly), but of having all the technological pieces developed to the stages needed to bring about the innovation. The latter point is often missed in considering what are seen to be desirable developments such as those associated with solar cell conversion of the light from the Sun, economically, into electricity on the large scale. (See Chapter 12 for a fuller discussion of energy generation and sustainable development.) Subsidies can sometimes accelerate this process, though making optimum judgements to 'back the winner' are fraught with difficulties, particularly if these are inappropriately influenced through lobbying, political activism or media pressure. Indeed, inappropriate subsidies may well divert effort from potentially more productive areas, particularly if the slow pace of development is associated with fundamental or conceptual blocks to progress. Subsidies are likely to be more effective, however, by promoting customer uptake once a product has gone into production, thereby accelerating the delivery of the economies of scale through quickening the learning process. Generally, however, industry cycles can be slow, especially if the costs of implementing change are high and the technological developments uncertain. For large investments, the degree of risk must be acceptable and this may change with time.

9.3 INVESTMENT AND RISK

The issue of the financial risk associated with a technological venture (whether or not the broader economic climate is favourable or depressed) can simply and readily be exemplified in Figure 9.3. At any point, companies will have a

[v] Bearing in mind the timescales generally needed to introduce new technologies, it may well be that greater focus is needed on the politically and socially more fraught approach of limiting material and energy consumption, as this, in principle at least, could be achieved more quickly.

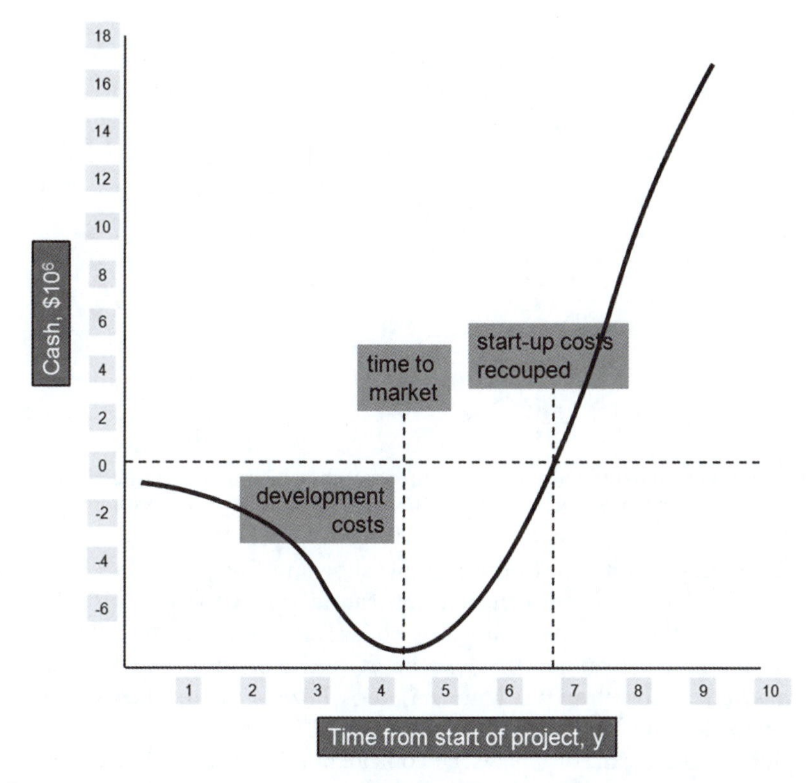

Figure 9.3 Idealised representation of the negative and positive flows (US$ million) with time (y) of cash for a development project for product manufacture showing the costs of initial development, the time at which the product first appears on the market and the time at which start-up costs are recouped from sales.

portfolio of projects overlapping in time. An important management focus is the optimal organisation of such a portfolio over time. Figure 9.3 plots the flow of cash (the difference, positive or negative, between the expenditure incurred by a company and the income received from its sales) during the process of development and introduction of a new product. Over the first few years, resources are committed to the R&D needed to define and understand the technology to bring the product to market on the required scale (and to have designed, tested and constructed the process for manufacturing it). These are the development costs, being spent in the expectation that financial benefits will subsequently accrue. Up to the point that the product is available for sale (4–5 years in the example shown), there will have been no income to offset the expenditure. If at any point an insuperable problem arises and the project is abandoned, then the expenditure is lost ('written off'). As soon as the product is put on the market, sales income will be generated and some of this can begin to offset the initial costs. However, several years may elapse before the total initial investment is recouped and the net cash flow (income – expenditure) for the

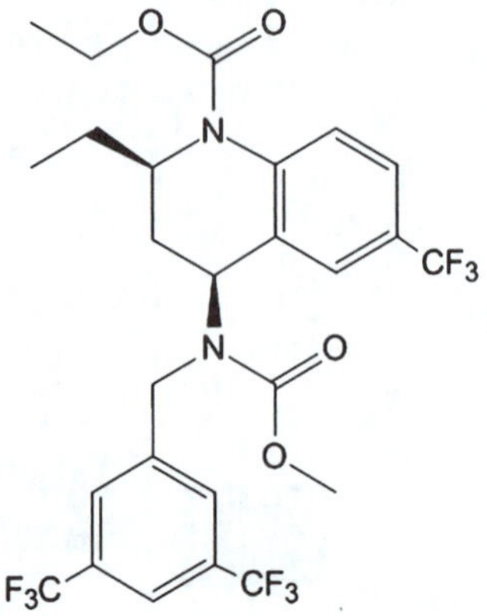

Figure 9.4 Candidate pharmaceutical, torcetrapib, which failed during clinical trials after >US$500 million had been spent on its 17-year development.[8]

project becomes positive.[vi] In the example shown in Figure 9.3, this takes over six years. The development of new drugs can take much longer with development costs of hundreds of millions of dollars. It is salutary to note the substantial amounts of money that can be at risk. For instance, the candidate pharmaceutical, torcetrapib (Figure 9.4), took Pfizer 17 years and over US$500 million to develop.[8] It failed during Phase III[vii] clinical trials and was not introduced. There can, of course, be coincidental benefits from such expenditure (*e.g.* learning about a new manufacturing process or the requirements of a new medical treatment), but these are unlikely ever to match the money lost.

Baekeland's dictum at the head of this chapter is relevant in making judgements between options for more sustainable technological development. The concept, design and evaluation phase of a project to build a chemicals plant may represent as little as 10–15% of the total cost of the project and 20% of the time to completion. Nevertheless, once detailed decisions about defining the

---

[vi] In an example unlikely to be repeated, ICI's investment in its pharmaceuticals business took >20 years to become 'cash positive' after the decision taken in the mid-1930s to develop it. The economic and commercial (as well as health) benefits that accrued from such long-term risk-taking were evident when ICI demerged its multi-billion pound pharmaceuticals operation in 1993 to form Zeneca plc (now part of AstraZeneca).

[vii] Drug development involves a series of steps[9] (running in parallel with process development):

- Pre-clinical investigations using animals
- Phase I: tests in healthy human volunteers to check safety and dose needed
- Phase II: tests for efficacy in patients
- Phase III: clinical trial in comparison with standard treatment or placebo
- Submission of data for regulatory approval
- Launch of product
- Phase IV: monitor clinical benefit.

The process may take 10 years or more. Each phase is more complex and costly than the previous one. Failure may occur at any time.

process and engineering design have been made at the end of this phase, they and the capital cost of construction and the plant operating costs that flow from them are irrevocably locked in. The attendant risks—significant as they are even when dealing with technology, products, feedstocks and markets that are broadly understood—are magnified when the additional uncertainties arising from major feedstock changes, energy costs and political decisions arising from concerns about climate change have to be factored in.

9.4 PRODUCT DEVELOPMENT

The basic elements of the complex process of product development are illustrated for the development of a new agrochemical in Table 9.2. Of the relative development costs of key activities, only 13% are devoted to exploratory chemical research for the identification of a candidate product (the 'lead'). Associated biological R&D (important, as the project is seeking to develop an agrochemical to control a pest) takes *ca.* 13% and field biology designed to assess efficacy and related environmental factors *ca.* 24%. To meet regulatory requirements in different countries, it is important to examine the toxicology, processes of degradation in the environment and the ultimate fate of breakdown products; this takes a further 23%. Identifying the appropriate means of delivering the active ingredient through formulation development adds 7%. Development of the chemical process to make the active ingredient and to compound it into the formulation may take 15%. The remaining 5% is incurred in such administrative matters such as patenting and planning.

Laboratory chemistry plays a lead role in only a part of the process. Chemistry plays a more supportive (but still important) role in other activities where collaboration with other disciplines (biology, engineering, formulation technology, analytical science, to name but four) is necessary.

The fine chemicals sector (of which agrochemicals are a part) is dependent on a succession of new products and typically spends *ca.* 20% of its sales on R&D, both to support products and to develop new ones. Carpenter[10a] estimates that,

Table 9.2 Key components of the development of a new agrochemical (taken from ref. 10a), their relative proportions of development costs. Those most dependent on chemistry are highlighted in bold, those in which chemistry plays a supporting role, in italics.

Activity	*Percentage of total development cost*
Chemical research	13
Biological R&D	13
Field biology: efficacy/environmental	24
Toxicology and fate	23
Formulation development	7
Process technology	15
Administration: patents/planning	5
Total	**100**

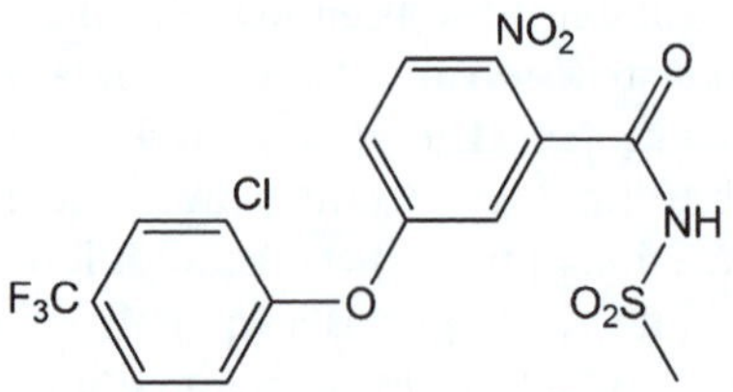

Figure 9.5 The herbicidal diphenyl ether, 5-[2-chloro-4-trifluoromethylphenoxy)]-*N*-(methylsulfonyl)-2-nitrobenzamide (CAS No. 72178-02-0).

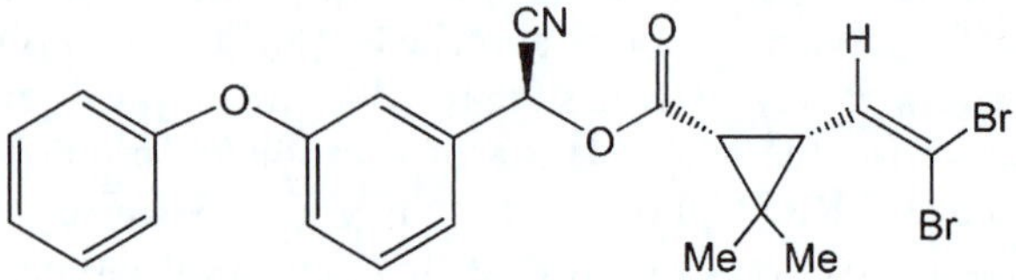

Figure 9.6 The widely-used synthetic insecticidal pyrethroid, deltamethrin: (1*R*,3*R*)-3-(2,2-dibromoethenyl)-2,2-dimethylcyclopropane carboxylic acid, (*S*)-cyano(phenoxyphenyl)methyl ester (CAS No. 52918-63-5).

for every product successfully marketed, about 5000 new compounds are screened, tested, investigated and found to be wanting at various stages of development. It is self-evident that the development costs of the 4999 failures must be recouped from the sales of the one successful product.

The time from the initial discovery of a 'lead' to the launch of a formulated and fully tested product can be about 10 years. As shown by the examples in Figures 9.5 and 9.6, agrochemicals are complex molecules requiring multi-stage syntheses. For obvious reasons, the companies that wish to market these materials have to satisfy regulatory authorities which specify ever more stringent tests regarding toxicology and environmental impact. With many consumer products containing compounds for which full safety testing data are not available, the European Commission seeks to produce such information through its REACH (Registration, Evaluation, Authorisation and Restriction of Chemicals) programme. However, a recent analysis[11] (Web 2) suggests that complying with REACH may have the unfortunate and unintended consequence of requiring 20 times more animals to be used in toxicological testing than initially estimated.

9.5 PATENTING

Agrochemicals are 'effect' chemicals some characteristic of which produces a useful effect (*e.g.* colour, adhesion or the control of a pest) that differentiates its impact (in kind or degree) from other materials. Such effects, and their superiority to what has been available before, can form the basis of a patent. A patent is a form of intellectual property[viii] with legal status. The basic consequence of

[viii] Other forms include copyright, know-how and trade-marks.

being awarded a patent is that the 'owner' has for a limited period of time (10–20 years, the patent lifetime) a monopoly on the exploitation of the invention (with strict rules defining what an invention and an inventor is).

Patent protection (whose validity can be tested in a patent court) begins when a patent is granted and it is a difficult judgement when to seek such protection (as the patent, when granted, becomes a public document). This should be done early enough to avoid someone else securing patent protection (as it is then they who would have the monopoly). However, a significant fraction of the life of the patent can be taken up with the process of development, meeting regulatory requirements and bringing the product to market. Once the patent has expired, others are free to use it. At this point, in the case of pharmaceuticals, a 'generic' manufacturer (who made no contribution to the costs of development) would be free (subject to meeting regulatory requirements) to make and market it.[ix] See the Bibliography for sources concerned with intellectual property and patenting.

9.6 APPLICATION OF PROCESS ENGINEERING AND CHEMISTRY

Chemical technology can be made less waste-producing by addressing the process steps (rather than the chemical steps) carried out on the large scale. Fitting what goes on in a chemical reactor into the bigger picture of an operating chemical plant is the concern of a series of interrelated academic disciplines (and the professionals who practice them) which includes process chemistry, process engineering, reaction engineering and chemical engineering. Chemists seeking to apply the principles of green chemistry should consult the texts in the Bibliography for a more detailed treatment of the basic principles of process and reactor engineering necessary to anticipate the constraints (and opportunities) of carrying out chemical transformations on the large scale.

Achieving optimum efficiency may require a process to be operated in a conventional continuous or batch reactor. There are a number of variants of these two basic approaches. In principle, a batch reactor (Figure 9.7a) has many of the characteristics (agitators, heating and cooling jackets or coils, ports for adding reactants and auxiliaries) of a laboratory reactor, though as we saw in Sections 8.10.17 and 18, important differences exist that affect reaction efficiency and waste formation. Batch reactors have flexibility of operation necessary if a number of products are required on a relatively small (by industrial standards) scale. A continuous stirred tank reactor (Figure 9.7b) differs from a batch reactor in that material is continuously added and removed, with conditions adjusted so that the concentrations of all the

[ix] Facilitating the process by which generic pharmaceuticals producers can challenge the patent of a brand name drug is designed to reduce the cost of some prescription drugs. However, reducing the timescale over which the company that originally developed it can generate revenue from its sale (and thereby recover the costs of its development) may have the unintended consequence of limiting the development of new treatments.[12a] See ref. 12b for an additional consequence of substituting high-cost branded subscription pharmaceuticals by generics, which sometimes arises from associated changes in minor ($<0.1\%$) impurities.

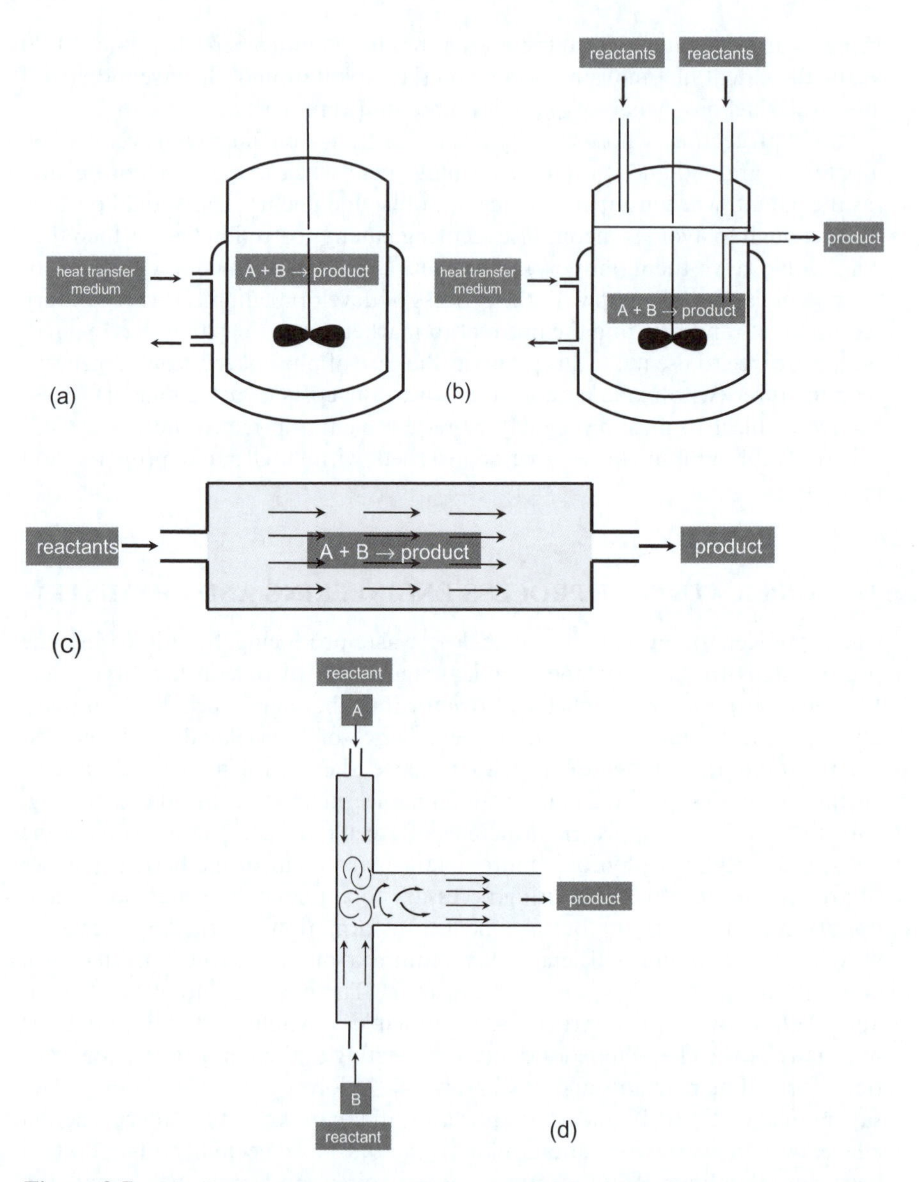

Figure 9.7 Idealised representation of the essential features of: (a) a batch reactor; (b) a continuous stirred tank reactor (CSTR); (c) a flow reactor; and (d) a T-piece mixer.

components in the tank are constant. This, like other continuous processing, requires a series of integrated unit operations (stills, separators, driers, *etc.* linked by pipework, pumps, valves and storage vessels) designed to separate the desired product from the streams exiting the reactor, to recycle unreacted feedstock, to process by-product and to treat waste. The chemistry occurring in the reactor, therefore, must be understood in sufficient detail to enable both the design of these process elements and their effective and safe operation.

Examples of the benefits in processing efficiency and safety that may arise from the working together of chemists and engineers are discussed in the following sections which consider:

- reaction sequence
- process intensification and microreactors (to increase productivity or limit inventory)
- clever contactors (to improve heat and mass transfer when mixing reagents and solutions together)
- novel stimuli
- distributed manufacture (to avoid transporting hazardous materials)
- industrial ecology.

9.7 REACTION SEQUENCE

The sequence in which chemical reactions are carried out can have profound effects on the hazards that the large-scale process can pose. Simply changing the order in which reaction steps are carried out can change process safety. A consequence of the phase-out of the insecticide DDT (see Section 2.8) was the need for a replacement. Among those developed was carbaryl (1-naphthyl methylcarbamate), manufactured by Union Carbide by the route shown in eqn (9.4) and eqn (9.5). Methylamine was reacted with phosgene to give the intermediate, methyl isocyanate, which was then added in a second step to 1-naphthol to give the required product.

Union Carbide Process

$$MeNH_2 + C(O)Cl_2 \longrightarrow \underset{\text{methyl isocyanate}}{MeNCO} + 2\,HCl \tag{9.4}$$

$$MeNCO + \underset{\text{1-naphthol}}{C_{10}H_7OH} \longrightarrow \underset{\text{carbaryl}}{C_{10}H_7OC(O)NHMe} \tag{9.5}$$

Alternative Route 1

$$C(O)Cl_2 + \underset{\text{1-naphthol}}{C_{10}H_7OH} \longrightarrow \underset{\text{1-naphthyl chloroformate}}{C_{10}H_7OC(O)Cl} + HCl \tag{9.6}$$

$$\text{1-naphthyl-OC(O)Cl} + MeNH_2 \longrightarrow \text{1-naphthyl-OC(O)NHMe} + HCl \quad (9.7)$$

Alternative Route 2

$$\underset{N,N'\text{-dimethylurea}}{MeNHC(O)NHMe} + \text{1-naphthol (OH)} \longrightarrow MeNH_2 + \text{1-naphthyl-OC(O)NHMe} \quad (9.8)$$

Methyl isocyanate is a volatile liquid, bp 39 °C. In a major industrial disaster in Bhopal, India, in December 1984, about 40 tonnes of methyl isocyanate escaped after water leaked into a storage tank causing a violent reaction to occur. The immediate death toll was about 3500, with as many as 10 000–15 000 dying thereafter. Of the several factors that contributed to this disaster, two are critical to the current discussion, namely, the use of such a hazardous intermediate and its storage in such large quantities. First, it is surprising, therefore, to realise that the formation of methyl isocyanate can be avoided altogether by simply changing the order of addition of the reagents. Addition of phosgene to 1-naphthol gives 1-naphthyl chloroformate, which can be converted to carbaryl by reaction with methylamine (Alternative Route 1: eqn 9.6 and eqn 9.7). Secondly, why was it necessary to produce and store so much methyl isocyanate? Process design, including novel reactors and using the concept of process intensification (Section 9.9), could well have 'engineered' an inherently safer process. Such considerations could also reduce the 'inventory' of phosgene, itself a material of concern.

Replacing phosgene with a less hazardous alternative might be considered to be an attractive target. Indeed, the use of *N*,*N*′-dimethylurea would seemingly avoid phosgene (Alternative Route 2: eqn 9.8). While this looks attractive at first sight, we have to ask how the dimethylurea itself is made: the most likely route is *via* a related reaction between phosgene and monomethylamine, which proceeds *via* methyl isocyanate as a reactive (and non-isolated) intermediate. While finding routes to useful compounds that avoid the use of phosgene or isocyanates is an active area of research (Section 9.11), none so far has been scaled up to industrial production.

9.8 MIXING AND MASS TRANSFER

We now turn to factors that affect the size of reactors used on an industrial scale. We have already considered briefly the issue of heat transfer

(Sections 8.10.17 and 8.10.18) when considering scale-up of a chemical reactor from the laboratory scale. A related issue is the question of mass transfer. It is obvious that, if materials are not in contact or not mixed together, they cannot react. Mixing does not usually present an insuperable problem in laboratory-scale chemical synthesis. However, serious problems[x] can arise when carrying out chemical change on the large scale. This can be illustrated by some simple examples.

When a reaction is carried out in the laboratory that involves the mixing of solutions of A and B dissolved in the same solvent (or in solvents that are miscible), it is usually assumed that the mixing rate to give a homogeneous solution will be quicker than reaction rate. We tend not to worry about what happens during the period which elapses before the mixed solution become homogeneous. However, we would certainly be wrong to do this for reactions on the larger scale. Even with the vigorous mechanical agitation that can be achieved in a typical industrial batch reactor, complete mixing of large volumes of reactants to give a homogeneous solution may occur so slowly that the reaction proceeds at a rate governed by the speed of mixing rather than by the rate of reaction assuming instantly achieved homogeneity. The reaction is then described as being ‘mass transfer limited’.

There are two parts to Figure 9.8, which plots the rate of reaction against agitation rate (represented by the frequency of rotation of the agitator). If it was possible to achieve instantaneous mixing, then the linear horizontal plot would be seen, associated with so-called ‘intrinsic’ kinetics of the reaction (*i.e.* the reaction rate that would be governed solely by the rate constant for the reaction and the concentration of reactants). As mixing on the large scale can take a finite time, then during the period in which the system is inhomogeneous, the reaction can proceed only at a rate governed by the rate at which the materials are brought into contact with one another on the molecular scale rather than by the intrinsic kinetics. As the mixing rate is increased (by enhanced agitation brought about by increased stirrer speed), a point is reached at which mixing is sufficient to remove the mass transfer limitation and the rate is then governed by the chemistry rather than the physical processes of mixing. Agitation at any higher rate does not bring about any further improvement.

In the laboratory, waiting a little longer for the reaction to proceed may not be considered all that important. However, industrially, the productivity of a reactor is a key factor in determining process economics and is a matter of critical importance.

However, even in the laboratory, we should not assume that we can always ignore the consequences of slow mixing. A paper[14] published in 1987 provides a cautionary warning. Benzoylation of a large excess of 1,2-diaminoethane with benzoyl chloride occurs in two steps, giving the mono and then the dibenzoyl

[x] An example of an apparently trivial factor that can turn out to be very important is illustrated as follows: the yield of a formylation reaction carried out in a pilot plant[13] nearly doubled (from 38% to 70%) when a butyl lithium solution in hexane was added beneath the surface of the reaction mixture rather than above it, as specified from laboratory studies, which themselves showed >70% yield.

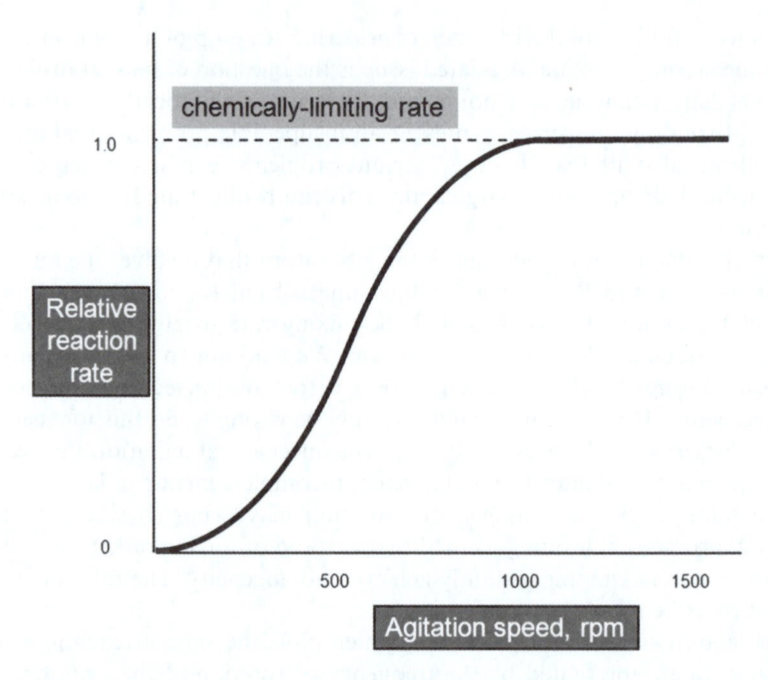

Figure 9.8 Variation of observed reaction rate as a function of mixing speed, demonstrating mass transfer limitations and the approach to the limit set by intrinsic reaction kinetics.

compound (eqn 9.9 and eqn 9.10). The rate constants for these individual steps are known and the concentrations of the two products expected on the basis of intrinsic kinetics can be calculated. Under the conditions of the reaction with a five-fold excess of the diamine and the reaction being carried out in dichloromethane at −78 °C, this calculation suggests 90% mono- and 10% dibenzoylamine should be formed. However, because we have two sequential reactions, unusual effects arising from local concentration gradients and micro-inhomogeneities result, as reaction occurs as mixing proceeds.

$$\begin{aligned} &H_2NCH_2CH_2NH_2 + C_6H_5C(O)Cl \\ &\quad \rightarrow C_6H_5C(O)NHCH_2CH_2NH_2 + HCl \end{aligned} \tag{9.9}$$

$$\begin{aligned} &C_6H_5C(O)NHCH_2CH_2NH_2 + C_6H_5C(O)Cl \\ &\quad \rightarrow C_6H_5C(O)NHCH_2CH_2NHC(O)C_6H_5 + HCl \end{aligned} \tag{9.10}$$

In fact, Jacobson *et al.*[14] observed (Table 9.3) that the percentage of the dibenzoylamine was significantly different from that expected. At high concentrations of reactants, 99% of the stoichiometrically possible dibenzoyl was formed even though the diamine was in overall excess. Even at concentrations of each starting material a factor of 20 lower (such that the intrinsic kinetics would be 400-fold slower), an amount of the dibenzoyl compound higher than

Table 9.3 The effect on the yield of the dibenzoylated product of carrying out the reaction of benzoyl chloride with a five-fold excess of 1,2-diaminoethane (eqn 9.9 and eqn 9.10; –78 °C in dichloromethane) using two different sets of initial reactant concentrations. Data from ref. 14.

1,2-diaminoethane concentration (mol L^{-1})	*PhC(O)Cl concentration (mol L^{-1})*	*Yield of dibenzoyldiamine (%)*
0.67	0.4	99
0.033	0.02	22

that predicted on the basis of intrinsic kinetics was observed. Jacobson *et al.* suggested that the first acylation occurs so rapidly at the interface between the drop of the acyl chloride solution and the diamine solution that the monoamide is available for a second acylation before mixing can lead to a completely homogeneous solution.

Turning now to a related situation in which reactants (that would react together rapidly if in homogeneous solution) are introduced into a reactor in different and immiscible solvents. Suppose, for instance, that A and B react to give C, but that selectivity to C is poor because of a further reaction of the desired product C to give a by-product D. It is sometimes possible to avoid or reduce the formation of D by putting A and B into immiscible solvents—A in solvent 1, B in solvent 2. The reaction is now governed by mass transfer between the two phases, something that is dependent on the interfacial area. It may now be unacceptably slow, but can be speeded up by increasing the interfacial area by breaking the mass of the two solvents into much smaller droplets through agitation; again, this can lead to an approach to intrinsic kinetics. Whether or not C reacts further to D will depend on whether the initial product C is retained in one solvent and the reagent, A or B, with which it reacts to form the by-product, D, is retained in the other.

A graphic historical example[15] of mass transfer limitation is given by the production of the important commercial liquid explosive, nitroglycerine or glyceryl trinitrate (made safe as dynamite when mixed with the mineral, kieselguhr). Nitroglycerine is formed when glycerol and a mixture of concentrated nitric acid and oleum ('fuming' sulfuric acid: sulfuric acid + sulfur trioxide) react together. These two liquids are not completely miscible and the reaction is mass transfer limited. However, the reactants may become more miscible as heat of reaction is evolved and the reaction temperature increases. Temperature control is critical, with the possibility of a runaway reaction if the temperature is allowed to rise too far and too quickly. In an early example of process monitoring, an operator was required to watch the temperature of the reaction and to dump the contents of a reactor into a quench solution if the temperature rose above 30 °C. The operator was provided with a one-legged stool that required him to remain awake to stay seated!

9.9 PROCESS INTENSIFICATION

The problem of immiscibility combined with the avoidance of a runaway reaction associated with contacting large volumes of the reactants can neatly be avoided by bringing only small amounts of glycerol and nitric acid/oleum together at one time by effecting rapid and intense mixing in a flow reactor (shown in simplified form in Figure 9.7d), the use of which allows the reaction temperature to be safely controlled and the volume of material reacting at any one time to be kept to a minimum.

This is an example of the means process chemists and process engineers have devised to reduce the constraints of heat and mass transfer. In addition, processes can be operated more safely by keeping to a minimum the inventory of reactive chemicals at any one time. Such approaches, which enable 'intrinsic' kinetics (the chemistry) to be revealed safely by improved mixing or 'contacting' (the engineering) are known collectively as 'process intensification'[16] and the concept can be extended to any development at the reaction (or molecular), reactor (or meso-) or process (or macro-) scales that makes processing more effective.

In a chemical manufacturing process, the reactor/separator usually represents only *ca.* 20% of total capital cost of the chemical plant (Figure 9.9). The

Figure 9.9 Chemical plant. Image copyright Light & Magic Photography 2010. Used under license from Shutterstock.com.

other 80% represents the cost of associated unit operations, the pipework, pumps and valves that connect them and the associated costs of installation and infrastructure. Reducing the volume of the reactor by a factor of 10^2–10^3 can significantly reduce some of the cost of this 80%. On the other hand, if the volume of streams to be processed is unchanged and the composition of the process stream generated is the same, then the costs of the ancillary processing equipment will not be much reduced (unless, of course, this also can be intensified). However, if the reactor achieves sufficient selectivity to the target, it may be possible to use the product without further processing. In addition, unit operations can be combined. In either case, significant capital savings may be possible. Reactive distillation (Section 9.9.3), which combines the chemical reactor and the distillation step in a single unit operation, is an example of the latter. Process intensification, therefore, aims at increased efficiency (reduced costs) by both miniaturisation and integration.

9.9.1 Microreactors

The primary motivation of process intensification is to achieve significant savings (25–50%) in costs. (The risks associated with applying radical changes such as process intensification may not be justified by more modest savings.) There are, however, additional benefits. Smaller reactors can reduce equipment costs sufficiently to permit 'distributed manufacture', in which material is produced where it is required rather than being centrally produced and transported to where it is needed. In addition, there are improvements in safety associated with reduced quantities of material being processed at any one time. The tragedy of Bhopal would have been on a much smaller scale had the amount of methyl isocyanate made and stored been only 40 kg rather than 40 tonnes.

Two further possible safety benefits arise when carrying out fast and exothermic reactions in microreactors[17] in which at least one reactor dimension is in the sub-millimetre range. Very rapid heat transfer may quench the exotherm which may otherwise lead to a runaway reaction. Alternatively, such an arrangement may suppress explosions by encouraging the scavenging at reactor walls of radicals responsible for propagation, allowing processes to be operated safely with potentially explosive mixtures. A microreactor for the direct combination of dihydrogen and dioxygen over a heterogeneous catalyst to give hydrogen peroxide has been investigated[18] on these grounds, as mixtures of air containing hydrogen in the range 5–74% (v/v) are explosive. Such a process would also eliminate the need to concentrate aqueous hydrogen peroxide as required in the conventional process (p. 180 *et seq.*).

Furthermore, savings can arise from reductions in working capital, *i.e.* money tied up in the volumes and values of materials at various stages of the process. In addition, it can be seen that the nature of the scale-up process has now changed. Rather than building a bigger unit, it becomes possible simply to put together multiples of the single unit. This so-called 'number-up' or 'scale-out' rather than 'scale-up' brings with it less uncertainty in carrying out processing on the larger scale. However, the instances where such an approach has been implemented are so far limited. Engineering the integration of a large number of small units brings its own problems associated with the controlling of flow and the avoidance of blockages. Reported examples include the production of dichlorine using modularised membrane reactors, phosgene and hydrogen cyanide.

The development of intensified processes usually (as we see below) requires innovative reactor design that may not have been tested on the large scale. Such developments usually carry risks of the unexpected, something that those seeking to recoup major investments as quickly as possible are loath to take. Nevertheless, the use of microreactors is an active area of research (see Bibliography), as we shall see when considering very recent developments in the synthesis of the analgesic, ibuprofen (Section 10.8.6).

9.9.2 Spinning Disc Reactor

We have already noted some of the problems with classical batch reactors ('stirred pot'):

- poor mixing
- slower formation of product
- lower productivity
- possibility of by-product formation.

A novel alternative takes the form of the spinning disc reactor (Figure 9.10). Reactants (liquid or in solution) are impinged at the centre of the surface of a spinning disc, which can be heated or cooled as appropriate. Centrifugal forces

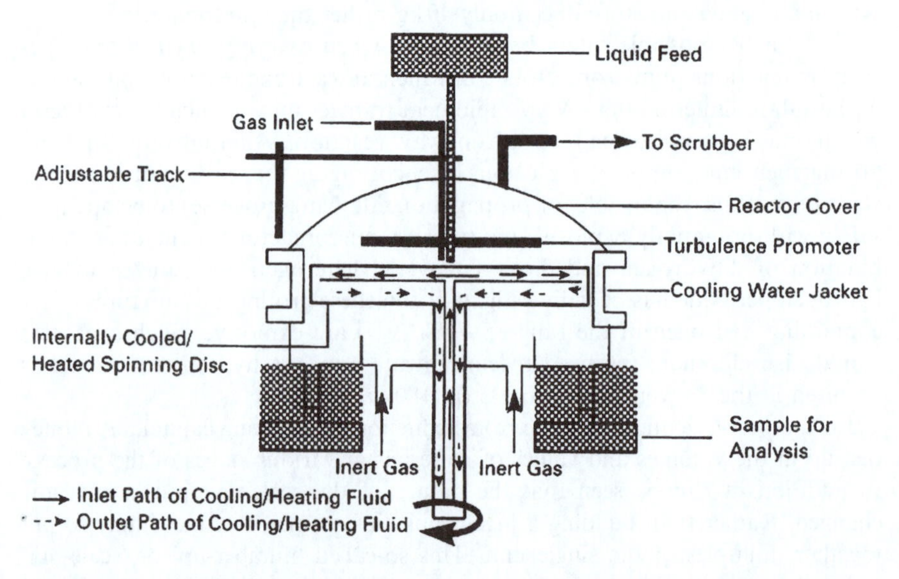

Figure 9.10 A spinning disc reactor. Reproduced from ref. 16a with the permission of Wiley-Blackwell.

Table 9.4 Comparison of conditions and outcomes for the zinc salt-catalysed rearrangement of α-pinene oxide in a batch reactor and on a spinning disc reactor[16a,19]

Conditions for α-pinene oxide rearrangement	*In a batch reactor*	*On a spinning disc reactor*
Feed (g/h)	1	100
Conversion (%)	50	85
Selectivity (%)	80	75
Processing time (s)	3600	~1

spread the liquid into a thin film that flows to the edge of the disc from which it is then ejected and collected. The process is characterised by good heat and mass transfer, low inventory (and is therefore inherent safer), and relies on rapid reaction with short residence times. The technology is untried on the industrial scale.

α-pinene oxide —[$ZnBr_2$ or $Zn(OTf)_2$ on SiO_2]→ campholenic aldehyde (9.11)

This idea has been tested out on the small scale[19] using a reaction (eqn 9.11) in which the natural product, α-pinene oxide (2,7,7-trimethyl-3-oxa-tricyclo[4.1.1.0(2,4)]octane) is converted to campholenic aldehyde [1-(2,2,3-trimethylcyclopent-3-enyl)ethanone], a component of flavours and fragrances. The conventional process, carried out in a batch reactor and catalysed by zinc bromide, displays poor selectivity to the desired product. A significant quantity of waste is generated during the separation of the catalyst. A redesigned process employing a spinning disc[16a] used $Zn(OS(O)_2CF_3)_2$ as catalyst immobilised on SiO_2, itself attached to the disc surface and spinning at 100–1500 rpm (corresponding to a residence time 0.2–1.2 s). The intense mixing and rapid heat transfer results in a much increased throughput. Table 9.4 shows the benefit of the spinning disc reactor (with increased conversion and much reduced processing time).

9.9.3 Reactive Distillation

In principle, combining reaction and separation into a single unit operation can save[20a] between 15 and 80% capital and operating costs.[xi] For instance, a 1985 estimate (cited in ref. 20b) suggested that the use of reactive distillation in the

[xi] It is as well to point out that, sometimes, adding a step can be beneficial, as noted for the manufacture of ethanediol on p. 229 *et seq.*

production of the gasoline additive, methyl tert-butyl ether (MTBE), had saved >3000 GW in electricity compared with alternative processes. In 2006, 146 such commercial-scale processes were said to be in operation,[20a] including a 200 000 $t\,y^{-1}$ plant for the production of methyl acetate from methanol and acetic acid (which requires precise control of feed rates to avoid product contamination with stoichiometric excesses).

$$2\,C_6H_5CH_3 \rightleftharpoons C_6H_6 + 1,2\text{-}, 1,3\text{- and } 1,4\text{-}C_6H_4(CH_3)_2 \qquad (9.12)$$

However, not all processes are amenable to such developments. An evaluation[20b] of combining disproportionation of toluene (eqn 9.12) over a fixed bed catalyst with distillation suggested that the economic benefits (based on 2002 costs and prices when the petrochemicals market was depressed) were 'equivocal'.[xii] The conventional process (vapour phase, 400–425 °C, 20–25 bar) is equilibrium limited (toluene conversion 45–50%), with reactive distillation offering the possibility of increasing this closer to 100%. However, this would require two-phase operation and the column operating at a pressure of 30 bar. This would add to costs that would offset savings elsewhere, as well as eliminating the formation of a relatively high value by-product, ethylbenzene. Were the reactive distillation column to be operated instead at 15 bar, costs could be reduced by 33%, though this would require a catalyst active at *ca.* 220 °C. That such a catalyst was unavailable was known at the beginning of the project, highlighting the potential benefits of interdisciplinary working!

9.10 NOVEL STIMULI

Chemistry that is brought about by heating reactants up and then cooling them down is inefficient because, generally, there is also a need to heat up a solvent and the reactor as well. What if only the reactants were subjected to the reaction stimulus, avoiding the wasteful necessity of treating non-reactants as well? Furthermore, as heat transfer by conduction is a slow process, can a different stimulus to reaction be introduced in ways that are instantaneous? Are there ways in which just sufficient energy (and no more) can be introduced where (and when) it has the desired chemical effect on a single reactant rather than heating an entire reactor with its inventory of reactants, products and solvent?

Among the reaction stimuli that have been tried (some in combination) are:

- Light: photochemistry[21a–h]

[xii] It is relatively rare to find such 'negative' findings described in the academic literature as such results tend not to be published. Indeed, the absence of data where one might expect them may have information content! In addition, because of the commercial sensitivity of such process data much information is retained by companies to avoid making disclosures that could be of use to a competitor. This makes the task of analysing competing claims about the relative performance of technologies and evaluating the present state of innovative developments particularly difficult.

- Microwaves: microwave chemistry[22a–m] (Web 3)
- Electrons: electrochemistry[23a–e]
- Ultrasound: sonochemistry.[24a–e]

These are all major areas of chemistry in their own right. A Bibliography of additional works can be found at the end of this chapter for those wanting to find out more about the fundamentals. The prime purpose here is to examine these methods in detail sufficient to assess their contribution and potential in achieving more efficient less waste-producing technologies, particularly for chemicals production.

9.10.1 Photochemistry

We are concerned with that aspect of photochemistry that involves the synthesis or transformation of compounds by irradiation with electromagnetic radiation, usually in the ultraviolet (UV) and visible regions. A quantum of light must be absorbed by a molecule, producing an excited intermediate species that is activated for reaction. Bringing about a reaction using light has the advantage, at least, that the photon is a 'residue-less' reactant which leads to no corresponding co-product or by-product (apart perhaps from a little heat) that must be separated. Not all the light energy is used in activation nor does each absorbed photon bring about reaction. Of course, the overall process by which artificial light is generated in the first place will be waste producing and the associated waste formation must be accounted for in assessing the overall benefits that the application of photochemistry might bring.

The advantage of light is that it can be switched on and off as required. The sustainability of a chemical process might be enhanced if sunlight, though intermittent and variable, could be used directly[21a]—and not simply in the process of biomass generation through photosynthesis (Chapter 11). There is also no doubt that novel chemistry may be accessible that might not be available by more conventional means, or which may lead to improved selectivities or to reaction under milder conditions.

On the other hand, the overall efficiency in the conversion of electrical energy (usually fossil-sourced) into photons is low (*ca.* 9%[21a]). Conventional light sources, other than lasers, produce light with a range of wavelengths, many of which produce no useful chemical outcome and will be wasted. The efficiency with which light energy is converted into chemical energy is measured by the quantum yield (*i.e.* the number of molecules transformed divided by the number of photons absorbed). A quantum yield of unity represents the situation in which each molecule that absorbs a photon undergoes the expected chemical change. Quantum yields of much less than unity are known in which only a small fraction of the photons brings about the expected chemical change. Some reactions are photocatalysed[21b] or photo-initiated such that the quantum yield is greater (often much greater) than unity.

The relatively small number of industrial processes that use photochemistry is probably associated with the need for specialised equipment such as lamps

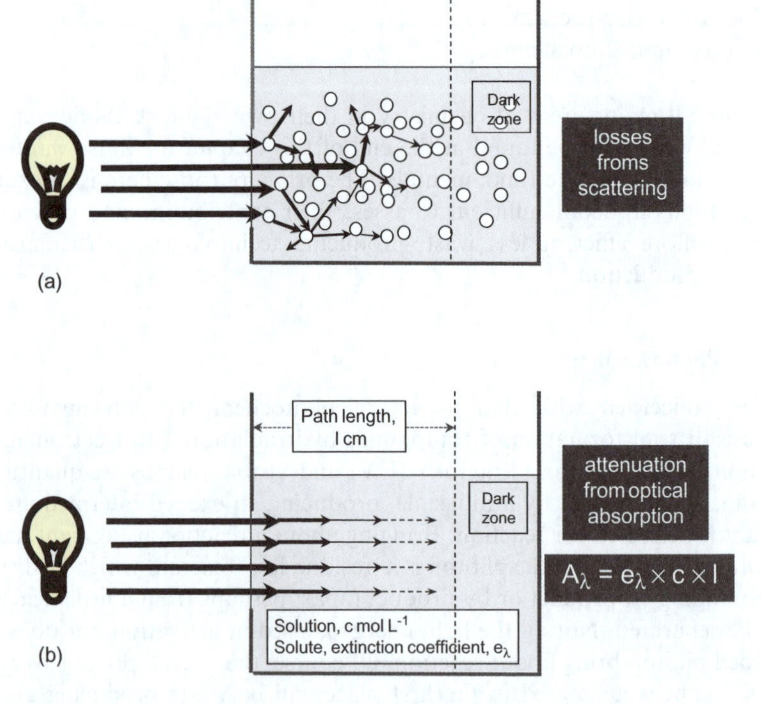

Figure 9.11 Limits of photoreaction arising from: (a) scattering as a result of turbidity or the presence of bubbles; and (b) optical absorption.

and photoreactors, and the need for the optically transparent windows that allow the light from the source to enter the reactor to remain free from fouling. This remains a major challenge. In addition, the basic laws of spectroscopy tell us that the light intensity decreases according to the inverse square of the distance from the source and that absorption of light is governed by the Beer–Lambert Law dependence on extinction coefficient of the components of the medium, their concentrations and the length of the optical path. It is evident from Figure 9.11 that the depth-dependence of light absorption will place a constraint on reactor design.[xiii] There is little point for any dimension of the photoreactor to be significantly greater than that at which light absorption becomes negligible. Such constraints require a different design of reactor for each photoreaction to avoid dark zones and to maximise the volume of the vessel that is subject to irradiation. In addition, this 'effective' depth will be further minimised by scattering associated with turbidity (associated with a precipitate formed during the reaction), or the presence of gas bubbles resulting from agitation or introduced as a reagent or produced as a product.

The use of photochemistry on an industrial scale is not widespread. Examples of photochemical transformations that have been operated on the large

[xiii] A recent review of large-scale solar photocatalytic reactors identifies eight basic designs.[21c]

scale[21d,e] include the Toray process for caprolactam production (used as an intermediate in the manufacture of nylon-6) operated on 100 000 t y^{-1} scale and a process for making vitamin D_3.

$$\text{cyclohexane} + NOCl \xrightarrow{h\nu\ 535\ nm} \text{cyclohexanone oxime} + HCl \qquad (9.13)$$

$$\text{cyclohexanone oxime} \xrightarrow{H_2SO_4} \varepsilon\text{-caprolactam} + H_2O \qquad (9.14)$$

$$\text{cyclohexanone} + [NH_3OH]_2[SO_4] + NH_3 \longrightarrow \text{cyclohexanone oxime} + [NH_4]_2[SO_4] \qquad (9.15)$$

In the Toray process, cyclohexane is converted to cyclohexanone oxime (eqn 9.13) by photolysing NOCl to NO$^\bullet$ and Cl$^\bullet$ at 535 nm provided by low-cost, low-pressure (but short-lived) mercury lamps that deliver light in the range 350–600 nm. Shortwave radiation in the mercury discharge results in fouling of the light-transmitting surfaces. This is suppressed by doping the mercury with thallium iodide. The oxime is then subjected to the Beckmann rearrangement (eqn 9.14) to give ε-caprolactam. While this process avoids the formation of significant quantities of waste ammonium sulfate generated by the conventional process (eqn 9.15), little (if any) of the 4.2 Mt (2004) world production of ε-caprolactam is now produced photochemically.

Previtamin D_3 is produced by photo-induced ring-opening of 7-dehydrocholesterol by weak 280, 289 and 297 nm emissions from a mercury lamp (Scheme 9.1). The product undergoes a subsequent photoreaction to give tachysterol with a quantum yield greater than that with which the product is formed. Optimum conversion of 7-dehydrocholesterol is therefore limited to 30–50%. The previtamin is thermally isomerised to the more stable vitamin D_3. The process suffers from a low quantum yield, with *ca.* 80 kWh energy (kg of product)$^{-1}$ being consumed[21e] and must be carried out in dilute solution. However, while no viable thermally activated process is currently available, the hydroxylated derivative, alfacalcidol, forms the same active metabolite as is formed from vitamin D_3 but is easier to produce (also *via* a photochemical process) (Scheme 9.2).

Seeking to use solar energy directly in chemical conversions has a long history[21e,f] as has an awareness of the difficulties of doing so on a large scale.[21a,e,f] Replacing the artificial light with sunlight in the production of ε-caprolactam

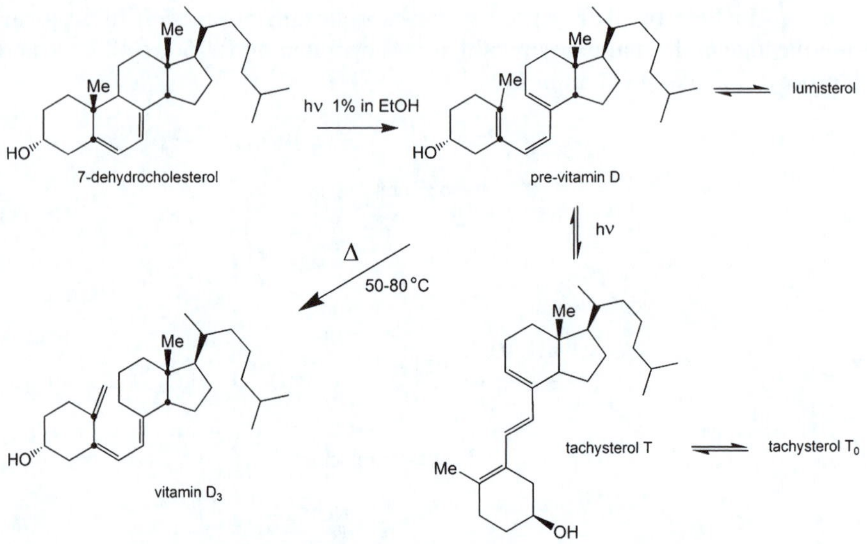

Scheme 9.1 Production of vitamin D_3 from the thermal isomerisation of pre-vitamin D formed photochemically from 7-dehydrocholesterol (showing additional by-products formed from the pre-vitamin).[21d]

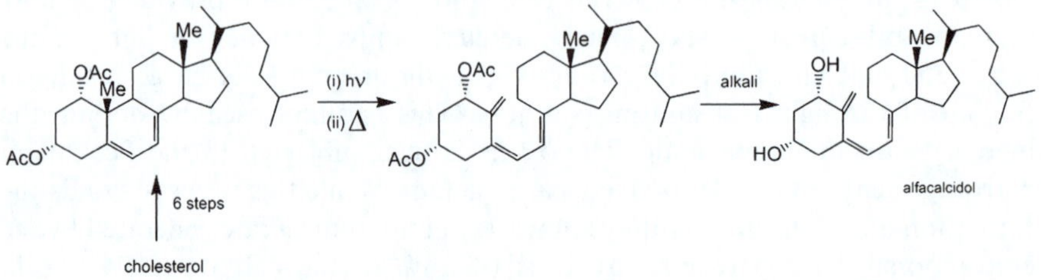

Scheme 9.2 Production of hydroxylated vitamin D_3 (alfacalcidol) in a sequence of reactions from cholesterol (showing the photochemical step leading to the acetic ester of alfacalcidol).

would avoid the CO_2 emitted from electricity generation, estimated[21g] to be 1.5–2.5 kg CO_2 (kg ε-caprolactam)$^{-1}$ depending on the energy 'mix'. Apart from the variability and intermittency of solar energy and its dependence on meteorological conditions, location (latitude) and altitude, its collection and utilisation are themselves important engineering challenges even before seeking to carry out a chemical transformation. The photochemistry possible using sunlight will be dependent on available wavelengths (realistically limited to 350–700 nm) and its concentration to achieve economical space–time yields to minimise the size (and cost) of the photoreactor.

In addition to direct photoreaction, a photo-induced reaction can be accelerated by a catalyst. It has been estimated[21b] that the productivity of photoreactors (combined with photocatalysts) (0.05–0.1 mol m^{-3} s^{-1}) needs to be improved 100–1000 fold before such approaches become commercially attractive.

Much chemistry research is currently devoted to bringing about the oxidation of water to mimic the processes of photosynthesis, particularly the role of the so-called oxygen-evolving complex (which has a Mn_4Ca cluster at its core). Major challenges stand in the path of the development of the 'artificial tree'. Nevertheless, these have stimulated a range of fascinating catalytic and coordination chemistry studies.[21j,k]

9.10.2 Microwaves

Microwaves are electromagnetic radiation in the frequency range 300 MHz to 30 GHz. Chemical applications[22a–c] use 2450 MHz radiation to avoid interfering with telecommunications. Materials dissipate microwave energy by:

- dipole rotation—in which molecules align their permanent or induced dipole with the electric field component of the radiation
- ionic conduction—through the migration of dissolved ions with the oscillating electric field, leading to frictional heating.

Interest in using microwaves to drive chemical synthesis was stimulated by papers by Gedye *et al.*[22d] and Giguere *et al.*[22e] in 1986,[xiv] who reported a 10–50 fold reduction in reaction times compared with conventional reflux for a series of reactions (Table 9.5).

There is some dispute as to whether the effect arises simply from localised heating, or whether there might also be a 'microwave' effect.[xv] However, there is no doubt that there are operational benefits since the energy is introduced remotely, with no contact between source and reagents. Energy input starts and stops instantaneously and heating rates (typically 1–2 °C s^{-1}) are much greater than for conventional heating[22b,c] (Figure 9.12) (though this requires the presence of at least one component capable of coupling with microwaves, an effect dictated by the dielectric properties of the chosen medium). The use of sealed vessels allows temperatures greater than the normal boiling point of the reaction solvent to be reached. When coupled with forced cooling, reaction times can be much reduced without loss of yield.

[xiv] Chemistry-related application of microwaves can be traced back to the 1950s. Studies of microwave-enhanced enzymatic hydrolysis of cellulose materials[22f] and sulfide mineral dissolution in acid[22g] appeared in 1984 and 1985, respectively. The former work is now of greater significance as efficient processing of the lignocellulosic component of biomass is a major hurdle in developing more renewable sources of chemical feedstock.

[xv] This ongoing controversy appears close to having been resolved in a recent report by Kappe *et al.*[22h] Temperature measurements in a reaction vessel made of silicon carbide which is strongly microwave absorbing, chemically inert and thermally conducting show that SiC shields microwave-absorbing materials contained within it whilst rapidly heating them *via* contact with the SiC surface. For a series of 18 reactions, the results (conversion, purity profile, product yields) were the same for both reactions in Pyrex vials and SiC vials irradiated with microwaves.

Table 9.5 Reaction times and yields for a hydrolysis, an oxidation, an esterification and a nucleophilic substitution subjected to microwave irradiation compared with similar data for reactions carried out under conventional reflux.[22d]

	Reaction time (min)		*Product yield (%)*		*Recovered starting material (%)*	
Reaction	*reflux*	*microwave*	*reflux*	*microwave*	*reflux*	*microwave*
$PhC(O)NH_2 \rightarrow PhC(O)OH$	60	10	90	99	5	2
$PhCH_3 \rightarrow PhC(O)OH$	25	5	40	40		
$PhC(O)OH + MeOH \rightarrow PhC(O)OMe$	480	5	74	76	19	11
$4\text{-}NCC_6H_4O^- + PhCH_2Cl \rightarrow PhCH_2OC_6H_4(4\text{-}CN)$	720	3	72	74		

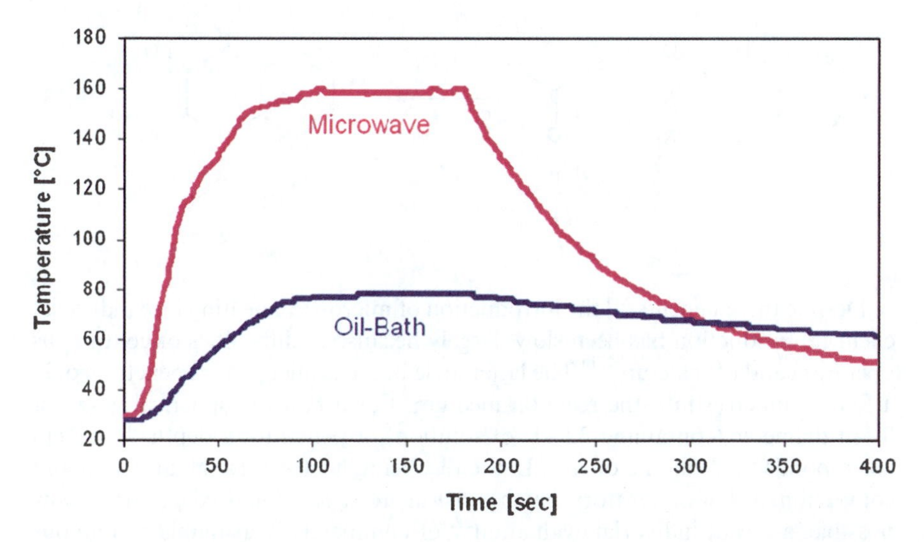

Figure 9.12 Comparison of the heating and cooling rates of a conventionally-heated and a microwave-irradiated reactor. Reproduced from Ref. 22b with the permission of Macmillan Publishers, Copyright 2006.

Greatest use so far has been made in laboratory investigations on the small (gram) batch scale. Its application in drug discovery (in conjunction with high-throughput methods) can result in significant improvements in productivity.[22b] For example,[22b] a three step synthesis of 4-acyl-1,2,3,4-tetrahydroquinoxalin-2-ones (2) from 1-fluoro-2-nitrobenzene (1) (eqn 9.16–9.18) using conventional (optimised) processing gave an overall yield of 4% in 37 working days, whereas using a microwave method the same product was obtained in 88% overall yield in just two working days. Because of the much faster reaction times when microwaves are used, a higher degree of optimisation can be achieved.

NO_2 F **1** + H_2N C(O)OMe R_1 → NO_2 NH R_1 C(O)OMe (9.16)

NO_2 NH R_1 C(O)OMe $\xrightarrow{H_2 \text{ or } HCO_2^{\ominus},\ Pd/C}$ H N O N H R_1 (9.17)

$$\text{(quinoxalinone)} + R_2COCl \longrightarrow \mathbf{2} \tag{9.18}$$

Despite these benefits,[xvi] the introduction of microwave heating into industrial chemical production has been slow, largely because of difficulties of continuous operation and of scale-up.[22i,j] The latter arise because microwaves penetrate only a few centimetres into the reaction medium. For a half-full spherical vessel of 0.5 m diameter (containing 32.5 L of solution), a penetration depth[xvii] of 2 cm corresponds to a volume of *ca.* 7 L, the rest being heated through agitation and convection. While laboratory batch scale-up to a few hundred grams is now possible, a recent industrial evaluation[22i] of commercially available continuous microwave reactors using a standard reaction (the isomerisation of aryl *O*- to *S*-thiocarbamates) concluded that there was currently '*no obvious solution to the problem of scale-up*'. This may not be the final word.[22l,n]

On the other hand, DuPont has investigated the production of hydrogen cyanide (currently manufactured from methane and ammonia in the presence of oxygen over a Pt-Rh catalyst at 900 °C) by the reaction of methane and ammonia (eqn 9.19) at 1200 °C over alumina using microwave heating.

$$CH_4 + NH_3 \rightarrow HCN + 3\,H_2 \tag{9.19}$$

The process may permit distributed manufacture to occur at the point where HCN is used, thereby reducing a significant transportation hazard.

Two additional aspects of the use of microwaves are important for considerations of sustainability:

- energy usage and
- applicability for processing biomass.

9.10.2.1 Energy Efficiency. Greater energy efficiency would appear self-evident in that heating rates and times are so much higher and shorter,

[xvi] Another benefit[22i] (less obvious to a laboratory chemist) arising from avoiding the hot surfaces necessary for contact heating is the prevention of degraded material from solids deposited on the heat transfer surface above the solvent line (as may occur during agitation or reagent addition).

[xvii] The penetration depth, D_P (at which incident microwaves are attenuated by ~37%), is given approximately by $D_P \approx (\lambda_o \sqrt{\varepsilon'})/2\pi\varepsilon''$, where λ_o is the wavelength of the microwave radiation (12.2 cm at 2.45 GHz), ε' and ε'' are the relative permittivity and loss factor, respectively, of the solvent. For water, $\varepsilon' = 77.4$ and $\varepsilon'' = 9.48$.[22k] D_P can be calculated to be 1.8 cm.

respectively. Indeed, simple[xviii] laboratory-scale studies[22n] that compare the electrical energy consumed in providing heating *via* microwaves or from an oil bath suggest that the Diels-Alder reaction between anthracene and maleic anhydride (eqn 9.20) carried out on the 10 mmol scale in toluene gave similar yields (93% oil bath/180 min/reflux; 90% microwaves/2 min/sealed tube) consuming 17.7 and 1.7 kWh mol^{-1} respectively, for conventional and microwave heating.

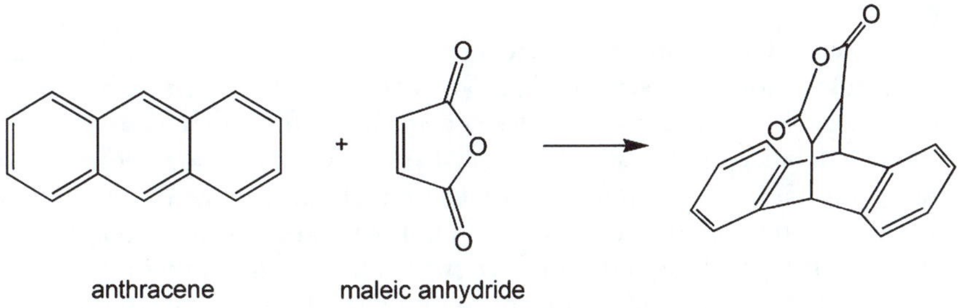

anthracene + maleic anhydride → (9.20)

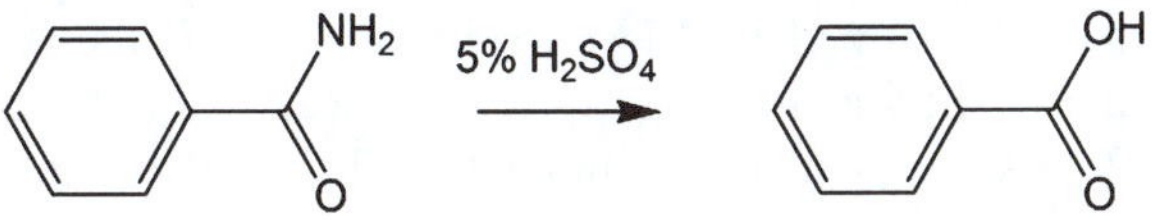

5% H_2SO_4 (9.21)

Similar results are seen when the hydrolysis of benzamide (eqn 9.21) using aqueous sulfuric acid is carried out similarly. However, carrying out the microwave-heated reaction in an open vessel (same mass of reactant, same volume of medium), which limits the temperature to 100 °C, sees the power consumption values reversed (18.3 *vs.* 48.2 kWh mol^{-1} for conventional and microwave heating). An autoclave heated to the temperature achieved in a microwave-heated sealed tube (180 °C) (albeit on a smaller scale) consumes 31.3 kWh mol^{-1}.

9.10.2.2 Processing Biomass. Maximum utilisation of biomass as a renewable feedstock (see Chapter 11) will require identification of an efficient means of breaking down lignocellulose. Depolymerising cellulose to glucose is important in making downstream products such as bioethanol from biomass. The hydrolysis of cellulose is a difficult reaction to initiate and under conventional conditions the initial sought-after products are more labile than the starting material. While advantage may be taken of the rapid heating and cooling that microwaves provide to promote hydrolysis the problems of efficient scale-up and continuous operation remain.

[xviii] While the transformations selected may be straightforward, a proper interpretation of the results requires a full appreciation of the precise circumstances of, and the conditions used in, each experiment.

9.10.3 Electrotechnology

Electrochemistry has been used on an industrial scale in two major conventional technologies, namely:

- the production and refining of metals such as for aluminium from bauxite
- the high-tonnage production of dichlorine, dihydrogen and sodium hydroxide from the electrolysis of brine.

Development of modularised electrochemical cells has permitted distributed manufacture allowing local production that limits the transportation of hazardous materials. Because of reduced demand for dichlorine and its impact on the availability (and cost) of sodium hydroxide production, processes to produce the latter without the co-production of the former (so-called 'salt splitting' of materials such as sodium sulfate) are under investigation. Aspects of electroplating and cathodic protection, while relevant to prolonging the useful life of structures and products, are not considered further. Battery and fuel cell technology are briefly discussed in Chapter 12.

Electrochemistry can offer the flexibility of a wide range of accessible potentials that can be tuned to effect oxidations and reductions that either avoid the need for specific stoichiometric chemical oxidants and reductants (see Section 7.9.4), or can be combined with redox couples present in catalytic quantities. Importantly, water can be used as a safe, cheap and environmentally benign reaction medium, though this can cause problems when seeking to separate products and to process aqueous wastes (because of the very high latent heat per kg of water). In principle, as electrons are relatively cheap to deliver (costing £8–10 $kmol^{-1}$; ref. 23a) they can be considered cost-effective relative to analogous chemical oxidants and reductants. However, this does not take account of the additional costs needed to carry out the reaction electrochemically and losses associated with voltage and faradaic inefficiencies.

$$\Delta G^\circ = -nFE^\circ \quad (9.22)$$

An additional benefit when considering the thermodynamic efficiency of electrochemical processes is the absence of entropic losses in the relationship between free energy change and electrochemical potential (eqn 9.22). Compared with conventional heat-engine derived power generation, those based on interfacial transfers (*e.g.* electrochemistry or photovoltaics) are not limited by Carnot efficiency (p. 153) and can, in principle, be of much greater thermodynamic efficiency.[23c,d] Interestingly, the conversion of nitrogen and hydrogen to ammonia can be carried out at atmospheric pressure (and at 570 °C) using a solid electrolyte fuel cell that avoids the thermodynamic requirement of elevated pressure (p. 127). Electrical work is used to achieve the transformation much in the way that water can be electrolysed to hydrogen and oxygen at ambient temperature and pressure acrylonitrile + $e^{\ominus}$ → $CN^{\bullet -}$ $\xrightarrow{H^{\oplus}}$ $CH_3CH_2CN^{\bullet}$ $\xrightarrow{e^{\ominus}}$ $^{\ominus}CH_2CH_2CN$ → $NCCH_2CH_2CH_2CHCN^{\ominus}$ $\xrightarrow{H^{\oplus}}$ $NCCH_2CH_2CH_2CH_2CN$ (adiponitrile)

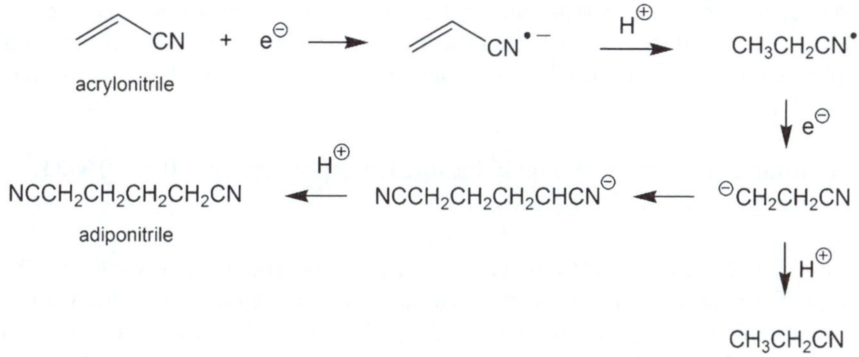

$^{\ominus}CH_2CH_2CN$ $\xrightarrow{H^{\oplus}}$ CH_3CH_2CN

Scheme 9.3 Stages in the electrohydrodimerisation of acrylonitrile for the industrial production of adiponitrile.

when, at equilibrium, the calculated mole fractions of H_2 and O_2 are only ca 10^{-27} (ref. 23e).

The downsides to electrochemical processing include the need to add a supporting electrolyte to achieve adequate conductivity (this may be difficult to bring about if this needs to be achieved in the presence of high organic reactant concentration). The careful choice of electrode materials is a major consideration both in terms of chemical (and electrochemical) inertness, but also in terms of electrochemical performance. There may also be difficulty in separating the required product from the reaction medium—and the need to recover and recycle the supporting electrolyte, which itself must be chemically inert. Electrochemical cells involve two electrode processes, one at the anode the other at the cathode. The product that is sought may be produced at the one, and an unwanted co-product may be formed at the other, thus limiting overall process efficiency.

There are some electrosynthetic processes operated industrially. For example, electrohydrodimerisation of acrylonitrile to adiponitrile (Scheme 9.3) developed by Monsanto, is operated on the 200 000 $t\,y^{-1}$ scale at Decatur, Alabama.

Acrylonitrile reduction occurs at negative potentials where hydrogen evolution would be expected. This is inhibited through the use of cadmium, a high hydrogen-overpotential cathode material. O_2 is produced at anode. The process consumes 2.4 $kWh\,kg^{-1}$.

Sequeira and Santos[23b] list over 70 processes for electrochemically transforming organic materials that they say are '*commercially available*'.

9.10.4 Sonochemistry

Sonochemistry takes advantage of two phenomena associated with the interaction of ultrasound and liquids. The first involves the various consequences of cavitation,[24] that results from 'insonation' of a liquid with sound waves in the kHz range. Cavitation is the generation, growth and collapse of bubbles in a liquid, with cavity lifetimes of the order of 10^{-4} s. Rapid bubble collapse has two effects:

- instantaneous (though highly localised) temperatures of 1000–10 000 K
- pressures of 500–5000 atmospheres,

capable of generating both reactive intermediates and local turbulence,[24b,c] which if occurring close to a surface can lead to surface damage or activation. Both of these effects can be exploited synthetically,[24e–g] although they also require the design of special reactors.

Cavitation can also be achieved through vibration induced in a steel blade by the flow of a liquid across it (in other words, a liquid 'whistle'), an effect exploited in the food industry to bring about emulsification or homogenisation. Stimulated mass transfer ('acoustic streaming') can result in the absence of cavitation when a probe oscillates in a liquid at MHz frequencies (so-called megasonics). Such phenomena can lead to reduced experimental reaction times and can eliminate induction times. However, questions remain concerning the operability of such stimuli for large-scale chemicals production and the energy consumed ($kWh\,kg^{-1}$) compared with alternative approaches. This may limit the technology to a small number of specialised or high-value applications.

9.11 INHERENT SAFETY AND INHERENT WASTE MINIMISATION

The concept of inherent chemical safety has been an important factor in the design of chemical processes, urgently stimulated by the terrible occurrences in Bhopal in 1984 and from earlier events in the UK such as the Flixborough disaster in 1974 and the Piper Alpha disaster in 1976. One of the industrial leaders in this aspect of chemical engineering was Trevor Kletz.[15] Interestingly, many of the principles of inherent safety have parallels with efforts to minimise waste and to improve efficiency in material and energy usage.[1]

We have already seen the importance of minimising the volume of reactors using ideas of process intensification. Limiting 'buffer' stocks of intermediates made in a process that are consumed in a later step reduces the risks to both safety and the environment. However, the overall sequence of processes—producing and using such intermediates as they are needed—requires careful management for efficient and timely operation ('just-in-time' production). In addition, distributed manufacture can avoid the hazards, losses, emissions and costs associated with transportation.

Inherent safety has some additional principles that indirectly can assist process operability and minimise emissions. By removing unnecessary complexity, fewer processing steps are needed. Having fewer unit operations will

$$NaCl \rightarrow Na^{\oplus} + 0.5\,Cl_2 + e^{\ominus}$$

$$CO + Cl_2 \rightarrow C(O)Cl_2$$

$$RNH_2 + C(O)Cl_2 \rightarrow RN{=}C{=}O + 2\,HCl$$

Scheme 9.4 Basic steps in the production of isocyanates from phosgene.

$$2\,CH_3OH + CO + 0.5\,O_2 \rightarrow (CH_3O)_2CO + H_2O$$

$$2\,CH_3OH + C(O)Cl_2 \rightarrow (CH_3O)_2CO + 2\,HCl$$

$$(CH_3O)_2CO + RNH_2 \rightarrow RNHC(O)OCH_3 + CH_3OH$$

$$RNHC(O)OCH_3 \rightarrow RN{=}C{=}O + CH_3OH$$

Scheme 9.5 Alternative process using dimethyl carbonate for the production of isocyanates.

require less pipework and fewer pumps to connect them, which should as a consequence mean less to leak and maintain. The development of reactive distillation (Section 9.9.3) combining a chemical reactor and a distillation unit into a single operation is an example.

Substitution or replacement of hazardous materials—as suggested by the third green chemistry principle (Section 8.10.3)—by more benign ones is a principle of inherent safety that can be extended to the use of alternatives with limited impact on the environment. One such material is phosgene, a chemical intermediate that continues to be of central importance to many chemical technologies. It is currently used to make isocyanates from amines as shown in Scheme 9.4. Isocyanates are extensively used in manufacturing urethanes and polyurethanes. Dimethyl carbonate, $(MeO)_2CO$, has been proposed as an alternative reagent (Scheme 9.5), though its relatively low reactivity requires the presence of a catalyst. An industrial production route to dimethyl carbonate based on phosgene would simply move the hazard to somewhere else in the manufacturing chain.[xix] In addition, an N-methylation reaction (Scheme 9.6)

[xix] Eni-Chem operates a non-phosgene route to dimethyl carbonate producing *ca.* 12 000 t y^{-1} *via* the liquid phase (120–140 °C, 20–40 bar) oxidative carbonylation of methanol: 4 CH_3OH + 2-CO + $O_2 \rightarrow$ 2 $(CH_3O)_2CO + H_2O$, using a copper(I) chloride slurry catalyst.[25a] While inherently more attractive, it is still problematic as dioxygen used as a feedstock poses an explosion risk. Carbonylation of methyl nitrite avoids this and a further 3000 t y^{-1} of dimethyl carbonate is produced this way.

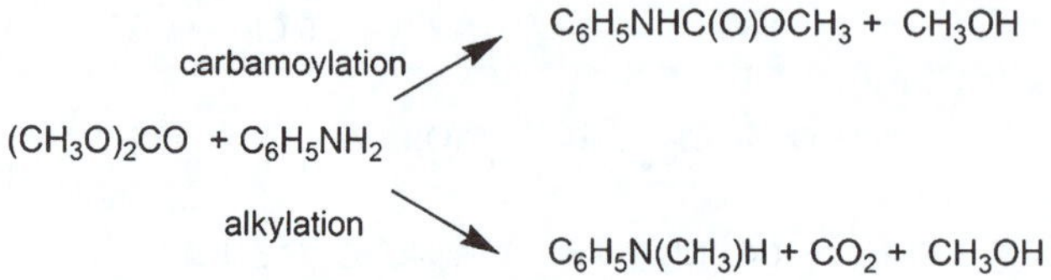

$$(CH_3O)_2CO + C_6H_5NH_2 \xrightarrow{\text{carbamoylation}} C_6H_5NHC(O)OCH_3 + CH_3OH$$

$$(CH_3O)_2CO + C_6H_5NH_2 \xrightarrow{\text{alkylation}} C_6H_5N(CH_3)H + CO_2 + CH_3OH$$

Scheme 9.6 Competition between *N*-methylation and carbamoylation in the reaction of arylamines with dimethyl carbonate.

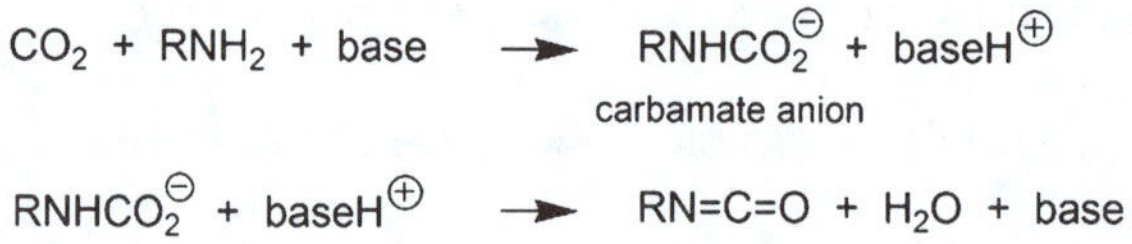

$$CO_2 + RNH_2 + \text{base} \rightarrow \underset{\text{carbamate anion}}{RNHCO_2^{\ominus}} + \text{baseH}^{\oplus}$$

$$RNHCO_2^{\ominus} + \text{baseH}^{\oplus} \rightarrow RN{=}C{=}O + H_2O + \text{base}$$

Scheme 9.7 Target process for the production of isocyanates from carbon dioxide.

competes with carbamoylation when dimethyl carbonate reacts with the arylamines that represent *ca.* 90% of isocyanate production, rendering the use of dimethyl carbonate of little or no current industrial value.[25b] However, recent research,[25c] has from laboratory studies now identified a heterogeneous catalyst (gold on cerium dioxide) capable of achieving *N*-carbamoylation of 2,4-diaminotoluene by dimethyl carbonate with 95% selectivity at 99% conversion. In addition, the methylating ability of dimethyl carbonate has been exploited[25d] in the continuous synthesis of ethers from alcohols (*e.g.* 1-octanol) using supercritical carbon dioxide as the reaction medium and acidic γ-alumina, giving the best selectivities to 1-methoxyoctane (and minimising dehydration to 1-octene and transesterification of dimethyl carbonate to methyloctyl and dioctyl carbonate).

Ideally, carbon dioxide should be used[25e] as shown in Scheme 9.7. Unfortunately, the direct reactions are found to be too slow to be technologically of interest. The search for better catalysts continues. An indirect route in which carbon dioxide is converted to dimethyl carbonate is shown Scheme 9.8. CO_2 is reacted with ethylene oxide to give ethylene carbonate (see p. 230) from which dimethyl carbonate could be obtained by transesterification. However, the latter step is highly equilibrium limited.

9.12 PROCESS INTEGRATION AND INDUSTRIAL ECOLOGY

A characteristic of the modern chemical industry is the degree to which it is integrated. A benefit of such integration is the (relative) efficiency with which materials are used. While this may be so, the consequence is that there will be inherent resistance to wholesale structural change such as that needed to move to technologies based on renewable biomass-based feedstocks. It is inevitable

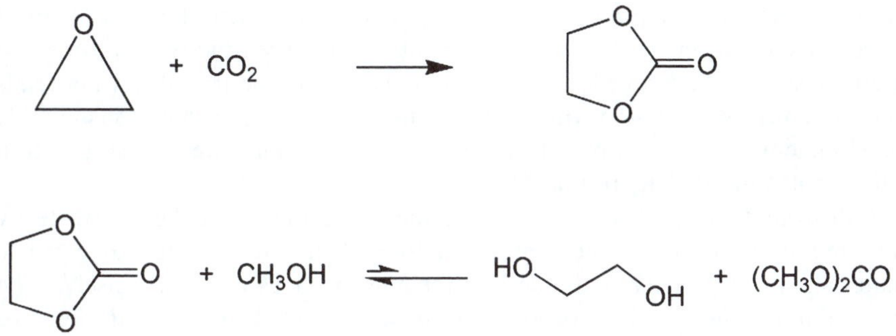

Scheme 9.8 reaction: ethylene oxide + CO_2 → ethylene carbonate; ethylene carbonate + CH_3OH ⇌ $HOCH_2CH_2OH$ + $(CH_3O)_2CO$

Scheme 9.8 A possible route to dimethyl carbonate from carbon dioxide.

that the process of transition will generally occur piecemeal over a relatively protracted period. On the other hand, the construction of completely new communities with the objective of being carbon neutral overall can, if the appropriate resources and political and organisational will are in place, bring about change more quickly. Such opportunities are limited; the Masdar City development[26] in Abu Dhabi may be one (Web 4) notable also for proposals to introduce an energy-based parallel currency (the 'ergo').[26c]

While technological solutions are necessary (if not sufficient) to achieve sustainable development, simply improving the efficiency of production, while important in itself, is also not sufficient. There is a need to consider ways to limit the absolute quantity of waste, taking account of both its volume and impact.

The key technological question posed by the challenges of sustainable development then is: how do we extract all usable energy from any material or from any process using sufficient of it to aim for zero environmental impact accepting, from thermodynamic arguments, that there is a theoretical maximum work potential in any set of material or energetic inputs into a process?

We have stressed the importance of accounting for all material and energy inputs to a process by identifying and quantifying all outputs including those that arise from the extraction of the primary raw materials (the 'cradle') and from ultimate disposal at the end of its useful life (the 'grave'). The search for a beneficial use for a waste, or the recycling of recovered material at different stages within the life-cycle, has largely been driven by economic considerations within a company or an operation. Concerting such activities between businesses and companies with wider environmental motivations in mind is a much more recent development.

Industrial ecology,[xx] a term popularised in an article[28] in *Scientific American* in 1989, brings together two ideas that some believe (like sustainable development) are incompatible and are in contradiction to one another: the one, industry, linked to the activities of humankind, the technosphere or

[xx] If you explore industrial ecology (IE) further, you should be aware of the related concept of ecological modernisation (EM). Deutz[27] explores the relationships between the two suggesting that '*EM emphasises economic development and technological advances within a policy framework. IE, by contrast, emphasises inter-firm cooperation and voluntary compliance inspired by eco-efficiency savings*'.

anthrosphere, and the other, ecology, wholly concerned with the natural world or ecosphere. Nevertheless, some see parallels[xxi] between the two that may be useful in seeking to reconcile industrial activity with the source of raw materials and the sinks for wastes on which it relies (in much the same way as sustainable development seeks to improve the standard of life of an increasing population while protecting the environment).

Industrial ecology therefore seeks to maximise the utilisation of waste by thinking in terms of an ecological analogy. What if the inevitable waste materials and products at the end of their useful lives could, themselves, be used as inputs into other processes and products? What if the design of processes and products could take this waste utilisation into account? Could we develop an industrial system that conforms to this ideal? Indeed, are there examples where this can be seen to have been demonstrated? Those wanting to explore industrial ecology (also known as 'industrial metabolism') further may want to consult ref. 30a,b,c. There is also a journal devoted to this topic (Web 5).

Ecology is the scientific study of the distribution of, abundance of, and the interactions between, organisms and their natural environment. Industrial ecology envisages:[30b]

> '*an industrial ecosystem, analogous in its functioning to a community of biological organisms and their environment. . . . In the industrial ecosystem, each process and network of processes must be viewed as a dependent and interrelated part of a larger whole. The analogy between the industrial ecosystem concept and the biological ecosystem is not perfect, but much could be gained if the industrial system were to mimic the best features of the biological analogue*'.

The essential question, therefore, is: can we close material and energy cycles to minimise waste, with residues from one process becoming feedstocks for another? Consideration of a petrochemical complex, with its multiple flows of intermediates between processes leading from the raw material input at one end and the products coming out of the other, suggests that some aspects of chemical technology already display elements of this ecological model.

Three basic types of system have been identified. The first is a linear one-way flow.[30b] This 'reservoir–use–disposal' sequence characterises the LeBlanc process and its effect on towns such as Widnes at the end of the 19th century (Figure 2.8). The second typifies current processes that are characterised by a

[xxi] Ehrenfeld[29] asks us carefully to distinguish between the use of the ecosystem as an analogy and as a metaphor. He also believes natural scientists need a changed perspective when they address the complexity of (and the social and environmental dimensions of) the challenge of sustainability. As in many other aspects of sustainable development, there continues to be debate about what the underpinning philosophical basis of industrial ecology should be. Ehrenfeld criticises the more 'technocentrist' view of the subject as outlined by leading proponents such as Allenby. He argues that more attention should be given to more recent ideas of ecosystem behaviour, based on emergence and complexity. While being aware of this debate, we nevertheless summarise the topic as presented by Allenby and Graedel in their text.[30b]

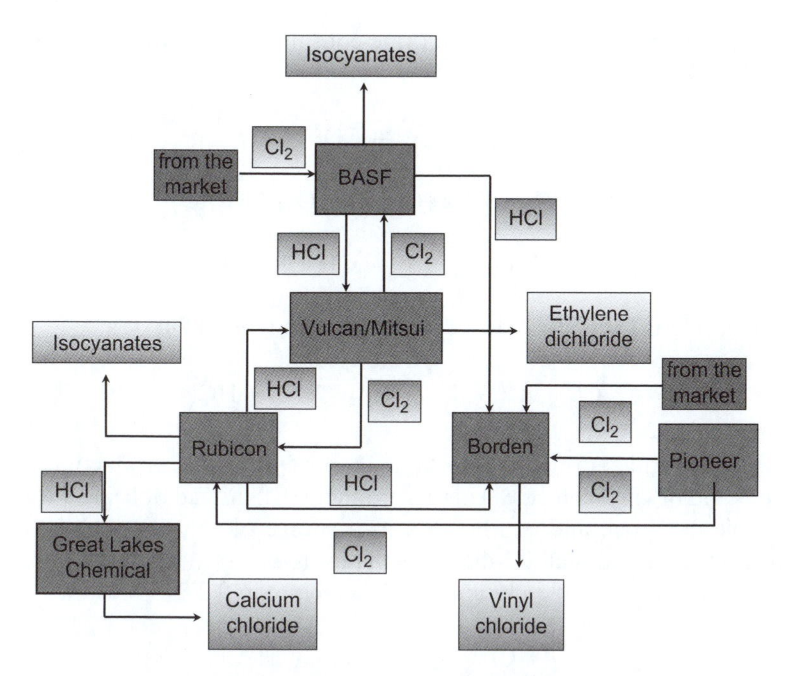

Figure 9.13 Integrated product chain (in 1998) for isocyanates, ethylene dichloride, vinyl chloride and calcium chloride showing a series of interchanges of key reactants, hydrogen chloride and dichlorine. Adapted from ref. 31a,b.

certain amount of recycling in which virgin material is supplemented with some reused or recycled material, with the final waste disposed of outside the system. Manufacturing processes that assemble complex products from a number of individual components are able to implement this type of development. The production of mobile phones, cars and printer/photocopies falls into this category, with the design of these products (and the materials from which they are made) now taking into account the need for subsequent disassembly for reuse ('remanufacturing'). However, the demands and consequences of assembly and disassembly of something like an electronic device are qualitatively different from the production and separation of a similarly complex chemical formulation or composite.

An example more relevant to chemicals processing (Figure 9.13) shows the relative flows associated with exchanges of products and wastes between companies engaged in the different but related chemical processing of chlorine-containing or chlorine-using products.[xxii] Product chains associated with the manufacture of organic isocyanates and vinyl chloride can be effectively integrated, though not perfectly because of the different scales on which the

[xxii] This arrangement is more properly described as 'industrial symbiosis', a form of industrial ecology operating at the inter-company level.

materials are produced. The dichlorine-consuming chemistry is shown in eqn (9.23–9.25):

$$CO + Cl_2 \rightarrow COCl_2 \tag{9.23}$$

$$RNH_2 + COCl_2 \rightarrow RNCO + 2\,HCl \tag{9.24}$$

$$CH_2{=}CH_2 + Cl_2 \rightarrow CH_2ClCH_2Cl \tag{9.25}$$

Vinyl chloride is formed *via* dehydrochlorination (eqn 9.26):

$$CH_2ClCH_2Cl \rightarrow CH_2{=}CHCl + HCl \tag{9.26}$$

Hydrogen chloride is thus an inevitable co-product in the production of isocyanates and of vinyl chloride. This co-product HCl, instead of being a waste, is a valuable feedstock and can be used as a source of chlorine such as in the production of additional 1,2-dichloroethane in a catalysed oxychlorination, which is carried out in the presence of dioxygen (eqn 9.27):

$$2\,CH_2 = CH_2 + 4\,HCl + O_2 \rightarrow 2\,CH_2ClCH_2Cl + 2\,H_2O \tag{9.27}$$

The production of methyl chloride using methanol as a starting material (eqn 9.28) is also an important large-scale chemical process using hydrogen chloride. In both of these latter examples, the co-product is water.

$$MeOH + HCl \rightarrow CH_3Cl + H_2O \tag{9.28}$$

In principle, it should be possible to build a network with a degree of integration within and between independent companies to maximise the utilisation of the primary raw materials input at the front-end of these processes to their mutual benefit. It should be remembered that any integration of this type has hitherto grown up as the result of the operation of market economics rather than from conscious planning.[31] The challenge for the wider integration of industries, as envisaged by proponents of industrial ecology is whether such conscious planning is, in fact, possible (and if so, who should be the planners?).

The third type of system represents the hypothetical ideal that would see the only input into the industrial system being solar energy with all waste recycled. Parallels between this and the technosphere-ecosphere model examined in Chapter 6 should be apparent.

It is often difficult to appreciate in specific cases whether there is an advantage or not from any proposed integration. In such instances, the advantages or otherwise can only be gauged by comparing relative benefits using aggregated data.

Industrial ecology is, therefore, seen by its proponents as providing a holistic approach to redesigning industrial activities by modelling industrial systems on ecological principles. Industrial ecology is concerned with complex patterns of material flows within and outside the industrial system. This could be critical for the long-term evolution of groups of technologies designed to bring about the transition from the current (unsustainable) industrial system to a more viable industrial ecosystem. At the very simplest level, there is little point in seeking to match an industry requiring or providing only a small amount of material with one that requires or provides a great deal. An efficient system will seek to balance these flows as far as possible, and a moment's thought suggests that this will not be a straightforward process. Whether this will arise inevitably from the motivation of potential partners through better co-ordination of action, communication and information or whether this requires stimulation *via* political or regulatory action to bring about more sustainable long-term industrial system evolution is a moot point.

The most widely cited example of industrial ecology in operation is a series of industries in a 'symbiotic' (close and long term, see Web 6) association with one another in Kalundborg, Denmark (ref. 32). These inter-related industries or operations (Figure 9.14) include a power station, an oil refinery, a biotech company, plasterboard production, district heating and soil remediation in which material and energy are exchanged between them to their mutual advantage and, importantly, with a benefit to the environment in that greater use is made of the raw materials and feedstocks than would otherwise have been the case. The claimed benefits[xxiii] include:

- reduced consumption (oil 45 000 t y^{-1}, coal 15 000 t y^{-1}, water 600 000 m^3 y^{-1})
- reduced emissions (CO_2 175 000 t y^{-1}, SO_2 10 200 t y^{-1})
- better waste utilisation (sulfur 4500 t y^{-1}, gypsum 90 000 t y^{-1}, fly ash 130 000 t y^{-1}).

A total of US$75 million was invested to bring about these exchanges, with an average five-year payback. These figures represent a measure of the real benefit such a system can bring about. Interestingly, the system was self-organised rather than imposed from above through government action. The difficulty is, therefore, to make such developments happen elsewhere among other industries.

It is worth bringing together at this point some of the aspects of industrial ecology of which those that have studied it in more detail are critical.[34] There is no single perfect answer to the challenges of sustainable development and industrial ecology is no exception. There may be circumstances in which industrial ecology works, like at Kalundborg, but this is likely not to be universally applicable because of specific or local factors. Historically, 'dematerialisation' reduces unit costs of production. Reducing costs cheapens products,

[xxiii] Interestingly, while the Kalundborg example is widely used to highlight the benefits of industrial ecology or industrial symbiosis, only in the cases of water and steam utilisation have the claimed benefits been independently assessed and reported in the peer-reviewed literature and then only in 2006.[33] However, more detailed analyses of the benefits of related schemes are now beginning to appear.

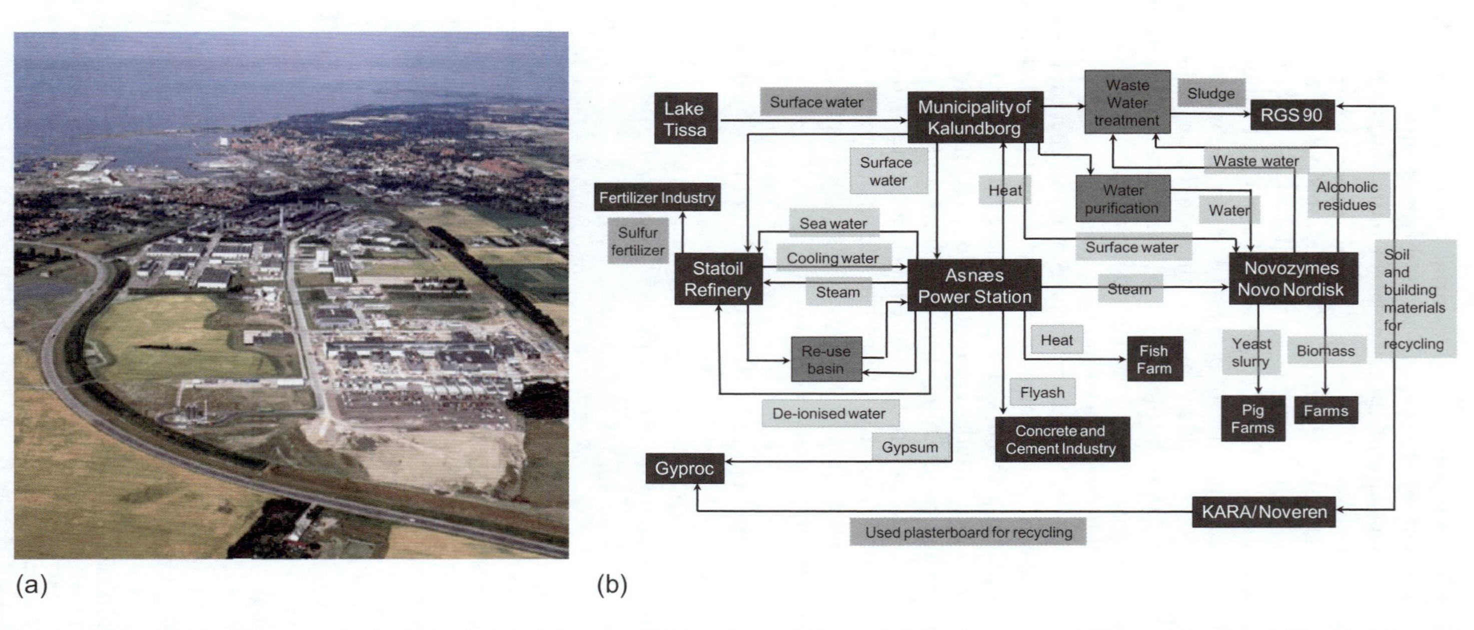

Figure 9.14 (a) Aerial photograph of the industrial zone of Kalundborg, Denmark (photo courtesy of Novo Nordisk A/S); (b) Schematic representation of the inter-relationship of component industrial operations that make up the symbiotic system (adapted from www.symbiosis.dk).

increases demand and intensifies extraction of material and energy use. There are no physical limits to the resulting accelerated circulation of capital in contrast to limitations on extraction of primary resources and their replacement limitations on which will become only too evident.

Innovation may stimulate competition and, in its turn, further innovation. It might be expected that the new technology would quickly displace the old. However, as we have previously seen, older technologies will have paid off all initial development costs and investments. The operation of the older technology can continue, particularly if a demand remains for its product. The price that the newer technology must charge for its own product may be significantly higher than that of the old because of the need to recover start-up costs and because the benefits of learning have yet to be realised. If the service provided to the user by the two products is similar (even if the benefit to the environment is quite different), it is even possible that the output from the mature (more waste-producing) operations need not necessarily be displaced or reduced. Political action may then be needed, such as the subsidy currently being offered to citizens in the UK and elsewhere (Web 7) to replace cars more than 10 years old—not only does this provide an economic stimulus, it replaces more-polluting cars with less.

Industrial ecology will be difficult to implement as it requires the availability of information to identify wastes capable of being used downstream and then the bringing of businesses interested in the exchange together optimally. O'Rourke and his colleagues[34] ask: will there ever be enough information available to make such decisions? There will inevitably be a trade-off between different (possibly conflicting) outcomes that will require not just technical but also economic, social and political value judgements that businesses will not be in the best position to make which could affect different groups who would have a legitimate interest in them. It also depends on putting a monetary value on environmental 'goods' (*e.g.* the protection of wilderness or biodiversity, reduction in ozone depletion) as a means of using a common and widely understood metric. Many find this idea unpalatable, but it is the only way to 'price the unpriceable', a challenge addressed by the disciplines of environmental economics or ecological economics (see Section 2.10).

There are clear differences between products manufactured through assembly of components and those obtained *via* chemical processing and formulation. Having said that, the integration needed to minimise waste is already practised in high volume chemicals processing and is an additional reason why Sheldon's E-factor for volume chemicals production is lower than that for fine chemicals.

There are, of course, limits to the extent to which wastes can be used. There is almost certainly an ultimate waste for which use or reuse is impossible (particularly if more material or energy would be consumed in bringing the waste back into use than is represented by the material or energy content of the waste). In the jargon, there are limits to 'waste cascading'. A good example is the limited use to which carbon dioxide can be put. As we have seen, CO_2 is at the end of the carbon chain, being fully oxidised. Its conversion into carbohydrate by photosynthesis relies on the availability of 'free' solar energy. CO_2 can be used as a product in its own right (*e.g.* in the carbonation of fizzy drinks) or as a monomer or reagent in a limited number of chemical processes. However, these represent only a trivial fraction of the global anthropogenic production of CO_2. Perhaps a difficulty with industrial ecology that does not come first to mind is the possibility that such an integrated and interdependent system becomes more difficult to change when changes are possible or are needed. Such integration may lock in component parts that have become sub-optimal choices. As a consequence, this may inhibit the development of new technology.

REFERENCES

1. (a) P. T. Anastas and J. B. Zimmerman, *Environ. Sci. Technol.*, 2003, **37**, 94A;(b) W. McDonough, M. Braungart, P. T. Anastas and J. B. Zimmerman, *Environ. Sci. Technol.*, 2003, **37**, 434A;(c) D. T. Allen, *Chem. Eng.*, 2007, **114**, 36;(d) M. Mendez, *Chem. Eng.*, 2007, **114**, 41.
2. (a) D. O. Hall and K. K. Rao, *Photosynthesis*, Cambridge University Press, Cambridge, 6th edn, 1999;(b) D. W. Lawlor, *Photosynthesis*, BIOS Scientific Publishers, Oxford, 3rd edn, 2000.
3. H. O. Pritchard, *Energy Environ. Sci.*, 2009, **2**, 815.
4. M. W. Jack, *Bioresour. Technol.*, 2009, **100**, 6324.
5. G. Huppes and M. Ishikawa, *Ecol. Econ.*, 2009, **68**, 1687, citing M. Hirooka, *Innovation Dynamism and Economic Growth: A Nonlinear Perspective,* Edward Elgar Publishing, Cheltenham, 2006.
6. C. Christ, *Production-Integrated Environmental Protection and Waste Management in the Chemical Industry*, Wiley-VCH, Weinheim, 1999, pp. 132–135.
7. D. Cressey, *Nature*, 2010, **463**, 422.
8. J. L. LaMattina, *Drug Truths: Dispelling the Myths about Pharma R&D*, Wiley, Hoboken, 2009, pp. 23–37.
9. T. Y. Zhang, *Chem. Rev.*, 2006, **106**, 2583.
10. (a) K. J. Carpenter, *Chem. Eng. Sci.*, 2001, **56**, 305; (b) J. H. Atherton and K. J. Carpenter, *Process Development Physicochemical Concepts*, Oxford University Press, Oxford, 1999.
11. T. Hartung and C. Rovida, *Nature*, 2009, **460**, 1080.
12. (a) M. J. Higgins and S. J. H. Graham, *Science*, 2009, **326**, 370; (b) T. Laird, *Org. Process Res. Dev.*, 2009, **13**, 1037.
13. (a) H.-J. Federsel, *Acc. Chem. Res.*, 2009, **42**, 671; (b) H.-J. Federsel, *Org. Process Res. Dev.*, 2000, **4**, 362.
14. A. R. Jacobson, A. N. Makris and L. M. Sayre, *J. Org. Chem.*, 1987, **52**, 2592.
15. Taken from T. A. Kletz, *Plant Design for Safety: A User-friendly Approach*, Taylor and Francis, New York, 1990, p. 26.
16. (a) R. J. J. Jachuck, in *Handbook of Green Chemistry and Technology*, ed. J. H. Clark and D. Macquarrie, Wiley-Blackwell, 2002, ch. 15, p. 366; (b)

T. Van Gervan and A. Stankiewicz, *Ind. Eng. Chem. Res.*, 2009, **48**, 2465; (c) A. Stankiewicz, *Chem. Eng. Res. Des.*, 2006, **84**, 511.
17. (a) K. F. Jensen, *Chem. Eng. Sci.*, 2001, **56**, 293; (b) S. Becht, R. Franke, A. Geißelmann and H. Hahn, *Chem. Eng. Process.*, 2009, **48**, 329; (c) K. Jähnisch, V. Hessel, H. Löwe and M. Baerns, *Angew. Chem., Int. Ed.*, 2004, **43**, 406.
18. Y. Voloshin and A. Lawal, *Chem. Eng. Sci.*, 2010, **65**, 1028.
19. M. Vicevic, R. J. J. Jachuck, K. Scott, J. H. Clark and K. Wilson, *Green Chem.*, 2004, **6**, 533.
20. (a) G. J. Harmsen, *Chem. Eng. Proc.*, 2007, **46**, 774; (b) E. H. Stitt, *Chem. Eng. Sci.*, 2002, **57**, 1537.
21. (a) P. Esser, B. Pohlmann and H.-D. Scharf, *Angew. Chem., Int. Ed. Engl.*, 1994, **33**, 2009; (b) T. Van Gerven, G. Mul, J. Moulijn and A. Stankiewicz, *Chem. Eng. Process.*, 2007, **46**, 781; (c) R. J. Braham and A. T. Harris, *Ind. Eng. Chem. Res.*, 2009, **48**, 8890; (d) I. R. Dunkin, in *Handbook of Green Chemistry and Technology*, ed. J. H. Clark and D. Macquarrie, Wiley-Blackwell, Oxford, 2002, ch. 18, p. 416; (e) M. Fischer, *Angew. Chem., Int. Ed.*, 1978, **17**, 16; (f) G. Ciamician, *Science*, 1912, **36**, 385; (g) A. Schönberg and A. Mustafa, *Chem. Rev.*, 1947, **40**, 181; (h) K.-H. Funken, F.-J. Müller, J. Ortner, K.-J. Riffelmann and C. Sattler, *Energy*, 1999, **24**, 681; (i) V. Balzani, A. Credi and M. Venturi, *ChemSusChem*, 2008, **1**, 26; (j) X. Sala, I. Romero, M. Rodríguez, L. Esriche and A. Llobet, *Angew. Chem., Int. Ed.*, 2009, **48**, 2842; (k) S. W. Kohl, L. Weiner, L. Schwartsburd, L. Konstantinovski, L. J. W. Shimon, Y. Ben-David, M. A. Iron and D. Milstein, *Science*, 2009, **324**, 74.
22. (a) C. O. Kappe, *Angew. Chem., Int. Ed.*, 2004, **43**, 6250; (b) C. O. Kappe and D. Dallinger, *Nat. Rev. Drug Discov.*, 2006, **5**, 51; (c) C. R. Strauss, *Org. Process Res. Dev.*, 2009, **13**, 915; (d) R. Gedye, F. Smith, K. Westaway, H. Ali, L. Baldisera, L. Laberge and J. Rousell, *Tetrahedron Lett.*, 1986, **27**, 279; (e) R. J. Giguere, T. L. Bray, S. M. Duncan and G. Majetich, *Tetrahedron Lett.*, 1986, **27**, 4945; (f) H. Ooshima, K. Aso, Y. Harano and T. Yamamoto, *Biotechnol. Lett.*, 1984, **6**, 289; (g) F. Smith, B. Cousins, J. Bozic and W. Flora, *Anal. Chim. Acta*, 1985, **177**, 243; (h) D. Obermeyer, B. Gutmann and C. O. Kappe, *Angew. Chem., Int. Ed.*, 2009, **48**, 8321; (i) J. D. Moseley, P. Lenden, M. Lockwood, K. Ruda, J.-P. Sherlock, A. D. Thomson and J. P. Gilday, *Org. Process Res. Dev.*, 2008, **12**, 30; (j) A. Stadler, B. H. Yousefi, D. Dallinger, P. Walla, E. Van der Eycken, N. Kaval and C. O. Kappe, *Org. Process Res. Dev.*, 2003, **7**, 707; (k) C. Gabriel, S. Gabriel, E. H. Grant, B. S. J. Halstead and D. M. P. Mingos, *Chem. Soc. Rev.*, 1998, **27**, 213; (l) M. H. C. L. Dressen, B. H. P. van de Kruijs, J. Meuldijk, J. A. J. M. Vekemans and L. A. Hulshof, *Org. Process Res. Dev.*, 2010, **14**, 351; (m) T. Razzaq and C. O. Kappe, *ChemSusChem*, 2008, **1**, 123; (n) F. Bergamelli, M. Iannelli, J. A. Marafie and J. D. Moseley, *Org. Process Res. Dev.*, 2010, **14**, 926.
23. (a) K. Scott, in *Handbook of Green Chemistry and Technology*, ed. J. H. Clark and D. Macquarrie, Wiley-Blackwell, Oxford, 2002, ch. 19,

p. 433; (b) C. A. C. Sequeira and D. M. F. Santos, *J. Braz. Chem. Soc.*, 2009, **20**, 387; (c) K. Hemmes, G. P. J. Dijkema and H. J. van der Kooi, *Russ. J. Electrochem.*, 2004, **40**, 1284; (d) J. Kunze and U. Stimming, *Angew. Chem., Int. Ed.*, 2009, **48**, 9230; (e) G. Marnellos and M. Stoukides, *Science*, 1998, **282**, 98.
24. (a) K. S. Suslick, *Science*, 1990, **247**, 1439; (b) K. S. Suslick and D. J. Flannigan, *Annu. Rev. Phys. Chem.*, 2008, **59**, 659; (c) H. Xu, N. G. Glumac and K. S. Suslick, *Angew. Chem., Int. Ed.*, 2010, **49**, 1079; (d) P. R. Gogate, *Chem. Eng. Process.*, 2008, **47**, 515; (e) L. H. Thompson and L. K. Doraiswamy, *Ind. Eng. Chem. Res.*, 1999, **38**, 1215; (f) T. J. Mason and P. Cintas, in *Handbook of Green Chemistry and Technology*, ed. J. H. Clark and D. Macquarrie, Wiley-Blackwell, Oxford, 2002, ch. 16, p. 372; (g) T. J. Mason and J. P. Lorimer, *Applied Sonochemistry: The Uses of Power Ultrasound in Chemistry and Processing*, Wiley-VCH, Weinheim, 2002.
25. (a) N. Keller, G. Rebmann and V. Keller, *J. Mol. Catal. A: Chem.*, 2010, **317**, 1; (b) R. Juárez, A. Corma and H. García, *Top. Catal.*, 2009, **52**, 1688; (c) R. Juárez, P. Concepción, A. Corma, V. Fornés and H. García, *Angew. Chem., Int. Ed.*, 2010, **49**, 1286; (d) P. N. Gooden, R. A. Bourne, A. J. Parrott, H. S. Bevinakatti, D. J. Irvine and M. Poliakoff, *Org. Process Res. Dev.*, 2010, **14**, 411; (e) T. E. Waldman and W. D. McGhee, *J. Chem. Soc., Chem. Commun.*, 1994, 957.
26. (a) M. W. Johnson and J. Suskewicz, *Harvard Bus. Rev.*, November 2009, 53; (b) D. Reiche, *Energy Policy*, 2010, **38**, 378; (c) S. Sgouridis and S. Kennedy, *Energy Policy*, 2010, **38**, 1749.
27. P. Deutz, *Geogr. J.*, 2009, **175**, 274.
28. R. A. Frosch and N. E. Gallopoulos, *Sci. Am.*, September 1989, **261**, 94.
29. J. R. Ehrenfeld, *J. Clean. Prod.*, 2004, **12**, 825.
30. (a) T. E. Graedel, *Environ. Sci. Technol.*, 2000, **34**, 28A; (b) T. E. Graedel and B.R. Allenby, *Industrial Ecology*, Prentice Hall, Upper Saddle River, NJ, 1995; (c) T. E. Graedel and B.R. Allenby, *Industrial Ecology and Sustainable Engineering*, Prentice Hall, Upper Saddle River, NJ,2009.
31. D. T. Allen and D. R. Shonnard, *Green Engineering: Environmentally Conscious Design of Chemical Processes*, Prentice-Hall, Upper Saddle River, NJ, 2002, ch. 14.
32. M. R. Chertlow, *J. Ind. Ecol.*, 2007, **11**, 11.
33. N. B. Jacobsen, *J. Ind. Ecol.*, 2006, **10**, 239.
34. D. O'Rourke, L. Connelly and C. P. Koshland, *Int. J. Environ. Pollut.*, 1996, **6**, 89.

BIBLIOGRAPHY[xxiv]

Y. Jin, D. Wang and F. Wei, *Chem. Eng. Sci.*, 2004, **59**, 1885.
B. Gagnon, R. Leduc and L. Savard, *Environ. Eng. Sci.*, 2009, **26**, 1459.

[xxiv] These texts supplement those already cited as references.

M. A. Abraham, Ed., *Sustainability Science and Engineering: Defining Principles*, Elsevier, Amsterdam, 2006.
D. T. Allen and D. R. Shonnard, *Green Engineering. Environmentally Conscious Design of Chemical Processes*, Prentice Hall, Upper-Saddle River, NJ, 2002.
O. Levenspiel, *Chemical Reaction Engineering*, Wiley, New York, 2nd edn, 1999.
I. S. Metcalfe, *Chemical Reaction Engineering: A First Course*, Oxford University Press, Oxford, 1997, Oxford Primer Series.
R. H. Perry and D. W. Green, *Perry's Chemical Engineering Handbook*, McGraw-Hill, Columbus, OH, 7th edn, 1997.
T. Wirth, *Microreactors in Organic Synthesis and Catalysis*, Wiley-VCH, Weinheim, 2008.
P. A. Roussel, K. N. Saad and T. J. Erickson, *Third Generation R&D: Managing the Link to Corporate Strategy*, Harvard Business School Press, Boston, MA, 1991.
P. Weaver, L. Jansen, G. van Grootveld, E. van Spiegel and P. Vergragt, *Sustainable Technology Development*, Greenleaf Publishing, Sheffield, 2000.
C. Fussler with P. James, *Driving Eco Innovation: a Breakthrough Discipline for Innovation and Sustainability*, Pitman Publishing, London, 1996.
P. Bamfield, *Research and Development in the Chemical Industry*, VCH, Weinheim, 1996.
C. A. Heaton, *An Introduction to Industrial Chemistry*, Leonard Hill, Glasgow, 1984.
D. O'Connell, *Inside the Patent Factory: The Essential Reference for Effective and Efficient Management of Patent Creation*, J. Wiley & Sons, Chichester, 2008.
J. H. Atherton and K. J. Carpenter, *Process Development: Physicochemical Concepts*, Oxford University Press, Oxford, 1999, Oxford Primer Series.
I. K. Bradbury, *The Biosphere*, Wiley, Chichester, 2nd edn, 1998.
B. Wardle, *Principles and Applications of Photochemistry*, J. Wiley & Sons, Chichester, 2009.
D. T. Sawyer, A. Sobkowiak and J. L. Roberts, Jr., *Electrochemistry for Chemists*, Wiley, New York, 2nd edn, 1995.

WEBLIOGRAPHY

1. www.direct.gov.uk/en/Nl1/Newsroom/DG_180003
2. www.altex.ch/resources/rovida_hartung_altex_3_09.pdf
3. www.maos.net
4. www.masdar.ae/en/home/index.aspx
5. http://www3.interscience.wiley.com/journal/118902538/home
6. http://en.wikipedia.org/wiki/Symbiosis
7. www.direct.gov.uk/en/Motoring/BuyingAndSellingAVehicle/AdviceOnBuyingAndSellingAVehicle/DG_177693

All the web pages listed in this Webliography were accessed in May 2010.

CHAPTER 10

Catalysis

'This new force, which was unknown until now, is common to organic and inorganic nature. I do not believe that this is a force entirely independent of the electrochemical affinities of matter; I believe, on the contrary, that it is only a new manifestation, but since we cannot see their connection and mutual dependence, it will be easier to designate it by a separate name. I will call this force catalytic force. Similarly, I will call the decomposition of bodies by this force catalysis, as one designates the decomposition of bodies by chemical affinity analysis.'

Jöns Jacob Berzelius, 1836

Of all the sub-disciplines of chemistry, catalysis is the one whose fundamental study is most intimately linked with its technological application.[i] Evidence for this should already be apparent from the various contributions of catalysis to waste minimisation in chemicals processing already mentioned. The purpose of this chapter, having first introduced the concept, is to explore further and exemplify the central importance of catalysis to the efficiency of large-scale chemicals production—past, present and future.

A catalyst is a chemical component of a reacting system that changes the rate at which chemical equilibrium is attained without itself being consumed in the reaction being catalysed. Types of catalyst and catalysis are many and varied, including:

- homogeneous catalysis
- heterogeneous catalysis

[i] The interplay between technological development and academic catalysis research is well exemplified in two recent papers[1] which discuss the industrially important Wacker process for the catalytic conversion of ethylene to acetaldehyde and fundamental investigations into the role of the homogeneous palladium(II) catalysts used.

Chemistry for Sustainable Technologies: A Foundation
By Neil Winterton

Published by the Royal Society of Chemistry, www.rsc.org

- asymmetric catalysis
- supported catalysis
- biocatalysis
- phase transfer catalysis
- organocatalysis
- electrocatalysis
- photocatalysis (Section 9.10.1)
- hybrid catalysis
- combinatorial catalysis
- environmental catalysis.

Each is an important independent area of study and application. Several of these areas have been met, incidentally, so far. Most will be introduced or further illustrated during the course of this and the remaining chapters on the use of renewable feedstocks for chemicals manufacture and energy generation.

A key attribute of catalysis, in addition to its important contribution to waste minimisation, is the availability it provides of routes to compounds that are difficult, if not impossible, to achieve non-catalytically. A good example, important industrially, is olefin metathesis catalysis[2]—a transformation that involves the exchange of carbons (and their attached substituents) linked by carbon–carbon double bonds (eqn 10.1) and discussed in detail in Section 8.7. The understanding of the catalytic process and the purposeful design of catalysts have led to new transformations of both scientific and technological importance, discoveries that merited the award of the Nobel Prize for Chemistry in 2005 (Web 1).

$$2\,CH_2{=}CHMe \rightarrow CH_2{=}CH_2 + MeCH{=}CHMe \tag{10.1}$$

10.1 CATALYSIS, KINETICS AND THE CATALYTICALLY ACTIVE SPECIES

Catalysis is a kinetic phenomenon. Catalysts enhance reactivity by lowering the activation energy of a reaction. This can be represented by consideration of the two reaction co-ordinates[ii] in Figure 10.1a,b. Figure 10.1a shows the (uncatalysed) rate-determining step from A + B *via* the transition state, $X^{\ddagger}$, with an activation energy $E_a(1)$. Figure 10.1b shows the corresponding catalysed process in which an intermediate, E, is formed from A + B in the presence of the catalyst, involving two new transition states, $Y^{\ddagger}$ and $Z^{\ddagger}$. The activation energy for the process leading to the intermediate, $A + B \rightarrow Y^{\ddagger}$, $E_a(2)$ must be less than that for the uncatalysed reaction.

The alternative representation that shows a simple reduction in the activation energy for the rate-determining step, $A + B \rightarrow X^{\ddagger}$, without the formation of an intermediate is, strictly, not an indication of catalysis at all; rather it is a

[ii] See cautionary comments in footnote xi, p. 148.

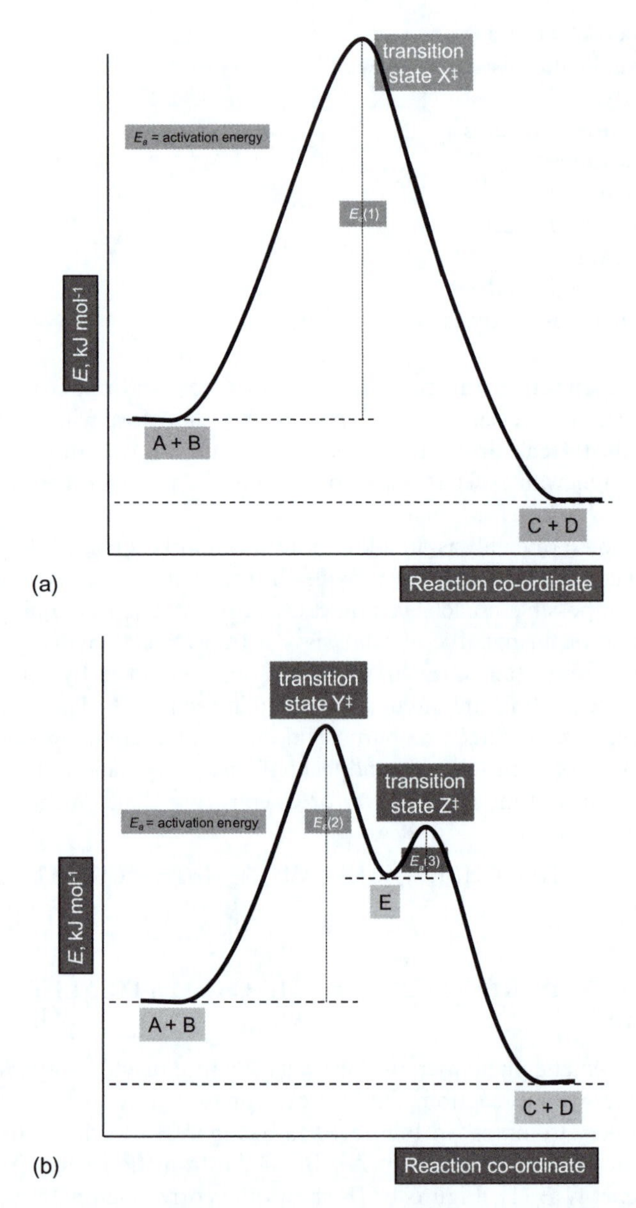

Figure 10.1 Reaction profile for molecules A and B reacting to give products C and D: (a) in an uncatalysed process; and (b) via an intermediate, E, as a result of the intervention of a catalyst the consequence of which is to lower the barrier between reactants and products ($E_a(1) > E_a(2)$).

consequence of a medium effect on a reaction resulting either from stabilisation of the transition state, $X^{\ddagger}$, or destabilisation of the reactants. It is important to distinguish between catalysis and medium effects, always bearing in mind that both may be operating in the chemical transformation of interest.

In general, catalysts affect the step in the sequence of chemical transformations from feedstock to required product that is rate-determining (otherwise, obviously, the overall rate would be unaffected). However, if the catalyst produces an intermediate of greater stability than that of the required product, then such a material (rather than the target) will accumulate. An example of this effect (discussed further in Section 10.5.2) is provided in ref. 3, in which an unwanted hydroxylamine intermediate accumulates (in addition to the desired amine) in the catalysed hydrogenation of *N*-cyclohexyl-*N*-methyl-2-nitrobenzenesulfonamide. The effectiveness of a catalytic transformation is governed, therefore, not simply by the activation energy of the rate-determining step but by the difference between this and the energy of the most stable intermediate. The same arguments apply to suggest that the reaction profile for the mechanistic cycle of the catalyst itself (see Scheme 10.17, for example) should similarly have no very high nor very low points. This is illustrated in Figure 10.2, where reaction profiles are shown for two hypothetical catalytic cycles that differ in the free energies of the intermediates and the height of the activation barriers between them.

It is also important to note that the precise nature of the entity responsible for the catalytic effect is not, without independent evidence, readily inferred from the initial nature of the putative catalytic material (the 'pre-catalyst')

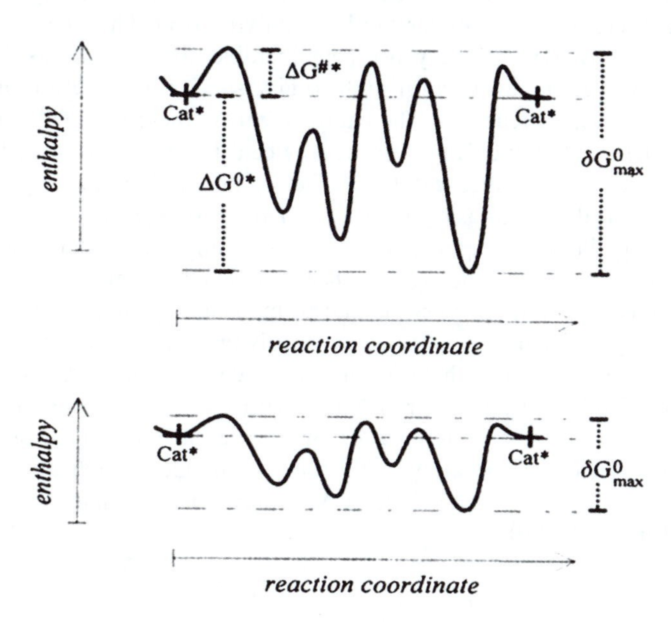

Figure 10.2 Two related reaction profiles (differing only in the heights of barriers and depth of wells) for a series of intermediates formed from a pre-catalyst in a catalytic cycle (in which the active catalytic species undergoes a series of transformations that ultimately results in its regeneration) and showing the importance of the relative free energy difference between them to the overall kinetic effect. Reproduced from ref. 4, with permission from Elsevier.

Table 10.1 Some general advantageous and disadvantageous characteristics considered to be typical of homogeneous and heterogeneous catalysis.

Homogeneous catalysis: catalyst contained in the same phase/medium as the catalysed reaction		*Heterogeneous catalysis: catalyst contained in a phase/medium different from the one in which the reagents are found*	
Advantages	*Disadvantages*	*Advantages*	*Disadvantages*
Very high rates Mild conditions Good selectivity Less subject to heat and mass transfer limitations Less affected by poisons	Difficult to separate and recycle May contaminate product Tend to be short-lived Difficult to reactivate	Readily separated especially if solid Easily recovered and recycled Easily reactivated Often very long-lived (many years) Suitable for use in continuous processes	May be mass transfer limited Surface area dependent Can be deactivated by poisons Batch-to-batch variation

added at the beginning of the reaction. The so-called 'active' species is often formed in a sequence of reactions from the pre-catalyst. It is one of the prime challenges of catalyst studies to identify what the true catalytically active species is and, thereby, seek to understand its catalytic effect. The true catalyst may be generated only slowly. It may also be vulnerable to side reactions that divert it down some additional reaction path to give species that are no longer catalysts for the desired reaction. The catalytic entity, therefore, will usually not have an indefinite lifetime. It may eventually be transformed to something with a much reduced or changed catalytic effect or no catalytic activity at all—at which point it is described as being 'deactivated' or 'poisoned' (if an identifiable component is largely responsible). Understanding the latter phenomena, therefore, is important to the process chemist in efforts to maintain the desired catalytic activity as long as possible and to manage any changes in activity.

Most catalysts for industrial-scale chemicals production fall into two basic types: homogeneous and heterogeneous. Each has advantages and disadvantages (Table 10.1), with variants and hybrids developed for specific needs. Generally, however, the preference in large-scale chemicals processing is for heterogeneous catalysts. Catalysis for biotechnological application is sufficiently distinct for it to be considered separately. Environmental catalysis is addressed in Section 10.8.

10.1.1 Heterogeneous Catalysts

Heterogeneous catalysts[5] are contained in a phase or medium different from the one in which the reagents are found. Heterogeneous catalysts are usually solids and are in contact with reactants in the gaseous or liquid (or solution) phase, often at high temperatures and pressures. Their chemistry is intimately bound up with the phenomena of adsorption and desorption, and associated surface phenomena. It is no coincidence that our understanding of heterogeneous catalysis has both motivated, and benefited from, development of a panoply of surface analytical techniques.[5b]

The activity of heterogeneous catalysts is dependent on characteristics such as porosity and surface area. Some solids (*e.g.* solid acids or bases) have intrinsic catalytic activity (with the physical form designed to maximise performance); other solids can be essentially inert with the catalytically active species distributed on or in them (giving **supported catalysts**); yet others will display both types of activity.

The key strength of heterogeneous catalysts is that they can be shaped (for efficient contacting with reactants), fluidised (for which mechanical stability is required) and separated. They are more easily recovered and recycled. They tend to be more readily reactivated and are often very long-lived (as much as several years).

Figure 10.3 displays schematically the essential features of a heterogeneous reaction. The basic individual steps involve diffusion of the reactants to the solid surface or the pore, adsorption at or near the active site, physisorption and chemisorption, followed by the catalysed chemical change at the active site and subsequent desorption and diffusion into the reaction medium. For porous materials, transport between the bulk medium and the active site may be relatively slow or unfavoured, leading to mass transfer limitations on the overall rate of reaction.

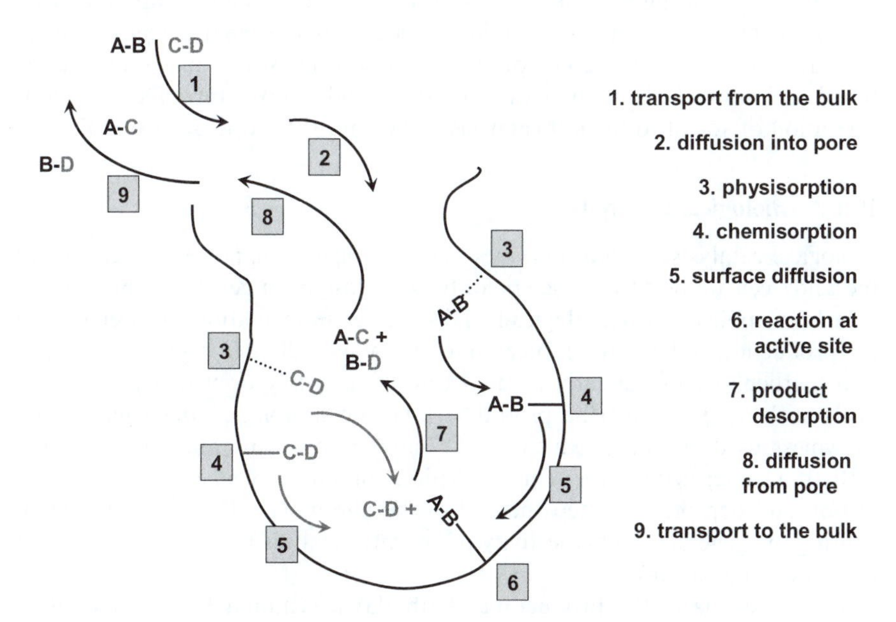

Figure 10.3 Essential features of a heterogeneous reaction, A–B + C–D → A–C + B–D, taking place in a porous material.

Heterogeneous catalysts can be deactivated by poisons though, as seen with 'coke' formation (Section 10.3.5), it is sometimes possible to reverse this. Because of their nature and the way heterogeneous catalysts function, activity can be dependent on the catalysts' support and there can be batch-to-batch variation associated with the process by which the catalytic species is introduced to the support material.

10.1.2 Homogeneous Catalysts

Homogeneous catalysts[6] are contained in the same phase or medium as that in which the catalysed reaction is carried out. This is usually the liquid phase. Homogeneous catalysis is thus concerned with the role of the reaction medium in its impact not only on the reaction chemistry, but also on the nature and behaviour of the catalytically active species. Homogeneous catalysis is generally (but not invariably) characterised by very high rates and the need for relatively mild conditions; it is less subject to mass transfer limitations than heterogeneous catalysts and less affected by poisons. On the other hand, homogeneous catalysts can be difficult to separate and recycle, leading to high costs associated with catalyst losses. While homogeneous acid or base catalysts are usually robust, those derived from transition metal co-ordination complexes can be short-lived and difficult to reactivate. On the other hand, the wide range of metals and ligands available provides scope for optimisation of activity, once found.

Much research is undertaken in attempts to have the best of both of these worlds by 'heterogenising' homogeneous catalysts by various means. These include incorporating a homogeneous catalyst into an insoluble (and therefore readily separable) polymeric scaffold and methods that enable ready catalyst separation from homogeneous solution using solvents whose properties can be tuned and controlled by changes in pressure and temperature. Detailed consideration of so-called '**hybrid catalysis**' is beyond the scope of this book.

10.1.3 Biological Catalysts

Biological catalysts[7] such as enzymes—either isolated, supported or retained in the whole cell, derived from bacteria or fungi, wild-type or genetically modified—have been applied biotechnologically. Historically, fermentation has been used in the production of beer, wine, bread and cheese as well as in chemicals production, particularly in the early manufacture of naturally occurring pharmaceuticals such as penicillin (the prototype β-lactam antibiotic) and vancomycin. Enzymes, used in the production of 7-aminocephalosporanic acid (Section 8.10.8), an intermediate for semi-synthetic antibiotics, avoid the wasteful protection and deprotection steps needed in the alternative chemical route, highlighting the generally high selectivity of enzyme catalysis to the desired product from a chosen substrate.

Sometimes, as in the production of riboflavin (vitamin B_2, Scheme 10.1), biocatalysis can bring about directly in a biosynthetic pathway what a chemocatalytic route requires several separate steps to achieve. But while in the

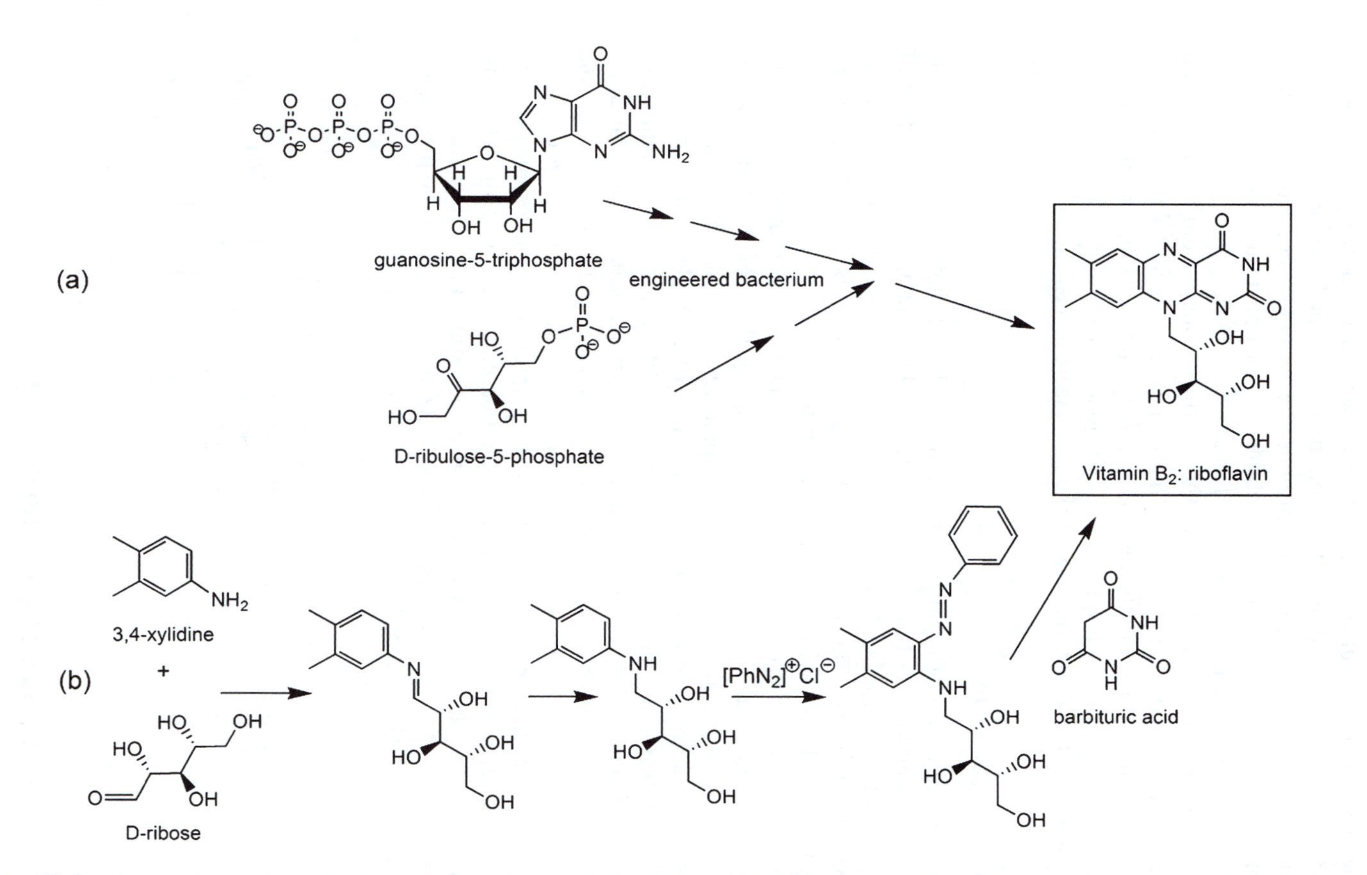

Scheme 10.1 Comparison of production processes to riboflavin (vitamin B_2): (a) a direct biocatalytic route using a genetically modified organism; and (b) a chemocatalytic route involving several individual stages.

case of vitamin B_2 significant savings compared with classical chemistry can be made, other processes suffer from relatively low rates, generate significant levels of additional biomass and must function in aqueous media at neutral pH. The particular benefits of biocatalysis[7] provide the basis of some technologically viable sustainable processes using vegetable oils, carbohydrates, lignocellulosic and other biomass-derived renewable feedstocks.

However, initially at least, most of the products sought from biomass (*i.e.* a hydrophilic and oxygen-rich feedstock) are likely to be hydrophobic and oxygen-poor. This will require the design, discovery and development of catalysts able to work in both hydrophilic and hydrophobic environments or, as in the case of a recent research report,[8] at the interface between the two.

10.1.4 Organocatalysis[9]

The catalytic effects of organic compounds on the rates of chemical reactions have long been known. Indeed, von Liebig noted in 1860 the catalytic effect of acetaldehyde on the hydration of cyanogen, $(CN)_2$, to form oxamide, $H_2NC(O)C(O)NH_2$. Furthermore, there are parallels between the role of enzymes and the behaviour of organocatalysts.

More recent interest in this topic is associated with efforts to promote reactions using small organic molecules that avoid the need to employ expensive metal-based catalysts. Industrial applications are currently modest—largely as a consequence of the high catalyst (10–20 mole%) concentrations needed. However, the classic example (Scheme 10.2), in which L-proline promotes a sequence of reactions, including an asymmetric intramolecular aldol reaction leading to the Wieland–Miescher ketone from readily available starting materials,[9d–e] has been carried out on the large scale to provide an intermediate needed in steroid synthesis and to prepare taxol. Recent research results[9f] have shown that the addition reaction (eqn 10.2; the conjugate addition of an aldehyde to a nitroolefin) can give good yields and high enantiomeric excess (see eqn 10.6) in the presence of as little as 0.1 mol% of the peptide H-Pro-Pro-Glu-NH_2 (Figure 10.4).

0.1 mol% catalyst
9:1 $CHCl_3$: *i*PrOH
N-methylmorpholine
RT, 48h

NO_2 NO_2 O O (10.2)

10.2 CATALYSIS IN THE ENVIRONMENT

Bearing in mind the universal applicability of catalysis to any chemical process, it will be no surprise that evidence of catalysed processes is to be found everywhere

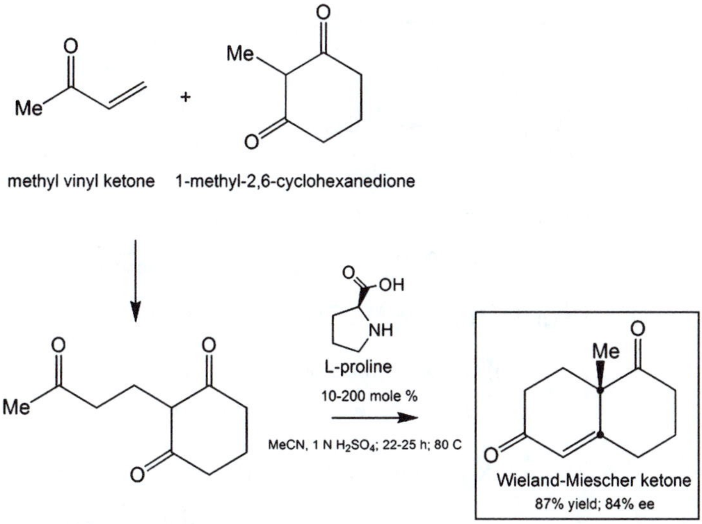
Me
Me
O
O
O
+
methyl vinyl ketone
1-methyl-2,6-cyclohexanedione
Me
O
O
O
OH
NH
L-proline
10-200 mole %
MeCN, 1 N H2SO4; 22-25 h; 80 C
Me
O
O
Wieland-Miescher ketone
87% yield; 84% ee

Scheme 10.2 Organocatalysis by L-proline of the asymmetric intramolecular aldol reaction to form a Wieland–Miescher ketone.

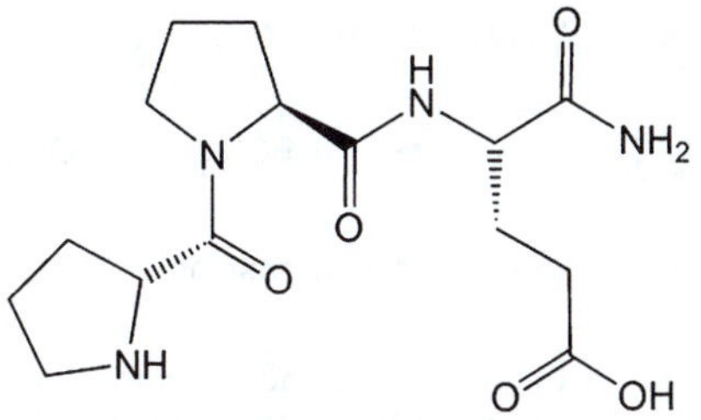
N
H
N
NH2
O
O
O
NH
O
OH

H-Pro-Pro-Glu-NH_2

Figure 10.4 A tripeptide (Pro = proline, Glu = glutamic acid) organocatalyst used (as the trifluoroacetate salt) in the synthesis of γ-nitroaldehydes[9f] (eqn 10.2).

in the natural world from the creation and destruction of ozone in the stratosphere (Scheme 10.3) to the role of enzymes in all life-processes such as the biological fixation of nitrogen[11] discussed in Section 8.10.6. It is also entirely possible that key steps that transformed molecular precursors into early forms of life ('abiogenesis') involved chemical catalysis. Seeking to replicate these so-called 'pre-biotic' processes is currently an exciting area of research (Section 8.1),[12] following the classic (non-catalytic) experiments of Stanley Miller and Harold Urey in which (racemic) amino acids were formed from mixtures of ammonia, methane, water and hydrogen subjected to simulated lightning. One fundamental and unresolved question is the origin of the so-called 'homochirality' of biology, as natural proteins consist of only L-amino acids, not their D form. The exquisite

$$O_{2\,(g)} + h\nu \xrightarrow{\lambda < 242\ nm} 2\,O_{(g)}$$

$$O_{2\,(g)} + O_{(g)} + M \rightarrow O_{3\,(g)}$$

ozone formation

$$O_{3\,(g)} + h\nu \xrightarrow{\lambda\ 200\text{-}300\ nm} O_{(g)} + O_{2\,(g)}$$

$$O_{3\,(g)} + O_{(g)} \rightarrow 2\,O_{2\,(g)}$$

ozone decomposition

(a) Chapman reactions

$$O_{3\,(g)} + X_{(g)} \rightarrow XO_{(g)} + O_{2\,(g)}$$

$$XO_{(g)} + O_{(g)} \rightarrow X_{(g)} + O_{2\,(g)}$$

catalysed decomposition

$$O_{3\,(g)} + O_{(g)} \rightarrow 2\,O_{2\,(g)}$$ net reaction

$X = NO^{\bullet},\ N_2O,\ Cl^{\bullet},\ Br^{\bullet},\ H^{\bullet},\ OH^{\bullet}$

$$CFCl_{3\,(g)} + h\nu \rightarrow {}^{\bullet}CFCl_{2\,(g)} + Cl^{\bullet}_{(g)}$$

(b) Ozone decomposition

Scheme 10.3 Basic steps in the formation and decomposition of ozone in the stratosphere (based on ref. 10): (a) the 'Chapman' reactions for the photogeneration and decomposition of ozone; and (b) catalysed decomposition of stratospheric ozone by species, X, such as chlorine atoms, present naturally or anthropogenically (*e.g.* from the photolysis of long-lived chlorofluorocarbons).

selectivity of biochemical transformations has motivated efforts first to understand and then to mimic the role of enzymes, cells and organisms in synthetic or engineered catalytic systems. Of particular interest are efforts to synthesise chiral intermediates biotechnologically and to effect biochemical transformations of renewable feedstocks, both of which we explore further below.

10.3 MEASURING CATALYST PERFORMANCE

The most important empirical measures of catalyst technical performance are:

- turn-over number (productivity)
- turn-over frequency (activity).

These parameters, however, give no indication of the identity of the catalytically active species, no insight into the mechanisms of catalytic action nor even whether or not all of the added pre-catalyst is involved in the catalytic reaction. These are all investigated using the full range of physical and analytical techniques.[5–7]

10.3.1 Catalyst Productivity

Catalyst productivity is measured using the turn-over number ('ton'), the amount of target material produced per unit of catalyst. It can be expressed on a mole (eqn 10.3) or mass (eqn 10.4) basis—so it is important to note the units used.

When the identity of the active catalyst is unknown, the quantity used in the denominator is the mass or number of moles of the catalyst precursor, or (in the case of processes involving metal catalysts), the mass or number of gram atoms of the metal present.

$$\text{ton} = \frac{\text{mole product}}{\text{mole catalyst}} \tag{10.3}$$

or

$$\text{ton} = \frac{\text{mass product}}{\text{mass catalyst}} \tag{10.4}$$

A catalyst with a turn-over number of 10 000 (on a weight basis) would give 1 000 000 g (1 tonne) of product from 100 g catalyst. It is evident that for a catalyst costing £100 g^{-1} the cost of the catalyst per tonne of product would be £10 000. The higher the cost of the catalyst, the higher must be the turn-over number to achieve the same cost of catalyst used per tonne of product.

Because of the very high value of some pharmaceutical products, it is sometimes acceptable to use catalysts with quite low turn-over numbers, but even here they must usually be >100, particularly if the unit cost of the catalyst itself is very high. Typically, tons of >1000 are required for other high-value products and $>50\,000$ for higher volume products.

10.3.2 Catalyst Activity

The measure of catalyst activity is the turn-over frequency ('tof'). This represents the number of moles (eqn 10.5) or mass of product produced per unit of catalyst per unit time.

$$\text{tof} = \frac{\text{mole product/hour}}{\text{mole catalyst}} \tag{10.5}$$

or

$$\text{tof} = \text{ton}\,\text{h}^{-1} \text{ (note that 'ton' here is turn-over number.)}$$

A tof of 1000 is equivalent to 0.1 mol% catalyst giving 100 moles of product in one hour. Values $>500\,\text{h}^{-1}$ are required for small-scale processes and $>10\,000\,\text{h}^{-1}$ for the large scale.

10.3.3 Reactor Productivity

Ton and tof are of central importance in industrial chemicals manufacture as they determine the productivity of a reactor or process. Productivity (or space-time yield, STY) is a critical metric in assessing process economics (Section 9.3) as it dictates the size of reactor (and hence cost) needed to produce the required quantity of material.

To deliver a defined quantity of product (of the desired quality) in a set time, a high tof catalyst, generally, will require a smaller reactor compared with that required for a low tof catalyst. High tof catalysts are often relatively short-lived (being subject to poisoning). Such catalysts require frequent reactivation or replacement. The associated plant shut-downs clearly limit plant availability and overall process productivity.

Other important metrics or technical characteristics of catalysts are their selectivity and lifetime.

10.3.4 Selectivity

Ideally, a catalyst should be specific, a term restricted to those catalysts that are perfectly, or 100%, selective to the desired product. Catalysts typically are not specific. An acceptable selectivity will be dependent on particular circumstances, though they are usually required to be better than >95% selective to enable the value of the additional product to offset the extra costs associated with the purchase, manipulation and recovery of catalyst.

Selectivity may take the form of chemo-, regio-, stereo- or enantioselectivity.[13a]

10.3.4.1 Chemoselectivity. Chemoselectivity[13b] (Web 2a) describes the extent to which a chemical reagent, when free to react with a number of functional groups, reacts preferentially with one of them. Greater chemoselectivity is likely the greater the difference between the chemical characteristics of the functional groups concerned.

For example, the reduction by sodium borohydride of the C=O in the conjugated enone (Scheme 10.4), in which the C=C is also susceptible to hydrogenation, can be carried out chemoselectively. No saturated alcohol is formed if cerium(III) chloride is present, though stereoselectivity is poor. A similarly selective reduction of a ketone in the presence of an aldehyde group is ascribed[13d] to the conversion of the aldehyde into a geminol diol coordinated to cerium(III).

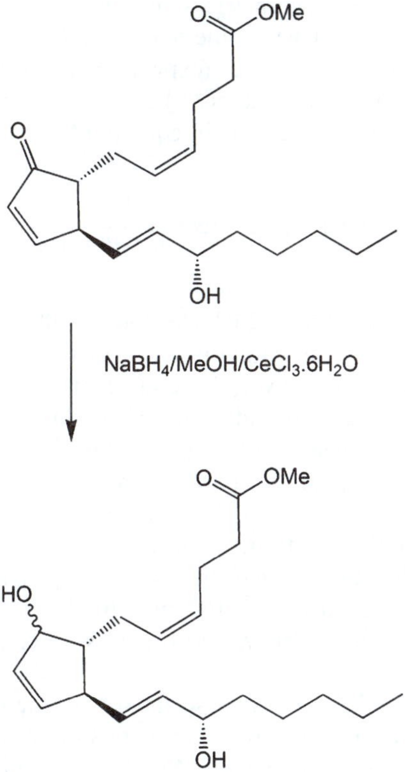

Scheme 10.4 reaction: NaBH$_4$/MeOH/CeCl$_3$.6H$_2$O

Scheme 10.4 Chemoselective reduction of a ketone function in a conjugated enone.[13c]

10.3.4.2 Regioselectivity. Regioselectivity (Web 2b) is demonstrated when, in a reaction, one process of bond making or breaking in a molecule occurs preferentially to all other possible processes.

A reaction outcome can favour one regioisomer over another as a consequence of differences in reaction mechanism arising from a change of reagents. For instance, using 70% perchloric acid, the acid-catalysed addition of a benzenethiol to an alkene produces[13e] RCH(SPh)CH$_3$, whereas radical addition (using azodiisobutyronitrile, Me$_2$C(CN)N=NC(CN)Me$_2$, as initiator) gives the other product,[13f] RCH$_2$CH$_2$SPh.

10.3.4.3 Stereoselectivity. Stereoselectivity (Web 2c) is the preferential formation of one stereoisomer over another. When the stereoisomers[iii] are enantiomers, the phenomenon is called **enantioselectivity** (Section 10.5.3); when they are diastereomers, it is called **diastereoselectivity**.

[iii] Enantiomers are non-superimposable mirror images of one another. Diastereomers are stereoisomers that are non-superimposable non-mirror images of one another.

The effectiveness of an asymmetric synthesis is described in terms of its enantiomeric excess (ee) or diastereomeric excess. The ee of a mixture of two enantiomers is the absolute difference (expressed as a percentage) between the mole fraction of the two enantiomers (*R*) and (*S*) or (in terms of the molar concentrations of the two isomers) as in eqn (10.6):

$$\text{Enantiomeric excess (ee)} = \frac{([R]-[S])}{([R]+[S])} \times 100 \qquad (10.6)$$

For a mixture of 75% (*R*) and 25% (*S*), the ee will be 50%.

It should be noted that as catalysis is a kinetic phenomenon, the value of ee may change with time over the course of the reaction as reactants are converted to products. In addition, if a chiral precursor is reacted with a racemic substrate, then the maximum yield of the desired diastereomeric form of the product can be no more than 50%. (This is often a limitation of enzyme catalysis as, usually, only one enantiomer of a racemate is a substrate.) If both enantiomers of the racemic substrate can equilibrate with one another then it may be possible, by dynamic kinetic resolution (DKR),[14] to raise this, in principle, to 100%, with consequential improvements in reaction efficiency (and the possible elimination of a process step).

A maximum yield of 50% is also associated with the use of diastereomeric crystallisation, favoured industrially because of its simplicity. (1*S*, 4*S*) sertraline [the active pharmaceutical ingredient (API) of the anti-depressant, Zoloft] can be separated from the other three diastereomers by crystallisation with (*R*)-mandelic acid (Scheme 10.5). Concurrent DKR is not possible because the temperature at which the racemisation catalysts were active was too high in the chosen solvent to permit crystallisation of the mandelate salt. After separation of the (1*S*, 4*S*) sertraline (*R*)-mandelate, the waste amine [(1*S*, 4*R*), (1*R*, 4*S*) and (1*R*, 4*R*)] was racemised at the amino-chiral centre (*via* the imine intermediate **A**) using $[Ir(\eta^5\text{-}C_5Me_5)I_2]_2$ and at the methine chiral centre using *t*-butoxide.[14b]

The development of chiral stationary phases for use in liquid chromatography has allowed racemic mixtures to be resolved. However, the additional costs for equipment and the large volumes of organic solvents used as elutants are such that the use of chromatographic separation on the industrial scale will only be attractive if the procedure allows a shorter sequence of reactions to be employed and the solvent used in the separation can be readily recycled. This is the case[14c] for the manufacture of chiral *exo*-2-norbornyl thiourea (Figure 10.5), the starting material in the synthesis of an API (**B**) that may form the basis of a treatment for type 2 diabetes. The thiourea could be obtained in high optical purity either by diastereomeric salt formation of *exo*-2-aminonorbornane with (*R*)-(+)-*N*-(1-phenylethyl)phthalamic acid (Figure 10.5) or by a chromatographic separation using methanol as the mobile phase.

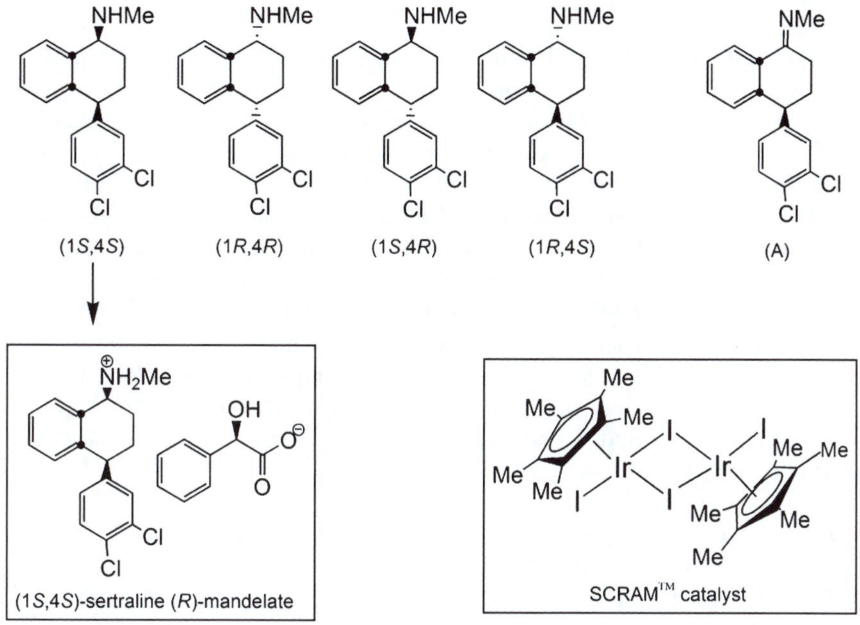

NHMe NHMe NHMe NHMe NMe
Cl Cl Cl Cl Cl Cl Cl Cl Cl Cl
(1*S*,4*S*) (1*R*,4*R*) (1*S*,4*R*) (1*R*,4*S*) (A)

(1*S*,4*S*)-sertraline (*R*)-mandelate

SCRAM™ catalyst

Scheme 10.5 Separation of (1*S*, 4*S*)-sertraline from (1*S*, 4*R*), (1*R*, 4*S*) and (1*R*, 4*R*) diastereomers by crystallisation with (*R*)-mandelic acid. After separation, the waste amines can be racemised at two centres using the iridium catalyst (SCRAM) for one—*via* the imine intermediate (A) — and *t*-butoxide for the other.[14b]

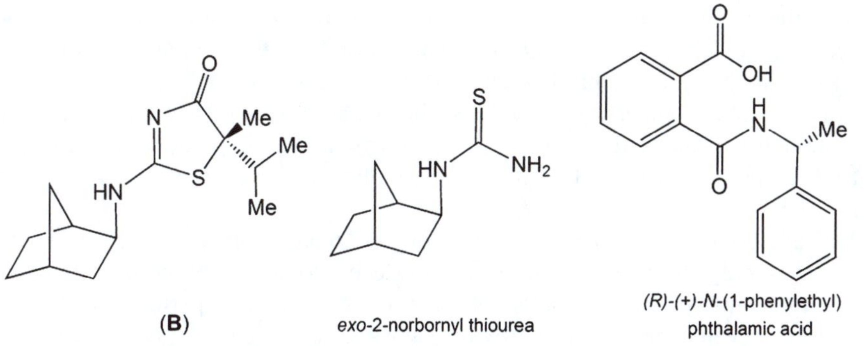

(B) *exo*-2-norbornyl thiourea *(R)-(+)-N-*(1-phenylethyl) phthalamic acid

Figure 10.5 The preferred production of the precursor, chiral *exo*-2-norbornyl thiourea, to the active pharmaceutical ingredient, **B**, is effected using large-scale chiral chromatography rather than *via* the diastereomeric salt formed between the phthalamic acid and *exo*-2-aminonorbornane.[14c]

10.3.5 Lifetime

Catalyst lifetime (the time taken for the catalyst to lose activity, productivity or selectivity sufficient to render the process uneconomic or inoperable) determines times between shut-downs and can span values from hours to decades. In the case of some short-lived catalysts and where no alternatives are available, additional process steps are included to effect in-process reactivation. An example is fluid catalytic cracking (FCC) used in processing the high-boiling high-molecular weight components of crude oils. The acid zeolite catalyst used is rapidly deactivated by the deposition of carbonaceous material or 'coke'. This is burned off in a separate catalyst regenerator. For a FCC unit processing 75 000 barrels of oil each day, it is necessary to circulate over 50 000 tonnes of catalyst between the reactor and regenerator in the same time (Web 3).

To be attractive an *industrial* catalyst must also meet some techno-commercial criteria that we note at this point for completeness including preferably a requirement to be:

- easy to recover
- capable of ready reactivation
- tolerant to functional groups
- commercial available
- free for use without licensing or royalty costs.

10.4 CATALYSIS AND SUSTAINABILITY

In Section 3.2 we concluded that efforts to reduce our burden on the Earth's ecosystems rely on significant improvements in the burden per unit of economic activity; in other words, the reduction in the impact of processing material extracted from the environment per unit of economic value obtained. While there has been a historic trend (particularly in the developed and the recently emerged industrial economies) that has reduced this by about a factor or 2–2.5 over the last 30 years, it is now recognised that this historical trend must be accelerated. This can only be achieved consensually through technological development. Catalysis has had an important role to play in past improvements in technological efficiency. It will also be central to the development of more sustainable chemical technologies[15] as efforts are made to dematerialise and transmaterialise them. The latter will involve the greater use of renewable resources, the focus of Chapter 11.

While past environmental benefits of new technology have largely arisen coincidentally from the economic and technological motivation for their development rather than having been their dominant impetus, nevertheless there has been more recently an increasing focus on pollution prevention and waste minimisation. This has come about because of the need to comply with an increasing number of legal and regulatory obligations, not only relating to the process of production (which will be under the manufacturer's control), but to product use and its ultimate fate or disposal after use (which may not). The continuing contribution of catalytic technology to these improvements can be seen as anticipating the ninth green chemistry principle, which states that 'selectively catalysed processes are superior to stoichiometric processes'. This was discussed in Chapter 8 where, indeed, we saw that the atom efficiencies of catalysed reactions were significantly higher than those of corresponding reactions involving stoichiometric reagents.

The differences in Sheldon's E-factor for different sectors of the chemical industry (Table 6.1) become more understandable when we consider the different use made of catalysts in the different sectors. The better relative performance (using this comparison) of the petrochemical and commodity sectors is, in some measure, associated with the greater use (in fact, the central and essential use) of catalysts compared with the fine chemical and pharmaceutical sectors, where the use of catalysts is less.

10.5 CATALYSIS IN INDUSTRY

Evidence for the current importance of catalytic technology can be seen from the following data:

- about 80% of chemical processes employ catalysts
- 35% of annual global GDP (global GDP in 2008 was US$60.6 trillion; Web 4) depends in some way or other on catalysis
- the production and sale of catalysts was a US$11 billion y^{-1} business in its own right in 2005
- 130 catalysts were commercialized in the USA alone during the 1990s.

Most of these developments came with direct and indirect environmental benefits.

We have already compared the characteristics of the different chemical sectors and it is instructive to examine the way in which catalysts might be used in each.

10.5.1 Commodity Chemicals

Commodity chemicals such as sulfuric acid, ammonia, chlorine, sodium hydroxide, nitric acid, benzene, ethene, methanol, poly(ethylene)[16,17] are produced on a scale of millions of tonnes per year. Processes to make them, in the main, employ catalysts (chlorine and sodium hydroxide production being an exception).

High-volume production is carried out in dedicated plant operated continuously. The construction of such production units and associated infrastructure can involve the investment of hundreds of millions of dollars of capital. Bearing in mind this money will be spent before any product can be made and the investment recouped, there has to be a high probability that the technology will work as expected. For this reason (as we explored in

Chapter 9), operators of such technology are slow to introduce so-called 'step-change' innovations (*i.e.* those that involve radical new processes, unproven on the very large scale), preferring to carry out evolutionary changes to the existing process.

An example of the latter might be the development of an improved catalyst that can be introduced with a minimum of plant or process modification. This might, for example, enhance the selectivity to the required product by as little as 1% but, when seen against the annual production of (say) 500 000 t of a product selling at £250 t^{-1}, this can represent increased production of 5000 t y^{-1} generating additional sales income of £1.25 million y^{-1}.

Such developments arising from the R&D carried out by chemical companies provide commercial advantage and their nature is generally kept confidential (a general problem when attempting to review industrial applications of catalysis). Some developments, however, may be of sufficient significance to be considered worth protecting by patents (Section 9.5), particularly if royalty income can be made by licensing the technology to another company.

10.5.2 Fine Chemicals, Pharmaceuticals and Agrochemicals

Fine chemicals, pharmaceuticals and agrochemicals are manufactured in smaller volumes say in the range 1 to 10 000 t y^{-1}. Such 'effect' chemicals generally have a relatively short life as marketed products. This arises because new compounds are continually under development that display improvements in efficacy or safety compared with the current materials, or are capable of producing the same effect more cheaply. Effect chemicals may also be protected by patents. Patents only give protection for a limited time and can be circumvented if the effect can be produced in a novel and non-obvious way or, demonstrably, in a significantly better way.

A critical factor in the development of an effect chemical is the so-called 'time-to-market'—the time taken to make the chemical available for sale after the decision is made to offer it. As patent lifetimes are of finite length, the objective is to get the product to market (and to satisfy all the regulatory requirements) as soon as possible so as to maximise the time in which the product can be sold before someone else is free to offer it. At the end of a patent's life, its protection is lost. It is then possible for another company (which does not have to recoup the initial R&D costs that led to its discovery) to make the compound (a so-called 'generic' product) in competition with the developer.

The high E-factor in the production of fine chemicals and pharmaceuticals raises the question of the degree to which the volume of waste could be reduced by the greater user of catalysts in their manufacture. The relatively limited implementation of catalytic (or other innovative process) developments in large-scale production of effect chemicals (beyond the use of 'traditional' catalysts such as Raney nickel for hydrogenations) arises largely from pragmatic considerations.[18] Any development applied while bringing an effect chemical to market must work when required. An unproven novel development is likely to be judged to represent a risk that might delay the programme of development and introduction unacceptably. As the production of pharmaceuticals and other complex organic molecules usually involves multi-step syntheses *via* intermediates containing a number of functional groups usually carried out in multipurpose batch reactors, the ability to achieve E-factors similar to those of the commodity sector may be unrealistic, even with the greater use of catalysts.

To be effective, a catalyst must be selective (hopefully specific, *i.e.* 100% selective) in only catalysing the reaction of the target functional group and leaving unaffected all the other functional groups. A traditional approach to ensuring only the desired functional group reacts has been to convert the group where reaction is not wanted into an inert derivative—so-called 'protection'. After the target functionality has been transformed the protected group can be 'deprotected'. This process of protection/deprotection is extremely wasteful and its avoidance, as expressed in the eighth green chemistry principle (Section 8.10.8), would improve the atom efficiency (AE) of many production routes to fine and pharmaceutical chemicals.

An example is provided by the selective reduction of a nitro group to an amino group in the conversion of the nitroaryl-substituted allyl ester to the aminoaryl-substituted allyl ester (eqn 10.7).

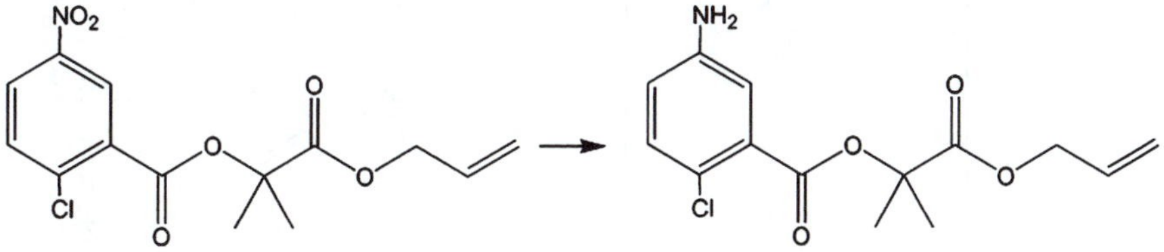

(10.7)

It is worth noting that improved selectivity to a sought-after product can also be achieved by the suppression of by-product formation. This is exemplified in Scheme 10.6 for the reduction of $-NO_2$ to $-NH_2$. On treatment with 20 bar H_2/5% Pd-C/AcOH at 120 °C, *N*-cyclohexyl-*N*-methyl-2-nitrobenzenesulfonamide (**A**) gives 41% of hydroxylamine intermediate (**B**). The accumulation of this intermediate can be limited to $<1\%$ in presence of ammonium vanadate, which catalyses the conversion of the intermediate to the desired product (**C**).

10.5.3 Enantioselectivity and Asymmetric Catalysis

The drug thalidomide is enantiomeric (Figure 10.6). In the 1960s, it was prescribed to pregnant women as a racemic mixture. Unfortunately, the (*S*)-enantiomer is teratogenic[iv] and thalidomide was withdrawn. The (*R*)-enantiomer is the active pharmaceutical and has a satisfactory therapeutic ratio

[iv] A teratogen is an agent that causes defects in the growing foetus during pregnancy.

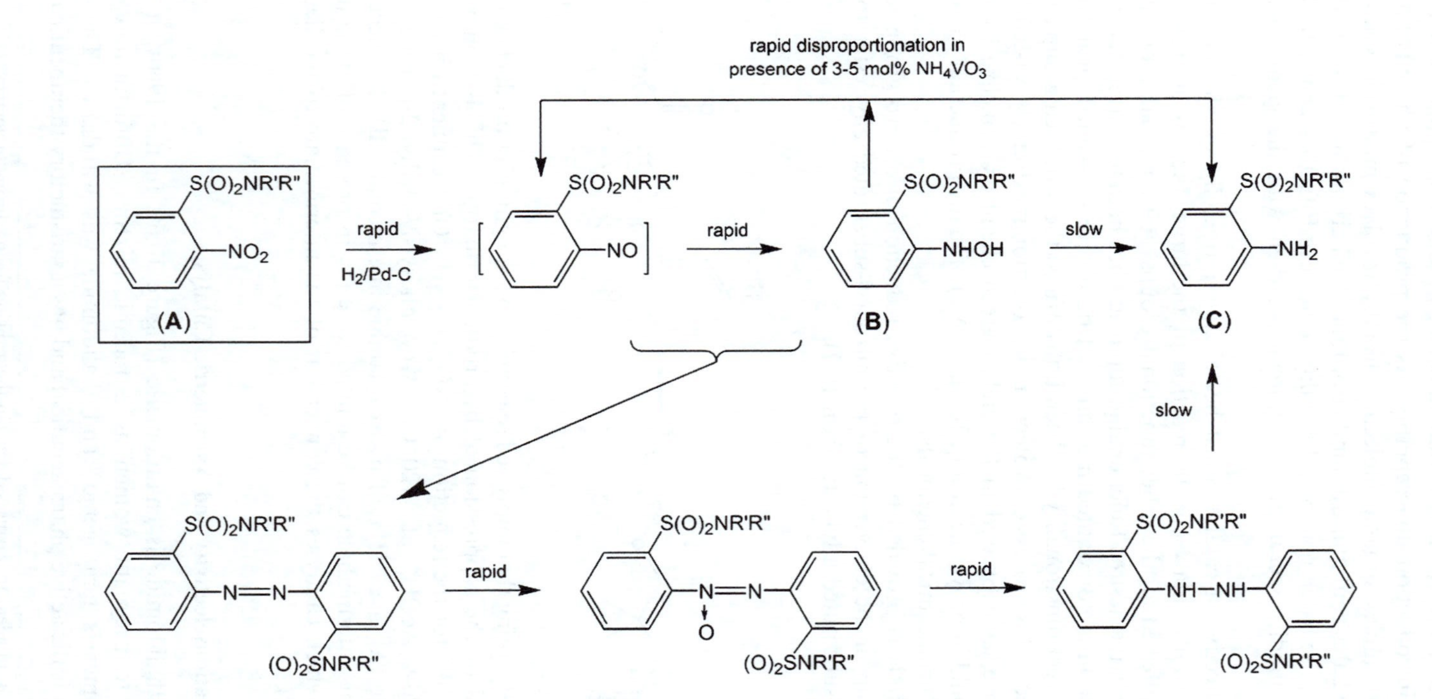

Scheme 10.6 Catalytic conversion of the nitro-group in *N*-cyclohexyl-*N*-methyl-2-nitrobenzenesulfonamide (**A**) to an amino group (**C**) revealing the series of reactions that enable selectivity to the desired product to be improved through the use of ammonium vanadate to suppress the accumulation of the intermediate (**B**).[3]

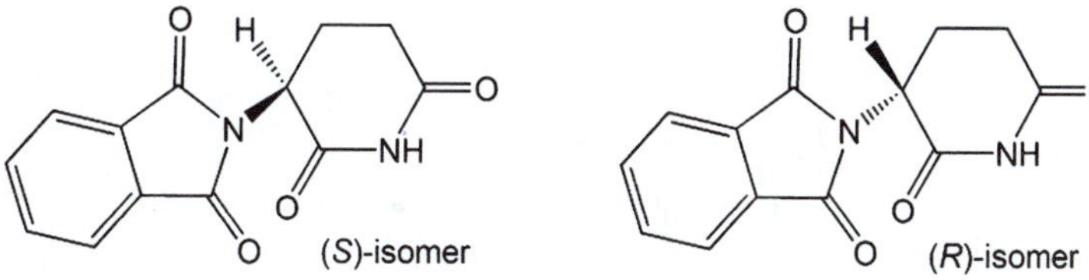

Figure 10.6 (*S*)- and (*R*)-thalidomide.

for other treatments.[v] It is, therefore, now a requirement that a chiral pharmaceutical contains only the enantiomer or diastereomer that is the pharmacologically active component. In the case of thalidomide, studies of its metabolic transformation and breakdown revealed, however, that the safe (*R*)-form is epimerised to the teratogenic (*S*)-form in the body.

The need for such enantiomeric and diastereomeric selectivity has motivated an intense research effort into chiral synthesis, much of it into chiral catalysis and chiral catalysts. This approach is well-exemplified by enantioselective hydrogenation, a method used in the synthesis of the amino acid, L-DOPA (3,4-dihydroxy-L-phenylalanine), a treatment of (among other things) Parkinson's Disease.[19] The Nobel Committee awarded one-fourth of the 2001 Prize in Chemistry (shared with Ryoji Noyori and K. Barry Sharpless) to William S. Knowles (Web 8) for his work on chirally-catalysed enamide hydrogenation, the most important being the synthesis of L-DOPA. The process is shown in Scheme 10.7 with [Rh(cod)L_2] as the hydrogenation pre-catalyst. The current technology[20] uses a Rh-dipamp catalyst (Figure 10.7) giving 95% ee (Section 10.3.4), with a ton of 10 000–20 000 and tof of 1000 h^{-1}.

Because of the unforgiving requirements of industrial and commercial operation (high ee + high ton + high tof), relatively few other enantioselective catalysts are used on the large scale. A Rh-binap catalyst is used industrially to produce L-menthol (Scheme 10.8) (97.6% ee, ton = 400 000, tof = 1 300 h^{-1}). The largest volume process currently operated[21] makes 10 000 t y^{-1} of the herbicide, (*S*)-metolachlor. The initial process (Scheme 10.9) produced a racemic mixture. However, using an iridium-xyliphos catalyst (Scheme 10.10), the imine intermediate (**D**) formed from 2-methyl-5-ethylaniline and methoxyacetone is asymmetrically hydrogenated to (**E**) with 80% ee, ton = 1 000 000 and tof = 180 000 h^{-1}.

10.5.4 Feedstock Changes

Catalytic technology develops in response to changes in feedstocks. Over the past 50 years or so, there have been changes of feedstock from coal [alkyne (ethyne) based chemistry] to oil [alkene (ethene) based chemistry] to natural gas [alkane (methane and ethane) based chemistry].[22] As the feedstock is changed,

[v] While thalidomide has more recently been found to be an effective treatment for a serious complication arising from leprosy, the World Health Organization does not recommend its use because of difficulties of foolproof patient surveillance (Web 7).

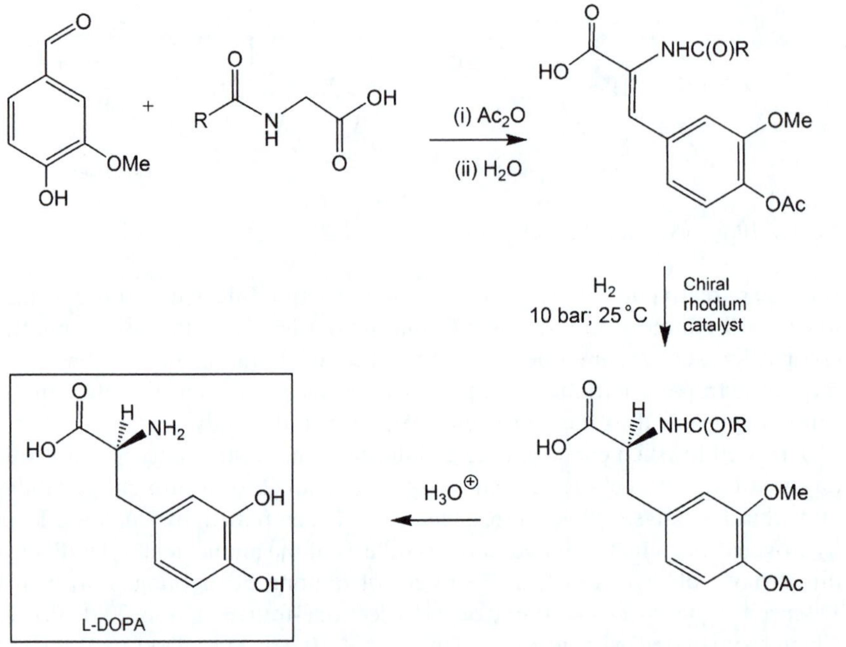

Scheme 10.7 The industrial process to L-DOPA, an enantioselective catalysed ene-amide hydrogenation using a chiral rhodium pre-catalyst, $[Rh(cod)L_2]BF_4$ (cod = 1,5-cyclooctadiene; examples of L_2 are shown in Figure 10.7).

the process chemistry and process engineering must be changed as well. For these reasons, the transition from one feedstock to another may take many tens of years.[vi] Even the transition from propene to propane in the manufacture of acrylic acid is not straightforward. Most of the world's 2.7×10^6 t annual production of acrylic acid is currently manufactured (in 87% yield) *via* a two-step process from propene (eqn 10.8 and eqn 10.9), the first carried out at 320 °C and the second at 210 °C.

$$CH_2{=}CHCH_3 + O_2 \rightarrow CH_2{=}CHC(O)H + H_2O \quad (10.8)$$

$$CH_2{=}CHC(O)H + 0.5\,O_2 \rightarrow CH_2{=}CHC(O)OH \quad (10.9)$$

In a life-cycle comparison of this process with an alternative, single-step, process from propane (eqn 10.10), the latter is predicted[23] to have a lower environmental impact and break even economically taking into account raw

[vi] Indeed, this process may never be entirely complete, particularly when special circumstances (*e.g.* the impact of global politics) may deny access of a country to a raw material supply. During the apartheid era in South Africa, there was a partial trade embargo that restricted the supply of oil. South Africa was, therefore, obliged to develop its own technology to use its indigenous coal as the basic raw material for making transportation fuels, something it continues to do today.

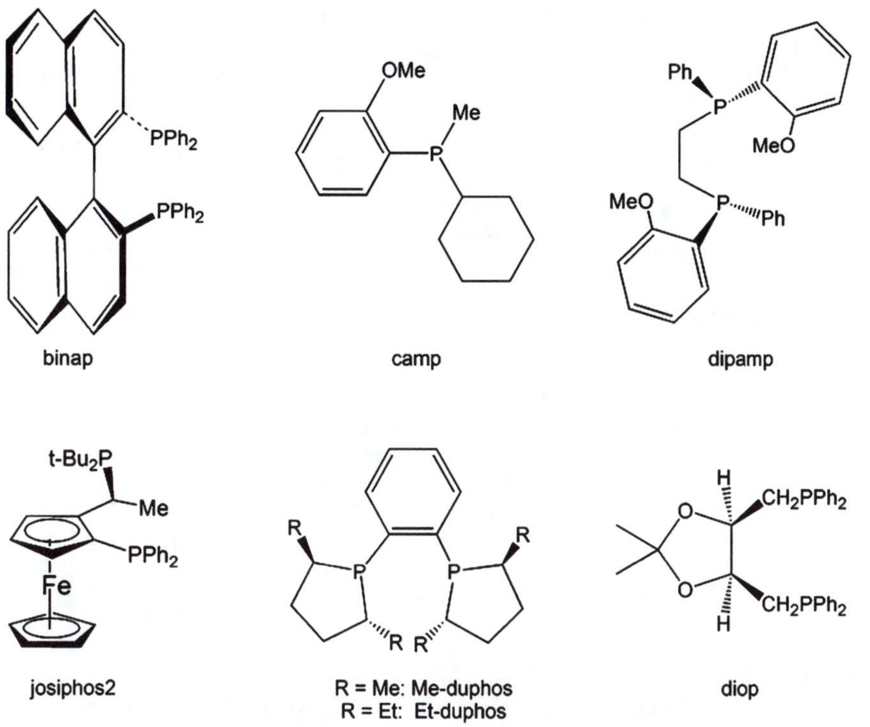

Figure 10.7 Representative examples of chiral bidentate ligands, L_2, used in enantioselective hydrogenations such as that shown in Scheme 10.7.

material costs) if the yield is 59–60%. The best yield reported for a mixed-metal oxide catalyst is 48% (at 380 °C). It is believed[23] that such evaluations can be used to guide the research and development of new catalysts for this transformation.

$$CH_3CH_2CH_3 + 2\,O_2 \rightarrow CH_2{=}CHC(O)OH + 2\,H_2O \qquad (10.10)$$

From a global perspective, however, the next step is to begin the transition to renewable feedstocks (*e.g.* biomass) as the basis of future product chains that meet the chemical and energy needs of society. This is covered in the next chapter.

10.5.5 Catalyst Discovery and Combinatorial Catalysis

Competitive pressure and academic curiosity[1] combine to ensure that catalyst research and development continues its search for innovations able to bring about enhanced process efficiency and cost improvements. Some of this is motivated by questions of waste reduction or the utilisation of a renewable feedstock as a route to a new 'platform' chemical; some applications have resulted from the discovery of wholly new transformations that cannot be

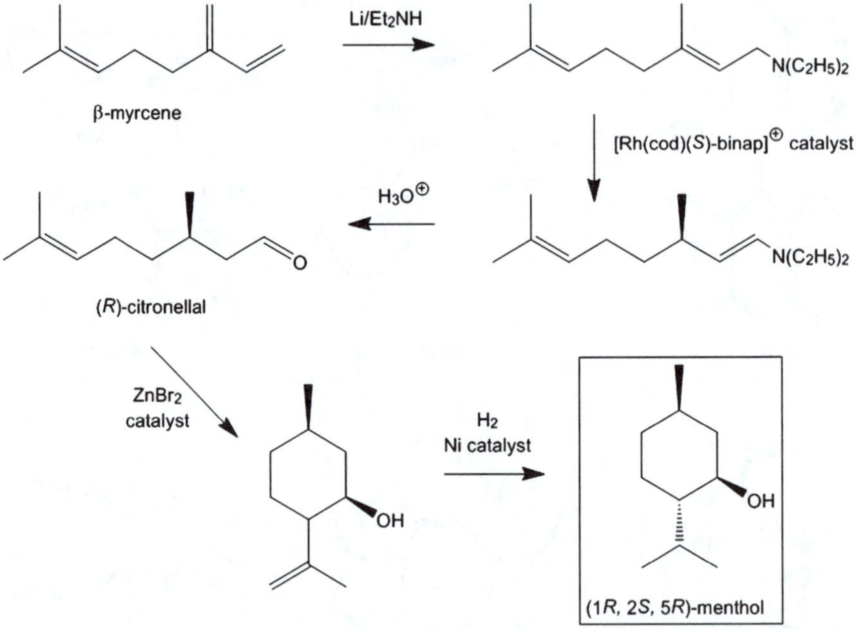

Scheme 10.8 Industrial production of L-menthol in which a chiral pre-catalyst, $[Rh(cod)(S)\text{-binap}]^+$, is used for the asymmetric isomerisation.

achieved without the intervention of a catalyst. Some useful and interesting examples are provided in the following sections and in subsequent chapters.

An issue of some importance is that, despite all of the advances in understanding of catalysts and the way they work, it is impossible fully to design *a priori* a new catalyst to carry out a novel transformation. Understanding and experience will certainly get part of the way towards the target. However, much industrial catalyst discovery and development involves a process called 'screening' in which large numbers of candidate catalysts, usually variants of compositional or structural themes, are tested on the small scale to find the most active material (the 'lead') for further study and possible development. A major innovation of the last decade has been the use of automated high-throughput or combinatorial methods (Web 9) to screen thousands of materials for activity. Purposeful discovery using such methods (see, for example, ref. 24) is now a feature of many areas of chemical technology, not simply in catalysis.

10.5.6 Catalysis and Economics

The benefits of catalysis (and the corresponding reductions in waste) have been driven in the past primarily by competitive pressure or the seeking of economic or commercial advantage. This will continue to be so. Much will depend (as it has always done) on process economics. On the other hand, economics of chemicals production will, increasingly, be influenced by the requirement to

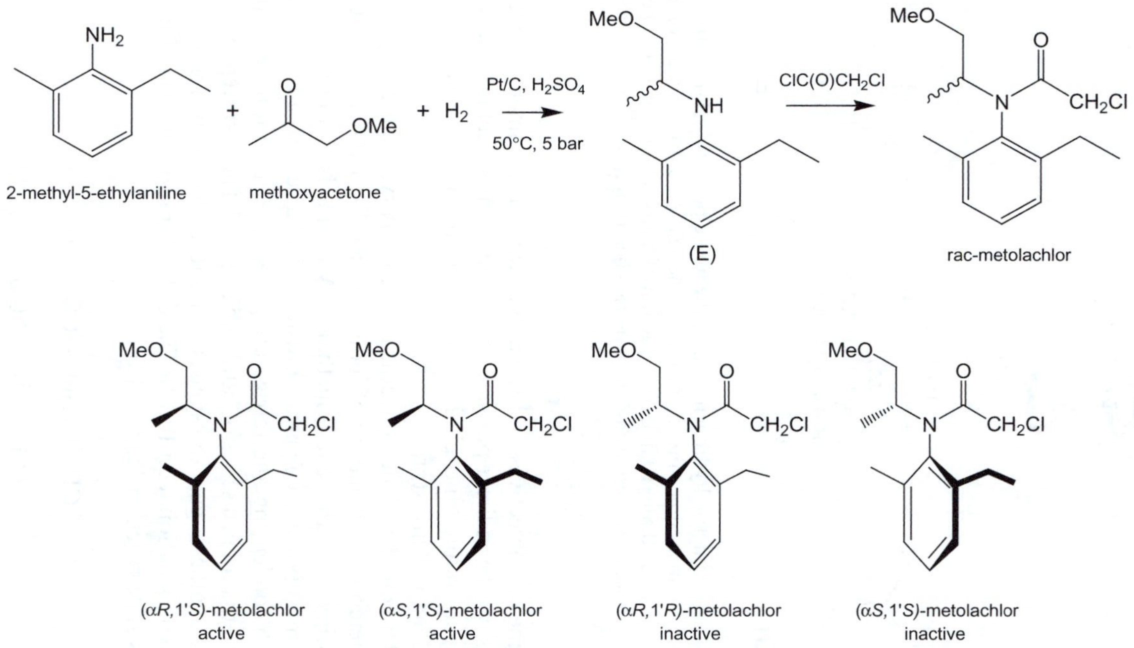

Scheme 10.9 The herbicide metolachlor, initially produced as a racemic mixture comprising active and inactive isomers, *via* the intermediate (**E**) from 2-methyl-5-aniline and methoxyacetone.[21]

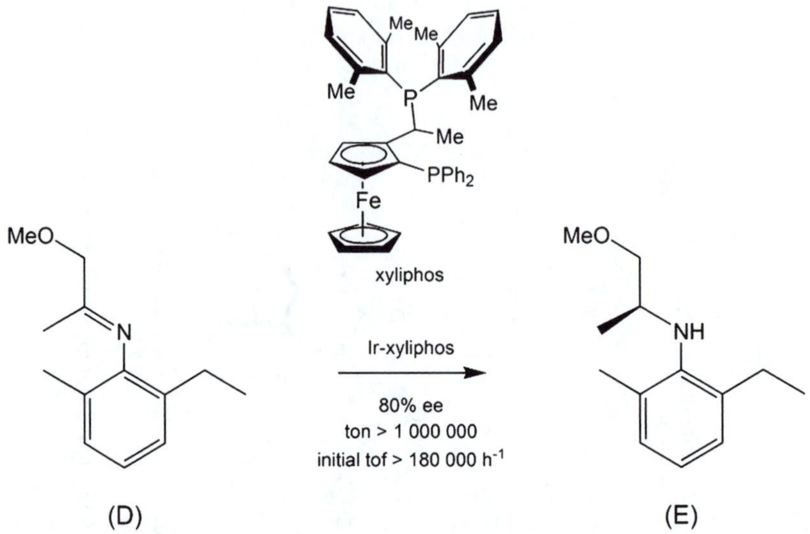

Scheme 10.10 Asymmetric hydrogenation of the imine precursor (**D**) from 2-methyl-5-ethylaniline and methoxyacetone (Scheme 10.9) leads to the key intermediate (**E**) in the production of metolachlor, currently the largest volume process using asymmetric chemocatalysis.[21]

meet environmental objectives and to comply with environmental regulation. On top of this will be indirect pressure to develop cleaner technology resulting from public perceptions and choices made by consumers, and more direct pressure from activist investors.

Some of the complex issues governed by chemistry, engineering and economics that determine the nature of industrial development can be illustrated by the processing of methane and methanol. Methane is a component of natural gas and an important chemical feedstock. However, because of the costs associated with condensing and refrigerating methane, it is more convenient to convert it to methanol in a series of catalysed steps.[25a] The first step involves the production of synthesis gas, or 'syngas' for short, via steam reforming (eqn 10.11) or partial oxidation (eqn 10.12). The CO : H_2 ratio can be modified using the water–gas shift reaction (eqn 10.13):

$$CH_4 + H_2O \rightarrow CO + 3\,H_2 \tag{10.11}$$

$$2\,CH_4 + O_2 \rightarrow 2\,CO + 4\,H_2 \tag{10.12}$$

$$CO + H_2O \rightarrow CO_2 + H_2 \tag{10.13}$$

Treatment of syngas with a copper–zinc oxide catalyst, supported on alumina, produces methanol with high selectivity. Any hydrogen in excess of the stoichiometry needed for eqn (10.14) is consumed (eqn 10.15) by added CO_2.

$$CO + 2\,H_2 \rightarrow CH_3OH \tag{10.14}$$

$$CO_2 + 3\,H_2 \rightarrow CH_3OH + H_2O \tag{10.15}$$

As syngas can be made from coal, it is not surprising that, where coal is most readily available such as in China, its conversion to methanol is seen as economically attractive. As we shall see in Chapter 11, the production of syngas from biomass feedstocks suggests that methanol will also be an important platform chemical in the transition from fossil feedstocks. In addition, methanol can be dehydrated catalytically [MTG (methanol-to-gasoline) or GTL (gas-to-liquids) processes] over microporous crystalline aluminosilicate (zeolite) catalysts[25b–d] such as ZSM-5 to give hydrocarbon mixtures suitable for use as a transportation fuel.

At first sight it seems strange that methane cannot be converted to methanol directly, something that methane monooxygenase is able to achieve in nature. However, the reason why the use of methane as a chemical feedstock is problematic is that the C–H bond in methane is significantly more resistant to chemical change (to form CH_3X) than the C–H bond of the product, CH_3X.[25e] Because of this, the latter can react further, and more readily, to give CH_2X_2 (eqn 10.16 and eqn 10.17). As it is very difficult to stop the reaction after the first step, such processes are poorly selective to CH_3X.

$$CH_4 + X_2 \rightarrow CH_3X + HX \tag{10.16}$$

$$CH_3X + X_2 \rightarrow CH_2X_2 + HX \tag{10.17}$$

Achieving improved selectivity to the monosubstituted product is, therefore, an important fundamental challenge, one that has been elegantly solved by Periana and colleagues.[25f] They were able, selectively and in good yield, to form methanol from methane (as shown in eqn 10.18–10.21) in 102% sulfuric acid at 220 °C using the platinum pre-catalyst, (η^2-bipyrimidyl)dichloroplatinum(II) (Figure 10.8).

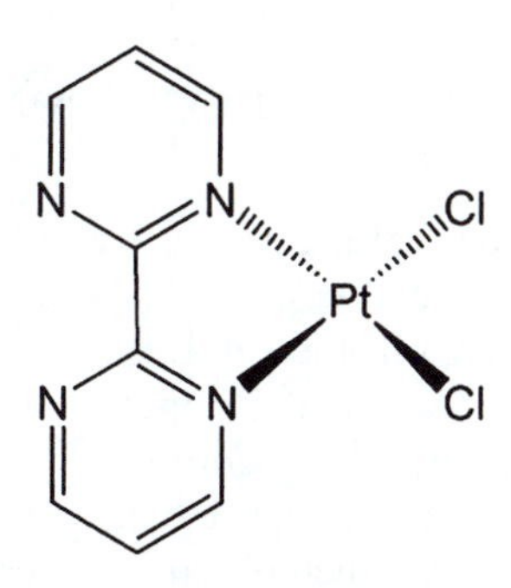

Figure 10.8 (η^2-bipyrimidyl)dichloroplatinum(II) pre-catalyst used in selective conversion of methane to methanol.[25f]

$$CH_4 + H_2SO_4 + SO_3 \rightarrow CH_3OSO_3H + H_2O + SO_2 \quad (10.18)$$

$$CH_3OSO_3H + H_2O \rightarrow CH_3OH + H_2SO_4 \quad (10.19)$$

$$SO_2 + 0.5\,O_2 \rightarrow SO_3 \quad (10.20)$$

The net reaction is given below:

$$CH_4 + 0.5\,O_2 \rightarrow CH_3OH \quad (10.21)$$

The reaction shows *ca.* 90% conversion of methane, with 81% selectivity to methyl bisulfate which arises from the *ca.* 100-fold larger rate constant for the reaction of methane compared with that for methanol. Unfortunately the desired product, methanol, can only be recovered by hydrolysing the methyl bisulfate, thereby diluting the concentrated sulfuric acid which must then be reconcentrated to bring about further reaction. The process is, simply and unhappily, an expensive way of diluting sulfuric acid and exemplifies well the severe challenges associated with translating novel catalysis discovered in first-rate research into industrial processes that are both commercially and sustainably viable.

10.5.7 Catalyst Recovery

Both the metal and the ligand that make up many transition metal catalysts are expensive. This and the prevention of metal contamination of ultimate products combine to make catalyst recovery an important technocommericial objective. As the cost of the ligand often is greater than that of the metal, research designed to recover both using techniques such as nanofiltration has been described for the ruthenium–binap asymmetric hydrogenation catalyst.[26] The growing research area of organocatalysis (Section 10.1.4) seeks to develop catalysts that avoid the use of metals altogether, though clearly, the organocatalysts themselves require separation and recovery.

10.6 WASTE REDUCTION AND PREVENTION THROUGH CATALYSIS

Waste reduction associated with catalyst use (when compared with the equivalent non-catalysed process) can arise directly from improved efficiency (and associated reduced costs) in the catalysed transformation and indirectly both from reductions in energy used to operate a process and from capital saved in smaller (and/or fewer) unit operations.

Using catalysts enables stoichiometric reagents to be avoided (and the inevitable stoichiometric co-products). Catalysts can also improve reaction selectivity to the desired product, thereby improving feed utilisation and reducing or eliminating the need for the separation and treatment of waste.

Inhibitors (to be distinguished from catalyst poisons) can also improve selectivity by suppressing a transformation leading to an undesirable product or intermediate.

By increasing reaction rates, it is possible that reactor productivity can be enhanced. This can lead either to higher throughput of material per unit time or to the need for a smaller reactor. The latter will produce consequential benefits, including the elimination of unnecessary pipework and related infrastructure and the reduction of the quantity of material being processed in the reactor at any one time (its 'inventory'), thereby contributing to greater inherent safety. Catalysed reactions can be carried out under milder conditions as catalyst efficacy is enhanced, permitting the use of lower temperatures and pressures compared with the analogous uncatalysed reaction (if one exists). The use of milder conditions will lead to a reduction in the energy expended in heating and in compression. If the reactor can be made smaller because of its greater productivity, then the capital required for its construction will be lower. Operating at lower temperatures and pressures generally also requires less capital. If a product can be made sufficiently pure in the reactor (because the catalysed reaction can be made sufficiently selective that negligible quantities of by-products are formed) that it does not require a subsequent extraction or distillation stage, then it is possible to avoid the need for the relevant unit operation (a still or an extractor) and the associate costs can be saved.

Catalysts with longer lifetimes are beneficial as, other things being equal, less downtime will be needed as the frequency with which the catalyst must be replaced or reactivated is reduced. During process shut-downs and start-ups, the process is not operating optimally and additional by-products and associated waste may be formed.

10.7 CATALYSIS AND WASTE AS FEEDSTOCKS

An objective of developing more sustainable chemical processes is to identify ways of making use of waste products that otherwise would have to be treated (sometimes with catalysts and often at substantial cost) before they are released into the environment. In addition, ways will need to be found to recover components from manufactured products at the end of their useful life so that they can be reused. An example of each of these is found below.

10.7.1 Nitrous Oxide

The conventional manufacturing process for phenol begins with benzene, which is alkylated catalytically (usually using propene and phosphoric acid) to cumene (iso-propylbenzene).[27] Cumene is then hydroperoxidised with dioxygen to give phenol with acetone as the obligatory co-product. As we will see below, nitrous oxide (N_2O) is a by-product from the process for the manufacture of adipic acid ($HO_2CCH_2CH_2CH_2CH_2CO_2H$) from cyclohexane (Scheme 10.11). N_2O is a vigorous oxidant and can oxidise benzene directly over a metal-loaded zeolite

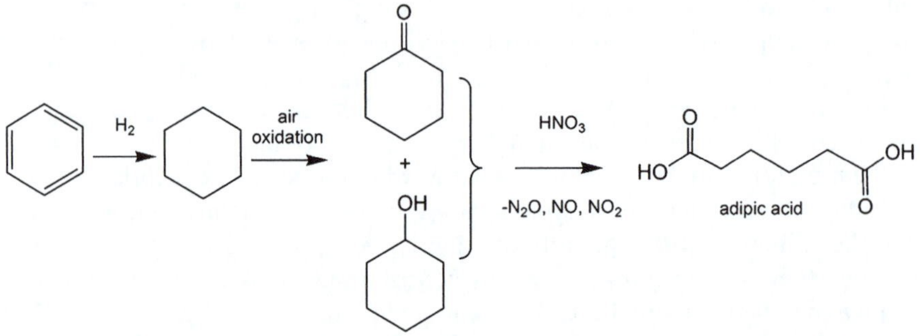

Scheme 10.11 Formation of by-product nitrous oxide, N_2O, in the oxidation of a mixture of cyclohexanone and cyclohexanol ('KA oil') leading to adipic acid.

heterogeneous catalyst, with only dinitrogen as co-product.[28] This provides an attractive means of beneficially using a material which formerly was a waste that required processing (it is implicated in both ozone depletion and global warming). The oxidation to phenol requires excess benzene, involves a conversion from benzene which is low but of high selectivity, and high conversion of N_2O. Catechol and *p*-benzoquinone (from hydroquinone) are minor by-products.

While such integration has the obvious benefit of making use of something that would otherwise involve a cost to dispose of, the downside is the need, to meet demand, for a reliable source of N_2O supply not linked to adipic acid production.[vii] Expanding the use of N_2O as an oxidant in so-called 'standalone' processes (*i.e.* that do not have to be linked with adipic acid production) would require an independent route to the production of N_2O. Research and development effort is directed to identifying a better process of manufacture than the thermal decomposition of ammonium nitrate (eqn 10.22). The front runner at the moment appears to be a catalytic process involving the oxidation of ammonia with dioxygen (eqn 10.23).[28]

$$[NH_4]NO_3 \rightarrow N_2O + 2\,H_2O \tag{10.22}$$

$$2\,NH_3 + 2\,O_2 \rightarrow N_2O + 3\,H_2O \tag{10.23}$$

10.7.2 Recycling of Polymers

In many respects, a typical process waste is relatively easy to deal with. It is usually part of a process stream of known and roughly constant composition.

[vii] Operators of chemical processes want to avoid being dependent on a single source of supply of a critical raw material, particularly one whose long-term availability may be questionable. Users prefer to source critical raw materials from more than one supplier, not only for such security-of-supply reasons but also to avoid being held to commercial ransom. Similar thinking lies behind government decisions to establish multiple sources of oil and gas to supply a country's energy needs.

It is produced in a chemical plant which, in principle, can be extended or modified to treat the waste. More difficult is the treatment of domestic waste. This will be a highly complex set of very variable and largely unknown components that are difficult to separate. The individual components are themselves complex assemblies of materials that, in all probability, have been designed to be robust and as a consequence are difficult to disassemble for reprocessing. The collection of domestic waste and its separation into useful streams are both energy intensive processes. Simple life-cycle considerations suggest that, for there to be a net environmental benefit, the energy saved in recycling waste taken to a central disposal point (*e.g.* a bottle bank) should always be greater than that consumed in individual journeys to the disposal point.

The challenges are well-illustrated when considering the recycling of synthetic carpet material. In the US alone, *ca.*1.5 Mt y^{-1} carpet is disposed of, a significant fraction containing polyamide-based fibre. In addition to recovery of the component chemicals, three alternatives are possible:

- Landfill—in which slow anaerobic biodegradation can occur, the gas from which can be collected and used to generate electricity. Unfortunately, if the gas is not collected, then the burden of greenhouse gases in the atmosphere will be increased.
- Recycle—though the product recovered inevitably will have suffered a significant loss of quality that will reduce its value.
- Incineration—from which the net energy value of the carbon content can be recovered.

While incineration is the most practical treatment, it will deliver to the atmosphere as CO_2 the fossil carbon from the petrochemical feedstock used to make the components of the carpet. The use of non-fossil-sourced feedstocks to make the polymer in the first place would clearly reduce, if not negate, this particular concern.

Recovery of the main component of synthetic carpet material, the polymer that forms the fibre, involves first separation of polymer from the rest of the carpet and the removal of plasticisers, dyes, fillers, stabilisers, adhesives and anti-static agents (as well as domestic dirt). This, in itself, is a major engineering and chemical challenge. Once the polymer has been removed then, in principle, it could be depolymerised. Unfortunately, most polymers [poly(methyl methacrylate) is an exception] are decidedly unreactive towards depolymerisation. This, indeed, is why they are attractive in use. This general lack of reactivity may be circumvented by some drastic chemistry involving thermolysis/pyrolysis, oxidative degradation, solvolysis or hydrogenolysis.[29]

A widely used polymer is poly(ethylene terephthalate) (PET). In addition to the fibre uses discussed in Chapter 2, it is used extensively in drinks containers; in principle, much of this PET could be recycled. However, to improve the retention of pressurising gas and prevent the ingress of oxygen (both of which will degrade the bottle contents), PET bottles are treated with so-called barrier coatings to prevent such transfers. Such coatings would need first to be removed in a pre-treatment step. While the depolymerisation of PET is, chemically at least, fairly straightforward, practically it is less so. PET can be transesterified with either ethylene glycol, methanol or hydrolysed with water to produce a monomer, which after appropriate purification, could be introduced back into the PET manufacturing process. Again, it is worth noting that, if these materials were sourced from renewable feedstocks, then incineration would be by far the most attractive and efficient option.

By far the bigger challenge will be the depolymerisation of poly(alkylenes) such as poly(ethylene) or poly(propylene). This is not a major research area, almost certainly because the difficulty of reversing this chemistry would require a greater amount of energy to be consumed (and associated CO_2 emitted) than would be saved by the recovery of the carbon content.

Consequently, work has so far only been done on the laboratory research scale.[30] Hydrogenolysis leading to C–C cleavage has been described for low MWt poly(ethylene) (C_{20-50}) and poly(propylene) using a catalyst obtained by attaching a catalyst precursor to a silica surface—a 'support' for a heterogeneous catalyst. Treatment of an inorganic solid, silica–alumina [containing 25% (mol) Al)], with $ZrNp_4$ [(Np = neopentyl; $CH_2C(CH_3)_3$] converts surface hydroxyl groups (represented by $\equiv$SiOH) to $\equiv SiOZrNp_3$. On further treatment at 150 °C for 62 h under 10^5 Pa H_2, methane and ethane are formed with $\equiv SiOZrNp_3$ being converted to $(\equiv SiO)_3ZrH$, $\equiv SiH_2$ and AlH species at the surface. These are believed to be the catalytically active sites at which polyolefin reacts. For instance, on treatment with this catalyst at 190 °C for 15 h under 10^5 Pa H_2, 40% of the poly(propylene) (MWt 250 000), is hydrogenolysed to a mixture of alkanes made up of CH_4 (11.6%), C_2H_6 (5.7%), C_3H_8 (6.2%), C_4H_{10} (7.2%), C_5H_{12} (6%), C_6H_{14} (2.8%) and C_7H_{16} (0.5%).

Polymers are generally too large to diffuse into, and fit into, the micropores of typical crystalline zeolites so these have been of little use in transforming them. Much research has focused on modifying the conditions under which zeolites are synthesised to produce materials in which the pores have been enlarged or the thickness of the zeolite crystals reduced. An important step has recently been reported[31] in which sheets of the zeolite ZSM-5, a single unit cell in thickness, have been prepared. Not only are these materials active in methanol-to-gasoline conversion, they achieve this while the tendency to deactivation through the formation of 'coke' is significantly reduced. Importantly, in the context of this section, branched poly(ethylene) can be 'cracked' with 85% conversion using the new material compared with only 27% for a conventional ZSM-5 catalyst.

10.8 ENVIRONMENTAL AND SUSTAINABLE CATALYSIS

Bearing in mind that there are benefits to the environment resulting from the switch to catalysed processes or from the development of more active, more selective longer-lived catalysts, it may seem redundant to identify an aspect of catalysis as 'environmental' catalysis. However, this aspect has become much more a conscious focus of catalyst research, development and use and is now the subject of a number of useful texts (see Bibliography). The editor of one the earliest of these, John Armor, defined environmental catalysis as follows:

> '*Environmental catalysis refers to a collection of chemical processes that use catalysts to control emissions of environmentally unacceptable compounds. The term also encompasses the application of catalysis for the production of alternative less-polluting products, waste minimisation and new routes to valuable products without the production of undesirable pollutants.*'[32]

Catalysis can be applied to benefit the environment in a number of ways in addition to the development of a radical new technology.

10.8.1 Remediation

Waste that has already been emitted into the environment can undergo natural degradation by enzymes or other organisms. Clean-up processes ('remediation') have been designed that exploit this capability by identifying, selecting, modifying and growing bacteria present in the polluted soil that can be used for purposeful remediation.

10.8.2 'End-of-pipe'

End-of-pipe describes the treatment of a waste stream from a process that avoids the modification of the process itself simply be adding another processing unit to transform the waste. This may be applied (but at a cost) when the emissions from an established technology are considered unacceptable but no satisfactory technological substitute is at hand and the abandonment of the established technology without a substitute would be too disruptive. A good example is the auto-exhaust catalyst that is placed in waste gas streams from an internal combustion engine. This device,[33] a major technological triumph, is a heterogeneous catalyst supported on a carefully engineered material to fulfil a complex series of functions continually during the life of the car, by oxidising CO to CO_2 and reducing NO_x to nitrogen.

10.8.3 Retrofit

In many instances, when new processes are first implemented on the large scale they generally have not been developed to their full potential (as we saw when olefin polymerisation using Ziegler–Natta catalysis was first introduced; Section 9.1). This arises from two factors (at least), illustrated as follows:

- If a process to manufacture a material for which there is a demand has been made more efficient through the discovery and development of a new catalysed process, there is an understandable tendency to want to seek to benefit from it at the earliest possible opportunity (to begin to recoup the development costs, at the very least).

- After a new process has been commissioned further investigations may identify a better catalyst. Such improved catalysts can sometimes later be introduced into the process with the minimum of process modification (a retrofit) and can increase the productivity of the process or improve the specification of the product (perhaps by lessening the formation of a critical by-product). The latter is exemplified below (Section 10.8.7) in the context of the manufacture of acetic acid.

10.8.4 Regulation

While early developments in catalytic technology were driven largely by commercial considerations, more recent developments have come about as a consequence of legislation and regulation. The most far-reaching resulted from the application of the Montreal Protocol (Web 10), an international treaty signed in 1987 and which came into force[viii] in 1989. The Montreal Protocol sought to reverse stratospheric ozone loss by phasing out a series of halogenated organic materials—mainly chlorofluorocarbons (CFCs) and bromofluorocarbons. These materials were chemically inert and of low toxicity compared with the materials they had replaced (ammonia, hydrocarbons, carbon tetrachloride) and were ideally suited as refrigerants, aerosol propellants and fire-fighting chemicals. However, their very inertness led to their accumulation in the atmosphere (particularly their transport to the stratosphere) where they decomposed under the influence of ultraviolet (UV) radiation to add chlorine atoms that catalyse the decomposition of ozone. The products developed to replace these materials, the hydrofluorocarbons (HFCs), though chlorine-free, were manufactured from (non ozone-depleting) chlorinated precursors by specially (and rapidly) developed and engineered catalysed reactions with hydrogen fluoride.[34]

The requirement to reduce the sulfur content of fuels to ever lower levels (to avoid the consequences of emissions of sulfur oxides into the atmosphere, associated with smog formation and acid rain) through the process of hydrodesulfurisation (HDS), has been driven by specifications for transportation and other fuels set by different authorities. Some idea of the technical challenge can be seen from the reduction in permitted levels from 300–500 ppmw (parts per million by weight) total sulfur in diesel fuel in the early 2000s to 15–30 ppmw in 2006. Some 64 million tonnes of sulfur produced in 2005 were the by-product from the hydrogenolysis of thiols, sulfides, thiophenes (eqn 10.24) and related compounds present in oil refinery process streams. HDS takes place at 300–400 °C and 30–130 atmospheres, typically using a cobalt-modified MoS_2–alumina catalyst. The challenge in developing such catalysts is to effect the hydrogenolysis ever more

[viii] For such international agreements to have legal effect, they have to be endorsed by a minimum number of the governments or parliaments of the countries that wish to sign up to them. When this occurs, the agreement then 'enters into force'.

selectively while avoiding alkene hydrogenation this having a negative effect on the fuel characteristics of the desulfurised product.

$$\text{3-CH}_3\text{C}_4\text{H}_3\text{S} + 3\,\text{H}_2 \rightarrow (\text{CH}_3)_2\text{C}{=}\text{CH}(\text{CH}_3) + \text{H}_2\text{S} \quad (10.24)$$

The hydrogen sulfide is removed from the process stream by contact with an amine. After recovery ('stripping'), the H_2S is converted to either sulfur or sulfuric acid.

These general topics and the role of catalysts in improving environmental performance of chemical processes are further illustrated with the following examples.

10.8.5 Lazabemide Production

Lazabemide (Scheme 10.12), trialled as a treatment for Parkinson's disease, provides a spectacular example of the significant improvements that synthetic route modification and the use of catalysis can make to the production efficiency of manufacture of a fine chemical.[35a] The laboratory route begins with 2-methyl-5-ethylpyridine giving lazabemide in an eight-step reaction sequence involving two stoichiometric oxidations (the first involving the isolation of 2-methyl-5-pyridine carboxylic acid as the copper salt[35b]—not included as a reaction step) and protection–deprotection steps involving 1,2-diaminoethane and di(t-butyl) carbonate. The overall yield[35a–d] is only 8%. It is not surprising that the (idealised) laboratory route has an atom efficiency of only 17%.[ix] On the production scale,[35a,e] a palladium catalyst (ton = 3000) converts 2,5-dichloropyridine, carbon monoxide and 1,2-diaminoethane (Scheme 10.13) to lazabemide with essentially 100% AE (though traces of the metal must be removed following isolation). Closer study of the experimental description of the individual steps reveals how much waste is avoided by the industrial process (operated at 130 °C and 10 atm of CO).

10.8.6 Ibuprofen Manufacture

The classical example used to illustrate the waste-reducing character of catalytic processes compared with alternative routes relying on stoichiometric processes is that of the anti-inflammatory, ibuprofen {(2-[4-(2-methylpropyl)-phenyl]propanoic acid} (Scheme 10.14). The initial six-step manufacturing process from (2-methylpropyl)benzene involves stoichiometric reagents generating obligatory co-products with an overall atom efficiency of 46%. A three-step

[ix] Note that this assumes an ideal stoichiometry with 100% conversion and selectivity for each step, ignoring all the reaction auxiliaries that might be employed. It also assumes that the stoichiometries of the individual steps are known. Working out the idealised stoichiometry of the oxidation of an aromatic methyl group to a carboxylic acid using acidic potassium permanganate [assuming a reduction of Mn(VII) to Mn(II)], while requiring a little thought, would be a useful exercise. However, the reaction mass efficiency metrics developed by Andraos and others (discussed in Sections 7.8 and 7.9) are much more meaningful for addressing waste generation in such multi-step processes.

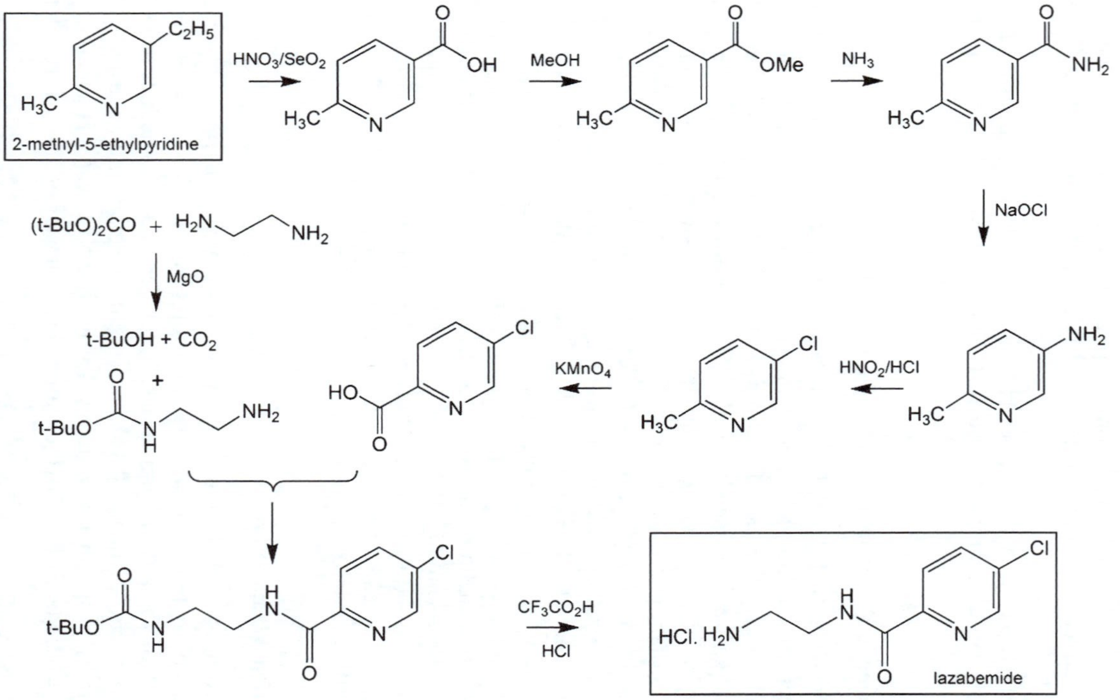

Scheme 10.12 Laboratory reaction sequence to lazabemide from 2-methyl-5-ethylpyridine with an overall yield of 8% and an atom efficiency of 17%.[35b]

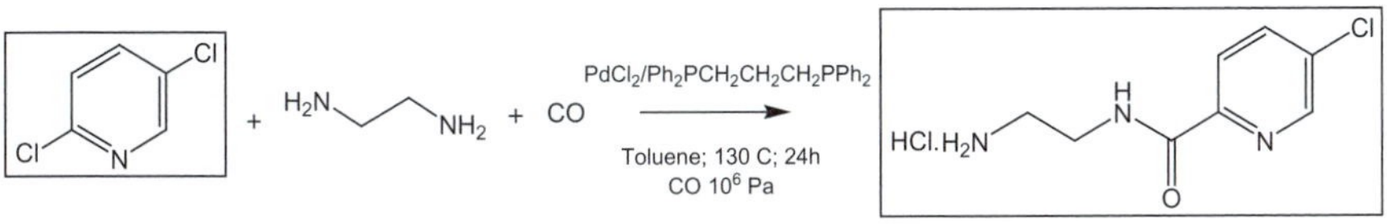

Scheme 10.13 Production route to lazabemide from 2-5-dichloropyridine with 100% atom efficiency.[35a,e]

production-scale batch process (Scheme 10.15) has replaced stoichiometric processes by catalytic, improving the idealised AE to 77%. Using a series of microreactors (Figure 10.9), a modified three-step process (Scheme 10.16) has been made continuous (albeit producing only 9 mg min^{-1} of crude product and a reduction in atom efficiency to 44%).

10.8.7 Acetic Acid Manufacture

One of the few large-tonnage processes that employs a homogeneous catalyst is the Monsanto process for the production of acetic acid by carbonylation of methanol using a rhodium catalyst (eqn 10.25).

$$\mathrm{MeOH + CO \rightarrow MeC(O)OH} \tag{10.25}$$

The original process was operated at 30–60 atm and 150–200 °C, with a selectivity from MeOH of 99%. The major catalytic species is believed to be $[Rh(CO)_2I_2]^-$. The catalytic cycle is shown in Scheme 10.17 and involves five basic steps:

1. MeI oxidative addition
2. Methyl migratory insertion
3. CO uptake
4. MeC(O)I reductive elimination
5. Hydrolysis.

At a water content of >8%, the rate-determining step is oxidative addition, whereas below this figure reductive elimination becomes rate-determining. A side reaction involving a rhodium-containing species catalyses the water–gas shift reaction in which carbon monoxide is oxidised by water with concomitant formation of dihydrogen (eqn 10.26). As a result acetaldehyde, also formed as a minor by-product, is reduced to ethanol (eqn 10.27). This can then enter a similar catalytic cycle to give the impurity, propionic acid (eqn 10.28).

$$\mathrm{CO + H_2O \rightarrow CO_2 + H_2} \tag{10.26}$$

$$\mathrm{CH_3CH(O) + H_2 \rightarrow CH_3CH_2OH} \tag{10.27}$$

$$\mathrm{CH_3CH_2OH + CO \rightarrow CH_3CH_2C(O)OH} \tag{10.28}$$

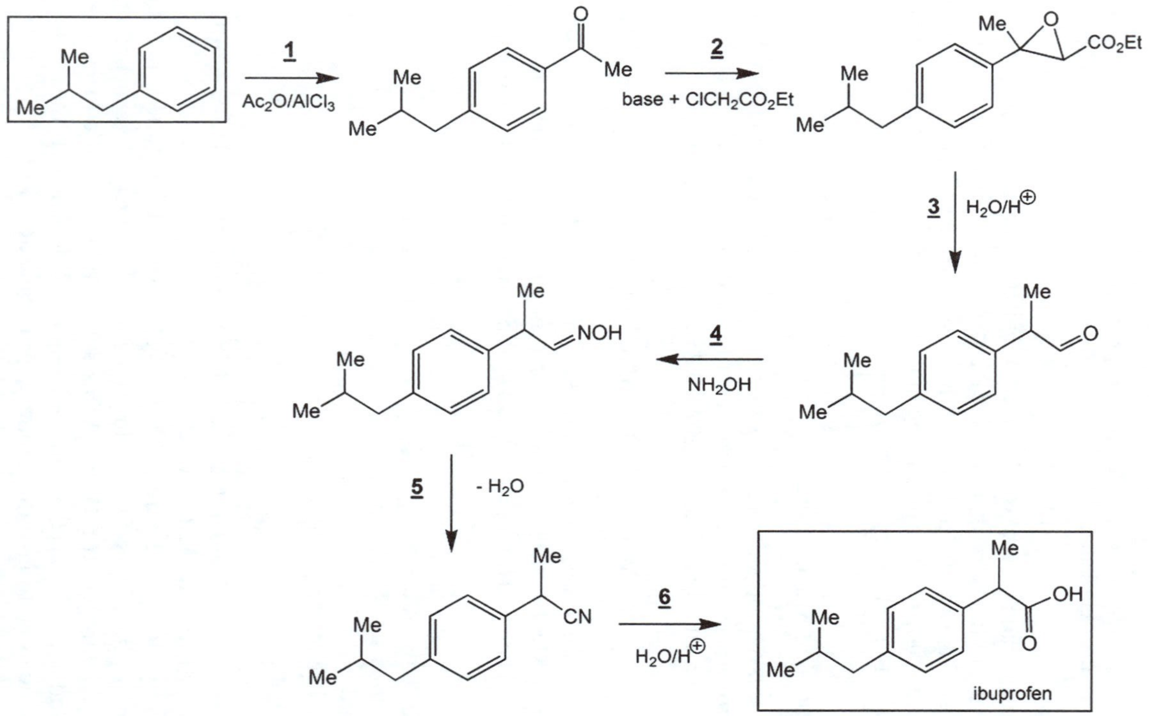

Scheme 10.14 Ibuprofen {2-[4-(2-methylpropyl)phenyl]propanoic acid} manufacture: process from (2-methylpropyl)benzene involving stoichiometric steps having an overall atom efficiency of 46%.

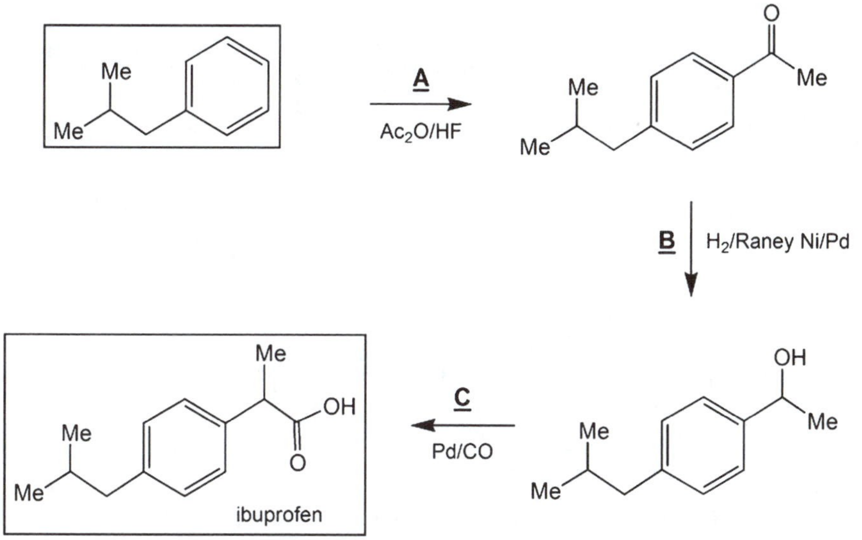

Scheme 10.15 Ibuprofen manufacture: process from (2-methylpropyl)benzene involving three catalysed steps with an overall atom efficiency of 77%.

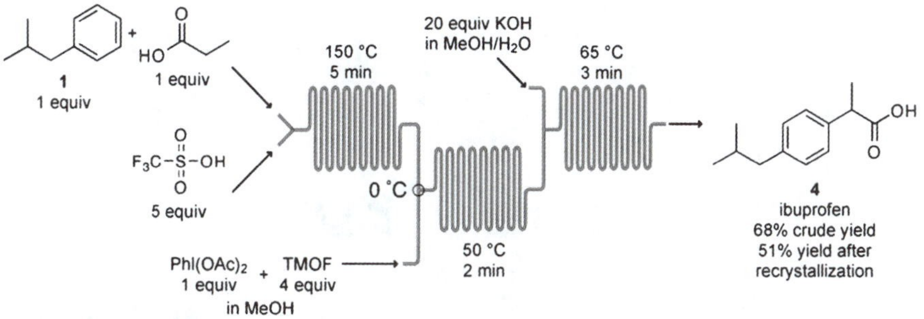

Figure 10.9 Small-scale continuous synthesis of ibuprofen in linked flow microreactors. Ref. 36: Copyright Wiley-VCH Verlag Gmbh & Co. KGaA. Reproduced with permission.

BP has made substantial evolutionary improvements to this process (now called the Cativa™ process[37]). The new process uses an iridium pre-catalyst instead of one based on rhodium. The motivation to change was the result of the very high cost of rhodium and the then relatively lower cost of iridium.[x] Being in the same transition metal group, the chemistry of iridium was expected to be similar to, but not identical with, that of rhodium. Indeed, a similar catalytic cycle is believed to operate, but the oxidative addition of MeI to the

[x]On 28 December 2007, the price of rhodium was US$6850 per troy ounce (1 troy ounce = 31.1034 g) and that of iridium US$450.

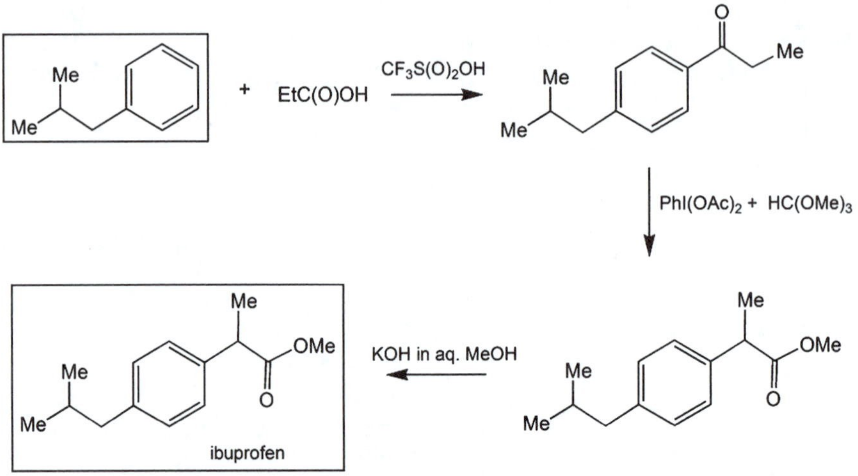

Scheme 10.16 Ibuprofen synthesis: continuous three-stage process[36] from (2-methylpropyl)benzene with an overall atom efficiency of 44%.

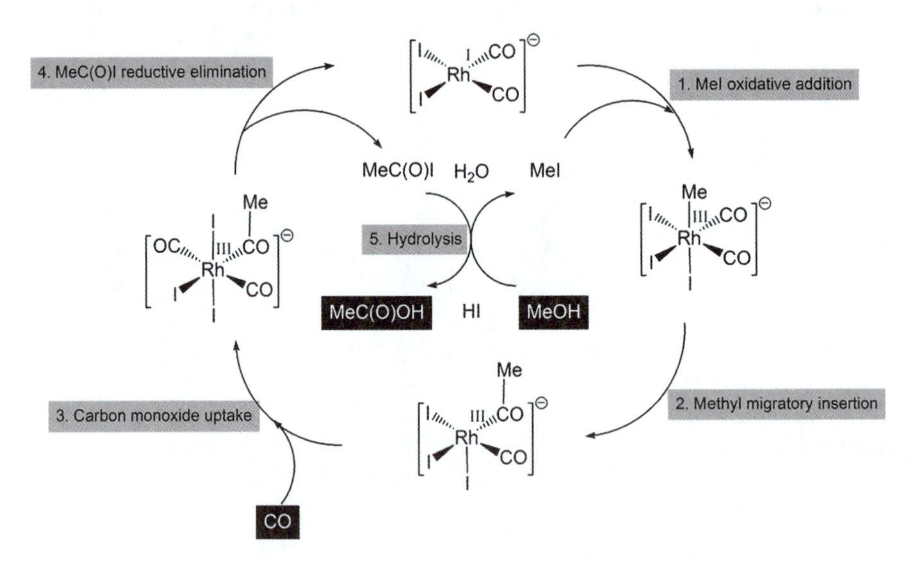

Scheme 10.17 Simplified homogeneous catalytic cycle for the production of acetic acid via methanol carbonylation. Based on ref. 37.

active catalytic species is now 150× faster than with Rh (and is no longer rate-determining). The rate-determining step is migratory insertion. The process operates with lower water concentrations, which have the benefit that less energy is needed to dry the product. The process is enhanced by the addition of promoters that facilitate iodide loss from catalyst. The process produces less organic iodide by-products. The water–gas shift reaction was found to be slower, resulting in a 70% reduction in CO_2 emissions and substantially less propionic acid. The net result is that the acetic acid produced can be used directly in the manufacture of vinyl acetate without further treatment. Overall, there are significant cost savings in addition to the reduction in waste.

10.8.8 Methyl Methacrylate Manufacture

Many of the factors that must be resolved and optimised to increase confidence that a radical technological development would, sufficiently, be both technically and commercially successful on the large scale to displace a well-established conventional technology are non-obvious, process-specific and time-consuming. This is well-illustrated in the following summary of a 14-year development by Lucite International (with academic partners[38]) that has resulted in the commissioning in late 2008 of a new plant for methyl methacrylate [$CH_2{=}C(CH_3)CO_2CH_3$, (MMA)] (Web 11) manufacture. MMA is conventionally manufactured on the 2.5 Mt y^{-1} scale predominantly *via* the ACH process from acetone *via* the cyanhydrin (eqn 10.29). The latter is then dehydrated and hydrolysed in the presence of sulfuric acid to give methacrylamide (as the bisulphate) and then MMA (eqn 10.30 and eqn 10.31).

$$(CH_3)_2CO + HCN \rightarrow (CH_3)_2C(OH)CN \tag{10.29}$$

$$(CH_3)_2C(OH)CN \rightarrow CH_2{=}C(CH_3)C(O)NH_2 \tag{10.30}$$

$$\begin{aligned} &CH_2{=}C(CH_3)C(O)NH_2 + CH_3OH + H_2SO_4 \\ &\quad \rightarrow CH_2{=}C(CH_3)CO_2CH_3 + [NH_4]^+ + [HSO_4]^- \end{aligned} \tag{10.31}$$

Despite its several apparent flaws such as the requirement for hydrocyanic acid and the co-production of ammonium bisulfate, ACH technology has remained attractive, despite the availability of several alternative catalytic routes all with much superior atom efficiencies. The search for an alternative process has been driven primarily by concerns over the availability of the primary feedstocks—acetone (dependent on phenol demand as a co-product from cumene oxidation) and hydrocyanic acid (dependent on production of acrylonitrile)—as well as the costs associated with the recovery of sulfuric acid.

An alternative technology, the methoxycarbonylation of ethene (eqn 10.32), was developed by Shell in the 1980s. This uses a palladium acetate (>£1 million t^{-1})/triphenylphosphine (*ca.* £10 kg^{-1})/*p*-toluenesulfonic acid catalyst in the liquid phase using methanol as solvent. The methyl propanoate is separated by distillation and fed to a gas-phase heterogeneously catalysed second step (eqn 10.33).

$$CH_2{=}CH_2 + CO + CH_3OH \rightarrow C_2H_5C(O)OCH_3 \tag{10.32}$$

$$C_2H_5C(O)OCH_3 + HC(O)H \rightarrow CH_2{=}C(CH_3)C(O)OCH_3 + H_2O \tag{10.33}$$

However, selectivities for the first step (*ca.* 95%, equivalent to 5 kt y^{-1} waste for a 100 kt y^{-1} plant) and ton [(3500 mol (g atom Pd)$^{-1}$] were inadequate for a commercial process. PPh_3 usage was 20 kg (t methyl propanoate)$^{-1}$, degrading to Ph_3PO, $[Ph_3PCH_3]^+$, $[Ph_3PC_2H_5]^+$ and $[Ph_3PC_2H_4C(O)C_2H_5]^+$. This was both wasteful and costly to purge from the process. Understanding provided by mechanistic investigations and studies of ligand chemistry identified 1,2-$\{(t\text{-Bu})_2PCH_2\}_2C_6H_4$ as a suitable replacement. While this provided much improved selectivity (99.95%, reducing waste from a 100 kt y^{-1} plant from 5000 to 50 t), productivity [ton = 10^7 mol methyl propanoate (g atom Pd)$^{-1}$] and activity (tof = 40 000 h^{-1}), its implementation required a cost-effective process for the ligand's own production. Laboratory-scale routes to the ligand from 1,2-$(BrCH_2)_2C_6H_4$ were available in the academic literature though these involved the use of pyrophoric $(t\text{-Bu})_2PH$ and expensive $Li[AlH_4]$. A modified route *via* 1,2-$(NaCH_2)_2C_6H_4$ and $(t\text{-Bu})_2PCl$ was therefore developed that was more suitable for the manufacturing scale.

Unfortunately, the solvent and the intermediate methyl propanoate form a roughly 50% (by weight) low-boiling azeotrope. The key product cannot, therefore, be separated by a single distillation. To achieve separation the liquid stream from the reactor is first flash distilled, with the methyl propanoate-rich distillate fed to a second still from which pure ester is obtained as a bottom product. The azeotrope overhead stream is recycled back to the reactor.

The technology was first subject to a three-year evaluation on a pilot plant operating at 0.5 kg h^{-1}. The results provided confidence to scale up by a factor 25 000 to invest in the 13 t h^{-1} plant recently commissioned in Singapore.

This project exemplifies well the nature of chemical technological development, the impossibility of foreseeing all the blocks to progress that may arise on the way, the length of time often needed to be confident of ultimate success and the rewards to those who stay the course.

10.9 CATALYSIS AND RENEWABLES

The exploitation of renewable feedstocks as raw materials in chemicals processing is addressed more fully in the next chapter, including, where appropriate, consideration of catalysis. However, it is convenient to introduce here some of the catalytic aspects to complete our survey, while keeping in mind that the challenges and timescales for the implementation of novel technologies, exemplified in the previous section, are the rule rather than the exception.

A range of primary raw materials will deliver biomass and biomass-derived material that will be subjected to a range of processing conditions, producing energy, process streams and platform intermediates. Further downstream processing will convert these to the components of everyday products for our health and well-being. Chemical intermediates from basic to commodity to fine chemical will, over time, be derived this way. However, the precise details—particularly the timescale over which this happens—will be a mix of science, technology, economics and politics.

Key, however, to making progress towards efficient renewables-based chemicals production will be the development of an understanding of the chemistry, biochemistry and processing of biomass-derived oxohydrocarbons to match that of hydrocarbon chemistry which has emerged from the requirements of technologies based on coal, oil and natural gas. The technological changes needed are well exemplified by Figure 10.10, which compares the nature of the processing conditions likely to be needed for compounds derived from biomass with those currently employed in technology based on fossil raw materials. Catalysis-related topics relevant to biomass and renewables processing have been the subject of a series of reviews.[39,40a–e] These are primarily of the academic literature.

Some chemical technologies that use non-fossil natural resources have survived competition from their fossil-derived equivalents. For instance, the conversion of glucose to sorbitan esters, the second largest group of carbocyclic acid-based surfactants, has been operated for over 50 years. As with many of these early processes, the catalysts are usually simple protic acids. Lactose or starch has also been used to make (*S*)-β-hydroxy-γ-butyrolactone [(*S*)-4,5-dihydro-4-hydroxy-2(3*H*)-furanone], a chiral intermediate, in a process employing sodium hydroxide as catalyst. Access to material of this sort, with chirality already built in, is potentially of great importance bearing in mind the difficulty of achieving this *via* conventional chemical means. A more recent development, which was recognised with a Presidential Green Chemistry Award in 1999 (Web 12), produces levulinic acid ($CH_3C(O)CH_2CH_2CO_2H$, 4-oxopentanoic acid), a relatively high-value product, in an acid-catalysed process from the cellulose waste or sludge that arises from the production of paper. This is the basis of the 'Biofine' process that has been operated on a 300 t y^{-1} scale, with suggestions (not independently confirmed) that this is being commercialised.

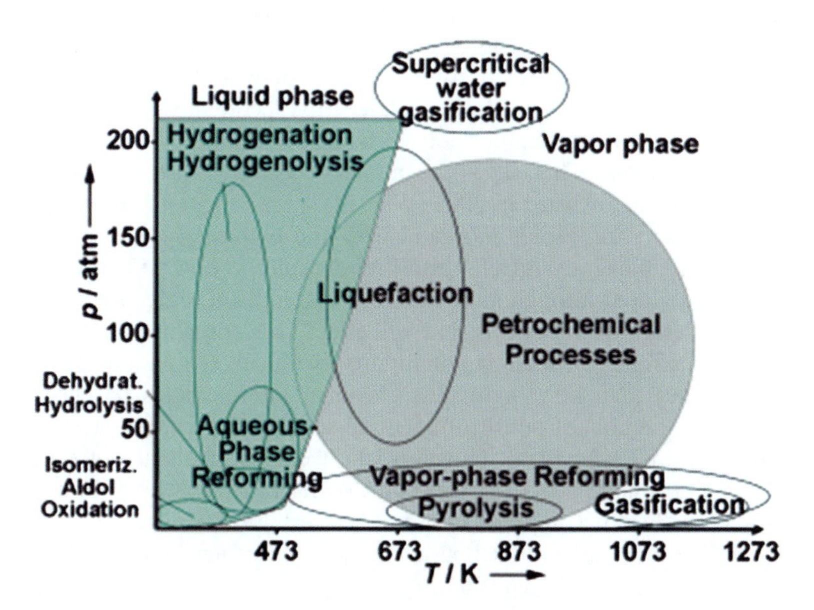

Figure 10.10 Differences in ranges of likely processing conditions for transforming petroleum- and biomass-based feedstocks. Ref. 39: Copyright Wiley-VCH Verlag Gmbh & Co. KGaA. Reproduced with permission.

$$RC(O)OCH_2CH\{OC(O)R\}CH_2OC(O)OR + 3\,MeOH \rightarrow 3\,RC(O)OMe + HOCH_2CH(OH)CH_2OH \quad (10.34)$$

$$HOCH_2CH(OH)CH_2OH \rightarrow HOCH_2CH(O)CH_3 + H_2O \quad (10.35)$$

$$HOCH_2CH(O)CH_3 + H_2 \rightarrow HOCH_2CH_2(OH)CH_3 \quad (10.36)$$

The production of biodiesel from the catalytic methanolysis of fatty acid triglycerides (eqn 10.34) obtained from a range of plant sources has conventionally used simple homogeneous base or acid catalysts. Solid heterogeneous catalysts for use in continuous transesterification processes is an active area of technical development.[41] A co-product from this reaction is glycerol, formed in large quantities (in excess of current or projected demand) as part of an impure process stream, *i.e.* 1 kg crude glycerol is produced for every 9 kg biodiesel. Glycerol is now the focus of attempts to convert it to useful products, again using catalysts. Glycerol can be dehydrated catalytically to acetol (1-hydroxypropanone) over a copper chromite ($CuCrO_2$) catalyst at 220 °C (eqn 10.35), with the conversion to product facilitated by reactive distillation (*i.e.* the product is removed by distillation as it formed), an example of process intensification discussed in Chapter 9. The acetol can then be hydrogenated (eqn 10.36) to 1,2-propylene glycol, a material for which there is considerable demand (currently satisfied from processes using fossil feedstocks). Suppes who developed the process[42] received the Presidential Green Chemistry Challenge Award in 2006 (Web 13).

Food crop waste biomass can be converted enzymatically to another chiral material, lactic acid, in a fermentation process. Lactic acid is starting material for a series of other useful products such as ethyl lactate and poly(lactic acid) (Scheme 10.18). The attraction of poly(lactic acid) is that it is a material that is 100% biodegradable. This development also received a Presidential Green Chemistry Challenge Award in 2002 (Web 14).

Instead of using traditional inorganic acid and base catalysts to bring these processes about, biocatalysis (ref. 7 and Bibliography) is now widely researched and increasingly used, though most often for materials of high value. This also includes the design of organisms that will effect such transformations. Nevertheless, the possibility of making commodity chemicals (*e.g.* adipic acid[43] and phenol[44]) rather than new high-value chemical intermediates has been suggested. The attraction of substituting fossil-sourced materials made in such large quantities is self-evident. However, the technical challenge of operating alternative biotechnological processes at the scale needed has yet to be met (and is explored further in Chapter 11). Some success has been achieved in converting glucose into *cis,cis*-muconic acid, which can then be converted using conventional hydrogenation technology into adipic acid. In such cases, the question of learning or experience which we considered in Chapter 9 becomes important, as the conventional petrochemical processes have been operated for decades whereas the newer biotechnological routes from glucose have only

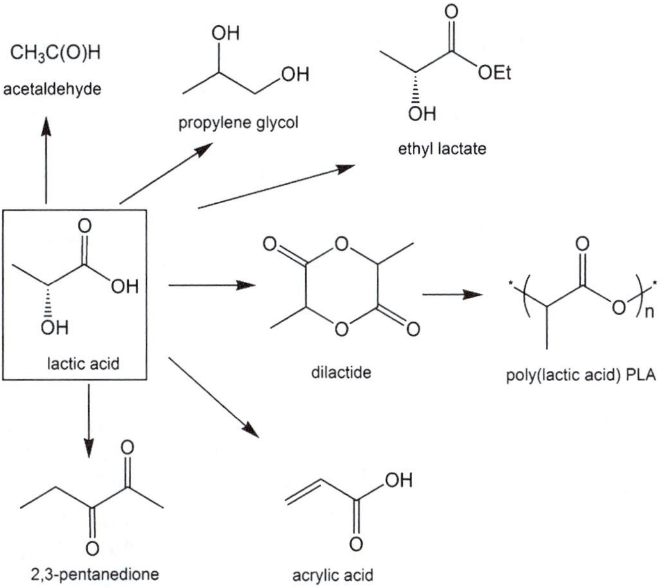

Scheme 10.18 A selection of products, some produced industrially on the large scale, accessible from the 'platform' chemical, lactic acid, itself obtained from food-crop waste *via* a biotechnological process.

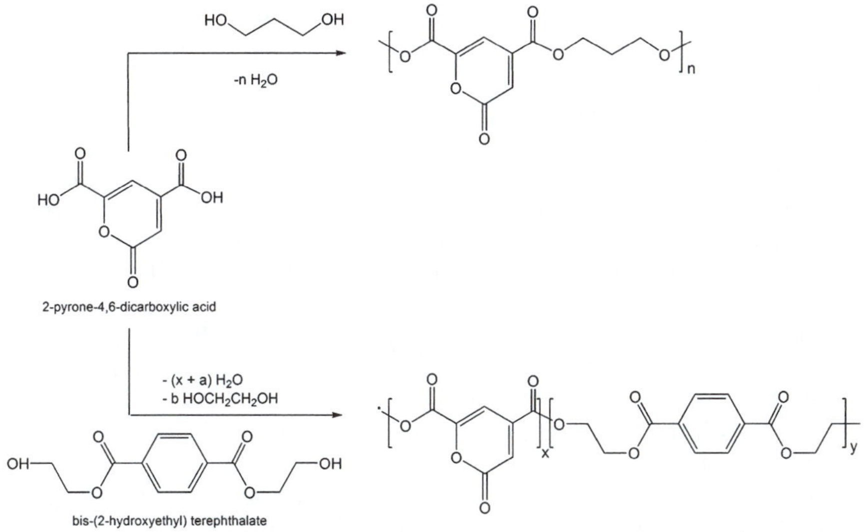

Scheme 10.19 Novel polyesters accessible from biomass-derived 2-pyrone-4,6-dicarboxylic acid including polymer derived wholly from biomass derived monomers, 2-pyrone-4,6-dicarboxylic acid and 1,3-propanediol.[45a,b]

recently been identified. As a consequence, the associated process chemistry is relatively undeveloped. From a technocommercial perspective, therefore, these newer processes are not very attractive; however, they may of course become so as experience is gained in operating them and the economies of scale realised. Such learning, in fact, may come from unexpected sources, *e.g.* when a new material is identified that has the properties or characteristics that meet someone's needs at a price considered acceptable. Who knows? Maybe recent reports[45a] of the microbial production of 2-pyrone-4,6-dicarboxylic acid (Scheme 10.19) and its use in making polyesters[45b] may be an example.

REFERENCES

1. (a) R. Jira, *Angew. Chem., Int. Ed.*, 2009, **48**, 9034; (b) J. A. Keith and P. M. Henry, *Angew. Chem., Int. Ed.*, 2009, **48**, 9038.
2. A. H. Hoveyda and A. R. Zhugralin, *Nature*, 2007, **450**, 243.
3. P. Baumeister, H.-U. Blaser and M. Studer, *Catal. Lett.*, 1997, **49**, 219.
4. C. Amatore and A. Jutland, *J. Organomet. Chem.*, 1999, **376**, 254.
5. (a) J. M. Thomas and W. J. Thomas, *Principles and Practice of Heterogeneous Catalysis*, Wiley-VCH, Weinheim, 1996; (b) G. A. Somorjai, H. Frei and J. Y. Park, *J. Am. Chem. Soc.*, 2009, **131**, 16589.
6. (a) B. Cornils, W. A. Herrmann, I. T. Horvath, W. Leitner, S. Mecking, H. Olivier-Bourbigou and D. Vogt, *Multiphase Homogeneous Catalysis*, Wiley-VCH, Weinheim, 2005; (b) B. Cornils and W. A. Herrmann, *Applied Homogeneous Catalysis with Organometallic Compounds: A Comprehensive Handbook*, Wiley-VCH, Weinheim, 2002.
7. K. Faber, *Biotransformations in Organic Chemistry*, Springer-Verlag, Heidleberg, 2004.
8. (a) S. Crossley, J. Faria, M. Shen and D. E. Resasco, *Science*, 2010, **327**, 68; (b) D. J. Cole-Hamilton, *Science*, 2010, **327**, 41.
9. (a) A. Berkassel and H. Gröger, *Asymmetric Organocatalysis: From Biomimetic Concepts to Applications in Asymmetric Synthesis*, Wiley-VCH, Weinheim, 2005; (b) S. Mukherjee, J. W. Yang, S. Hoffmann and B. List, *Chem. Rev.*, 2007, **107**, 5471; (c) P.I. Dalko and L. Moisan, *Angew. Chem., Int. Ed.*, 2004, **43**, 5138; (d) U. Eder, G. Sauer and R. Wiechert, *Angew. Chem., Int. Ed.*, 1971, **10**, 496; (e) Z. G. Hajos and D. R. Parrish, *J. Org. Chem.*, 1974, **39**, 1612; (f) M. Wiesner, G. Upert, G. Angelici and H. Wennemers, *J. Am. Chem. Soc.*, 2010, **132**, 6.
10. D. J. Jacob, *Introduction to Atmospheric Chemistry*, Princeton University Press, Princeton, NJ, 1999, ch 10.
11. K. Fisher and W. E. Newton, in *Nitrogen Fixation at the Millenium*, ed. G. J. Leigh, Elsevier Science, Amsterdam, 2002, ch. 1, p. 1.
12. (a) J. W. Szostak, *Nature*, 2009, **459**, 171; (b) M. W. Powner, B. Gerland and J. D. Sutherland, *Nature*, 2009, **459**, 239.
13. (a) B. M. Trost, *Science*, 1983, **219**, 245; (b) N. A. Afagh and A. K. Rudin, *Angew. Chem., Int. Ed.*, 2010, **49**, 262; (c) J.-L. Luche, L. Rodriguez-Hahn

and P. Crabbé, *J. Chem. Soc., Chem. Commun.*, 1978, 601; (d) J.-L. Luche and A. L. Gemal, *J. Am. Chem. Soc.*, 1979, **101**, 5848; (e) C. G. Screttas and M. Micha-Screttas, *J. Org. Chem.*, 1979, **44**, 713; (f) C. G. Screttas and M. Micha-Screttas, *J. Org. Chem.*, 1978, **43**, 1064.
14. (a) R. Noyori, M. Tokunaga and M. Kitamura, *Bull. Chem. Soc. Jpn.*, 1995, **68**, 35; (b) A. J. Blacker, S. Brown, B. Clique, B. Gourlay, C. E. Headley, S. Ingham, D. Ritson, T. Screen, M. J. Stirling, D. Taylor and G. Thompson, *Org. Process Res. Dev.*, 2009, **13**, 1370; (c) S. Caille, J. Boni, G. B. Cox, M. M. Faul, P. Franco, S. Khattabi, L. M. Klingensmith, J. F. Larrow, J. K. Lee, M. J. Martinelli, L. M. Miller, G. A. Moniz, K. Sakai, J. S. Tedrow and K. B. Hansen, *Org. Process Res. Dev.*, 2010, **14**, 133.
15. (a) J. M. Thomas and R. Raja, *Annu. Rev. Mater. Res.*, 2005, **35**, 315; (b) H. Arakawa, M. Aresta, J. N. Armor, M. A. Barteau, E. J. Beckman, A. T. Bell, J. E. Bercaw, C. Creutz, E. Dinjus, D. A. Dixon, K. Domen, D. L. DuBois, J. Eckert, E. Fujita, D. H. Gibson, W. A. Goddard, D. W. Goodman, J. Keller, G. J. Kubas, H. H. Kung, J. E. Lyons, L. E. Manzer, T. J. Marks, K. Morokuma, K. M. Nicholas, R. Periana, L. Que, J. Rostrup-Nielson, W. M. H. Sachtler, L. D. Schmidt, A. Sen, G. A. Somorjai, P. C. Stair, B. R. Stults and W. Tumas, *Chem. Rev.*, 2001, **101**, 953.
16. *Ullmann's Encyclopedia of Industrial Chemistry*, Wiley-VCH, Weinheim, 7th edn, 2009 (see Web 5).
17. *Kirk-Othmer Encyclopedia of Chemical Technology*, John Wiley & Sons, Hoboken, NJ, 5th edn, 2007 (a 27 volume set) (see Web 6).
18. H.-U. Blaser and M. Studer, *Appl. Catal., A*, 1999, **189**, 191.
19. W. S. Knowles, *Adv. Synth. Catal.*, 2003, **345**, 3.
20. H.-U. Blaser, F. Spindler and M. Studer, *Appl. Catal., A*, 2001, **221**, 119.
21. H.-U. Blaser, *Adv. Synth. Catal.*, 2002, **344**, 17.
22. R. Diercks, J.-D. Arndt, S. Freyer, R. Geier, O. Machhammer, J. Schwartze and M. Volland, *Chem. Eng. Technol.*, 2008, **31**, 631.
23. P. A. Holman, D. R. Shonnard and J. H. Holles, *Ind. Eng. Chem. Res.*, 2009, **48**, 6668.
24. (a) A. Hagemayer, B. Jandeleit, Y. Liu, D. M. Poojary, H. W. Turner, A. F. Volpe Jr. and W. H. Weinberg, *Appl. Catal., A*, 2001, **221**, 23; (b) H. W. Turner, A. F. Volpe Jr. and W. H. Weinberg, *Surf. Sci.*, 2009, **603**, 1763.
25. (a) J. M. Thomas and W. J. Thomas, *Principles and Practice of Heterogeneous Catalysis*, Wiley-VCH, Weinheim, 1996, ch. 8, p. 515; (b) J. M. Thomas and W. J. Thomas, *Principles and Practice of Heterogeneous Catalysis*, Wiley-VCH, Weinheim, 1996, ch. 8, p. 612; (c) C. S. Cundy and P. A. Cox, *Chem. Rev.*, 2003, **103**, 663; (d) A. Corma, *J. Catal.*, 2003, **216**, 298; (e) J. A. Labinger, *J. Mol. Catal., A: Chem.*, 2004, **220**, 27; (f) R. A. Periana, D. J. Taube, S. Gamble, H. Taube, T. Satoh and H. Fujii, *Science*, 1998, **280**, 560.
26. D. Nair, H.-T. Wong, S. Han, I. F. J. Vankelecom, L. S. White, A. G. Livingston and A. T. Boam, *Org. Process Res. Dev.*, 2009, **13**, 863.

27. W. L. Luyben, *Ind. Eng. Chem. Res.*, 2010, **49**, 719.
28. V. N. Parmon, G. I. Panov, A. Uriarte and A. S. Noskov, *Catal. Today*, 2005, **100**, 115.
29. W. Kaminsky and F. Hartmann, *Angew. Chem., Int. Ed.*, 2000, **39**, 331.
30. (a) V. Dufaud and J.-M. Basset, *Angew. Chem., Int. Ed.*, 1998, **37**, 806; (b) G. Tosin, M. Delgado, A. Baudouin, C. C. Santini, F. Bayard and J.-M. Basset, *Organometallics*, 2010, **29**, 1312.
31. (a) M. Choi, K. Na, J. Kim, Y. Sakamoto, O. Terasaki and R. Ryoo, *Nature*, 2009, **461**, 246; (b) A. Corma, *Nature*, 2009, **461**, 182.
32. J. N. Armor, *ACS Symp. Ser.*, 1994, **552** (Environmental Catalysis), iii.
33. (a) M. V. Twigg, *Appl. Catal., B*, 2007, **70**, 2; (b) R. M. Heck and R. J. Farrauto, *Appl. Catal., A*, 2001, **221**, 443.
34. L. E. Manzer and M. J. Nappa, *Appl. Catal. A*, 2001, **221**, 267.
35. (a) R. Schmid, *Chimia*, 1996, **50**, 110; (b) G. Nowlin, *US Patent* US2749350, 5 Jun. 1956; (c) R. Imhof and E. Kyburg, *US Patent* US4764522, 18 Aug. 1988; (d) J. Oehlke, E. Schrötter, S. Dove, H. Schick and H. Niedrich, *Pharmazie*, 1983, **38**, 591; (e) M. Scalone and P. Vogt, *European Patent* EP0582825 19 Feb. 1990.
36. A. R. Bogdan, S. L. Poe, D. C. Kubis, S. J. Broadwater and D. T. McQuade, *Angew. Chem., Int. Ed.*, 2009, **48**, 8547.
37. (a) J. H. Jones, *Platinum Met. Rev.*, 2002, **44**, 94 (see Web 15); (b) G. J. Sunley and D. J. Watson, *Catal. Today*, 2000, **58**, 293.
38. G. Eastham, *Catalysis: Fundamentals and Practice Summer School*, University of Liverpool, July 2009.
39. J. N. Chheda, G. W. Huber and J. A. Dumesic, *Angew. Chem., Int. Ed.*, 2007, **46**, 7164.
40. (a) P. Claus and H. Vogel, *Chem. Eng. Technol.*, 2008, **31**, 678; (b) R. Rinaldi and F. Schüth, *Energy Environ. Sci.*, 2009, **2**, 610; (c) G. W. Huber and A. Corma, *Angew. Chem., Int. Ed.*, 2007, **46**, 7184; (d) M. Stöcker, *Angew. Chem., Int. Ed.*, 2008, **47**, 9200; (e) Y.-C. Lin and G. W. Huber, *Energy Environ. Sci.*, 2009, **2**, 68.
41. Z. Helwani, M. R. Othman, N. Aziz, W. J. N. Fernando and J. Kim, *Fuel Process. Technol.*, 2009, **90**, 1502.
42. (a) C.-W. Chiu, M. A. Dasari, G. J. Suppes and W. R. Sutterlin, *AIChE J.*, 2006, **52**, 3543; (b) G. J. Suppes and W. R. Sutterlin, *World Patent* WO 2007053705, 10 May 2007.
43. (a) W. Niu, K. M. Draths and J. W. Frost, *Biotechnol. Prog.*, 2002, **18**, 201; (b) K. Li and J. W. Frost, *J. Am. Chem. Soc.*, 1998, **120**, 10545.
44. J. M Gibson, P. S. Thomas, J. D. Thomas, J. L. Barker, S. S. Chandran, M. K. Harrup, K. M. Draths and J. W. Frost, *Angew. Chem., Int. Ed.*, 2001, **40**, 1945.
45. (a) M. Nakajima, Y. Nishino, M. Tamura, K. Mase, E. Masai, Y. Otsuka, M. Nakamura, K. Sato, M. Fukuda, K. Shigehara, S. Ohara, Y. Katayama and S. Kajita, *Metab. Eng.*, 2009, **11**, 213; (b) T. Michinobu, M. Bito, Y. Yamada, M. Tanimura, Y. Katayama, E. Masai, M. Nakamura, Y. Otsuka, S. Ohara and K. Shigehara, *Polym. J.*, 2009, **41**, 1111.

BIBLIOGRAPHY[XI]

F. J. J. G. Janssen and R. A. van Santen, *Environmental Catalysis*, Imperial College Press, London, 1999.

R. A. Van Santen, P. W. M. N. van Leeuwen, J. A. Moulijn and B. A. Averill, *Catalysis: an Integrated Approach*, Elsevier, Amsterdam, 2nd edn, 2000.

R. H. Crabtree, *Green Catalysis*, Wiley-VCH, Weinheim, 2009.

P. Barbaro and C. Bianchini, *Catalysis for Sustainable Energy Production*, Wiley-VCH, Weinheim, 2009.

G. Rothenberg, *Catalysis: Concepts and Green Applications*, Wiley-VCH, Weinheim, 2008.

R. A. Sheldon, I. Arends and U. Hanefeld, *Green Chemistry and Catalysis*, Wiley-VCH, Weinheim, 2007.

H.-U. Blaser and E. Schmidt, *Asymmetric Catalysis on Industrial Scale: Challenges, Approaches and Solutions*, Wiley-VCH, Weinheim, 2004.

A. Liese, K. Seebach and C. Wandrey, *Industrial Biotransformations*, Wiley-VCH, Weinheim, 2006.

F. Cavani, G. Centi, S. Perathoner and F. Trifiró, *Sustainable Industrial Chemistry: Principles, Tools and Industrial Examples*, Wiley-VCH, Weinheim, 2009.

P. Grunwald, *Biocatalysis: Biochemical Fundamentals and Applications*, Imperial College Press, London, 2009.

WEBLIOGRAPHY

1. http://nobelprize.org/nobel_prizes/chemistry/laureates/2005/index.html
2. (a) www.iupac.org/goldbook/C01051.pdf
   (b) www.iupac.org/goldbook/R05243.pdf
   (c) www.iupac.org/goldbook/S05991.pdf
3. http://en.wikipedia.org/wiki/Fluid_catalytic_cracking
4. http://siteresources.worldbank.org/DATASTATISTICS/Resources/GDP.pdf
5. http://mrw.interscience.wiley.com/emrw/9783527306732/home/
6. http://eu.wiley.com/WileyCDA/WileyTitle/productCd-0471484946.html
7. www.who.int/lep/research/thalidomide/en/index.html
8. http://nobelprize.org/nobel_prizes/chemistry/laureates/2001/index.html
9. (a) http://en.wikipedia.org/wiki/Combinatorial_chemistry
   (b) http://en.wikipedia.org/wiki/High-throughput_screening
10. http://ozone.unep.org/Publications/MP_Handbook/Section_1.1_The_Montreal_Protocol/
11. www.luciteinternational.com/newsitem.asp?id = 124
12. www.epa.gov/greenchemistry/pubs/docs/award_entries_and_recipients1999.pdf

[xi] These texts supplement those already cited as references.

13. www.epa.gov/greenchemistry/pubs/docs/award_entries_and_recipients2006.pdf
14. www.epa.gov/greenchemistry/pubs/pgcc/winners/grca02.html
15. www.platinummetalsreview.com/pdf/pmr-v44-i3-094-105.pdf

All the web pages listed in this Webliography were accessed in May 2010.

CHAPTER 11
Chemicals from Biomass

'Where there's muck, there's brass'

Old Yorkshire adage

This chapter and the next one focus on the issue of renewable resources: to discuss, from a chemistry perspective, what these are and the degree to which (and on what timescale) we can move from our current heavy dependence on fossil-sourced materials for our food, energy and other material needs to technologies whose operation and whose use of raw materials are more sustainable (however this may be defined).

The debate about the economic, social and environmental benefits and costs of growing crops for use as biofuels and as sources of chemical products brings with it all the questions, perceptions and arguments that arise in seeking to understand and map out ways forward towards sustainable development. It is broadly accepted that we need to improve the rate at which we reduce our environmental burden per unit of economic activity, perhaps by a factor of at least four or as high as 10. The necessary acceleration of the pace and direction of change to bring this about represents a major challenge that can only realistically and humanely be achieved through the development of better technology; technology, moreover, that must be focused on the twin objectives of dematerialisation, that is 'do more with less' (the issue of efficiency and its limits is discussed in Chapter 6) and transmaterialisation ('change to something better'). The 'something better' includes the use of those resources that can be said to be 'renewable'. Chemistry is a central discipline[i] in developing technologies based on them.[1]

[i] When you have read this chapter, identify other disciplines that will make a contribution and what their role and importance will be.

Chemistry for Sustainable Technologies: A Foundation
By Neil Winterton

Published by the Royal Society of Chemistry, www.rsc.org

11.1 RENEWABLE RESOURCES

Whether or not we are at, or beyond, the limit of the world's carrying capacity, the ready availability of fossil resources such as oil and natural gas sufficient for our needs beyond the middle of the 21st century (a topic addressed in Chapter 12) cannot necessarily be relied upon. There are, in addition, non-oil raw materials[2] whose reserves are being consumed at rates faster than new sources can be identified. In the past, exploration motivated by such scarcity was usually able to find new reserves. Alternatively, an adapted technology was developed (or there was a preparedness to pay the higher price associated with scarcity) that extended the availability of existing sources or rendered uneconomic reserves economic.[ii]

Bearing in mind the timescale needed to bring novel technologies on stream, it is prudent to ask whether resources other than those that are petrochemical-derived can be used to power the world economy and to meet our nutritional, health and material needs. The hope is that, in the future, we can make greater use of resources that are continually being renewed.[1,3] Renewable resources are distinguishable from those that are from 'fossil', 'depletable' or 'exhaustible' sources in that the latter have (to varying degrees) diminishing availability. Some natural resources, on the other hand, are essentially inexhaustible such as sunlight, oxygen, carbon dioxide and water. Even so, their availability and distribution may be so variable that accessing and recovering them in the form and in the quantity we need may be problematic. Resources that can renew themselves (or can be quickly renewed by our own intervention) include biomass, a term that describes the complex and disparate set of organic materials generated in, or following, the processes of photosynthesis[4] occurring in plants, algae and certain bacteria, both on land and in water. Mammals such as ourselves are wholly dependent on such photosynthesising organisms for our food.

For the last 250 years or so, the energy and fuel needs of industrialised societies have been satisfied by using ancient geologic stores of solar energy by burning coal, oil and natural gas, as a consequence releasing historically captured carbon dioxide. The chemical industry processes less than 5% of crude oil production[2c,iii] compared with 70.6% to provide transportation fuel. While energy can be obtained from a number of alternative (and possibly sustainable) sources (as we will discover in Chapter 12), it is only from biomass that chemical and material products can be obtained that will substitute those currently derived from fossil sources. In the longer term (say in 20–30 years), we may be able to mimic the photosynthetic process itself on the industrial scale and synthesise a diversity of chosen chemical compounds directly from carbon dioxide, water and sunlight (and other essential materials).

[ii] This represents the operation of the 'market mechanism' discussed in Section 2.10. Now the issue is how, rationally, quantitatively, equitably and universally, to factor in the 'externalities' of the environmental impact of these activities. This is the province of the disciplines of ecological and environmental economics, and the practice of environmental accounting.

[iii] Other estimates give somewhat higher figures, depending on what is counted as a chemical and what is not.

11.2 BIOMASS, RENEWABILITY AND SUSTAINABILITY

The quantities of biomass produced annually are prodigious. The so-called net primary production of the biosphere,[5] organic matter derived from photosynthesis and biosynthesis, is estimated to be 105 GtC y^{-1} (1 GtC = 10^9 t of carbon). Until relatively cheap fossil resources became available biomass was the prime source of our energy, food and material needs. However, we should not automatically assume that the production of new biomass will be able to supply the needs of 6–9 billion of us (see Section 11.2.2 for a more detailed discussion). For instance, some useful (occasionally unique) naturally occurring materials may be present in such low concentrations in biomass from single plant sources that, to satisfy worldwide demand, the survival of individual species might be threatened (Section 11.6.11). Or, the rate of consumption may so exceed the rate of regeneration that a reservoir (*e.g.* timber from ancient woodland) may quickly be exhausted. In such cases, 'renewable' would certainly not mean 'sustainable'. As has occurred in the past, alternative technological options need to be developed. Now, the additional constraint of sustainability must be applied, namely:

- the resource consumption rate to meet our requirements should not exceed the replacement rate
- there must be no environmental impacts, either direct or indirect, which are unacceptable or irreversible.

The meaning of terms such as 'unacceptable' or 'irreversible' is a matter of opinion, judgement and debate.[iv]

Plant-derived biomass may be a renewable resource, but other factors[v] will determine the extent and the nature of its use in more sustainable technology. The first, more obvious, will require fast-growing plants or other organisms to be used (or developed using genetic modification); otherwise insufficient material will be available per unit time and far too great an area of land will be needed to provide the annual requirement.[vi]

[iv] Not surprisingly, there are many such opinions and judgements. A recent proposal[6a] seeks to place limits on seven key global parameters, including ocean acidification and species loss. While these are controversial and challengeable, the questions posed will have the benefit of focusing research, analysis and enquiry. (Indeed a series of responses quickly appeared[6b–h]). One purpose of this book is to help you question, test and understand the opinions and judgements of others and to search out and assess information to help you formulate your own (with a proper understanding of the limitations and uncertainties associated with this process).

[v] A recent paper[7] lists 15 criteria for the selection of crops for energy production that include: suitability for genetic improvement; high biomass accumulation; partition of nutrients to non-harvested biomass; high fraction of biofuel in harvested matter; suitable for growth on marginal lands; harvested material suitable for field-storage; high bulk density; high water-use and nitrogen-use efficiency; low potential as weed; high co-product potential, low cost of harvest; and potential for large-scale production.

[vi] You may see a parallel here between rate, productivity and reactor size (as discussed in Chapter 9).

Different crops have widely different productivities. A recent comparison of biomass sources of biodiesel[8a] uses three different metrics of productivity:

- yield (L oil ha^{-1})
- the land area [Mha, 1 hectare (ha) = $10^4\,m^2$] required to meet 50% of annual US needs for biodiesel
- the percentage of existing cropping land in the USA that this land area corresponds to.

For instance, soya beans (soybeans in the USA) produce 446 L of oil ha^{-1} and would need 594 MHa or 326% of current cropping land in the USA to meet 50% of US annual needs for diesel. Though not yet a developed technology, estimates[8b] that microalgae-derived biodiesel would need about 100-fold less land than soya beans to produce an equivalent quantity of biodiesel explains the current frantic interest in this source. On the other hand, more sceptical assessments[8c,d] suggest such claims are excessively optimistic.

The second becomes evident from life-cycle studies that show that much fossil-sourced energy or material is used in biomass production (fertilisers and pesticides), harvesting (fuel to drive agricultural equipment), processing and distribution (refrigeration[vii] and transport). As a consequence, carbon dioxide captured in vegetation and other organisms millions of years ago is released. While we might hope that the use of renewables should avoid such releases altogether, this will not generally be the case (and certainly not during the period of transition). It is clearly important to know how much fossil-CO_2 is released from the use of any renewables-based technology and for this to be minimised and reduced over time, both as technological innovation and learning takes place and as the use of fossil sources declines. This calculation should also take into account CO_2 emitted during the construction and operation of the renewable power plant and its decommissioning at the end of its working life. Emissions of any other greenhouse gases (*e.g.* nitrous oxide) that arise from the tillage and planting methods of conventional agricultural production need also to be taken into account. A recent analysis of such emissions[9] suggests that biomass production from perennial or self-seeding biomass sources may be preferred[10] to annual grain and oil-seed crops.

11.2.1 Technological Choices

The principles introduced and discussed in Chapter 8 may provide a starting point for deciding the optimum path for the development of chemical technologies based on renewables. However, it will become increasingly clear that they are largely irrelevant in guiding solutions to the fiendishly complex technological problems associated with relying on renewables for our needs. Even the chemical process metrics described in Chapter 7 can take us only so far in the absence of

[vii] About 50% of the carbon footprint of the UK supermarket Tesco is associated with refrigeration (Web 1).

data relating to, and even an appreciation of, far wider economic and social questions. Notwithstanding this, consideration and application of some the factors relating to chemical thermodynamics (particularly exergetic analysis[11]), reaction and process efficiency and process and business economics discussed in this book should help to identify the more worthwhile exploratory investigations.

The application of life-cycle methods (Section 6.9), while beset with arguments about the appropriate boundaries to use in making comparisons, nevertheless focus attention on much wider issues[12] than the chemistry and associated technology employed in processing biomass into energy and materials. First, wide-reaching thermodynamic analysis has highlighted the weaknesses in some apparently attractive sources of biomass linked with hitherto inadequately defined issues such as land-use changes and previously unrecognised sources of greenhouse gases. I note these difficulties and point to some recent reviews[13] that discuss them (without wanting to suggest that these necessarily represent the only, the correct or the final word).

11.2.2 Can We Rely Wholly on Renewables for All Our Needs?

Despite much research and comment, very little attention has been directed towards the central question: can we produce the entire food, material and energy requirements for the population, now and in the future, relying solely on the use of renewables? This 'wicked' problem has no simple 'yes' or 'no' answer: rather, asking it leads to a range of pertinent sub-questions that focus on the associated practicalities or the consequential impact of changes in land use, the impact on living systems generally and on local human populations in particular. Okkerse and van Bekkum have provided, in a short readable paper,[14] an initial evaluation of whether, in principle and in very simple terms, this is possible. They conclude that the possibility of total reliance on renewables for our needs cannot immediately be excluded. On the other hand, related publications[15] raise some practical and thermodynamic constraints (particularly in the context of biofuels production and photosynthetic productivity; indeed, one book is called *The Biofuels Delusion: The Fallacy of Large Scale Agro-biofuel Production*[15d]).

We have already concluded that it will simply not be possible to transform our society to one—last seen more than 200 years ago—in which our needs can be met from the natural world without recourse to modern technological intervention. While our capabilities and our understanding have developed during this time, it remains to be seen whether our foresight and forbearance have developed to the same degree (as the way forward is as much a political and social process as an economic or technological one). Okkerse and van Bekkum also introduce some of the additional questions that would lie behind a more in-depth analysis.

While many would challenge their 'back-of-the-envelope' analysis,[viii] it is nevertheless instructive (while we remain sceptical and open-minded) to follow Okkerse and van Bekkum's argument. They begin with two basic

[viii] Pickard has carried out an even shorter calculation.[16] He points out that the technique he uses to seek out the critical variables associated with a complex problem was famously employed by Enrico Fermi to sharpen the wits of his research students.

questions: 'What is the energy available to us? What are our needs for food, energy and materials?'

Their analysis is based on the consideration of energy from the Sun (Chapter 5). Total radiation emitted (in all directions) by the Sun is estimated to be 1.1×10^{17} EJ y^{-1} (1 EJ = 1 exajoule = 10^{18} joules) of which 2.8×10^{6} EJ y^{-1} is incident on the Earth. This is estimated to be *ca.* 3 000 times our current global annual energy requirement. In principle, at this global level, sufficient energy appears to be available. But can it be captured and used?

Solar energy can be captured either directly or indirectly. In the absence of artificial photosynthesis and efficient photochemical processing, we are reliant on biomass for our chemical, material and nutritional needs. Biomass can also sequester carbon (important in carbon dioxide mitigation if this can be made permanent) and provide heat and electricity *via* combustion technology. Will it be possible (and technically feasible) to feed 9–10×10^{9} people and provide them with chemicals and materials solely from the sun, CO_2, water and land?

There are some possible limits that are immediately evident. Will there be sufficient land able to deliver the required biomass? Will there be sufficient water to provide irrigation? (Recent work has noted the significantly higher 'water footprint'[17] for biomass production compared with conventional petrochemical-based technology.) It is also expected that the amount of agricultural land will reduce from 3.4×10^{9} ha in 1995 to 2.8×10^{9} ha by 2040.

How much biomass is produced will depend on the efficiency with which photosynthesis converts sunlight.[4,5a] The maximum theoretical photosynthetic yield (the amount of carbon that can be fixed photosynthetically) is estimated to be 6.6% of incident energy for white light. The practical maximum yield is species dependent,[4] being higher for one group of plants (2.4–3.2% for the so-called 'C_4' plants such as sugar cane and corn) than for another (1.7–1.9% for 'C_3' plants such as sugar beet). An average yield of 0.5% is the conservative value used by Okkerse and van Bekkum in their calculations. They cite estimates of annual biomass productivity in the range 30–70 t dry biomass ha^{-1} y^{-1} with world biomass production of 170×10^{9} t y^{-1}; 3% of this is cultivated or harvested, representing 5×10^{9} t y^{-1}.

Turning now to the food and materials needs to be met from biomass: food production for 10 billion in 2040 needing 2500 kcal day^{-1} will require 2.0×10^{9} ha. This is based on a standard daily diet of 480 g grain, 750 g milk and 80 g meat. If agricultural efficiency is such that 75% of the calculated practical yield could be achieved, then this land area could deliver 38 EJ y^{-1} of food. In addition, straw (32 EJ y^{-1}) and manure (16 EJ y^{-1}) would be co-produced. The remaining 0.8×10^{9} ha of agricultural land could yield 50 t ha^{-1} y^{-1}, or 40×10^{9} t y^{-1}, of biomass. In addition, forest land could deliver 4×10^{9} t y^{-1} and waste streams (*e.g.* from wood pulp production) a further 5×10^{9} t y^{-1}, making a total of *ca.* 50×10^{9} t y^{-1} additional biomass (*i.e.* after food requirements have been satisfied), which would be equivalent to *ca.* 25×10^{9} t y^{-1} carbon. The future demand for organic raw materials (currently 300×10^{6} t y^{-1}) can be estimated to be *ca.* $1\,000\times10^{6}$ t y^{-1} in 2040. At 20% conversion, this could be produced from 5×10^{9} t y^{-1} biomass. The remaining 45×10^{9} t y^{-1} could produce *ca.* 200 EJ y^{-1} energy.

On the face of it, these simple sums suggest that the energy from the Sun could provide all our food and chemical, and part of our energy needs. However, some qualifying statements need to be reiterated:

- The estimates provide an overall global view only and neglect regional variations and constraints, including associated differences in social and political conditions.
- The conclusion also relies on the validity of assumptions made about world population and its growth, and about per capita energy, nutritional and material requirements. It assumes that the appropriate technological developments needed for energy conversion/storage and for the production of chemicals from biomass will appear in a timely fashion to minimise the disruptive consequences of the transition.
- It is assumed there will be an adequate supply of water and other, possibly limiting, raw materials. Neither of these is necessarily a valid assumption.
- There will be a need for an appropriate infrastructure to link the new areas of supply to the areas of demand.

The issues of choice of crop, whether for food *vs.* fuel, has already entered (and polarised) the discussion of the production of bioethanol. In Europe, concerns about the environmental impact of the use of GM crops and the 'organic' farming movement (where low production efficiencies—and negligible nutritional benefits—relative to conventional agriculture are evident) are both factors affecting regional thinking and the associated political process particularly in regard to land use.

11.3 CHEMISTRY AND BIOMASS: AN OVERVIEW

Before discussing the type of chemicals and materials available from biomass, the chemical composition of its key components and the impact this has on the nature of the processing necessary to obtain transportation fuels and so-called 'platform' chemicals are described. This complex topic is represented schematically in Figure 11.1, which also highlights its role in the cycle involving atmospheric CO_2. Figure 11.1, like Figure 6.17, demonstrates the (transactional) relationship between the ecosphere and the technosphere showing, in more detail, the dependence of humankind on natural resources, whether fossil or renewable, for all our material needs. It highlights the different primary products that will be obtained from petrochemical raw materials and from biomass. Those from the latter then either must be transformed to give intermediates that can be processed conventionally or new processes (and associated product chains) must be developed able to use them directly.

The dominance of the petrochemical industry that arose during the first half of the 20th century means that $<5\%$ of today's chemicals are derived from biomass.[18] Historically, plants and animals were the source of many important and useful materials, including oils and fats, fibres, vegetable dyes and traditional

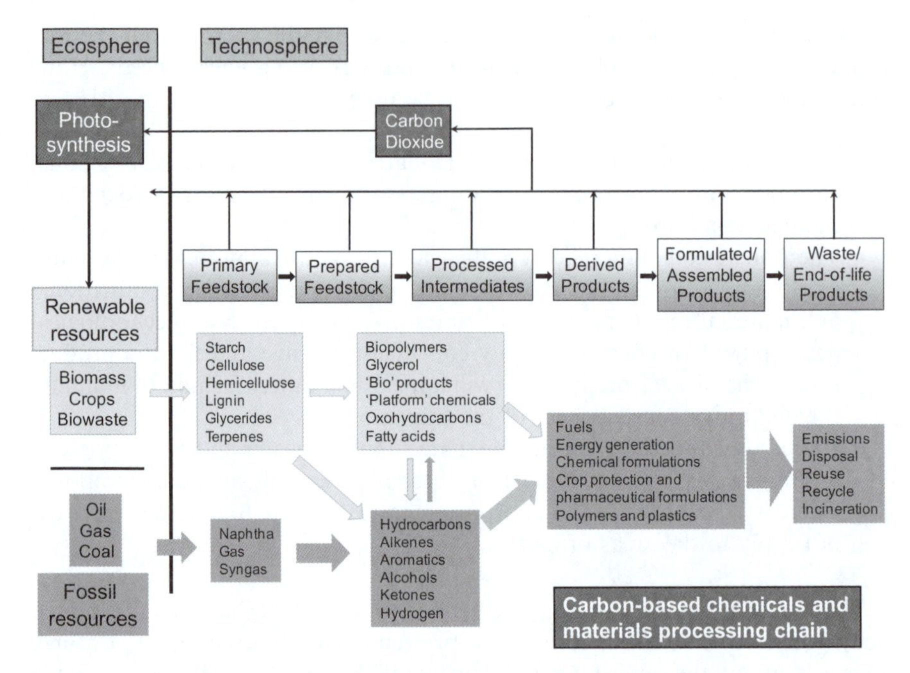

Figure 11.1 Processing stages in the organic chemical product chain highlighting the main products and their origin, the different primary and early-stage products available from fossil and new-biomass sources and their basic relationship with the ecosphere.

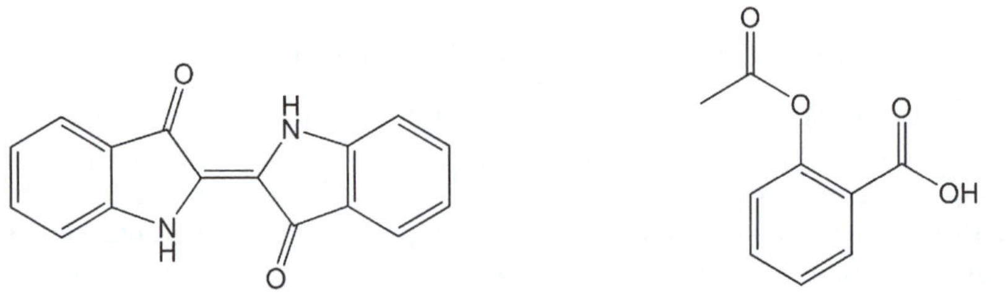

indigo

aspirin

Figure 11.2 Indigo and aspirin.

medicines. Dyestuffs such as indigo (Figure 11.2) and medicines such as willow bark extract—rich in salicylic acid (named from 'salix', Latin for 'bark') later developed into acetylsalicylic acid (2-acetoxybenzoic acid), better known as aspirin (Figure 11.2)—were initially isolated from plant materials. The origin of many of today's chemicals, particularly pharmaceuticals, can be traced back to natural products (see Bibliography) whose availability, purity and efficacy have been improved by chemical modification and chemical technology.

Major modern chemical technologies continue to process terpenes, vegetable oils and carbohydrates into important speciality products including oleochemicals, adhesives and surfactants. Because of the know-how and infrastructure associated with these industries, it is likely that they will provide the basis for early expansion of the wider use of renewables. The collaboration between Tate and Lyle, a sugar producer, and DuPont, a chemicals manufacturer, to develop and apply the technology to make 1,3-propanediol—to make synthetic polyesters (Section 11.6.10) from a renewable source (maize or corn)—is an example.

11.4 CHEMICALS FROM BIOMASS: THE NATURE OF BIOMASS AND ITS DERIVATIVES

Photosynthesis generates carbohydrates, a series of organic compounds produced by plants by capturing CO_2 present in the atmosphere (eqn 11.1), an important component of the carbon cycle (Section 5.4). Carbohydrates are the fuel that drives biochemical processes vital to life, including the production of all the components of 'biomass'—the 'stuff' of all living things.

$$n\,CO_2 + n\,H_2O + n \times 480\,kJ \rightarrow (CH_2O)_n + n\,O_2 \qquad (11.1)$$

Biomass, therefore, encompasses a group of chemically diverse complex biological materials, whose characteristics are dependent on the plant type that produces them and the function that such materials are synthesised to perform (see Bibliography). The major components include (in addition to water[ix]) lipids, steroids and triglycerides (Figure 11.3a), terpenes (Figure 11.3b) starch- and sugar-derived components (termed saccharides or polysaccharides), along with biopolymers such as energy storage materials, proteins and DNA, minerals and hormones.

Each of these components may itself be a complex mixture. For instance, lignocellulose (over 60% of terrestrial biomass) is made up of:

- cellulose—an unbranched polymer of *ca.* 10 000 glucose units (Figure 11.3c)
- hemicellulose—a complex, more readily hydrolysed, amorphous branched polysaccharide with 50–250 sugar units, mainly xylose (Figure 11.3d)
- lignin—a complex cross-linked polyphenolic material (Figure 11.3e,f) that, on its own, makes up *ca.* 30% of wood dried biomass.[19]

Glucose (a component of starch) and table sugar (sucrose, a disaccharide of glucose and fructose; Figure 11.3g) are easy to obtain from plant material on the industrial scale. It is not surprising, therefore, that these are used as starting

[ix] The water content of biomass can be substantial (freshly cut timber can contain 50% water) and is dependent on biomass type, its age and how it is stored. Care needs to be taken when discussing the weight or composition of biomass to establish whether this is before or after drying (and what the drying method might have been).

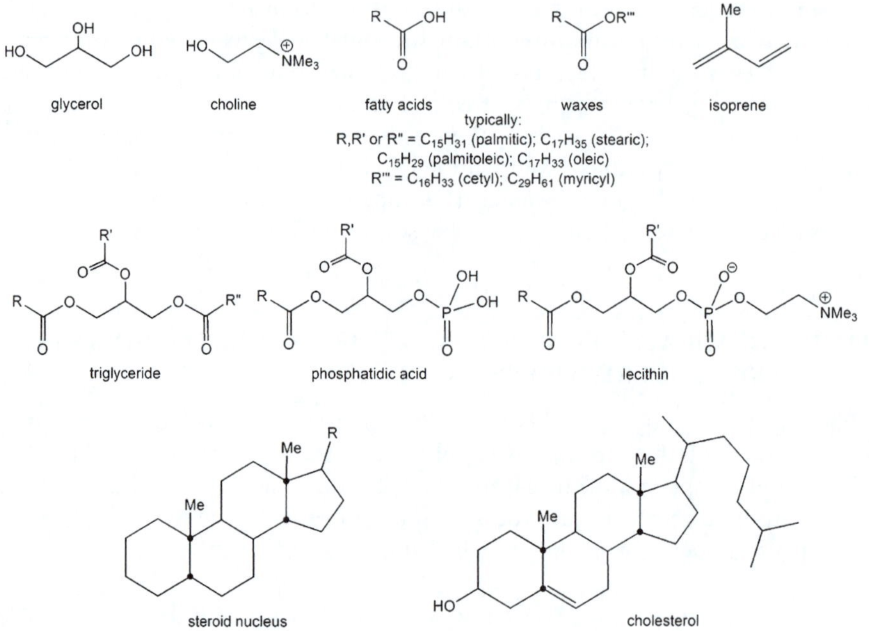

OH
HO
OH
glycerol
HO
NMe3
choline
R
OH
O
fatty acids
R
OR'''
O
waxes
Me
isoprene
typically:
R,R' or R'' = C15H31 (palmitic); C17H35 (stearic);
C15H29 (palmitoleic); C17H33 (oleic)
R''' = C16H33 (cetyl); C29H61 (myricyl)
triglyceride
phosphatidic acid
lecithin
Me
R
steroid nucleus
HO
cholesterol

Figure 11.3a Key biomass components: lipids, phospholipids and steroids.

camphene
β–pinene
α–pinene
pinane
myrcene
limonene
terpinolene

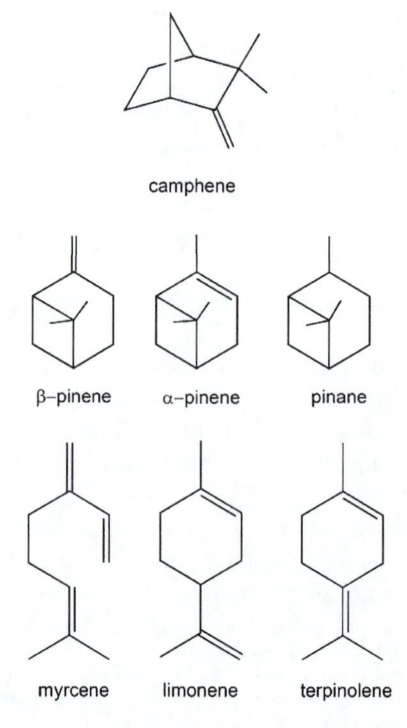

Figure 11.3b Key biomass components: terpenes.

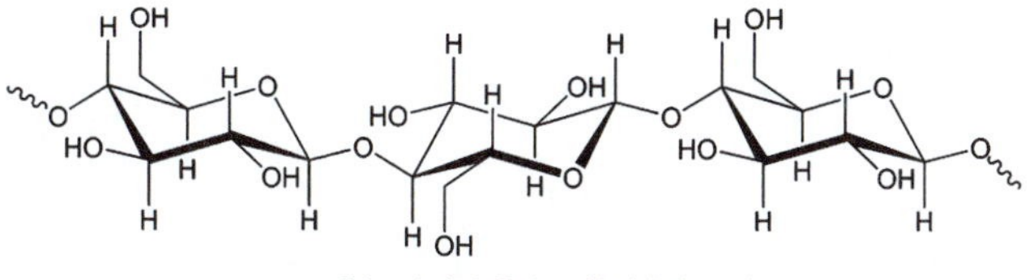

cellulose (poly-β-D-glucosido-1,4-glucose)

Figure 11.3c Key biomass components: cellulose.

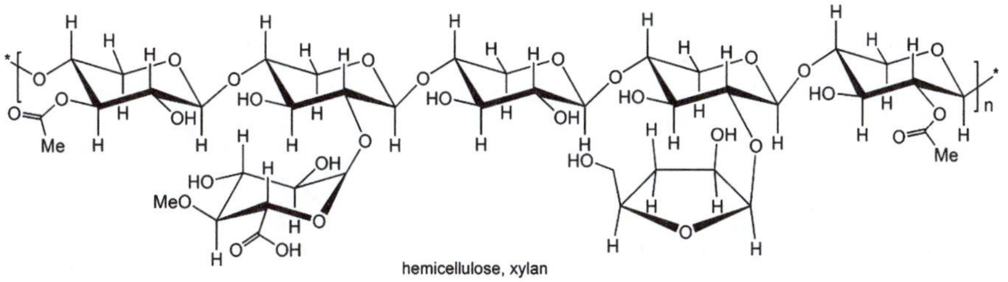

hemicellulose, xylan

Figure 11.3d Key biomass components: hemicellulose.

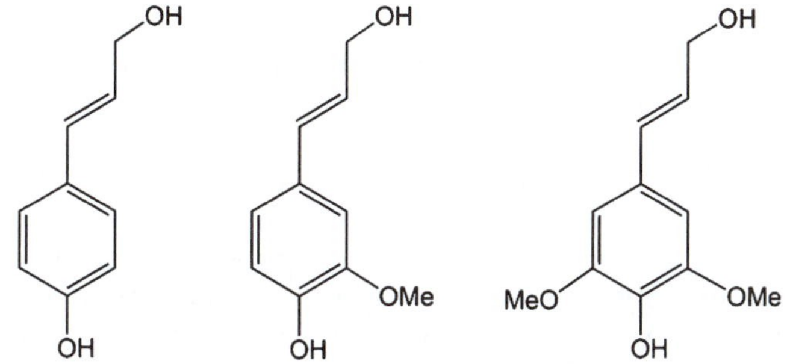

p-coumaryl alcohol coniferyl alcohol sinapyl alcohol

Figure 11.3e Key biomass components: lignin monomers. Lignin is a phenolic biopolymer (600–15 000 kDa) with a branched highly cross-linked phenylpropanoid framework (Figure 11.3f).

materials in the synthesis of existing chemicals. On the other hand, lignocelluloses, for reasons associated with their biological function, are particularly resistant to hydrolysis, making them difficult to process. Much research is currently being devoted to the relevant underlying chemistry seeking to improve their processibility to make a major source of biomass available for chemicals production. For completeness, a representation of what are believed to be the main features of the chemical structure[20] of coal is given in Figure 11.3h.

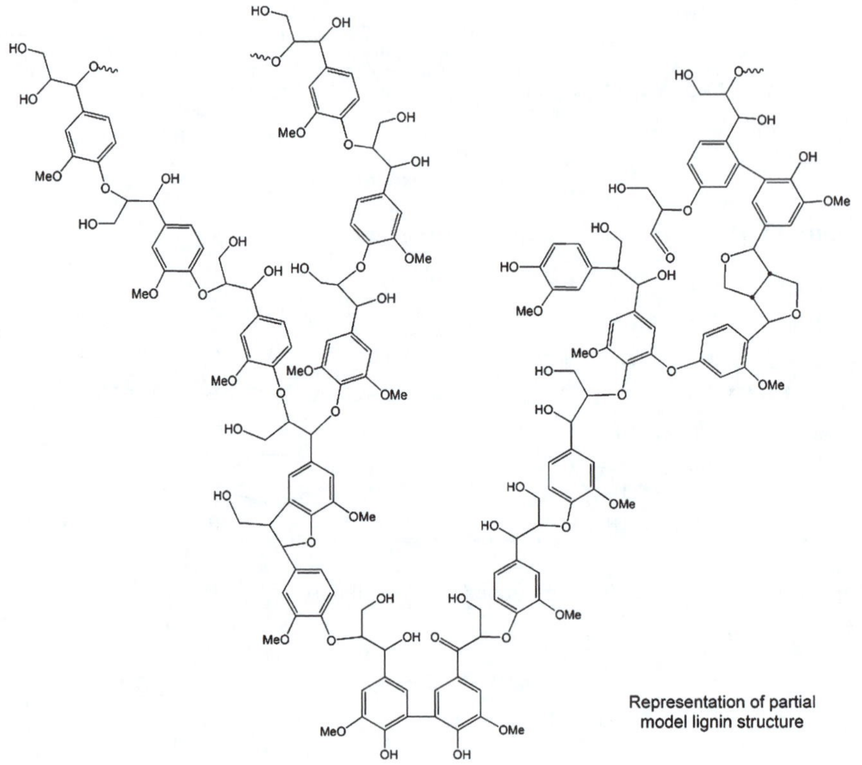

Representation of partial model lignin structure

Figure 11.3f Key biomass components: lignin.

The main purpose of processing biomass feedstocks is to generate energy and produce fuel for transportation. While biomass can be combusted directly (eqn 11.2), its energy content (given approximately[21] by eqn 11.3) per unit weight is low.

$$CH_xO_y + (1 + x/4 - y/2)O_2 = CO_2 + (x/2)H_2O \qquad (11.2)$$

$$\Delta H^0_{298} = -422.5 - 117.2x + 177.5y \ (\mathrm{kJ\,mol^{-1}}) \qquad (11.3)$$

This is significant from the sustainability point of view because biomass can be transported only over short distances[x] before more energy is consumed in the transportation than can be recovered from its combustion. This is one reason why there are currently 325 plants for processing sugar cane into sugar and ethanol in Brazil.[23]

The reason for this low energy content (other than the presence of significant quantities of water) is the high oxygen–carbon atomic ratio for the materials shown in Figure 11.3. This explains why early developments of biofuels have focused on biomass rich in components such as fatty acids or fatty acid esters

[x] One estimate[22] puts this value as low as 40–80 km.

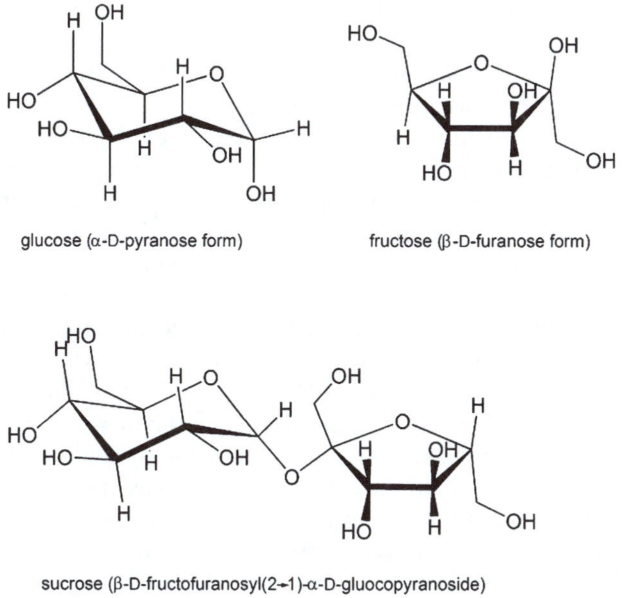

glucose (α-D-pyranose form)

fructose (β-D-furanose form)

sucrose (β-D-fructofuranosyl(2→1)-α-D-gluocopyranoside)

Figure 11.3g Key biomass components: representative sugars (saccharides), glucose, fructose and sucrose.

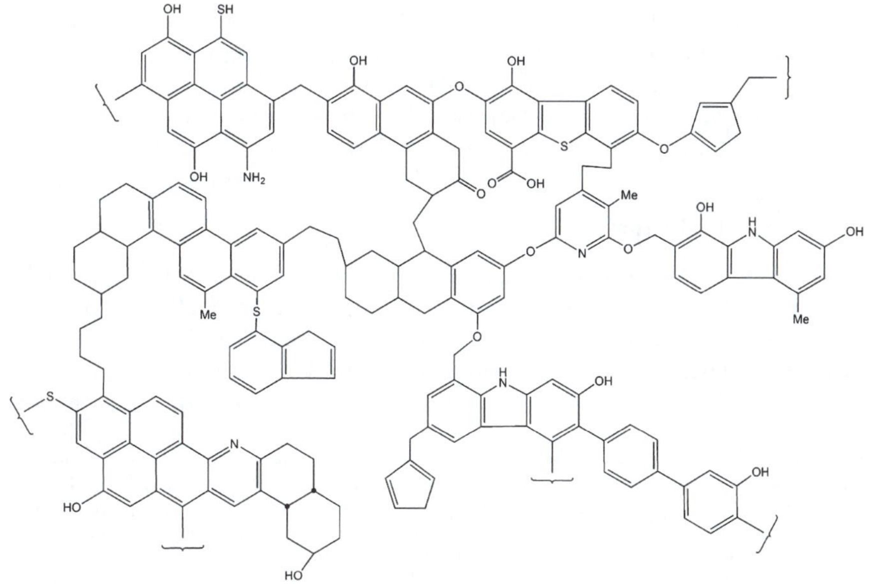

Figure 11.3h A representation of the chemical structure of coal.[20]

with relatively low oxygen–carbon atomic ratios. Simple thermochemical considerations show why this is so. Unfortunately, most biomass—particularly carbohydrate (the easiest to extract) and lignocelluloses (the largest component)—is oxygen rich. The chemistry of biomass processing is, therefore, largely concerned with transforming (economically and on the industrial scale) materials with high oxygen content into ones with lower oxygen content. This may be achieved by hydrogenation, hydrogenolysis, dehydration, decarbonylation and decarboxylation; this is reflected in the chemistry (Scheme 11.4) underlying the processes outlined in Section 11.7. As the following sections illustrate, the processing of biomass is concerned with bio- or chemo-catalysed transformations of ether, hydroxy, carbonyl, carboxy and glycoside functionalities in (in addition to those already mentioned) hydrations, hydrolyses, isomerisations, cyclisations, and aldol condensations.[24] A sequence of dehydrations/hydrogenations for the carbohydrate, fructose, leading ultimately to hexane is shown in Figure 11.4. Each step adds cost and uses up materials and energy with the potential for environmental impact. In addition, the technocommercial viability of any large-scale transformation will depend on assessments of reaction efficiency and selectivity, catalyst activity and recoverability and reactor operability and productivity (see Chapters 9 and 10).

11.5 CHEMICALS FROM BIOMASS: SOURCES OF BIOMASS

Chemical products and functional materials of various types can be obtained from biomass contained in purpose-grown non-food crops, surplus food commodities and from agricultural wastes and by-products such as straw, corn stover (plant material left after harvesting) and manure. All the biomass-derived precursors to biofuels and large-volume platform chemicals are oxo-hydrocarbons and it is, therefore, easy to overlook our dependence on the use of nitrogen fertilisers in crop growth for their production (and the associated perturbation of the nitrogen cycle). Wood is a major source of cellulose, lignin and waste from paper processing. Each of these is discussed further below.

11.5.1 Non-food Crops

Plant-seed oils[26] such as from rape, soya bean, jatropha, castor and jojoba provide triglycerides (tricarboxy esters of glycerol, Figure 11.3a). These are transesterified (Scheme 11.1) to fatty acid methyl esters (FAME) that are the basis of biodiesel[xi] for use as a transport fuel. In the process (Figure 11.5) the co-product, glycerol, is an impure waste stream. Its possible availability in high

[xi] The tendency to put 'bio' in front of every product that may have been produced from biomass is a widely used shorthand which I use in this book. In some instances, the product (*e.g.* diesel or synthesis gas) will be a mixture with a composition that is dependent on its source, whether renewable or petrochemical. In such cases, the distinction has meaning. However, for a single-component pure product such as ethanol or hydrogen, we should guard against encouraging the belief that the product is intrinsically different or that its biomass origin may indicate its superior sustainability. As we have seen, this is not always clear cut and can only be established after proper life-cycle or similar assessments.

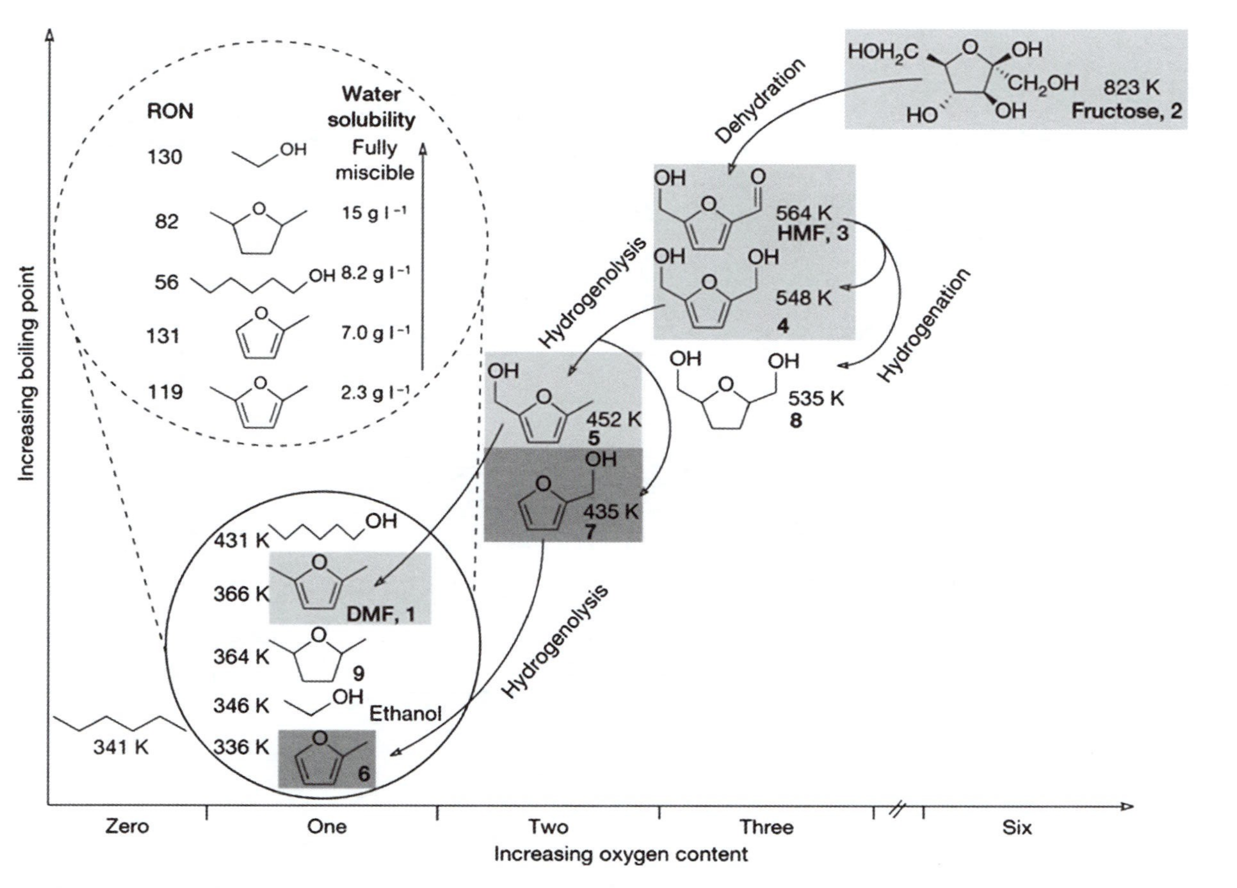

Figure 11.4 Effect of changes in oxygen content, as a result of chemical processing, on important physical properties of biomass-derived compounds, illustrated for a series of furans and tetrahydrofurans resulting from the dehydration of fructose and subsequent hydrogenation and hydrogenolysis. Reprinted from ref. 25 by permission of Macmillan Publishers Ltd, copyright 2007.

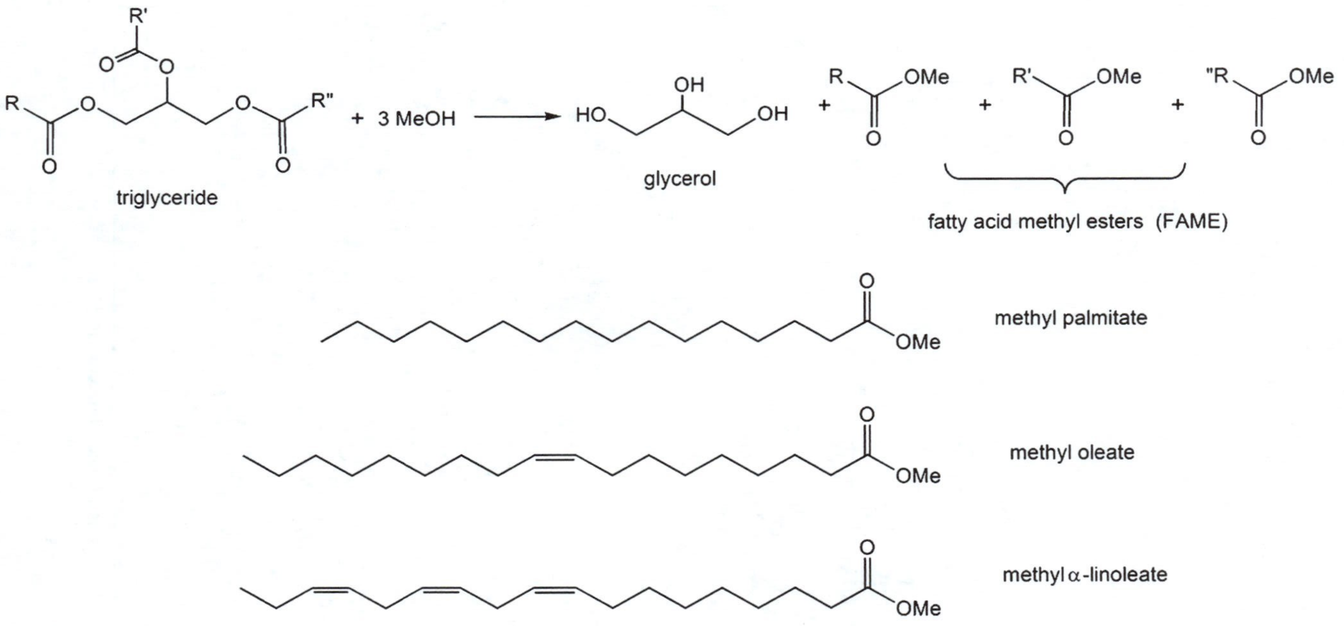

Scheme 11.1 Transesterification of triglycerides from plant seed oils to give fatty acid methyl esters (FAME), with some representative examples derived from palmitic, oleic and linoleic acids.

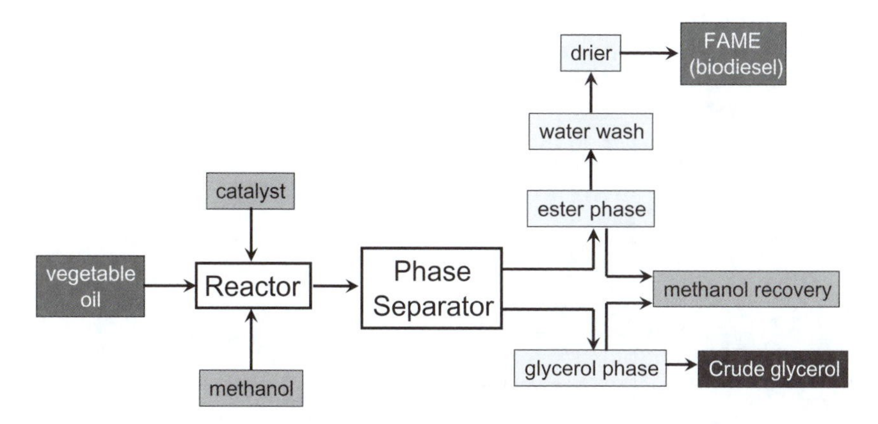

Figure 11.5 Basic flow diagram identifying important stages in the production of fatty acid methyl esters (FAME) from the transesterification of a vegetable oil using methanol, the recovery of methanol for recycle and the separation of crude glycerol. Based on Figure 2.48, Cavani *et al.*, Bibliography.

volume has stimulated much research and industrial interest in its use as a chemical feedstock (Section 11.6.10).

11.5.2 Food Crops

Maize (corn) and sugar cane are rich in starch and have long been used as sources of glucose and sucrose. In a process involving reduction and dehydration which has been operated for over 50 years, glucose is transformed to sorbitan esters (Scheme 11.2), a major group of non-ionic surfactants (*e.g.* sorbitan tristearate) used in a variety of applications as emulsifiers, plasticisers and additives. Dimethyl isosorbide is seen as an attractive solvent because of its relatively low toxicity.

The growth in the use of (non-surplus) food crops as feedstocks for making bioethanol as a transportation fuel is believed to have affected food prices in countries that depend on them.[xii] While other factors were important in determining high food prices, a negative (and incorrect) public perception developed regarding all biofuel production. This serves to reinforce the importance of seeking out reliable information on such topics to provide an informed context. Indeed, data do exist to allow the relative importance to be assessed of the environmental benefits of competing biomass-based technologies. In Figure 6.21,[27] plots are shown of greenhouse gas emissions against a measure of their overall environmental impact for different technologies. The data are normalised to petrol (gasoline), which is given a value of 100% for each parameter. Some of the alternatives being proposed fall above or beyond this reference point. However, it should be remembered that, as these newer technologies are at the early stage of development, they have not been optimised. Their investigation, evaluation and development may well be worth continuing.

[xii] See however footnote p. 164.

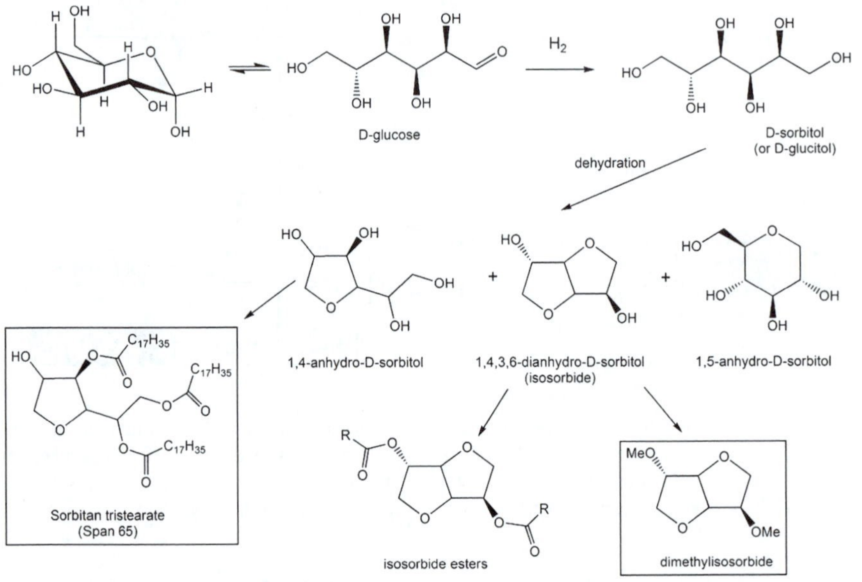

Scheme 11.2 Production of sorbitol from glucose and its further processing to the industrially widely used sorbitan esters and to other derivatives such as isosorbide esters and dimethylisosorbide, which are attracting more recent interest.

They may yet suit particular sets of circumstances and may yet make environmentally sound contributions to reductions in greenhouse gas emissions.

Four weakness of the use of annual crops such as maize and sugar cane in biofuel production should be noted:[9]

- the need to plant afresh each year;
- the need for the most productive land that might better be used to grow food-crops;
- the requirement of warm sunny conditions for productive growth; and
- the emission of significant quantities of nitrous oxide associated with their production.

Harvesting the annual growth of perennial or self-seeding grasses such as *Miscanthus giganteus*[28] and switchgrass[29] from relatively unproductive land in higher latitudes might address these weaknesses (though the costs and benefits remain a matter of contention[30a–c]). Such biomass could also be used in both energy and biofuel production.

11.5.3 Wood

Wood may or may not be sustainable, depending on its source and its management. Even when using sustainably produced wood, the pulp and paper industry generates large quantities of a range of waste material from its processing.[31a] The Kraft process treats wood pulp with aqueous alkaline sodium sulfide at 150–180 °C to separate cellulose (for papermaking) from lignin (Figure 11.3e,f). Lignin-containing wastes (amounting to *ca.* 72 Mt y^{-1} from the Kraft process[31b]) are used as fuel. Because such wastes are in excess of what is needed for energy generation in modern pulp mills, they are now considered as a potential chemical feedstock. Turpentine is also a volatile by-product from the Kraft process, from which today's flavour and fragrance industry[32] obtains its terpene feedstocks such as:

- α- and β-pinene {(1*S*,5*S*)-2,6,6-trimethylbicyclo[3.1.1]hept-2-ene, 6,6-dimethyl-2-methylenebicyclo[3.1.1]heptane}
- limonene (1-methyl-4-prop-1-en-2-yl-cyclohexene) (Figure 11.3b).

11.5.4 Macromolecules from Biomass

While biomass is an intractable, complex and variable solid it still may seem inefficient to take a material in which carbon–carbon and carbon–oxygen bonds have been laboriously built up during photosynthesis only to break them all down again to small molecules. Are there potentially useful functional components of biomass that can be extracted intact? One such example is cellulose, the glucose polysaccharide, the basis of film (cellophane) and fibre (rayon) produced in the historically important viscose process.[33] Cellulose was reacted with base and carbon disulfide to give a water-soluble xanthate ('viscose') which was extruded into acid, whereupon the cellulose—which could be formed into fibre (rayon) or film (cellophane)—was reformed (Scheme 11.3). The process produced high volumes of waste as well as used the highly flammable and toxic solvent, carbon disulfide. More modern processes dissolve cellulose directly with either *N*-methylmorpholine *N*-oxide monohydrate or urea to form cellulose carbamate.

The availability of useful macromolecules from whole-cell fermentation motivated an evaluation of poly(hydroxyalkanoates) (PHAs) used for energy storage in bacterial cells as biodegradable thermoplastics.[34] ICI pioneered fermentation technology (though used petrochemically-derived methanol as the carbon source) for the production of poly(hydroxybutyrate) (PHB) (Figure 11.6) in the 1980s. These materials remain of commercial interest, most recently to Metabolix, which was awarded a Presidential Green Chemistry Award in 2005 (Web 2) for its commercialisation of the manufacture of PHAs.

11.5.5 Algal Biomass

Algae are a particularly attractive alternative source of fatty acids and glycerides suitable for biofuel production.[35a] Algae (Web 3) are a group of diverse organisms which include seaweed (macroalgae) that are capable of photosynthesis but are classified as distinct from plants. Microalgae are single cell organisms, of which there are many thousand species, living either in marine or freshwater environments. Algae are capable of producing biomass more rapidly

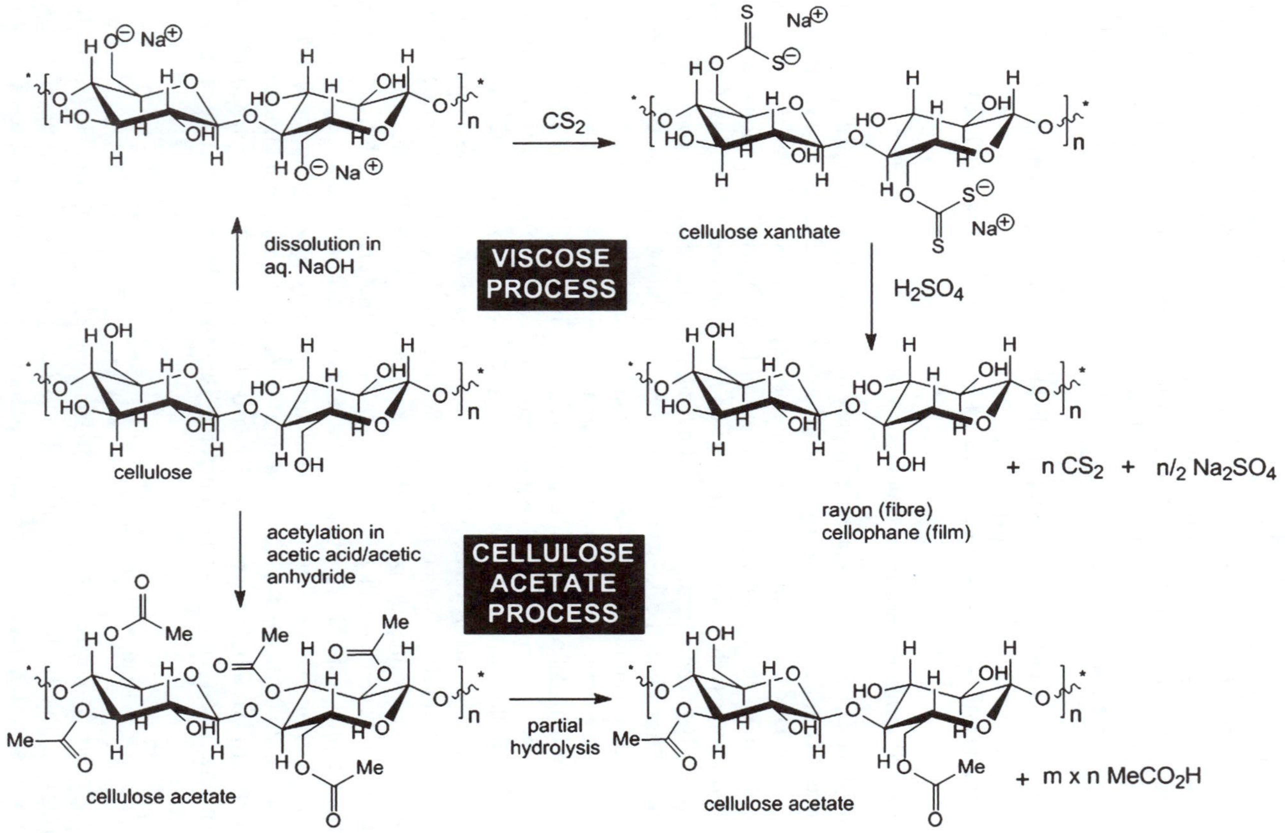

VISCOSE PROCESS

cellulose

dissolution in aq. NaOH

CS_2

cellulose xanthate

H_2SO_4

rayon (fibre)
cellophane (film)

$+ \ n\,CS_2 \ + \ n/_2\,Na_2SO_4$

CELLULOSE ACETATE PROCESS

acetylation in acetic acid/acetic anhydride

cellulose acetate

partial hydrolysis

cellulose acetate

$+ \ m \times n\,MeCO_2H$

Scheme 11.3 Historically important production of fibre and film from cellulose in the viscose and cellulose acetate processes.

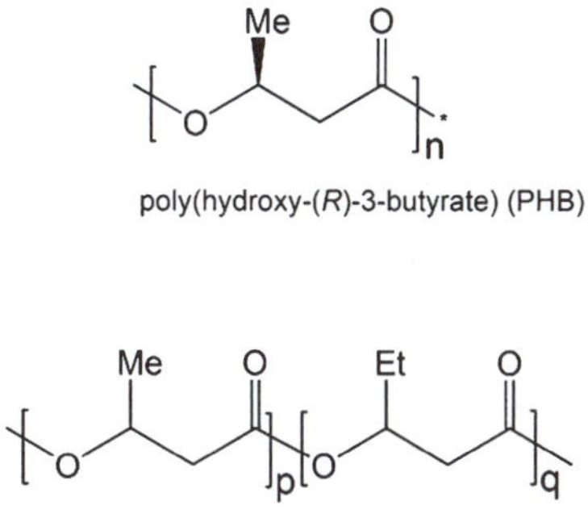

poly(hydroxy-(*R*)-3-butyrate) (PHB)

poly(hydroxybutyrate-*co*-hydroxyvalerate) (PHBV)

Figure 11.6 Two poly(hydroxyalkanoates), PHB and PHBV (where n and p + q are in the range 500–20 000), obtained from whole-cell fermentation and under development as biodegradable thermoplastics.

than terrestrial plants. However, they do so in aqueous suspension from which they must be subsequently separated. Their particular attraction is that algae productivity can be much higher than that of traditional oilseeds. According to Pienkos and Darzins,[35b] the soya bean oil yield in the USA in 2007, if converted to biofuel, would replace 4.5% total petroleum-derived diesel fuel. Were the same land area (*ca.* 26 million ha) used for algae cultivation (using a low estimate for likely productivity), *ca.* 60% of annual diesel would be replaced. While it is believed that land currently unusable for agriculture could be used, it would be necessary to place an algal reactor close to a power station to provide a constant and rich feed of CO_2.[36] Algae may grow in open ponds (cheaper, but difficult to exclude other organisms) or in closed photobioreactors (more expensive, but more controllable).[35c–e] The key problem would be the harvesting of algae (centrifugation is expensive; flocculation is slow) since 200–1000 kg of broth (containing 1–5 g algal biomass L^{-1}) must be processed to obtain 1 kg biomass. Even using the best current productivity projections, algal-derived oil can only compete when fossil oil is more than US\$110 bbl^{-1} (1 bbl = 158.99 L).

11.6 CHEMICALS FROM BIOMASS: BIOFUELS, COMMODITIES, SPECIALITIES AND 'PLATFORM' CHEMICALS

A wide range of chemical products, reflecting the diversity of biomass components and their associated chemistry (Scheme 11.4), is available from biomass both directly and indirectly. These may be characterised as target molecules (or derivable from them) or as co- or by-products that may be further processed.[37] This section exemplifies this diversity and includes end-product specialities, volume commodities (*e.g.* transportation fuels[38] and energy carriers) as well as 'platform' chemicals from which a range of derivatives may be accessible. These

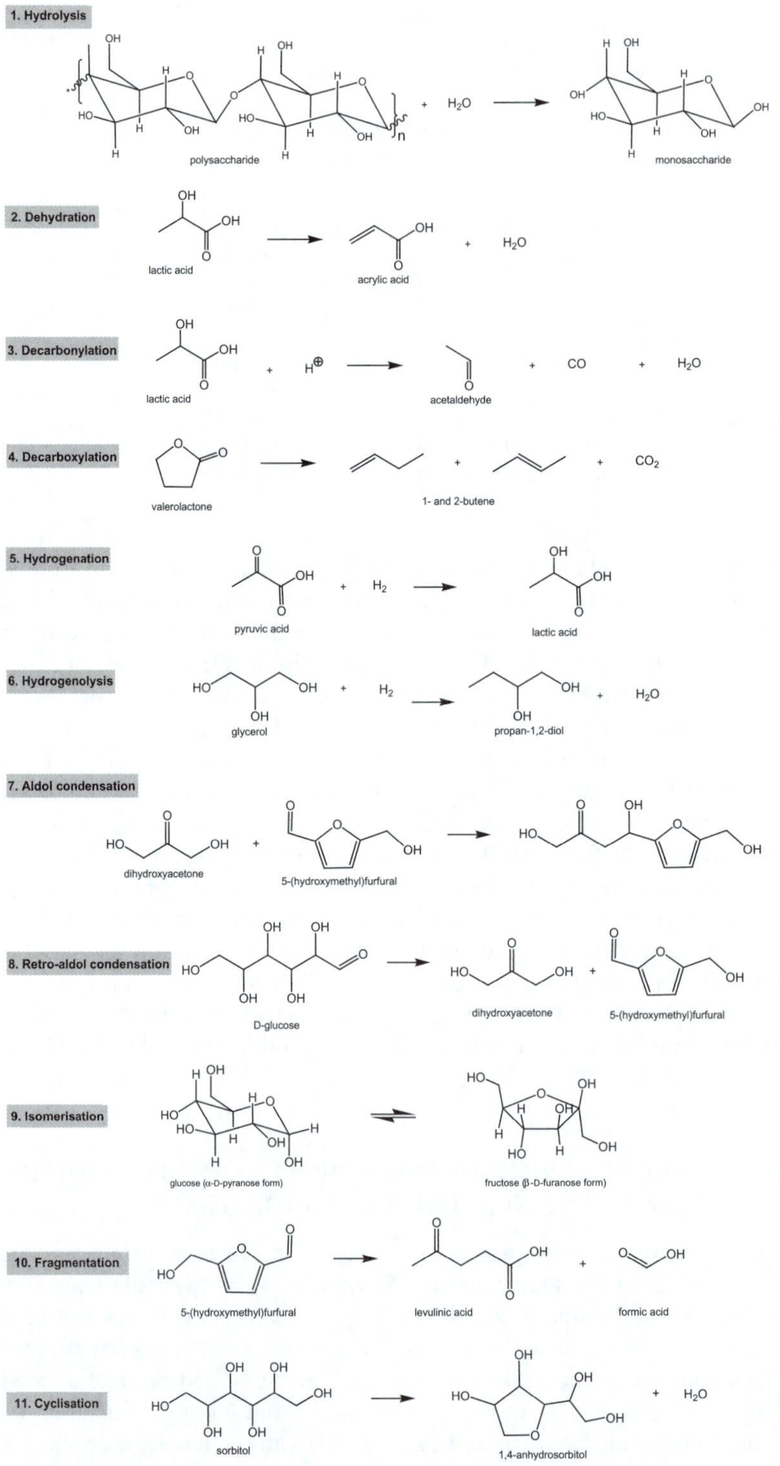

1. Hydrolysis

polysaccharide + H_2O → monosaccharide

2. Dehydration

lactic acid → acrylic acid + H_2O

3. Decarbonylation

lactic acid + $H^{\oplus}$ → acetaldehyde + CO + H_2O

4. Decarboxylation

valerolactone → 1- and 2-butene + CO_2

5. Hydrogenation

pyruvic acid + H_2 → lactic acid

6. Hydrogenolysis

glycerol + H_2 → propan-1,2-diol + H_2O

7. Aldol condensation

dihydroxyacetone + 5-(hydroxymethyl)furfural →

8. Retro-aldol condensation

D-glucose → dihydroxyacetone + 5-(hydroxymethyl)furfural

9. Isomerisation

glucose (α-D-pyranose form) ⇌ fructose (β-D-furanose form)

10. Fragmentation

5-(hydroxymethyl)furfural → levulinic acid + formic acid

11. Cyclisation

sorbitol → 1,4-anhydrosorbitol + H_2O

Scheme 11.4 Reaction types representative of biomass chemistry and of biomass-derived products.

will include bio-derived primary feedstocks to be used in existing chemical processing technologies and bio-derived intermediates that can substitute for important petrochemical analogues. The selection reflects the current state of play of a dynamic process of research, innovation, discovery and technological development the success of which will define the transition to more sustainable chemical technologies. Section 11.7 describes some of the basic processing technologies[39] that are likely to be used on the large scale and Section 11.9 the technological constraints likely to influence the final form of the chemical industry of the future.

11.6.1 Biodiesel

Triglycerides are found in animal fats and vegetable oils.[26] The seeds of the soya bean and rape contain *ca.* 20% and 40% by weight of triglycerides, respectively, which after extraction can be transesterified with excess methanol or ethanol, giving mixed methyl or ethyl esters of C_8–C_{22} fatty acids and glycerol (Section 11.6.10). The fuel characteristics, particularly the energy content or calorific value ($kJ\,kg^{-1}$) of the vegetable oils themselves (and the fatty acid esters derived from them), can be improved by a two-step hydrogenation–isomerisation process shown in Scheme 11.5. The particular advantage, when using vegetable-oil derived triglycerides as the feedstock, is the avoidance of glycerol as a by-product. This is also the basis of a process, recently commercialised (ref. 40; Web 4), that transforms tall oil fatty acids, by-products from the Kraft process for wood pulp manufacture.

11.6.2 Bioethanol

Fermentation of sugars released following enzymatic hydrolysis of starch from crops such as maize, sugar beet or sugar cane is currently the prime source of *ca.* $30\times10^6\,t\,y^{-1}$ of industrial bioethanol. The sequence of formal reactions from glucose *via* pyruvate and acetaldehyde (excluding the role of ADP and ATP) is shown in eqn (11.4) to (11.7), where NAD^+ and NADH are,

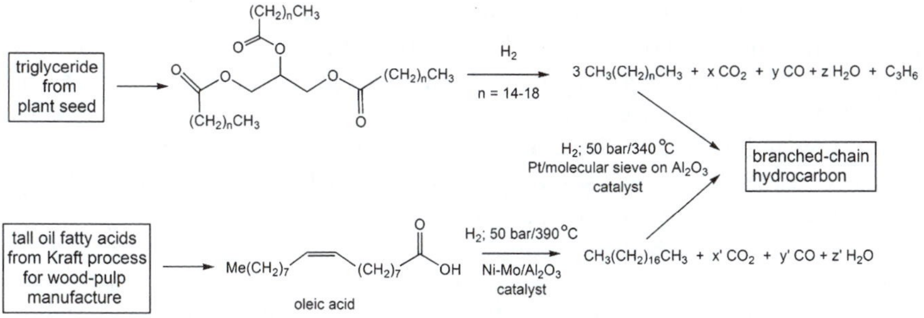

triglyceride from plant seed → $(CH_2)_nCH_3$ triglyceride → (H_2; n = 14-18) → 3 $CH_3(CH_2)_nCH_3$ + x CO_2 + y CO + z H_2O + C_3H_6

H_2; 50 bar/340 °C, Pt/molecular sieve on Al_2O_3 catalyst → branched-chain hydrocarbon

tall oil fatty acids from Kraft process for wood-pulp manufacture → $Me(CH_2)_7CH{=}CH(CH_2)_7COOH$ (oleic acid) → (H_2; 50 bar/390 °C, Ni-Mo/Al_2O_3 catalyst) → $CH_3(CH_2)_{16}CH_3$ + x′ CO_2 + y′ CO + z′ H_2O

Scheme 11.5 Hydrogenation of triglycerides and fatty acids, and subsequent isomerisation for biodiesel upgrading.[40] Note the avoidance of by-product glycerol formation from the triglyceride.

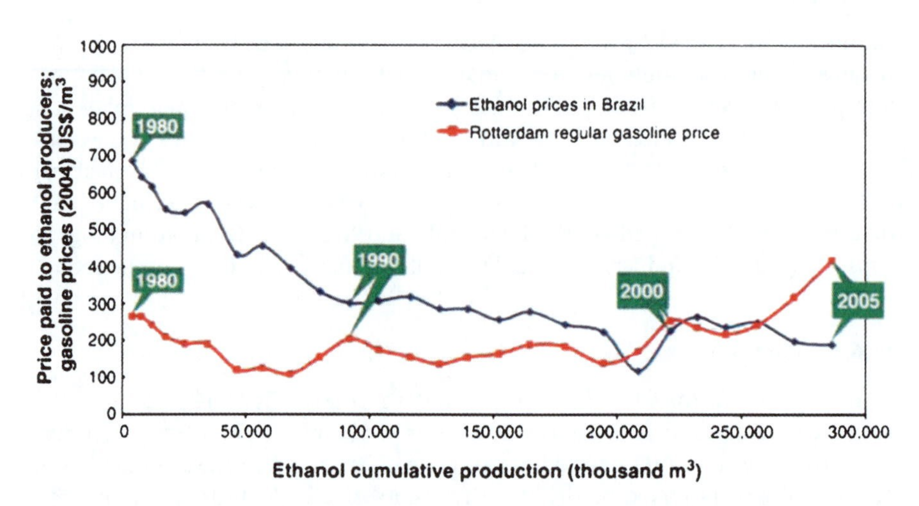

Figure 11.7 Price changes (in $US\$_{2004}$) for sugar cane derived ethanol during the period 1980–2005 compared with a gasoline price during the same period (as cumulative production increases to $300 \times 10^6\,m^3$). Reprinted from ref. 41 with permission of AAAS.

respectively, the oxidised and reduced form of nicotine adenine dinucleotide. Interestingly, the net reaction, eqn (11.7), has an atom efficiency (AE) of 51%.

$$C_6H_{12}O_6 + 2\,NAD^+ \rightarrow 2\,CH_3C(O)CO_2^- + 4\,H^+ + 2\,NADH \quad (11.4)$$

$$2\,CH_3C(O)CO_2^- + 2\,H^+ \rightarrow 2\,CH_3CHO + 2\,CO_2 \quad (11.5)$$

$$2\,CH_3CHO + 2\,NADH + 2H^+ \rightarrow 2\,C_2H_5OH + 2\,NAD^+ \quad (11.6)$$

$$C_6H_{12}O_6 \rightarrow 2\,C_2H_5OH + 2\,CO_2 \quad (11.7)$$

The benefits arising from the experience gained since the late 1970s associated with the production of bioethanol from sugarcane in Brazil can be seen in Figure 11.7. This compares the price in US\$ (2004) per m^3 paid to producers for ethanol and the gasoline price (in Rotterdam) over the period 1980–2005. As experience is gained (represented by cumulative production), the price of the new product generally decreases more quickly than that of the mature product.

The use of lignocellulosic biomass to manufacture ethanol is being given serious consideration because of the avoidance of food crop sources, though there is currently a distinct cost disadvantage[42] because of the need for delignification. There also appears to be no significant advantage for a thermochemical process to ethanol from lignocelluloses. The ethanol must be distilled and dried before it can be used as a fuel. Technical limitations on the use of ethanol as a transportation fuel have prompted the investigation of biobutanol as an alternative. Interestingly, carbon dioxide can be directly converted to isobutyraldehyde (and isobutanol) using a genetically engineered cyanobacterium.[43] Isobutyraldehyde is more volatile and less water-soluble than ethanol. While this would make it more readily separable from aqueous media used in fermentations, it would also require an additional processing step for its conversion to the corresponding alcohol.

Bioethanol can also be a feedstock, providing a non-petrochemical route to ethene by catalytic dehydration (eqn 11.8) and, from it, important downstream products such as ethylene glycol, ethylene oxide and vinyl chloride.

$$C_2H_5OH \rightarrow CH_2{=}CH_2 + H_2O \qquad (11.8)$$

11.6.3 2,5-Dimethylfuran

The deficiencies of ethanol as a transportation fuel (water miscibility and hygroscopicity, volatility, low energy density, poor storage stability, corrosiveness) can be addressed by converting carbohydrates into technically more attractive oxo-hydrocarbons. Water-immiscible products with higher energy density and with more suitable boiling points can be accessed[25] by a sequence of dehydrations and hydrogenations from fructose (see Figure 11.4). The first step is an acid-catalysed dehydration of fructose in aqueous hydrochloric acid from which 5-(hydroxymethyl)furfural (5-HMF) can be extracted into butanol. 5-HMF, with a boiling point of 291 °C, is too involatile for use as a transportation fuel. It can be further transformed in a series of steps over a copper–ruthenium catalyst into 2,5-dimethylfuran (bp 93 °C) and 2-methylfuran (bp 63 °C), which might be so used if they meet quite stringent technical criteria.

The availability of such materials on the large-scale might encourage their use as so-called 'platform' chemicals, *i.e.* be precursors for a series of downstream products. For example, 5-(hydroxymethyl)furfural could provide the series of derivatives shown in Scheme 11.6, any one of which (Section 11.6.10) might itself find large-scale use. However, yields of 5-HMF from carbohydrates such as fructose may well be currently too low to be economic.

11.6.4 Biosyngas

Syngas proper is currently obtained from fossil sources such as coal[xiii] and natural gas by partial combustion (eqn 11.9) or 'steam reforming' (eqn 11.10) over a long-lived Ni/MgO or Al_2O_3 catalyst operated at 15–30 bar and 825 °C. The ratio of H_2:CO in syngas is dependent on its source and on the process for making it, ranging from 0.5 for coal to 2 or 3 for methane. This is important, as different ratios are required depending on the downstream processing to be undertaken. This preferred ratio varies from 3 : 1 for H_2 and NH_3 production, 2 : 1 for MeOH production and Fischer–Tropsch synthesis (eqn 11.11) to 1 : 1 for hydroformylations, formaldehyde and CO production. Syngas from biomass

[xiii] In regions with large accessible reserves of coal available in the longer term, the development of 'clean' coal technologies for chemicals production (associated with carbon capture and storage) could build on well-established processes such syngas production and those that use syngas in Fischer–Tropsch and methanol-to-olefins (MTO) technology. For a recent review, see ref. 44.

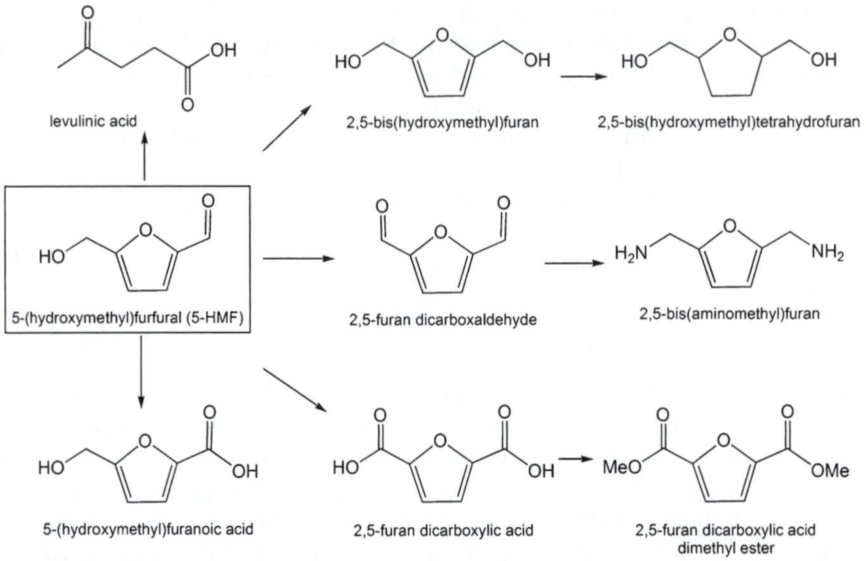

levulinic acid

2,5-bis(hydroxymethyl)furan

2,5-bis(hydroxymethyl)tetrahydrofuran

5-(hydroxymethyl)furfural (5-HMF)

2,5-furan dicarboxaldehyde

2,5-bis(aminomethyl)furan

5-(hydroxymethyl)furanoic acid

2,5-furan dicarboxylic acid

2,5-furan dicarboxylic acid dimethyl ester

Scheme 11.6 Compounds of technological interest accessible from 5-(hydroxymethyl)furfural (5-HMF).[37a]

gasification (Section 11.7.2) ('biosyngas') could be a source of hydrogen *via* the water–gas shift reaction following water addition (eqn 11.12). Or, it may be processed by Fischer–Tropsch chemistry[37b,38] into (second- or third-generation) hydrocarbon-type biofuels for use in internal combustion engines. Biosyngas contains H_2, CO, CO_2 and CH_4 and is, currently, unsuitable[45] for direct use (*i.e.* is not a 'drop-in' replacement) in current Fischer–Tropsch processes as the optimum Fischer–Tropsch catalyst is dependent on the precise $H_2 : CO$ ratio. The development of catalysts able to work with new syngas-type feedstocks is an active research area.

$$CH_4 + 0.5\,O_2 \rightarrow CO + 2\,H_2 \qquad (11.9)$$

$$CH_4 + H_2O \rightarrow CO + 3\,H_2 \qquad (11.10)$$

$$2\,H_2 + CO \rightarrow \text{`}{-}(CH_2){-}\text{'} + H_2O \qquad (11.11)$$

$$CO + H_2O \rightarrow CO_2 + H_2 \qquad (11.12)$$

$$2\,H_2 + CO \rightarrow CH_3OH \qquad (11.13)$$

Syngas provides a well-established route to basic chemical commodities such as methanol (eqn 11.13), from which transportation fuels can be obtained by the MTG (methanol-to-gasoline) process, as well as to other important intermediates including aromatics. Syngas is already used to build up bigger carbon-based molecules using the Fischer–Tropsch and related catalysed processes (eqn 11.11), yielding linear alkanes and alkenes. These synthetic products are much lower in sulfur- and nitrogen-containing impurities and in aromatics than their petrochemical equivalents. While the former would be a benefit in transportation fuel applications, the non-availability of basic aromatics such as benzene, toluene and xylenes (BTX) would be disruptive to the supply of a range of downstream intermediates and products.

11.6.5 Bio-oil

The thermochemical processing described in Section 11.7 delivers a complex mixture of oxygenated hydrocarbons known as 'bio-oil'. The crude product derived from woody biomass by pyrolysis, containing up to 40% by weight of oxygen and, having a water content of 15–30% and a pH of 2.5,[38] is a reactive corrosive mixture. This can be upgraded to give a more stable material with a higher energy density by catalytic cracking through a series of dehydration steps and hydrogen-transfer hydrogenations or by hydrotreating.

11.6.6 Biochar

'Biochar'[46a] represents the carbonaceous residue formed from either natural biomass burning or purposeful biomass processing. Biochar is produced naturally at a rate of $0.05–0.3\,GtC\,y^{-1}$ which, with an estimated global inventory in the soil of 80 GtC, provides a half-life in the soil of $10^2–10^3$ y.[46b] This estimate is important as it provides a measure of the degree to which atmospheric carbon could be permanently stored as biochar if returned to the soil and highlights the importance of defining the characteristics (or treatment prior to disposal) necessary for industrially produced biochar. (Its physical form suggests that other forms of disposal such as underground storage are likely to be more secure on a long-term basis than the more conventional carbon capture and storage.) Biochar could be an intentionally produced product associated either with chemicals production *via* biomass pyrolysis or from energy generation.[46c] As there would be no use for biochar, other than to sequester carbon, costs would be incurred in disposing of it (with limitations on transport distances to ensure net benefits in the reduction of greenhouse gases[46d]). Under conventional market conditions, biochar (a waste by-product) would have no value, so there would be no motivation to produce it. This could only be brought about by creating an incentive for its production that itself would require government action.

11.6.7 Biohydrogen

The use of hydrogen as a clean fuel and how it might be manufactured are considered in Chapter 12. Hydrogen can be obtained from biosyngas (Section 11.6.4) and, in principle, from the catalytic steam-reforming of bioethanol (eqn 11.14, ref. 47a) (as water is a reactant, its complete separation from fermentation-derived ethanol would not be required). However, carbon is produced from CO in competing reactions (eqn 11.15 and eqn 11.16) leading to catalyst deactivation.

$$CH_3CH_2OH + 3\,H_2O\,(+173.4\,kJ\,mol^{-1}) \rightarrow 6\,H_2 + 2\,CO_2 \qquad (11.14)$$

$$2\,CO \rightarrow CO_2 + C\,(+172.5\,kJ\,mol^{-1}) \qquad (11.15)$$

$$CO + H_2 \rightarrow H_2O + C\,(+131.4\,kJ\,mol^{-1}) \qquad (11.16)$$

H_2 could also be produced directly *via* microbial fermentation[47b] though such processes are inefficient (eqn 11.17) with typical maximum yields in the range 1–3 mol H_2 (mol sugar)$^{-1}$. A theoretical yield of 12 mol H_2 (mol sugar)$^{-1}$ is possible (eqn 11.18) and 11.6 mol H_2 (mol sugar)$^{-1}$ has been achieved experimentally under optimum *in vitro* conditions.[47c]

$$C_6H_{12}O_6 + 4\,H_2O \rightarrow 4\,H_2 + 2\,HCO_3^- + 2\,CH_3CO_2^- + 4\,H^+ \qquad (11.17)$$

$$C_6H_{12}O_6 + 12\,H_2O \rightarrow 12\,H_2 + 6\,HCO_3^- + 6H^+ \qquad (11.18)$$

$$CO_2 + 4\,H_2(g) \rightarrow CH_4(g) + 2\,H_2O(l) \qquad (11.19)$$

Hydrogen could be formed in a chemocatalytic aqueous phase reforming (APR) reaction following eqn (11.18).[48] This is thermodynamically favoured ($K = ca.\ 10^8$ at 500 K.) Unfortunately, a competitive process (eqn 11.19) is even more thermodynamically favoured ($K = ca.\ 10^{10}$ at 500 K). Nevertheless, using a 3 wt% Pt/γ–Al_2O_3 catalyst, an aqueous solution of 1 wt% glucose, sorbitol, glycerol or ethylene glycol (with H_2O : C = 165), treated at 225–265 °C, 27–54 bar for 24 h, delivers 80 g H_2 (kg catalyst)$^{-1}$ h^{-1}. The carbon conversion to gaseous products (CO_2, CH_4, C_2H_6, *etc.*) is 50–84% from glucose and 83–99% from glycerol. Importantly, the rates of production (tof) are better than *via* enzymic routes.

11.6.8 Adipic Acid

Adipic acid (used in the production of the polyamide, nylon 6-6) provides an example of a current commodity chemical conventionally manufactured from benzene (*via* cyclohexane, cyclohexanol and cyclohexanone; Scheme 10.11) that might be produced from a renewable source. Indeed, a fermentation process (Section 10.9) starting with glucose has been investigated[49] using a genetically engineered *E coli* to give *cis,cis*-muconic acid (see Scheme 11.7). This produces 36.8 g L^{-1} in 22% yield from glucose after 48 h in a fermenter. A conventional hydrogenation efficiently converts *cis,cis*-muconic acid (97% conversion) to adipic acid over a 10% Pt on carbon catalyst.

The key point is that the cost of manufacture from glucose was estimated (at the time of publication of the research article) to be US\$2.46 kg^{-1} (this is before many other costs and margins are added on to give a selling price), whereas the market purchase price for adipic acid was, at the time, only US\$1.52 kg^{-1}. This

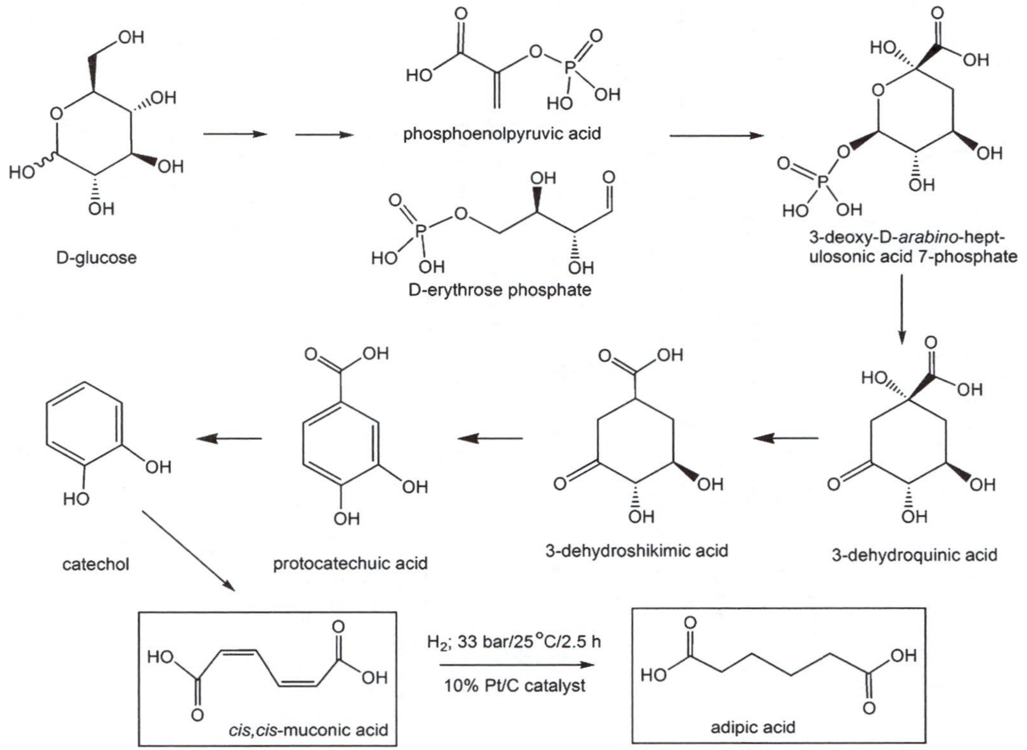

Scheme 11.7 A fermentation process using genetically modified organisms[49] for the conversion of glucose into a *cis,cis*-muconic acid, from which adipic acid can be readily obtained by chemocatalysis.

method of production was (and is) uneconomic.[xiv] No investment to operate the technology is likely to be made without a significant increase in the price of fossil-sourced adipic acid (such as might be associated with a sustained price of oil of $>$US\$140 bbl^{-1}) or a corresponding reduction in the cost of fermentation-derived adipic acid. The large fluctuations in the oil price make it difficult to forecast whether (or when) such technologies will be attractive at any point in the future. However, to the extent that scarcity drives up the price of raw materials (and of energy based on fossil sources), then the move to processes based on renewables is likely to continue, though the transition is likely to be slow unless it is stimulated in some way.

11.6.9 Bioaromatics

Terpenes are important natural products used as starting materials for some speciality chemicals.[32] However, while catalysed conversion of terpenes to aromatics is known, they are produced in quantities that are small in relation to global demand and are unlikely to be an important source. Lignin makes up *ca.* 20% of global biomass and is available as a waste from pulp and paper manufacture. If ethanol production from lignocellulosic biomass becomes more important, then a lignin-rich co-product stream will become available, providing a possible route to aromatics. However, efficiently transforming its complex cross-linked polymeric structure into aromatics is still a matter for laboratory investigation.[19b]

The production of aromatics from farm-grown biomass was demonstrated on the laboratory scale in a study motivated by a doubling of the oil price during the 1970s.[50] More recently,[51] catalytic fast pyrolysis using the zeolite ZSM-5 gave (though with coke formation) improved yields of aromatics including benzene, toluene, mixed xylenes and naphthalene (the latter is important in phthalic anhydride production).

A route to 'bio'-styrene (Scheme 11.8) has been proposed from 4-vinylcyclohexene[52a] involving a catalysed Diels–Alder cycloaddition of butadiene (itself formed from bioethanol *via* acetaldehyde and an aldol condensation over a MgO/SiO_2 catalyst[52b]). However, no large-scale process is likely to emerge except in the longer term because of the need to optimise several intermediate reaction steps.

11.6.10 'Platform' Chemicals

A study undertaken by the US National Renewable Energy Laboratory and published in August 2004[53a] suggests that the 'platform chemicals' shown in Figure 11.8 (accessible from carbohydrate sources) are key intermediates from which a range of downstream products, currently in demand, could be

[xiv] At a time when the 'impact' of publicly funded scientific research is an important determinant of how successful some see research to be, it should be stressed that the research described in ref. 49 was still worth doing as know-how, insight and a better definition of future research and technical targets would have been obtained in the process.

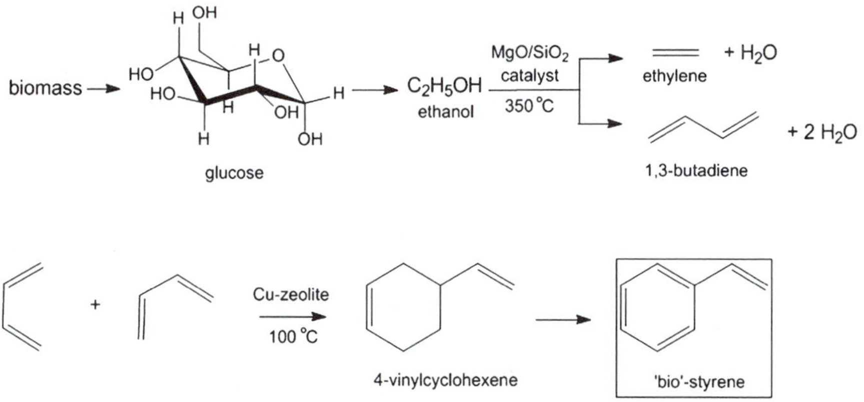

Scheme 11.8 A proposed reaction sequence[52a] for the production of styrene ('bio-styrene') in a series chemically precedented steps using intermediate products, ethanol, 1,3-butadiene and 4-vinylcyclohexane—all derivable ultimately from biomass.

manufactured. An update by Bozell and Petersen has recently appeared.[53b] A selection of those currently being highlighted is discussed in more detail below. Which of these become significant components of more sustainable globally-important chemical technologies remains to be seen.

11.6.10.1 2,5-Furandicarboxylic Acid. 5-(Hydroxymethyl)furfural (5-HMF) may be a component of processes leading to alternative biofuels. Its possible availability on the industrial scale is sufficient to stimulate widespread additional exploratory research interest (Scheme 11.6), building on much earlier work. It may be catalytically oxidised to the dicarboxylic acid or transformed directly to the dimethyl ester.[54] From either, bio-derived polyesters,[55] analogous to poly(ethylene terephthalate), can be obtained. Indeed, a parallel evaluation that uses a lignocellulosic starting point produces 2-pyrone-4,6-dicarboxylic acid, which may also form a polyester (Scheme 10.19).[55b] Yet more research suggests the furan series may lend themselves to particular classes of chemical reaction, such as their use as dienes in Diels–Alder reactions.[55c]

11.6.10.2 γ-Valerolactone. Carrying out the dehydration of a carbohydrate (sucrose) at the same time as a catalysed hydrogenation (using dihydrogen)[56a] leads (*via* a dehydration of 5-HMF) to levulinic acid (itself a platform chemical produced in the 'Biofine' process; Section 10.9) and formic acid and then (*via* a ruthenium-catalysed homogeneous reaction) to γ-valerolactone (Scheme 11.9) (with 4-hydroxypentanol and 2-methyltetrahydrofuran). γ-Valerolactone itself is a possible transportation fuel additive or a chemical starting material and has been evaluated both technically and economically, as a feedstock for dibutyl ketone.[56b]

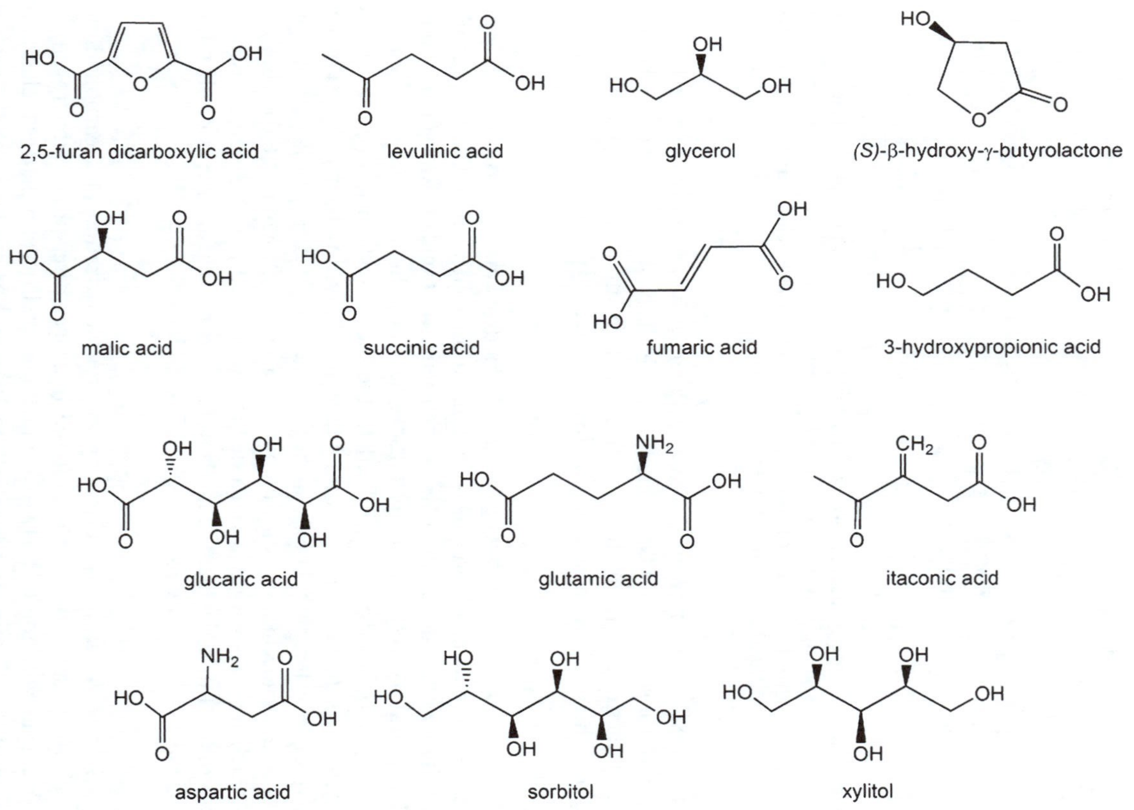

Figure 11.8 Fourteen potential 'platform' chemicals, accessible from carbohydrate sources, identified by the US National Renewable Energy Laboratory.[53a]

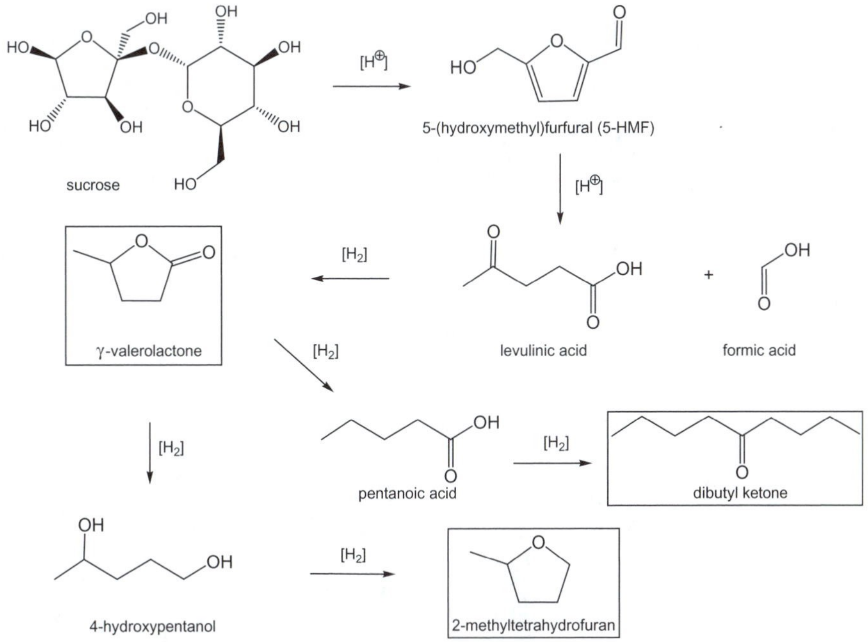

Scheme 11.9 Formation of γ-valerolactone, a possible transportation fuel, from sucrose, *via* levulinic acid, and its further reaction to 2-methyltetrahydrofuran[56a] and dibutyl ketone (DBK).[56b]

11.6.10.3 Glycerol. As long as oil-seed triglycerides are a major source of biodiesel fatty acid methyl esters produced conventionally, then glycerol will be ready available as a co-product. However, this ready availability does not necessarily mean that it will be a low-cost raw material. Bioglycerol is a component in a complex stream that contains methanol (a six-fold excess is used to effect economic conversions in the transesterifications), water, catalyst residues, free fatty acids and methyl esters, and unreacted glycerides. The crude waste is currently used as a concrete additive and in animal feed. Pure glycerol recovery is costly[57] so developments based on solid catalysts and reactive distillation[58] could yield a cleaner and more tractable source of glycerol. Despite this, Solvay are reportedly using glycerol from biodiesel production (as a substitute for increasingly expensive propene) in a process for the manufacture of epichlorohydrin (Scheme 11.10) that reduces both waste and water consumption.

Most chemistry research[59] (also discussed in Section 10.8) has focused on the use of pure, rather than waste, glycerol as a feedstock (Scheme 11.10). 1,3-Propanediol (1,3-PDO) may be produced from glycerol *via* a fermentation process and is used to make a polyester developed by DuPont. We concluded in Chapter 2 that, while the production and use of poly(ethylene terephthalate) (PET) as a synthetic fibre avoids the requirement for huge areas of land for the production of a natural material, wool, it is still produced from wholly petrochemical-derived feedstocks. A related material PTT, poly(trimethylene

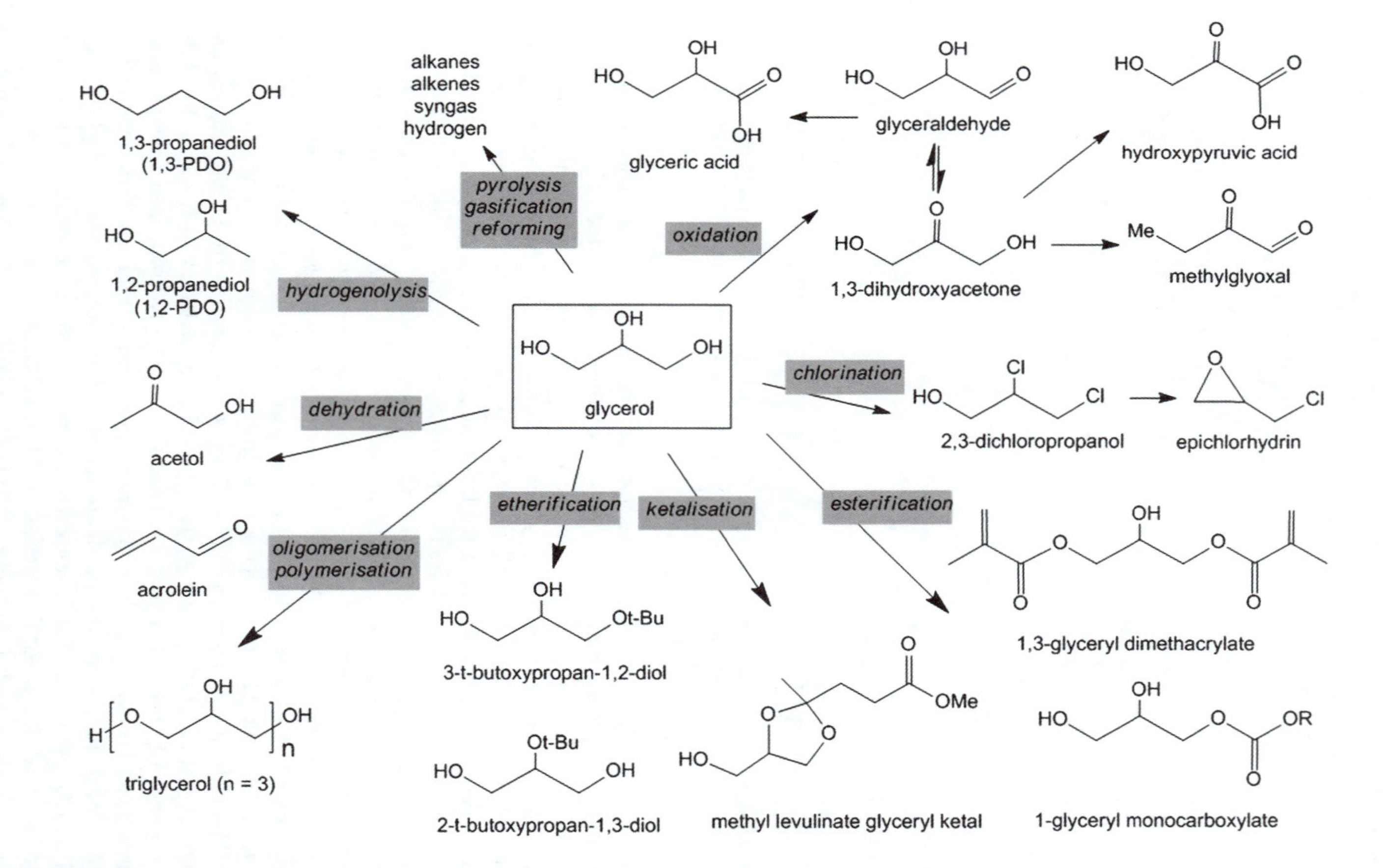

Scheme 11.10 Possible reaction products from glycerol exemplifying its potential as a platform chemical if low-cost material of appropriate purity becomes available from routes to biodiesel involving transesterification of triglycerides.

terephthalate) (trade name 'Sorona'), differs from PET in a single methylene unit in the diol monomer, 1,3-propanediol. However, the costs associated with the manufacture of 1,3-PDO through conventional technology (shown in Scheme 2.1) have inhibited the use of PTT. A recent life-cycle assessment of the fossil and biotechnological routes to 1,3-PDO[60] highlights the importance of including all of the impacts into the assessment. Recently, DuPont announced a one-step fermentation (biocatalytic) route (eqn 11.20) to 1,3-PDO that provides materials of a quality equivalent or superior to chemically produced material with claimed additional environmental benefits (Web 5):

$$\text{corn} \rightarrow \text{glucose} \rightarrow \text{glycerol} \rightarrow 1,3\text{-propanediol} \qquad (11.20)$$

A Bio-PDO™ plant, a joint venture with Tate & Lyle PLC, was commissioned in Loudon, Tennessee, in 2007 and the use of PTT fibre in carpet applications has been announced. Interestingly, routes to 1,3-PDO may also be accessible from glycerol derived from biodiesel manufacture *via* transesterification of triglycerides, previously mentioned in connection with 1,2-propanediol.

Bearing in mind that glycerol is a waste co-product from biodiesel manufacture, a change in FAME production technology that avoided its formation would be attractive to biodiesel producers. The results of an experimental design have very recently been disclosed[61] in which triglycerides are transesterified to FAME on reaction with supercritical methyl acetate (eqn 11.21). As glyceryl triacetate (or triacetin) is now the co-product, the mixture can be used as biodiesel directly.

$$\begin{aligned}&\mathrm{RC(O)OCH_2CH(OC(O)OR')CH_2OC(O)R'' + 3\,MeOAc \rightarrow}\\&\mathrm{RC(O)OMe + R'C(O)OMe + R''C(O)OMe + AcOCH_2CH(OAc)CH_2OAc}\end{aligned} \qquad (11.21)$$

11.6.10.4 Lactic Acid. Lactic acid production highlights an interesting contrast between established petrochemical and newer fermentation technology as a means of manufacture. A series of useful derivatives (a product 'chain') can be obtained from lactic acid (including a possibly economic process for acrylic acid when oil is at US\$55–60 bbl^{-1}). However, a solution to the problem of efficiently catalysing the dehydration reaction (eqn 11.22) is first required.

$$\mathrm{CH_3CH(OH)CO_2H \rightarrow CH_2{=}CHCO_2H + H_2O} \qquad (11.22)$$

The biotechnological approach to lactic acid uses corn syrup or molasses as feedstock and produces a technical-grade product. This process is slow (involving a 4–6 day fermentation) and is both energy-consuming (largely associated with the need to separate the desired product from aqueous solution) and waste-producing. Nevertheless, the product (and the associated downstream derivatives) is chiral. The classical chemical route is more efficient, producing little waste (see Scheme 11.11). However, it uses a hazardous reactant (HCN) and involves three distillations. On the other hand, the isolated product is of high

acetaldehyde + HCN —distill→ acetaldehyde cyanhydrin —H_2O/H_2SO_4→ crude lactic acid + $[NH_4]_2SO_4$

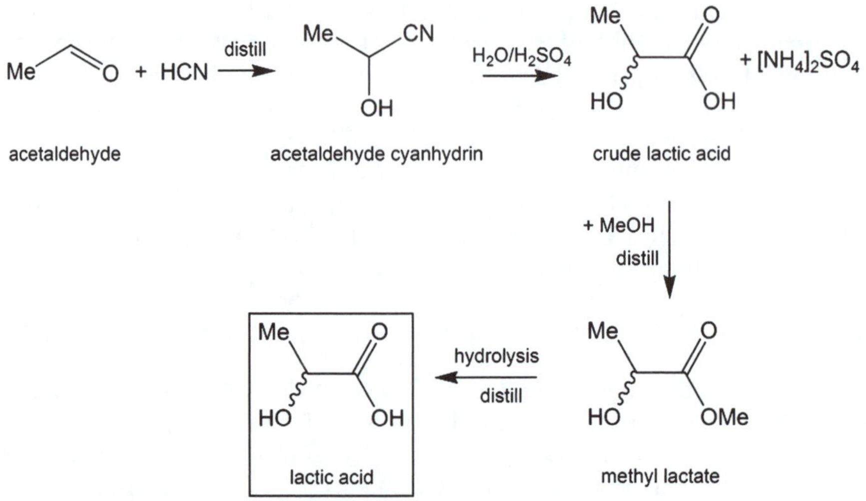

crude lactic acid —+ MeOH, distill→ methyl lactate —hydrolysis, distill→ lactic acid

Scheme 11.11 Classical chemical process for the efficient production of high-purity achiral lactic acid.

purity (though non-chiral). Clearly, the biotechnological route has quite a way to go to displace the conventional technology, except of course if chiral compounds are needed. Even here, though, the degree of purity of the product may be inadequate and some means of separating enantiomers from the purer non-chiral product may be more attractive. The higher economic value of the chiral product could well justify investment to develop a better fermentation technology that might result in the displacement of the petrochemical sourced route. The point here, of course, is that it is not always a straightforward judgement as to the optimum way forward. And, indeed, who should make the decision anyway?

Aqueous lactic acid can be converted catalytically to dilactide. After separation and purification, dilactide can be subjected to ring-opening polymerisation to give poly(lactide), a biodegradable thermoplastic (Scheme 11.12) marketed under the trade name Ingeo® (Web 6). This is manufactured by NatureWorks LLC in a 140 000 t y^{-1} plant in Nebraska, USA.

11.6.11 Specialities, Fine Chemicals and Pharmaceuticals

Antibiotics, steroids, vitamin C, dihydroxyacetone, sorbitan esters, sucrose esters, alkyl polyglucosides, alkyl glucamides and monoglycerides are all products that are (or have been) accessed from renewable starting materials. Some will have been displaced in the past by cheaper products from petrochemical sources. As the costs of using petrochemical-sourced feedstocks rise, the benefits of shifting back to renewables may begin to look more attractive.

Unfortunately, very few studies provide a proper side-by-side comparison between a petrochemical-sourced and a renewables-sourced process. Furthermore, there are very few reports that tell us of, and reasons for, abandoned (or

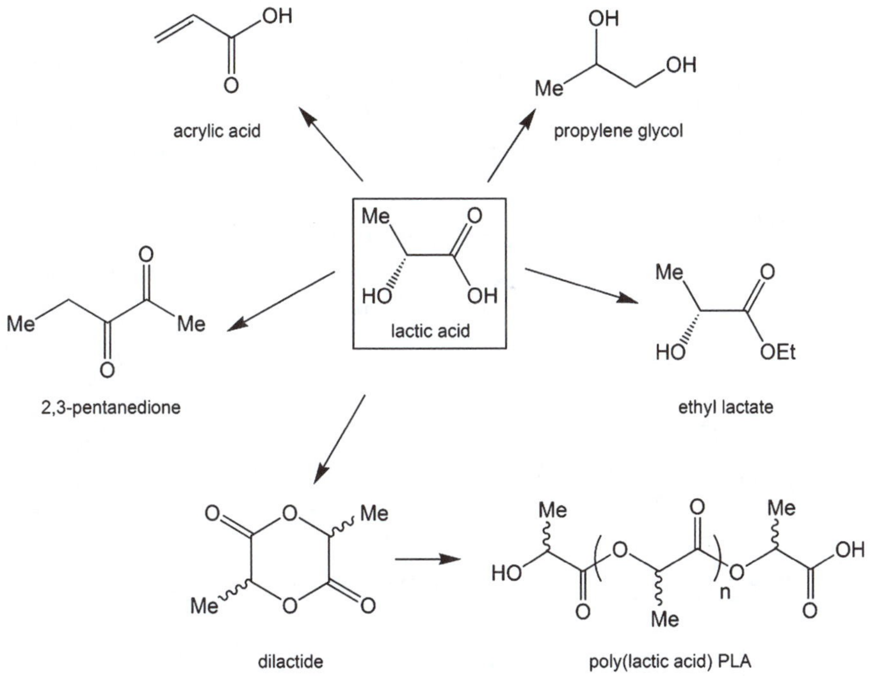

acrylic acid
propylene glycol
lactic acid
2,3-pentanedione
ethyl lactate
dilactide
poly(lactic acid) PLA

Scheme 11.12 Chiral lactic acid produced from corn syrup in a biotechnological process, including accessible downstream products such as dilactide and its conversion to a biodegradable thermoplastic, poly(lactic acid), marketed as Ingeo®.

delayed) investigations and developments. As this is an area of intense competition and commercial sensitivity, such publicly available information is generally in short supply.

An approach with the greater likelihood of earlier adoption (other things being equal) would be to make a product with a significantly higher value which might thus justify the additional costs associated with the use of a biotechnological process. Two examples are shikimic acid[62] and vanillin.[63]

11.6.11.1 Shikimic Acid. Shikimic acid, a natural product isolated from the Chinese star anise, is also accessible in a bioprocess from glucose (Scheme 11.13). Shikimic acid is used in synthetic routes to the anti-influenza drug, Tamiflu (Section 8.5).

11.6.11.2 Vanillin. Natural vanillin is obtained from glucovanillin formed when the beans of the orchid, *Vanilla planifola*, are cured. Unfortunately, this source (renewable, but hardly sustainable!) was insufficient to satisfy demand even as long ago as the 19th century when a synthetic route from eugenol [(a) in Scheme 11.14] was developed. This was subsequently displaced in the 1930s

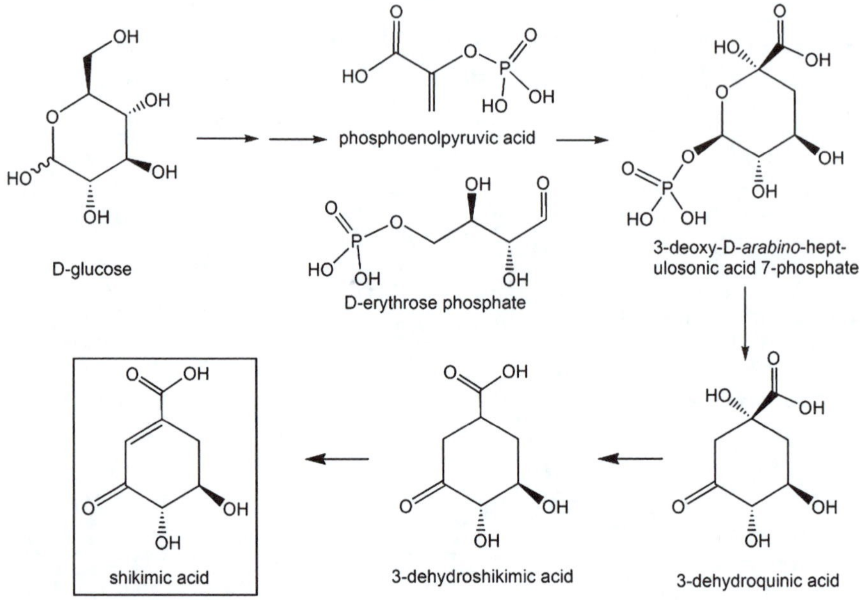

Scheme 11.13 A bioprocess leading to the important intermediate, shikimic acid, from glucose.

by a process, no longer operated, that used lignin from the waste generated in paper production.[63] The Norwegian company, Borregaard Industries, has developed a related process that yields 4 kg of vanillin from one tonne of wood. However, most of the 12 000 t annual production of this high value chemical (*ca.* £10 000 t^{-1}) comes *via* a petrochemical route [(b) in Scheme 11.14], though there are research reports that it may also be obtained by fermentation.[49b]

Because of the preference of some consumers for products from a natural source, there is a market for material produced by technology not relying on fossil feedstocks. As the vanillin example shows, satisfying such preferences may not always be sustainable. In fact, there is a spectrum of classification from 'natural' to 'nature-equivalent' to 'semi-synthetic' to 'synthetic'. The search for 'natural' or 'nature-equivalent' product has involved the use of recombinant *E. coli*[49b] which gives vanillic acid from glucose in the reaction sequence shown [(c) in Scheme 11.14], with the vanillic acid reduced in a subsequent step using the enzyme, aryl aldehyde dehydrogenase.

11.7 CHEMICALS FROM BIOMASS: BIOMASS PROCESSING

While there are many possible combinations of feedstocks and processing technologies (California's regulators considered no less than 12 different scenarios simply for producing fuel ethanol from corn[64]), it is probably fair to conclude that keeping the number of processing steps to a minimum will be one important factor in determining economic viability.

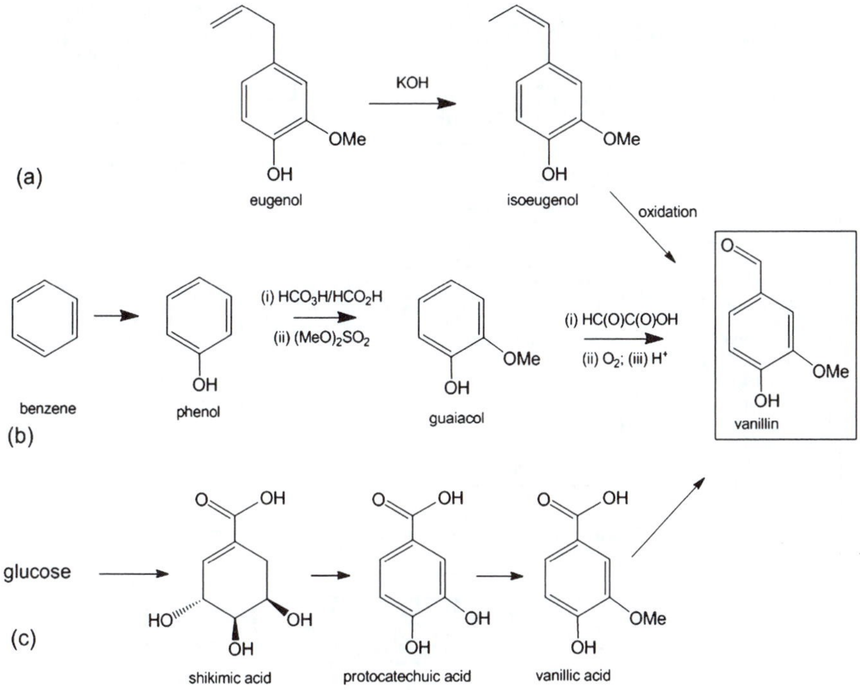

Scheme 11.14 Technologically important routes to the natural flavouring, vanillin: (a) an early process from the terpene, eugenol; (b) the petrochemical process from benzene; and (c) a research fermentation route from glucose via shikimic acid.[49b]

Manufacturing chemicals and energy carriers from biomass can be achieved in two basic ways (Figure 11.9):

- thermochemical processing
- water-based processing

The low volatility of many biomass components, their high reactivity, thermal instability and solubility in water makes aqueous phase processing conceptually appealing. The speed with which new technology for chemicals production from biomass can be introduced is dependent on the adequacy of our understanding of, and ability to control, the chemistry and chemical processing of oxohydrocarbons (particularly carbohydrate and lignin chemistry) compared with that of hydrocarbons (aliphatics and aromatics). Reliable technology must be developed suited to oxygenated rather hydrocarbon feedstocks, requiring a significant development in understanding of the technological characteristics of biomass in its various forms and the changes in existing processing methods and the development of new ones that will be necessary.

On the other hand, there is much expertise and existing processing technology that has grown up in processing petrochemical products. This is currently

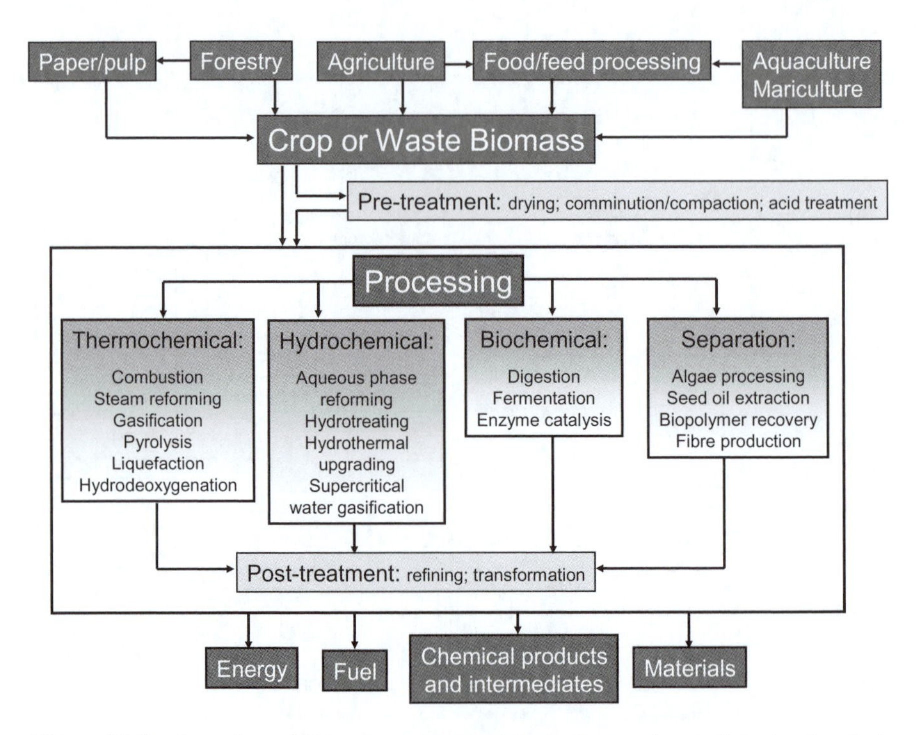

Figure 11.9 Overview of biomass sources and basic thermochemical, hydrochemical, fermentation and extraction processing needed to provide energy, fuel, chemical products and materials.

available. Conventional production of petrochemicals uses high-temperature, high-pressure, largely gas phase, catalytic processing in plant and equipment. These are also already available. However, oxohydrocarbon feedstocks from biomass, unlike the hydrocarbon components of oil and natural gas, are generally not suited to such conditions and would require additional processing to make them so. Figure 10.10 (taken from ref. 23) displays some of these technological options and the relevant operating conditions. The chosen option will depend on very many different factors, not all of which will be governed by chemistry. Should a particular chemical be derived from a defined crop-derived precursor? Or are we interested in developing a wholly new product chain? Do we want a route to an important petrochemical commodity such as ethene that is sourced from ethanol produced from a renewable raw material? Alternatively, the availability of new raw materials may require wholly novel processing technologies to be identified and developed. In such cases, time will be needed to gain the 'experience' needed to bring down unit costs to make the technology competitive enough to drive the shift from fossil feedstocks.

A further, conceptually simpler, alternative would be to transform all biomass to small primary building blocks in a manner analogous to that adopted when chemicals were largely derived from coal.[44] In this case, coke (the carbonaceous residue obtained from the destructive distillation of coal) and lime (calcium oxide) were heated in an electric arc furnace to give calcium carbide (eqn 11.23) from which acetylene, a key petrochemical feedstock, was obtained on reaction with water (eqn 11.24).

$$3\,C + CaO \rightarrow CaC_2 + CO_2 \quad (11.23)$$

$$CaC_2 + 2\,H_2O \rightarrow HC{\equiv}CH + Ca(OH)_2 \quad (11.24)$$

It will be necessary to convert biomass close to where it is grown and harvested into a similarly small number of basic chemical products (*e.g.* methanol) rather than transporting it long distances ($>100\,km$) to the equivalent of an oil refinery. This would avoid both the cost of (and emissions associated with) transporting large quantities of (solid) biomass and developing the necessary infrastructure. These basic chemical products could (in principle) be fed to existing processing plant, though the nature of such a bioproduced feedstock may be different in some crucial way (such as its impurity profile) that would prevent such a so-called 'drop-in' change.

The following processing methods are introduced briefly below:

- combustion (primarily for power generation)
- gasification[65a] (primarily for syngas)
- pyrolysis[65b] (giving bio-oils)
- hydrothermal upgrading, hydrolysis and fermentation.

Each is an active area of investigation. However, we should note that depending on its type, biomass may need to be subjected to any one of a number of physical or chemical pretreatments to render it more suitable for such processing.

11.7.1 Combustion

Chemically useful by-products from the generation of energy from biomass combustion (eqn 11.25, exemplified for a carbohydrate; related equations can be written for other biomass components) are likely to be limited in an optimally operated process. Indeed, the energy content of wastes and by-products from the transformation of biomass into chemicals is likely to be recovered in a linked power station rather than collected and transported to a central unit for further chemicals processing.

$$C_6H_{12}O_6 + 6\,O_2 \rightarrow 6\,CO_2 + 6\,H_2O \quad (11.25)$$

11.7.2 Gasification

Gasification of biomass delivers mixtures of H_2 and CO ('biosyngas'; Section 11.6.4) according to (the simplified) eqn (11.26).

$$C_6H_{12}O_6 + 1.5\,O_2 \rightarrow 6\,CO + 3\,H_2 + 3\,H_2O \quad (11.26)$$

11.7.3 Pyrolysis

Thermal treatment of biomass in the absence of oxygen can lead to both liquid and solid products, depending on processing conditions such as temperature, residence time and state of aggregation of the biomass. For instance, fast pyrolysis on very finely-divided feedstock at moderate temperatures with residence times of the order of seconds followed by rapid quenching of the volatiles to avoid depolymerisation can give high yields of so-called 'bio-oil'[66] (Section 11.6.5). A mechanistic scheme (Scheme 11.15) for solid cellulose pyrolysis shows[67a] that depolymerisation gives levoglucosan that can then be dehydrated and isomerised to other anhydrosugars that react further *via* fragmentation, retro-aldol condensations, decarbonylations and decarboxylations. Fifteen products, in addition to char (4.6–9.8%) and accounting for 50–88% of starting material, have been found[67b] to arise in the initial stages of fast pyrolysis of a series of mono-, di- and polysaccharides at 500 °C. In addition to anhydrosugars such as levoglucosan, the four most abundant products were formic acid (5.3–12.4%), glycolaldehyde (5.6–8.4%), 2-furaldehyde (1–8.4%) and 5-(hydroxymethyl)furfural (2.8–8.9%).

11.7.4 Aqueous Processing

Sugar fermentation is a widely operated technology, used industrially to manufacture ethanol and related materials. An alternative approach currently being researched treats carbohydrate separated from biomass in aqueous solution (rather than in the gas phase) to form hydrocarbons or oxohydrocarbons suitable for fuel use.

For instance, aqueous phase reforming of sorbitol (Scheme 11.16) leads ultimately to CO_2 and H_2. Repeated cycles of carbohydrate dehydration (over acidic SiO_2/Al_2O_3) and subsequent hydrogenation of the dehydrated intermediates can lead to alkanes with six or fewer carbon atoms. Typically, 4 wt% $Pt/SiO_2/Al_2O_3$ with 5 wt% sorbitol in water at 225 °C/33–53 bar and 265 °C/56–61 bar gives 76–95% of carbon as compounds in the gas phase and 8–27% in the liquid phase; $<5\%$ branched chain isomers are present and the alkane content of the gas phase is 53–63% (the rest being CO_2). Virent Energy Systems Inc. achieved a 2009 Presidential Green Chemistry Challenge award (Web 7) for the near commercialisation of such a process, claiming it would be competitive at a crude oil price of US\$60 bbl^{-1}. However, these processes are currently operated only on the exploratory stage (*ca.* 1 $L\,d^{-1}$), with a pilot plant producing 100 $L\,d^{-1}$ expected to be operational shortly.[68] Efforts to develop related technology to produce more useful higher alkanes focus on combining C–C forming reactions (*e.g.* aldol condensation) with dehydration/hydrogenation. The main catalytic challenge centres on the means of reducing the extent and impact of coke formation.

Related chemistry[48b] has been examined as a means of producing oxohydrocarbons (from fructose, itself derived from starch) as transportation fuels, focusing particularly on two compounds, 5-HMF and 2,5-dimethylfuran

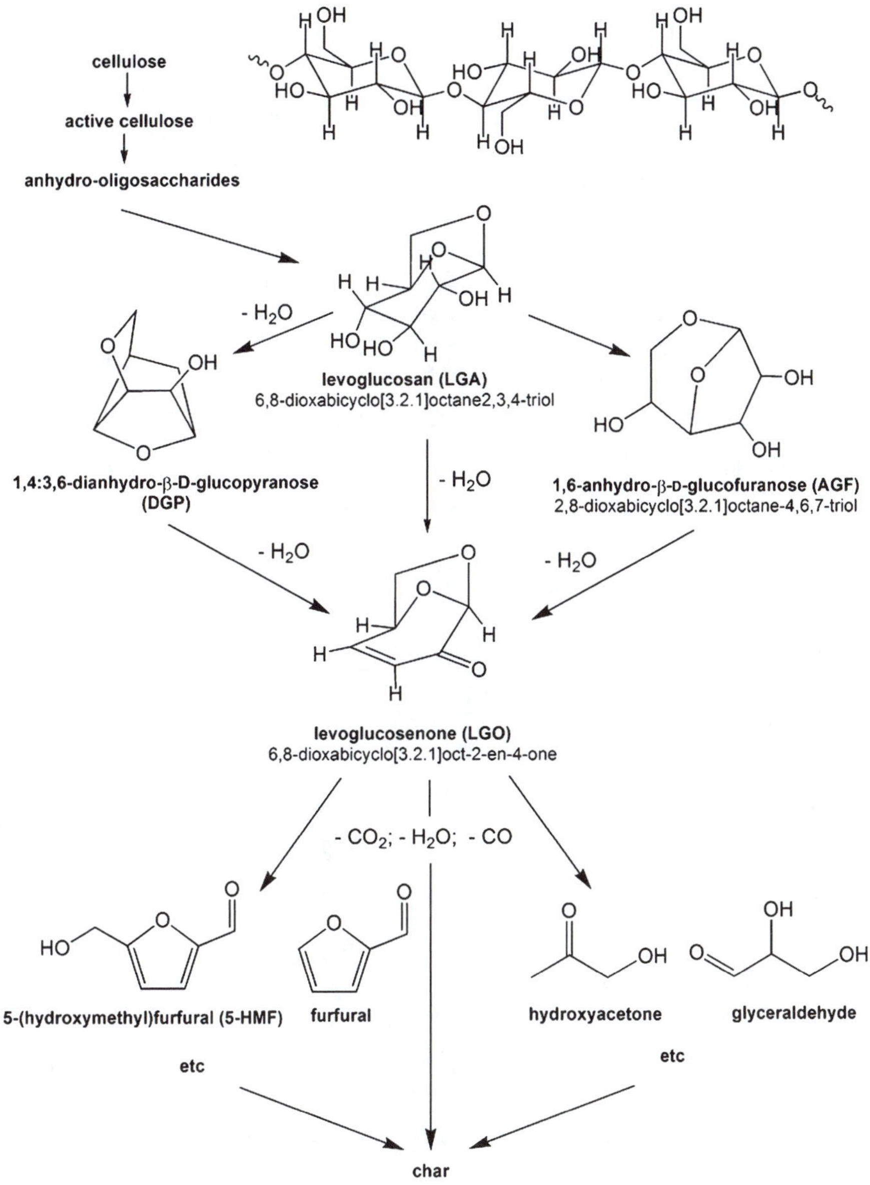

Scheme 11.15 Intermediates formed during the rapid thermal depolymerisation of cellulose. Redrawn from ref. 67a.

(DMF). The work highlights the trade-off—very important when considering matters from technological and technocommercial perspectives—between the amount of processing that is done and the nature of the product being sought. In this case, reasonably high volatility is important (other things

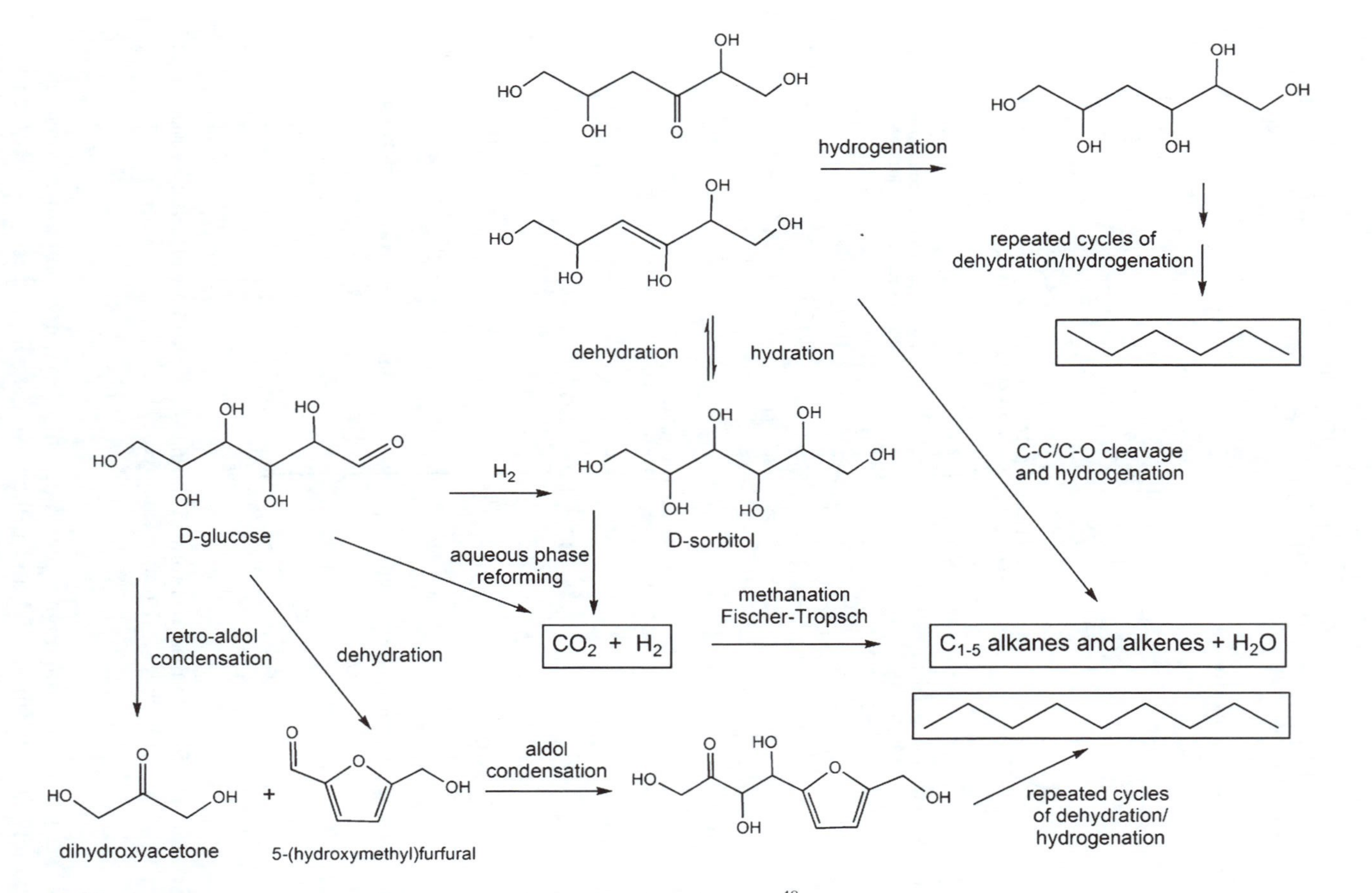

Scheme 11.16 Main processes involved in aqueous phase reforming of sugars.[48]

being equal) and the presence of large numbers of oxygens (particularly as OH) can result in high boiling points (as shown in Figure 11.4). So, the aim is to develop cost effectively, with the minimum of processing, an optimum product by dehydration and hydrogenation, making C–H bonds from C–O bonds.

11.8 TECHNOLOGICAL INTERDEPENDENCE AND INTEGRATION

While it may be feasible to produce some chemicals solely in a series of biotechnological transformations, the process and energy efficiencies that can be achieved by the use of high temperatures and pressures make it more likely that biomass-sourced chemicals production of the future will integrate, appropriately, both chemo- and bio-processing. For example, in an evaluation by DuPont[69] of industrial-scale production of 4-hydroxystyrene from glucose (Scheme 11.17), a genetically engineered microorganism was developed for the production of the intermediate, 4-hydroxycinnamic acid. Because the final product was toxic to the micro-organisms used to prepare the intermediate, the decarboxylation step used a conventional chemocatalysed process (using potassium acetate in dimethylacetamide).

11.8.1 The Biorefinery

Both energy and basic C_1 chemicals can, in principle, be co-produced from biomass using the appropriate integrated technology. This has led to the development of the 'biorefinery' concept[66,70] to parallel the idea of the oil refinery. The schematic generic representation shown in Figure 11.9 combines many of the components that have been suggested for inclusion in a biorefinery.[xv] While such a concept may be a model for the long term,

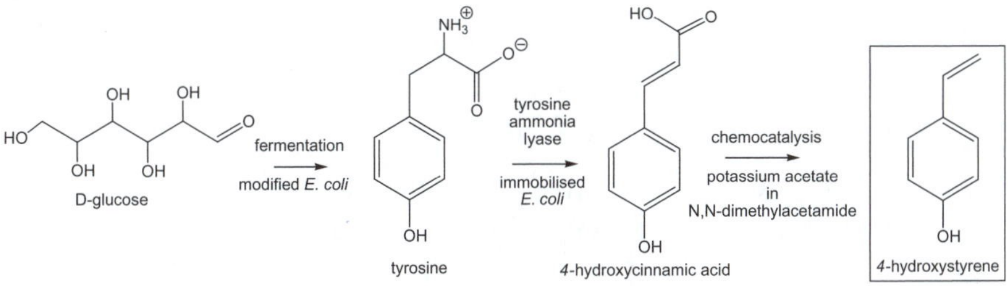

D-glucose → fermentation, modified *E. coli* → tyrosine → tyrosine ammonia lyase, immobilised *E. coli* → 4-hydroxycinnamic acid → chemocatalysis, potassium acetate in N,N-dimethylacetamide → 4-hydroxystyrene

Scheme 11.17 Combined biocatalytic and chemocatalytic process for the production of 4-hydroxystyrene.[69]

[xv] However, while the idea appears simple, the possible combinations of platforms, processes, feedstocks and products are complex and will be circumstance-specific based on the dominant input material and the desired output, whether energy, fuel or chemical products and intermediates. Under the aegis of the International Energy Agency, a common classification scheme for biorefineries has been proposed.[71]

developments in the immediate future are much more likely to be local, separate and independent as the availability of biomass supply and the demand for products, by-products and wastes for new processes and technologies are first explored and their economic and commercial viability determined. Inter-company 'symbiotic' arrangements, similar to those developed at Kalundborg (Section 9.12), might then grow around localised groupings of separate but complementary activities such as agriculture, wood- and biomass-processing, chemicals production and energy generation. On the other hand, bearing in mind the nature and scale of process streams available from lignocellulosic biomass (one lignin-based, the other cellulose-based), the integration of technologies for transforming both into high-value products on a single site is likely to be necessary for economic operation.

Biomass refining is analogous to the refining of oil—though the analogy is, in some important respects, a poor one. The latter involves the initial fractionation of crude oil by distillation into a series of high-volume product streams, including naptha, currently the prime carbon-containing feedstock for the chemicals industry. The biorefinery (Figure 11.9) would involve a number of basic processes for the initial treatment of diverse types of biomass integrated with the provision of the basic energy carriers, fuel and raw materials needed for society to function. An important possibility is that one output could be 'biochar'[46] (Section 11.6.6) or some other form of carbon that avoids the release of a certain proportion of the photosynthetically captured CO_2. The literature on biorefineries is now extensive and growing, highlighting both its importance and its complexity. A selection of reviews is provided in ref. 70.

A potential (and possibly unanticipated) benefit arising from the purposeful growth, harvesting, collecting and processing of biomass arises from the concentration of metals into the process residues. This bioaccumulation of metals from metal-containing soils into plant tissue ('phytoremediation') is very species dependent. The interesting possibilities that 'bioharvesting' or 'phytomining' of metals may have commercial potential (as sources of metals whose supply is seen as declining or which are becoming very much more expensive) have been suggested.[72]

11.9 TECHNOLOGICAL CONSTRAINTS

There are several key questions that will decide the pace with which any particular renewable biomass resource will be used to make chemicals and the strategies that will be adopted to bring this about, in the short and long terms, including the following:

To what extent can existing petrochemical feedstocks, intermediates and products, readily and directly, be substituted by those from renewables? That is, can we replace a petrochemically-sourced intermediate in a product chain with one generated from a renewable biomass feedstock? If this is technically feasible, how quickly can it be done? Can it be done economically? How can the net environmental benefits be assessed? Would subsidies and taxes have a beneficial or a negative effect on the change?

The period of transition to a more sustainable use of renewables, particularly for energy generation and the production of transportation fuels will be governed by a range of factors, not least the role played by governments in stimulating change. The change to maximise the use of renewables for chemicals production will be similarly influenced,[xvi] not least by the time needed to develop and introduce any new technology (Chapter 9). This period is unlikely to be less than 20 years, though it will be governed by how the end of the transition is defined. It is unlikely to be when petrochemical feedstocks are no longer used, as some fossil-sourced materials may remain essential even when we have maximised our move towards renewables.

One important consideration will be the ability of the supply to satisfy demand. This is of particular relevance when considering the utilisation of waste, which will initially be of low cost, though its price will increase as some uses are made of it. However, if the volume of waste is limited, then those wishing to use it, concerned it will not be available in the volumes required, will turn to alternative sources of supply. The avoidance of total dependency on a single supplier or source of supply is an important technocommercial consideration.

While the production of chemicals from petrochemical feedstocks may continue to be economically and commercially justified even as the scarcity of oil results in increased prices, its current close integration with, and dependence on, the thermochemical technologies for energy and fuel production may lead to changes primarily driven by a shift to the use of renewables for energy and fuel production. This could dictate which basic and feedstock chemicals (and, as a consequence, the downstream processes) become available. Alternatively, separate hydroprocessing operations that are primarily driven by chemicals (as opposed to biofuels) production could arise. The degree to which these two types of technology are integrated into 'biorefineries' (Section 11.8.1) (and their scale and location) will be determined largely by economics.

In principle, there are four main approaches to generate an entire product supply chain based on renewables. We can use:

- existing plant-derived feedstocks to make existing intermediates or products
- existing plant-derived feedstocks to make new intermediates and products
- new plant-derived feedstocks to make existing intermediates or products
- new plant-derived feedstocks to new intermediates or products.

Each of the pathways from biomass source to chemical product shown in Figure 11.1 or Figure 11.9 could define one or more of these supply chains. The particular choice will be very much circumstance-specific, circumstances that are impossible to predict. They will be influenced largely by economic factors that will determine whether the appropriate investments (and the rate of

[xvi] A relevant historical perspective on the development of biomass-derived chemicals is provided[73] in a study of the transition from coal-derived to petrochemical feedstocks in the UK in the period 1921–1967.

investment with time) will take place (and in a timely fashion). (We are already seeing, in 2009, some drawing back from investment in major renewables projects as the price of oil reduces from its peak of about US$140 bbl^{-1} to a level of US$50–70 bbl^{-1}. Such variations make it difficult to gain sufficient certainty in the growth of demand over the timescale of an investment on which to base reliable forecasts of costs and prices.)

An important constraint to the ready use of plant-derived material for large-scale continuous chemical production is the relatively slow growth of plants, which generally produce only one harvest per year. They are also vulnerable to climatic variability and pest infestation. They are highly complex structures whose processing on the very large scale needed to replace petrochemical feedstocks provides major engineering challenges. It is not surprising, therefore, that other biomass-producing organisms are under scrutiny, particularly those which generate biomass at a greater rate, and whose composition and form lend themselves to be more readily processed. Algae, which are fast-growing and rich in processible biomass, fall into this category[8,35,74] (Web 3). Fermentation processes (see Bibliography) using micro-organisms also take advantage of the very rapid growth of whole cells given the right conditions. A major current research area is using genetic modification to improve important characteristics of plants and other organisms relevant to large-scale processing. These include the efficiency of solar energy capture, growth rate, disease- and pest-resistance, and the ability to grow on marginal lands under conditions of low water or saline water availability.

To be commercially attractive (*i.e.* for investors to put large amounts of money at risk), the new feedstocks, products and processes need ideally to satisfy the following criteria, irrespective of the precise choice of technology used to produce them:

11.9.1 Performance and Technical Specification

To persuade a user to switch supply of an essential product from one based on petrochemical or non-renewable feedstocks to a material from a renewable source, the critical performance characteristics of an existing intermediate or chemical product derived from a renewable source must, at least, be equivalent to the fossil-based alternative. Users need to be assured that the characteristics of an intermediate or a product, important for their business, are not going to bring unexpected changes to a process or a final material.

11.9.2 Cost Differentials

There must also be a favourable cost differential compared with competing fossil-sourced products to induce the change, particularly to overcome any uncertainty about the performance of the substitute material. This may result either from the cost of the fossil-sourced material increasing over time because of scarcity of supply (or tax disincentive), or because of a lower cost of the material from a renewable source (which may, initially at least, be achieved through subsidies or other inducements), or from a combination of both. However, we have seen that the experience curve tends to disadvantage a new entrant into a marketplace because of the initial relatively high unit costs for a new product. The issue of timescale is an interesting one, as the process of substitution may be slow relative to the perceived need to limit future CO_2 emissions (and this may persuade governments to seek to speed up the substitution process). However, it should be borne in mind that the amount of CO_2 (and other greenhouse gases) emitted by the production of chemicals (as opposed to fuels and power generation) from petrochemical feedstocks is a relatively modest proportion of global anthropogenic greenhouse gas emissions.

11.9.3 Security of Supply

For a chemicals producer (particularly to satisfy the corresponding expectations of the putative user), there must be secure sources and a regular supply of plant-derived raw materials and intermediates. This may be challenging when one considers factors that fossil-sourced materials do not have to contend with. The supply of materials originating from biomass (particularly that grown sustainably, *i.e.* without consuming the historic stock of biomass present, for instance, in ancient woodland) is likely to be intermittent, associated with foreseeable seasonal factors that will affect growing and harvesting. Less predictable will be events (during both growth and storage) of the impact of disease and pest infestation, affecting both yield and quality. Any move to rely on biomass as the major primary source of raw materials for chemicals processing must, of course, itself take account of the effects of (and on) climate change.

11.9.4. Materials Handling and Transport

Petrochemical technologies have grown up that, for the most part (at least as far as so-called 'upstream' products are concerned), process liquids (of varying viscosity) and gases having a fairly narrow range of characteristics. Processes able to transform biomass will need to find ways of handling solid material of varying form and quality. Such variations will depend on the source of the biomass—whether this is a non-food crop (oil-seed, wood, grass or algae) specially grown, food crop (or other process) waste—as well as on the varieties grown and batch-to-batch variations associated with the site of growth and the nature of the growing season. In addition, there will need to be a suitable transport infrastructure (equivalent to an oil pipeline or oil tanker, but capable of handling solid biomass) to bring the harvested crop to the processing unit. An alternative would be to process the raw material close to the harvest site and convert it to a product that may more readily be transported and stored, and which would then be subject to further downstream processing. This may give rise to associated concerns regarding security and safety of supply.

REFERENCES

1. J. Dewulf and H. Van Langenhove, *Renewables-based Technology: Sustainability Assessment*, John Wiley & Sons, Chichester, 2006.
2. (a) R. Diercks, J.-D. Arndt, S. Freyer, R. Geier, O. Machhammer, J. Schwartze and M. Volland, *Chem. Eng. Technol.*, 2008, **31**, 631;(b) W. F. Pickard, *Global Planet. Change*, 2008, **61**, 285;(c) D. Cohen, *New Sci.*, 23 May 2007, p. 34;(d) A. Valero, A. Valero and A. Martínez, *Energy*, 2010, **35**, 989.
3. (a) M. Lancaster, *Green Chemistry: An Introductory Text*, Royal Society of Chemistry, Cambridge, 2002, ch. 6, pp. 166–209;(b) G. Centi and R. A. van Santen, *Catalysis for Renewables: from Feedstock to Energy Production*, Wiley-VCH Verlag, Weinheim, 2007.
4. (a) D. O. Hall and K. K. Rao, *Photosynthesis*, Cambridge University Press, Cambridge, 6th edn, 1999;(b) D. W. Lawlor, *Photosynthesis*, BIOS Scientific Publishers, Oxford, 2000.
5. (a) I. K. Bradbury, *The Biosphere*, Wiley, Chichester, 2nd edn, 1998;(b) C. B. Field, M. J. Behrenfeld, J. T. Randerson and P. Falkowski, *Science*, 1998, **281**, 237.
6. (a) J. Rockström, W. Steffen, K. Noone, Å. Persson, F. S. Chapin III, E. F. Lambin, T. M. Lenton, M. Scheffer, C. Folke, H. J. Schellnhuber, B. Nykvist, C. A. de Wit, T. Hughes, S. van der Leeuw, H. Rodhe, S. Sörlin, P. K. Snyder, R. Costanza, U. Svedin, M. Falkenmark, L. Karlberg, R. W. Corell, V. J. Fabry, J. Hansen, B. Walker, D. Liverman, K. Richardson, P. Crutzen and J. A. Foley, *Nature*, 2009, **461**, 472;(b) W. H. Schlesinger, *Nat. Rept.: Climate Change*, 2009, **3**, 112;(c) S. Bass, *Nat. Rept.: Climate Change*, 2009, **3**, 113;(d) M. Allen, *Nat. Rept.: Climate Change*, 2009, **3**, 114;(e) M. J. Molina, *Nat. Rept.: Climate Change*, 2009, **3**, 115;(f) D. Molden, *Nat. Rept.: Climate Change*, 2009, **3**, 116;(g) P. Brewer, *Nat. Rept.: Climate Change*, 2009, **3**, 117;(h) C. Samper, *Nat. Rept.: Climate Change*, 2009, **3**, 118.
7. R. J. Henry, *Plant. Biotechnol. J.*, 2010, **8**, 288.
8. (a) P. Spolaore, C. Joannis-Cassan, E. Duran and A. Isambert, *J. Biosci. Bioeng.*, 2006, **101**, 87;(b) K. Muffler and R. Ulber, *Chem. Eng. Technol.*, 2008, **31**, 638;(c) J. B. van Beilen, *Biofuels, Bioprod. Biorefin.*, 2009, **4**, 41;(d) E. Stephens, I. L. Ross, Z. King, J. H. Mussgnug, O. Kruse, C. Posten, M. A. Borowitzka and B. Hankamer, *Nat. Biotechnol.*, 2010, **28**, 126.
9. P. J. Crutzen, A. R. Mosier, K. A. Smith and W. Winiwarter, *Atmos. Chem. Phys.*, 2008, **8**, 389.
10. A. Monti, S. Fazio and G. Venturi, *Eur. J. Agron.*, 2009, **31**, 77.
11. M. Juraščík, A. Sues and K. J. Ptasinski, *Energy Environ. Sci.*, 2009, **2**, 791.
12. F. Cherubini, N. D. Bird, A. Cowie, G. Jungmeier, B. Schlamadinger and S. Woess-Gallasch, *Resour., Conserv. Recycl.*, 2009, **53**, 434.
13. (a) J. Dewulf, H. Van Langenhove and B. Van de Velde, *Environ. Sci. Technol.*, 2005, **39**, 3878; (b)T. W. Patzek and D. Pimentel, *Crit. Rev. Plant Sci.*, 2005, **24**, 327; (c)T. N. Kalnes, K. P. Koers, T. Marker and

D. R. Shonnard, *Environ. Prog. Sustain. Energy*, 2009, **28**, 111; (d) H. Kim, S. Kim and B. E. Dale, *Environ. Sci. Technol.*, 2009, **43**, 961.
14. C. Okkerse and H. van Bekkum, *Green Chem.*, 1999, **1**, 107.
15. (a) J. W. Ponton, *J. Cleaner Prod.*, 2009, **17**, 896; (b) D. A. Walker, *J. Appl. Phycol.*, 2009, **21**, 509; (c) J. U. Grobbelaar, *J. Appl. Phycol.*, 2009, **21**, 519; (d) M. Giampietro and K. Mayami, *The Biofuels Delusion: The Fallacy of Large Scale Agro-biofuel Production*, Earthscan, London, 2009.
16. W. F. Pickard, *Energy Policy*, 2010, **38**, 1672.
17. (a) R. Dominguez-Faus, S. E. Powers, J. G. Burken and P. J. Alvarez, *Environ. Sci. Technol.*, 2009, **43**, 3005; (b) A. Y. Hoekstra and A. K. Chapagain, *Water Resour. Manage.*, 2007, **21**, 35.
18. J. J. Bozell, *Clean*, 2008, **36**, 641.
19. (a) J. P. Lange, *Biofuels, Bioprod. Biorefin.*, 2007, **1**, 39; (b) M. Kleinert and T. Barth, *Chem. Eng. Technol.*, 2008, **31**, 736.
20. D. G. Levine, R. H. Schlosberg and B. G. Silbernagel, *Proc. Natl. Acad. Sci. U.S.A.*, 1982, **79**, 3365.
21. C. Posten and G. Schaub, *J. Biotechnol.*, 2009, **142**, 64.
22. P. A. Willems, *Science*, 2009, **325**, 707.
23. J. Goldemberg, *Energy Environ. Sci.*, 2008, **1**, 523.
24. J. N. Chheda, G. W. Huber and J. A. Dumesic, *Angew. Chem., Int. Ed.*, 2007, **46**, 7164.
25. Y. Román-Leshkov, C. J. Barrett, Z. Y. Liu and J. A. Dumesic, *Nature*, 2007, **447**, 982.
26. J. O. Metzger, *Eur. J. Lipid Sci. Technol.*, 2009, **111**, 865.
27. R. Zah, H. Böni, M. Gauch, R. Hischier, M. Lehmann and P. Wäger, *Ökobilanz von Energieproduckten: Ökologische Bewertung von Biotreibstoffen*, Empa, St Gallen, Switzerland, 2007, cited in J. P. W. Scharlemann and W. F. Laurance, *Science,* 2008, **319**, 43.
28. F. G. Dohleman and S. P. Long, *Plant Physiol.*, 2009, **150**, 2104.
29. M. R. Schmer, K. P. Vogel, R. B. Mitchell and R. K. Perrin, *Proc. Natl. Acad. Sci. U.S.A.*, 2008, **105**, 464.
30. (a) D. Tilman, J. Hill and C. Lehman, *Science*, 2006, **314**, 1598; (b) M. P. Russelle, R. V. Morey, J. M. Baker, P. M. Porter and H.-J. G. Jung, *Science*, 2007, **316**, 1567; (c) D. Tilman, J. Hill and C. Lehman, *Science*, 2007, **316**, 1567.
31. (a) G. Jiang, D. J. Nowakowski and A. V. Bridgwater, *Thermochim. Acta*, 2010, **498**, 61; (b) T. Voitl and P. R. von Rohr, *Ind. Eng. Chem. Res.*, 2010, **49**, 520.
32. K. A. D. Swift, *Top. Catal.*, 2004, **27**, 143.
33. D. Klemm, B. Heublein, H.-P. Fink and A. Bohn, *Angew. Chem., Int. Ed.*, 2005, **44**, 3358.
34. B. P. Mooney, *Biochem. J.*, 2009, **418**, 219.
35. (a) P. T. Pienkos and A. Darzins, *Biofuels, Bioprod. Biorefin.*, 2009, **3**, 431; (b) P. J. le B. Williams and L. M. L. Laurens, *Energy Environ. Sci.*, 2010, **3**, 554; (c) A. M. Kunjapur and R. B. Eldridge, *Ind. Eng. Chem. Res.*, 2010, **49**, 3516; (d) H. C. Greenwell, L. M. L. Laurens, R. J. Shields, R. W.

Lovitt and K. J. Flynn, *J. R. Soc. Interface*, 2010, **7**, 703; (e) A. L. Stephenson, E. Kazamia, J. S. Dennis, C. J. Howe, S. A. Scott and A. G. Smith, *Energy Fuels*, 2010, **24**, 4062.
36. D. E. Brune, T. J. Lundquist and J. R. Benemann, *J. Environ. Eng.*, 2009, **135**, 1136.
37. (a) A. Corma, S. Iborra and A. Velty, *Chem. Rev.*, 2007, **107**, 2411; (b) G. W. Huber, S. Iborra and A. Corma, *Chem. Rev.*, 2006, **106**, 4044.
38. R. Luque, L. Herrero-Davila, J. M. Campelo, J. H. Clark, J. M. Hidalgo, D. Luna, J. M. Marinas and A. A. Romero, *Energy Environ. Sci.*, 2008, **1**, 542.
39. G. W. Huber and A. Corma, *Angew. Chem., Int. Ed.*, 2007, **46**, 7184.
40. M. Jukka, A. Pekka and H. Elina, *US Patent* 2007006523, 11 Jan., 2007.
41. J. Goldemberg, *Science*, 2007, **315**, 808.
42. J. P. Lange, in *Catalysis for Renewables: from Feedstock to Energy Production*, ed. G. Centi and R.A. van Santen, Wiley-VCH Verlag, Weinheim, 2007, ch. 2, p. 21.
43. (a) S. Atsumi, W. Higashide and J. C. Liao, *Nat. Biotechnol.*, 2009, **27**, 1177; (b) J. Sheehan, *Nat. Biotechnol.*, 2009, **27**, 1128.
44. Y. Traa, *Chem. Commun.*, 2010, **46**, 2175.
45. M. M. Yung, W. S. Jablonski and K. A. Magrini-Bair, *Energy Fuels*, 2009, **23**, 1874.
46. (a) K. Kleiner, *Nat. Rept: Climate Change*, 2009, **3**, 72; (b) A. R. Zimmerman, *Environ. Sci. Technol.*, 2010, **44**, 1295; (c) J. L. Gaunt and J. Lehmann, *Environ. Sci. Technol.*, 2008, **42**, 4152; (d) K. G. Roberts, B. A. Gloy, S. Joseph, N. R. Scott and J. Lehmann, *Environ. Sci. Technol.*, 2010, **44**, 827.
47. (a) P. Ramírez de la Piscina and N. Homs, *Chem. Soc. Rev.*, 2008, **37**, 2459; (b) P. Westermann, B. Jørgensen, L. Lange, B. K. Ahring and C. H. Christensen, *Int. J. Hydrogen Energy*, 2007, **32**, 4135; (c) J. Woodward, M. Orr, K. Cordray and E. Greenbaum, *Nature*, 2000, **405**, 1014.
48. (a) R. R. Davda, J. W. Shabaker, G. W. Huber, R. D. Cortright and J. A. Dumesic, *Appl. Catal., B*, 2005, **56**, 171; (b) J. N. Chheda and J. A. Dumesic, *Catal. Today*, 2007, **123**, 59.
49. (a) W. Niu, K. M. Draths and J. W. Frost, *Biotechnol. Prog.*, 2002, **18**, 201; (b) K. Li and J. W. Frost, *J. Am. Chem. Soc.*, 1998, **120**, 10545.
50. (a) P. B. Weisz, W. O. Haag and P. G. Rodewald, *Science*, 1979, **206**, 58; (b) P. B. Weisz and J. F. Marshall, *Science*, 1979, **206**, 24; (c) B. G. Kyle, *Science*, 1980, **210**, 807.
51. T. R. Carlson, G. A. Tompsett, W. C. Conner and G. W. Huber, *Top. Catal.*, 2009, **52**, 241.
52. (a) J. van Haveren, E. L. Scott and J. Sanders, *Biofuels, Bioprod. Biorefin.*, 2007, **2**, 41; (b) S. Kvisle, A. Aguero and R. P. A. Sneedon, *Appl. Catal.*, 1988, **43**, 117.
53. (a) T. Werpy and G. Petersen, *Top Value Added Chemicals from Biomass: Volume 1: Results of Screening for Potential Candidates from Sugars and Synthesis Gas*, National Renewable Energy Laboratory, Golden, CO, 2004; (b) J. J. Bozell and G. R. Petersen, *Green Chem.*, 2010, **12**, 539.

54. O. Casanova, S. Iborra and A. Corma, *J. Catal.*, 2009, **265**, 109.
55. (a) A. Gandini, A. J. D. Silvestre, C. P. Neto, A. F. Sousa and M. Gomes, *J. Polym. Sci., Part A: Polym. Chem.*, 2009, **47**, 295; (b) M. Hishida, K. Shikinaka, Y. Katayama, S. Kajita, E. Masai, M. Nakamura, T. Otsuka, S. Ohara and K. Shigehara, *Polym. J.*, 2009, **41**, 297; (c)A. Gandini, D. Coehlo, M. Gomes, B. Reis and A. Silvestre, *J. Mater. Chem.*, 2009, **19**, 8656.
56. (a) H. Mehdi, V. Fábos, R. Tuba, A. Bodor, L. T. Mika and I. T. Horváth, *Top. Catal.*, 2008, **48**, 49; (b) A. D. Patel, J. C. Serrano-Ruiz, J. A. Dumesic and R. P. Anex, *Chem. Eng. J.*, 2010, **160**, 311.
57. M. Pagliaro, R. Ciriminna, H. Kimura, M. Rossi and C. Della Pina, *Eur. J. Lipid Sci. Technol.*, 2009, **111**, 788.
58. A. A. Kiss, A. C. Dimian and G. Rothenberg, *Energy Fuels*, 2008, **22**, 598.
59. (a) M. Pagliaro, R. Ciriminna, H. Kimura, M. Rossi and C. Della Pina, *Angew. Chem., Int. Ed.*, 2007, **46**, 4434; (b) C.-H. Zhou, J. N. Beltramini, Y.-X. Fan and G. Q. Lu, *Chem. Soc. Rev.*, 2008, **37**, 527.
60. R. A. Urban and B. R. Bakshi, *Ind. Eng. Chem. Res.*, 2009, **48**, 8068.
61. K. T. Tan, K. T. Lee and A. R. Mohamed, *Bioresour. Technol.*, 2010, **101**, 965.
62. S. S. Chandran, J. Yi, K. M. Draths, R. Von Daeniken, W. Weber and J. W. Frost, *Biotechnol. Prog.*, 2003, **19**, 808.
63. M. B. Hocking, *J. Chem. Educ.*, 1997, **74**, 1055.
64. J. J. Sheehan, *Curr. Opin. Biotechnol.*, 2009, **20**, 318.
65. (a) M. Balat, M. Balat, E. Kirtay and H. Balat, *Energy Convers. Manage.*, 2009, **50**, 3147; (b) M. Balat, M. Balat, E. Kirtay and H. Balat, *Energy Convers. Manage.*, 2009, **50**, 3158.
66. D. J. Hayes, *Catal. Today*, 2009, **145**, 138.
67. (a) Y.-C. Lin, J. Cho, G. A. Tompsett, P. R. Westmorland and G. W. Huber, *J. Phys. Chem. C*, 2009, **113**, 20097; (b) P. R. Patwardhan, J. A. Satrio, R. C. Brown and B. H. Shanks, *J. Anal. Appl. Pyrolysis*, 2009, **86**, 323.
68. K. Sanderson, *Nature*, 2009, **461**, 710.
69. F. S. Sariaslani, *Annu. Rev. Microbiol.*, 2007, **61**, 51.
70. (a) P. Gallezot, *Catalysis for Renewables: from Feedstock to Energy Production*, ed. G. Centi and R. A. van Santen, Wiley-VCH Verlag, Weinheim, 2007, Ch. 3, p53; (b) B. Kamm and M. Kamm, *Appl. Microbiol. Biotechnol.*, 2004, **64**, 137; (c) J. H. Clark, F. E. I. Deswarte and T. J. Farmer, *Biofuels, Bioprod. Biorefin.*, 2009, **3**, 72; (d) P. Gallezot, *Green Chem.*, 2007, **9**, 295.
71. F. Cherubini, G. Jungmeier, M. Wellisch, T. Willke, I. Skiadas, R. Van Ree and E. De Jong, *Biofuels, Bioprod. Biorefin.*, 2009, **3**, 534.
72. V. Sheoran, A. S. Sheoran and P. Poonia, *Miner. Eng.*, 2009, **22**, 1007.
73. S. J. Bennett and P. J. G. Pearson, *Chem. Eng. Res. Des.*, 2009, **87**, 1120.
74. (a) Q. Hu, M. Sommerfeld, E. Jarvis, M. Ghirardi, M. Posewitz, M. Seibert and A. Darzins, *Plant J.*, 2008, **54**, 621; (b) E. Waltz, *Nat. Biotechnol.*, 2009, **27**, 15; (c) L. Lardon, A. Hélias, B. Sialve, J.-P. Steyer and O. Bernard, *Environ. Sci. Technol.*, 2009, **43**, 6475.

BIBLIOGRAPHY[XVII]

B. G. Davis and A. J. Fairbanks, *Carbohydrate Chemistry*, Oxford University Press, Oxford, 2002.

T. K. Lindhorst, *Essentials of Carbohydrate Chemistry and Biochemistry*, Wiley-VCH, Weinheim, 3rd edn, 2007.

B. Kamm, P.R. Gruber and M. Kamm, *Biorefineries: Industrial Processes and Products: Status Quo and Future Directions*, Wiley-VCH, Weinheim, 2006.

J. R. Mann, R. S. Davidson, J. B. Hobbs, D. V. Banthorpe and J. B. Harbourne, *Natural Products: Their Chemistry and Biological Significance*, Longmans Scientific and Technical/J. Wiley and Sons, New York, NY, 1994.

P. F. Stanbury, A. Whittaker and S. J. Hall, *Principles of Fermentation Technology*, Butterworth-Heinemann, Oxford, 2nd edn, 1995.

J. D. Mauseth, *Botany: An Introduction to Plant Biology*, Jones and Bartlett, Boston, 4th edn, 2009.

E. F. Schumaker, *Small is Beautiful: a Study of Economics as if People Mattered*, Vintage, London, 1993.

W. Beckerman, *Small is Stupid: Blowing the Whistle on the Greens*, Gerald Duckworth, London, 1995.

F. Cairncross, *Costing the Earth*, The Economist Books, London, 1991.

F. Cavani, G. Centi, S. Perathoner and F. Trifiró, *Sustainable Industrial Chemistry: Principles, Tools and Industrial Examples*, Wiley-VCH Verlag, Weinheim, 2009.

WEBLIOGRAPHY

1. (a) www.bbc.co.uk/iplayer/episode/b00mvczy/You_and_Yours_01_10_2009/
   (b) www.endscarbon.com/_pdf/ends-carbon-supermarket-summary.pdf
2. www.epa.gov/greenchemistry/pubs/docs/award_entries_and_recipients2005.pdf
3. http://en.wikipedia.org/wiki/Algae
4. (a) www.hightech.fi/direct.aspx?area=htf&prm1=546&prm2=article
   (b) www.nesteoil.com
5. www.duponttateandlyle.com
6. www.natureworksllc.com
7. www.epa.gov/greenchemistry/pubs/docs/award_recipients_1996_2009.pdf

All the web pages listed in this Webliography were accessed in May 2010.

[xvii] These texts supplement those already cited as references.

CHAPTER 12

Energy Production

'The sun, with all those planets revolving around it and dependent upon it, can still ripen a bunch of grapes as if it had nothing else in the universe to do.'

Galileo

The importance and relevance of chemistry to the timely development of efficient technologies that will use renewable feedstocks should now be self-evident. In the last chapter we saw that sustainable energy generation and the use of renewables for chemicals production were, in some fundamental ways, bound up with one another. While this chapter deals with energy generation we should not lose sight of the importance of energy conservation, important of course intrinsically but also more pragmatically, as a consequence of the timescales on which new sources of energy generation can become available. Because of the focus of this book, a broad introduction only is provided that highlights aspects of energy generation in which chemistry has some role to play. (The Bibliography lists some texts that deal with the topic more deeply and completely. MacKay's splendid book[1a] can be particularly recommended, as can a shorter overview by Abbott.[1b]) To understand these technologies it is important to recall and to understand some basic underpinning fundamental chemical and physical principles and associated metrics (and, bearing in mind that different people use different metrics, how these may be converted).

We need first to define the unit of **energy**, bearing in mind that this is not the same as the thermodynamic term used in Chapter 6. It is closer in meaning to the term exergy, which (unlike energy) is not conserved and is a more practical measure of available energy. The SI unit is the Joule (J). For our purposes it is the work done continuously to produce one watt (W) of power for 1 second. One

Chemistry for Sustainable Technologies: A Foundation
By Neil Winterton

Published by the Royal Society of Chemistry, www.rsc.org

Joule is, therefore, one watt second and one kilowatt hour (kWh) is (1000×3600) or 3.6 MJ (megajoules). (In thermochemistry, one calorie is 4.184 J).

Energy and the units used to measure it need to be distinguished from **power**, which is the rate at which we use or generate energy such as per day or per year. A power of 1 joule per second is 1 watt second/second or 1 watt. An individual generates about 100 watts of heat. A country such as the UK has energy-generating capacity measured in billions of watts or GW (gigawatts). Globally, energy usage is discussed in terms of TW (terawatts; $1\,TW = 10^{12}$ watts) or EJ (exajoules; $1\,EJ = 10^{18}\,J$). The current global usage is about $500\,EJ\,y^{-1}$ or *ca.* 15 TW. A useful source of data is the Annual Energy Review undertaken by the US Department of Energy (DoE) (Web 1). Figure 12.1 (Web 2) illustrates the complexity of energy flows for the USA in 2008.

Figure 12.2 shows the historic sources of energy. In 2004, 80% of world energy demand was satisfied from fossil resources. It is estimated (Web 3) that the approximate timescales of known reserves are 42 years for oil, 60 years for natural gas and 122 years for coal. Important contributions are seen from hydroelectric (1.7%) and nuclear (6.3%) sources. Traditional biomass accounts for 8.5% but only *ca.* 3.4% from 'new' renewables including geothermal, wind, solar and 'new' biomass.

Figure 12.3 illustrates graphically our usage compared with the potential availability (without at this stage considering what may be practical). It can be seen at the top that our annual requirements are not vastly different from the known reserves (as listed above). Were it not for the evident link between climate change and the use of fossil fuels in energy generation (and other uses), we would at some point over the next century need to begin the transition, though over a period longer than we will be forced to do because of climate change concerns.

12.1 PRIMARY, SECONDARY, RENEWABLE AND SUSTAINABLE ENERGY

We first need to distinguish clearly between primary and secondary sources of energy, as this affects judgements about the categorisation of an energy source or energy-carrier as renewable or not, or sustainable or not.

12.1.1 Primary Sources

Primary sources of energy only require extraction or capture prior to their use. Such sources are reliant on:

- solar energy—ultimately derived from nuclear fusion processes
- geothermal energy—ultimately from nuclear fission processes in the Earth's crust (along with heat conducted from the Earth's core)
- potential energy associated with gravitational and centrifugal forces.

Fossil sources of energy (*e.g.* coal, oil and natural gas) fall into this category as, while they result from the capture of historic solar energy, once extracted can

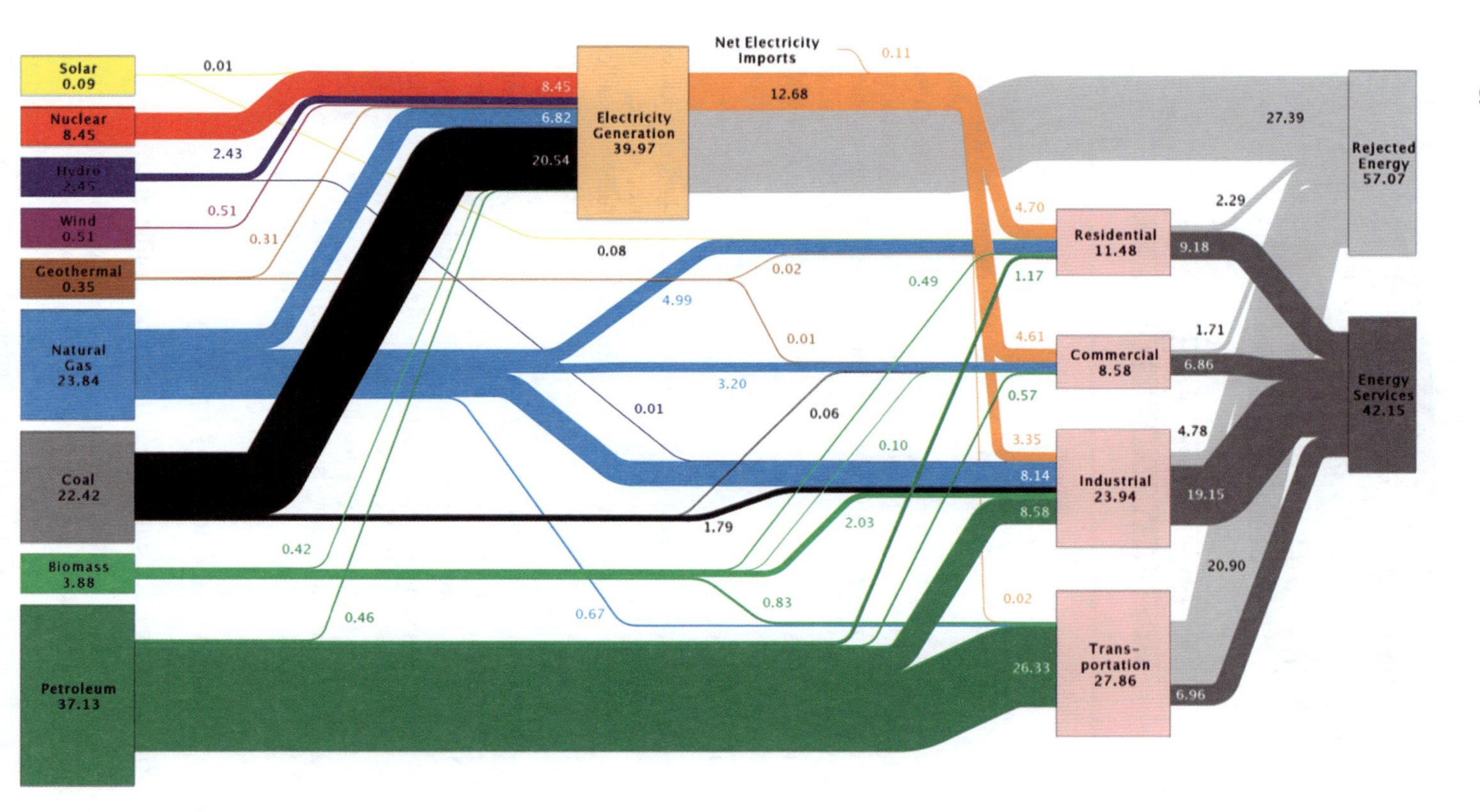

Figure 12.1 Energy flows in the US (data for 2008) highlighting the major sources and consumption of major economic sectors. Numerical data are given in quads, where 1 quad = 10^{15} British Thermal Units or 1.05506×10^{12} MJ. Source: Lawrence Livermore National Laboratory/US Department of Energy (Web 2).

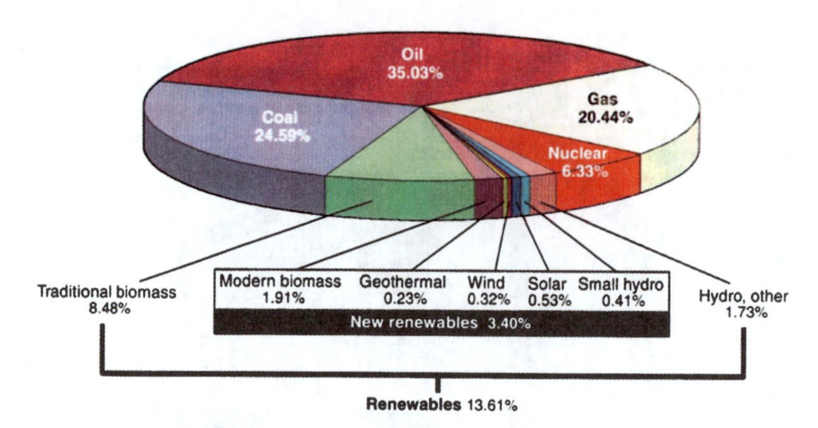

Figure 12.2 Sources of the total global energy supply (2004) of 470 EJ [or 11.2×10^9 toe (tons of oil equivalent)]. Reprinted from ref. 2 with permission from AAAS.

deliver energy with little further processing. Similarly, uranium and thorium must also be seen as primary sources, though significant processing is required for their extraction. To the extent that additional fuel may be formed during energy generation (so-called 'breeder' technology), nuclear power should be considered renewable—though others may argue that such technologies are not sustainable.

12.1.2 Secondary Sources

Secondary sources of energy are those that have to be transformed (with associated exergetic losses) from a primary source into some other form prior to use. This would include the generation of electricity or the production of fuels (also called energy 'carriers' or energy 'vectors') such as ethanol, hydrogen or petrol (gasoline).

Fuels such as hydrogen (or electricity itself) can be used in ways such that no polluting emissions are apparently generated at the point of use. In this limited sense they can be considered clean. However, they cannot be considered renewable, unless the primary source from which they are derived is itself renewable. For example, the hydrogen-powered car produces only water as a co-product when the fuel is combusted or fed to a fuel cell (see below). However, at the moment, most hydrogen is derived from fossil sources. Only when hydrogen can be obtained from a renewable primary source can itself be said to be renewable (and only then to the extent that no fossil resources are used in its production).

12.2 CONVENTIONAL SOURCES OF ENERGY

12.2.1 Fossil Energy Sources: Future Availability and Continued Use

Because of the timescale of the transition to large-scale and widespread implementation of more sustainable energy generation, it will be important to exploit conventional energy sources, including those derived from petrochemical sources.[4] Of course, this will need to be combined with improved energy

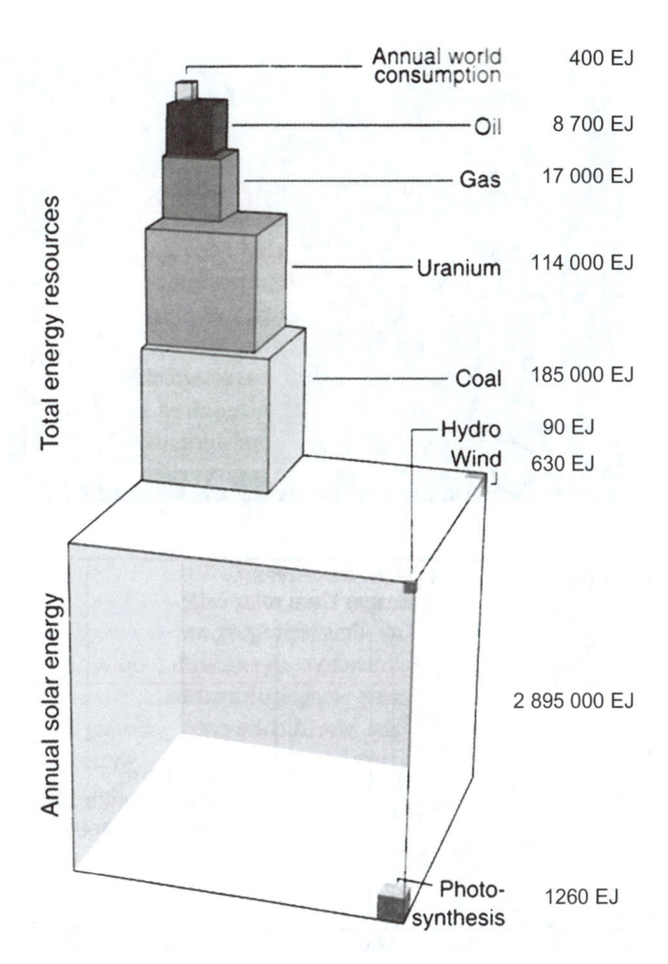

Figure 12.3 Estimates of the current requirement and availability of energy from a range of sources, fossil and renewable. Reproduced from ref. 3 with permission: Copyright Cambridge University Press. Numerical data from various sources.

generation—which is itself limited to the Carnot efficiency (Section 6.12)—and usage efficiency, and with efforts to limit the impact on the environment, particularly related to the control of greenhouse gas emissions.

The balance between consumption, the size of known reserves and the expectation of the discovery of additional reserves governs whether or not, and over what period, a resource will remain available. So it is with reserves of coal, oil and natural gas. Figure 12.4a (Web 4) shows the past use of hydrocarbon feedstocks and projections for future use, which takes a rough 'bell-curve' form that has given rise to the anticipation of a peak in availability, followed by a decline as reserves run out. The idea of 'peak oil' was first put forward by M. K. Hubbert in a prescient report[5b] (Web 5) published in 1956, from which Figure 12.4b is taken.

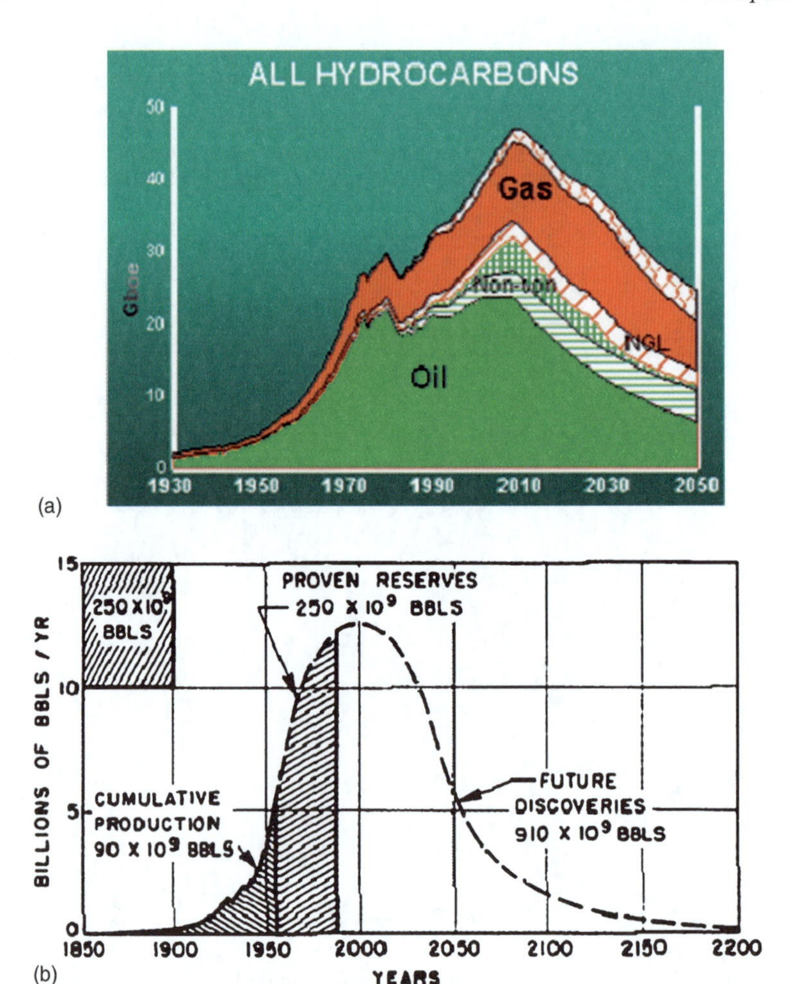

Figure 12.4 (a) Graph showing peaks in estimates of the production of various hydrocarbons. Taken from a presentation made by C. J. Campbell to a UK House of Commons All-Party Committee, 7 July 1999;[5a] and (b) Bell-curve representing growth and decline in the use and availability of petrochemical reserves taken from M. K. Hubbert's 1956 report.[5b]

The essential question of how close we are to this peak has been addressed in a recent review of over 500 studies.[5c,d] A more recent study of world crude oil production[5e] suggests, on the basis of 2005 world crude oil production and current recovery technologies, that world oil reserves are depleting at 2.1% per year. It also estimates world production to peak in 2014, with that from members of OPEC (Organisation of Petroleum Exporting Countries) in 2026. There are those who suggest the peak will be more like a plateau (as price increases moderate demand and slow the increase in consumption), making it difficult to predict when the downturn will occur. This adds to the difficulty of forecasting the point at which a new technology may become economic (and, hence, when it would be prudent to start investing in its large-scale development). The pressure added from concerns about the effect of emissions of carbon dioxide (and the fact that the costs associated with the impact of such emissions are not fully factored into the balance sheets of companies or state institutions involved in energy production, distribution and use) lessens confidence in the notion that market forces alone can deal satisfactorily with these uncertainties. Note also, from our consideration of the environmental burden and the trends in energy use efficiency (Figure 3.4) that an improvement of efficiency 2–5 times greater than that seen over the last 30–50 years is required. Combined, these represent formidable challenges, both technological and societal. The problem is that renewable energy sources are not immediately available on the scale or at a cost needed for rapid and widespread implementation. The timescale over which the deployment of new energy generation technology can be expected is shown in Figure 12.5. It seems likely that continued use of fossil sources will be necessary to lessen the economic disruption associated with the transition to renewables. Means to render such sources less impacting on the environment are, therefore, needed.[4]

In addition, during the early stages of development of a novel power generation technology, the proportion of the energy available to be extracted will usually be less than ultimately achievable as experience of its actual mode of operation on the large scale is gained. Such factors may partly be offset by the greater exploration, extraction and transportation costs and risks that will be associated with increasingly inaccessible (geographically or materially)[i] sources of fossil materials. This underlines the importance of comparing the beneficial impact and cost (both economic and environmental) of technologies from the point at which the raw material that provides the primary energy source is extracted all the way through to its final use.

12.2.2 Carbon Capture and Storage

CO_2 abatement (Box 6.2) is proposed to avoid (or at least limit[ii]) releases of carbon dioxide from energy generation to the environment. Options include sequestration, an available technology (though yet to be used to capture emissions from a full-scale energy-generating operation) involving a reversible reaction with ethanolamine ($H_2NCH_2CH_2OH$). Sequestration should capture CO_2 and contain it in such a way that it is permanently prevented from returning to the atmosphere. Methods currently proposed involve injection of

[i] These include the deep ocean, environmentally sensitive areas such as Antarctica, and oil shale or tar sands reserves. Extraction will be highly energy-consuming and this will increase costs. For instance, tar sands may be attractive commercially at an oil price in excess of US\$140 $bbl.^{-1}$ The dangers associated with such operations are highlighted by the consequences of the fatal explosion on the Deepwater Horizon drilling rig on 20 April 2010 and the ensuing oil leak.

[ii] The recovery of CO_2 from the exhaust gases of power stations will not be perfect. Direct emissions might be reduced by 80–90%,[7] but this figure might become only 65–70% when emissions of greenhouse gases associated with the energy expended in the process of capture are taken into account.

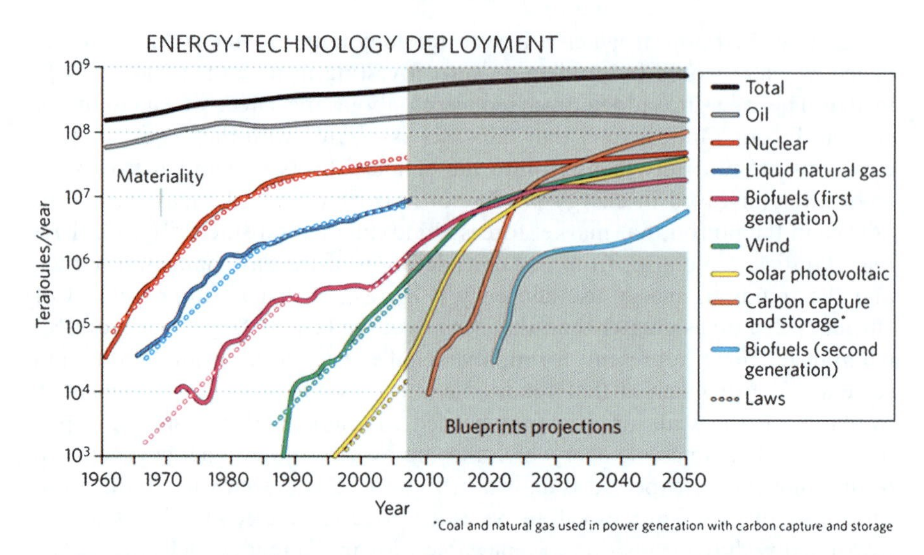

Figure 12.5 Timescales of deployment of primary energy-generating technologies capable of producing energy equivalent to 500 barrels of oil per day or 10^3 TJ y^{-1} to the point when they are, individually, capable of delivering 1% of world energy (10^6–10^7 TJ y^{-1}).[6] Reprinted by permission from Macmillan Publishers Ltd. Copyright 2009.

CO_2 under pressure into geological strata[iii] in which the understanding of its reactions (their nature, rates, extent and reversibility) with water and with mineral components becomes of great importance. This would only serve to limit the increase in atmospheric CO_2, though this is important in itself. Sequestration coupled with a process based on biomass could, in principle, reduce the atmospheric CO_2 mixing ratio. Converting biomass carbon into biochar would change the means of disposal to one that, on the face of it, is less associated with long-term risks of being re-emitted (perhaps in catastrophic amounts) if containment fails.

Sequestration technology, or the alternative feeding of CO_2-containing streams to an algal photobioreactor (Section 11.5.5), is likely to be of greatest applicability where waste streams high in CO_2 are available, as is the case in power generation. This is still estimated to be expensive. However, non-combustible dinitrogen makes up a large fraction of the waste gases from conventional technology which uses air for the combustion process. Combustion with pure O_2 instead of air would produce waste gases much richer in CO_2 and therefore more readily recoverable.[iv]

[iii] The Intergovernmental Panel on Climate Change (IPCC) estimates[8a] that a large fraction of CO_2 stored in suitable geological strata will be retained for $> 10^8$ years. Stable isotope ratios measured above a natural CO_2 reservoir formed 70–80 million years ago and sealed with mud rock suggest a migration of only 12 m in this period.[8b] The formation of carbon dioxide hydrate is also thought,[8c] from experimental studies, to reduce upward migration of CO_2.

[iv] A 30 MW research-focused pilot plant to capture and store CO_2 from the combustion of lignite in dioxygen (an 'oxyfuel' process) was commissioned in September 2008 by the Swedish company Vattenfall, which claimed (Web 7) that this was the first operation of its kind.

As photosynthesising organisms are able to remove carbon dioxide directly from the atmosphere, the question arises: why cannot this be done industrially? However, trees and plants are able to do this by harnessing directly the continuously available energy from the sun. Removal of CO_2 from the ambient air ('air capture'[9a,b]) has been proposed using a mixture of sodium hydroxide and slaked lime as the absorbent (eqn 12.1, the net reaction), giving calcium carbonate from which the CO_2 could be recovered using solar thermal energy (eqn 12.2). Heat could be recovered from the exotherm associated with the reaction with water ('slaking') shown in eqn (12.3).

$$Ca(OH)_2 + CO_2 \rightarrow CaCO_3 + H_2O \tag{12.1}$$

$$CaCO_3 \rightarrow CaO + CO_2 \tag{12.2}$$

$$CaO + H_2O \rightarrow Ca(OH)_2 \tag{12.3}$$

It has been estimated[9c] that 35 000 units would be needed (spaced sufficiently far apart so as not to be affected by local CO_2 depletion caused by an adjacent unit), each removing $250 \times 10^3\, t(CO_2)\, y^{-1}$,[v] costing very roughly US\$100–200 $t(CO_2)^{-1}$.[9a] This estimate is still seen to be optimistic.[9c]

Considering the time needed for the development and large-scale implementation of new more sustainable ways to generate energy, it is sensible to consider ways in which conventional technologies (particularly low emitters of CO_2) might be made more efficient (as well as improvements in the efficiency with which energy so produced is used). This would require technologies that were at stages of development from which widespread application could readily be achieved. A portfolio of projects to reduce CO_2 emissions by 1 $GtC\, y^{-1}$ and based on 15 such technologies was proposed in 2004[10] based on:

- improved energy efficiency and conservation
- a shift from coal to gas for energy generation coupled with carbon capture and storage
- the use of nuclear fission and renewables
- reduced deforestation
- improved agricultural practices.

An increase in district heating and combined heat and power (CHP) schemes (where electricity generation is close to centres of population) can make use of

[v] Each unit would therefore need to remove completely *ca.* 400 ppm of CO_2 from *ca.* 3×10 L of air, or 10 times this volume if only 40 ppm of the CO_2 were removed. It would be interesting to know how much energy would be consumed simply by pumping this volume of air through the capture unit and how much additional CO_2 would, thereby, be released. A recent estimate suggests that 1 000 250 000 $t\, y^{-1}$ capture units would consume as much energy as is produced by the entire annual nuclear power generating capacity of the USA.[9c] Clearly, if the process was driven solely by renewable energy, the additional CO_2 would be much reduced. On the other hand, these same 1000 units would require 135 000 1.5 MW wind turbines to drive them, roughly equivalent to current global installed capacity.[9c]

Table 12.1 Estimates from life-cycle studies[11] of the greenhouse gas emissions ($gCO_2e\,kWh^{-1}$) from energy generation technologies.

Energy source	$gCO_2e\,kWh^{-1}$
Wind	9–10
Hydroelectric	10–13
Anaerobic digestion biogas	11
Solar thermal (parabolic trough)	13
Biomass	14–41
Solar photovoltaics	32
Geothermal	38
Nuclear	66
Natural gas	443
Fuel cell	664
Oil	778
Coal	960–1050

waste heat that otherwise is lost in cooling towers (Figure 2.3). Improved insulation of dwellings (an applied materials chemistry problem) can result from better design and changes in materials used in construction.

The move from coal to oil to natural gas has resulted in a significant reduction in the mass of greenhouse gases (including CO_2) produced per kWh of electricity generated (Table 12.1). Had this switch not occurred and coal had continued to be the primary source of energy, it is entirely possible that the atmospheric concentration of CO_2 would have reached, much earlier, the value that now provides so much cause for concern. While incineration of municipal solid waste will make an additional, if modest, contribution as a source of energy this will, nevertheless, release CO_2 that originated in the petrochemical feedstock used to manufacture the material now to be incinerated. (This may nevertheless be preferable to the emissions of methane that would arise from the anaerobic biodegradation of such waste disposed of in landfill sites).

12.2.3 Hydroelectricity

Hydroelectric power generation occurs with no direct fossil fuel use, but is limited by the availability of suitable sites. Hydroelectricity recovers solar energy stored as the potential energy of dammed water that has resulted from precipitation and is recovered from its flow through turbines under the action of gravity. While there are no direct fuel costs, there are high initial capital costs associated with construction. However, once built, labour and other operating costs are low. A major advantage of hydroelectric power is its instant availability and the possibility of using off-peak or base-load electricity generation from other sources for pump-storage (in which water is pumped into a reservoir to generate a hydrostatic head) to meet peaks in electricity demand. There are downsides, however. Dam construction and reservoir creation can have negative effects on ecosystems as well as downstream. Dammed lakes can be sources of greenhouse gases (CH_4, CO_2) from the areas flooded or from material

Figure 12.6 The Three Gorges Dam located on the Yangtze River will generate 22.5 GW at full capacity when fully operational. Source: NASA Earth Observatory (Web 8a).

washed into the reservoir. There can be problems arising from the resulting social dislocation, not least because of the need to persuade (or force) large numbers of people to move. It is estimated that over one million people[12] have been relocated as a consequence of the building of the Three Gorges Dam on the Yangtze River in China (Figure 12.6).

New hydroelectric schemes are in train from modest-sized projects such as the Glendoe project (Web 9) in the UK to colossal projects such as the Three Gorges Project (Web 8b) which is damming the Yangtze River and will generate 22.5 GW (at full capacity[vi]) when it becomes fully operational in 2011. In contrast, the Glendoe project is intended to contribute just 0.1 GW to the UK's total current electricity generating capacity of 78 GW.[vii]

A hitherto under-utilised energy source arises from the free energy ('salinity gradient' energy) associated with the mixing of fresh water into sea water, estimated to be 2.2 kJ (L fresh water)$^{-1}$.[13] The Mississippi–Missouri river has a flow of *ca.* 16×10^{6} L s^{-1}, representing (a theoretical) source of 35 GW. The challenge is how to design a device to exploit such a renewable (solar energy

[vi] There is a difference between the anticipated or design energy generation capacity at full load and the energy actually produced following installation. The latter is generally smaller than the former. Likewise, there is a need to distinguish between 'peak' power and 'average' power. The ratio average : peak is the 'load factor'. This parameter is important when designing systems to meet demand and will be particularly so when intermittent sources such as wind are considered.

[vii] The Glendoe scheme was shut down shortly after it was commissioned in 2009 because of a major rock fall in one of its tunnels (Web 9c).

driven) source. This is very much in the design/experimental stage, with the first pilot plant being opened on 24 November 2009 at Tofte in Norway (Web 10). The technology is likely to be constrained by the same sort of mass transfer issues associated with carbon dioxide air-capture (as well as more practical concerns such as the removal of suspended matter in the feed waters and the long-term performance of the membranes used).

12.2.4 Nuclear Fission

Nuclear fission involves processes such as those shown in eqn (12.4) and eqn (12.5), with a range of technologies designed to capture the heat generated. The fission of 1 g of ^{235}U (natural abundance 0.7%, 99.3% ^{238}U) yields the same energy of *ca.* 2.7 t of coal.[14]

$$^{235}\mathrm{U} + \mathrm{n} \rightarrow {}^{236}\mathrm{U} \tag{12.4}$$

$$^{236}\mathrm{U} \rightarrow {}^{92}\mathrm{Kr} + {}^{141}\mathrm{Ba} + 3\mathrm{n} \tag{12.5}$$

The technology is well-developed and can reliably deliver a base-load supply of electricity. It is the only practical short-term option to enable a major shift from fossil fuel combustion needed to reduce CO_2 emissions. While an operating nuclear plant generates very little CO_2, the mining and processing of uranium ore and its extraction and enrichment are highly energy-intensive. From the analysis of recent life-cycle estimates (Table 12.1), nuclear power is associated with emissions of 66 g $CO_2e\,kWh^{-1}$, *i.e.* much lower than coal-fired (960 g $CO_2e\,kWh^{-1}$) and gas-fired (443 g $CO_2e\,kWh^{-1}$) plants, though higher than (intermittent) solar photovoltaic (32 g $CO_2e\,kWh^{-1}$) and onshore wind (10 g $CO_2e\,kWh^{-1}$). However, the technology is not without its own challenges—management of radioactive waste (a political not a technical problem), fuel recycling, decommissioning of obsolete plant—and the influence of its vocal and well-organised opponents.

The construction of 10 new UK nuclear power stations was announced by the Government on 10 January 2008[15] (Web 11b). There are currently 436 nuclear power stations operating globally (Web 11c), representing *ca.* 15% of world wide electricity-generating capacity, with a further 47 under construction and 133 planned. In the UK, we generate 14% of our electricity this way; in France it is 76%.

12.3 ENERGY FROM RENEWABLES

Meeting our future energy needs using more sustainable approaches will rely, to varying degrees, on a mix of the following technologies:

- wind and wave power
- photovoltaics (PV)
- tidal

Table 12.2 Estimates of global energy use (1 EJ = 10^{18} J) in 1990 and projections for global energy requirements in 2040 with the contributions from different energy sources, petroleum- and renewables-derived. Data from Shell, quoted in ref. 16.

Energy source	*EJ (1990)*	*EJ (2040)*
Coal	70	180
Oil	130	160
Gas	55	140
Hydro	18	50
Nuclear	17	60
Wind	–	70
Biomass	55	60
New biomass	–	120
Solar	–	130
Total	350	970

- geothermal
- nuclear fission and (possibly in the longer term) fusion
- biofuels/biogas.

To which we can add fuel cells and hydrogen.

Estimates of the use that might be made of these in 2040 are shown in Table 12.2, though the precise mix will be a consequence of the combined impact of many factors (local, regional, political, social and technocommercial).

Each of these sources has a combination of advantages and disadvantages such that no individual technology represents a global solution to meeting future energy needs. Relevant considerations include geography (*e.g.* the favourable location of Iceland close to sources of exploitable geothermal energy), economics and the timescale on which, individually, each will become a technologically practised reality, implemented widely and on the large scale.[6]

The appropriateness of an individual technology will not be based on vague descriptions of its 'sustainability', 'renewability' or 'greenness'. Realistic and robust solutions will be based on rational choices between carefully and fully analysed (and costed) options. Such rational comparisons must employ data and evidence[1a] rather than emotion, and ensure a careful definition of boundaries and the validity of metrics used in such analyses using methods such as life-cycle analysis (particularly to take account of the associated fossil energy use in those technologies promoted as renewable). However, such comparisons are 'wicked' problems in which the data are not uncontested and attitudes, values and aesthetics will inevitably influence final judgements.

Table 12.2 shows that in 2040 about 970 EJ of energy will be required (a range of estimates from different publications such as the data in Table 12.3 can be found that are based on different assumptions and methods of projection and estimation). It is, however, evident that fossil-sourced energy will continue to make a contribution alongside the growing use of renewables. Table 12.2

Table 12.3 Projections of the growth in global energy (EJ; 1 EJ = 10^{18} J) available from different renewable sources based on a 'business-as-usual' scenario, compared with estimates of capacity. Data from ref. 4.

	Estimates of Annual Contribution			
		Business as usual		
Renewable source	*Potentially economic*	*2000*	*2050*	*2100*
Primary energy				
Hydro	29-50	9	20	30
Biomass: traditional	100-300	45	70	90
Biomass: modern		7	50	120
Wind	250-600	0.1	20	90
Solar	1 500-50 000	0.2	8	30
Geothermal	500-5 000	0.3	2	20
Tidal/Wave	1.5	0.002	0.01	0.1
Total	**2 300-56 000**	**62**	**170**	**380**
Electricity generation				
Biomass		1	20	70
Solar		0.005	2	20
Geothermal		0.2	1	15

should also be read alongside Table 12.1 which includes estimates[11] of CO_2 emissions arising from these different sources. The imperative nature for the development of carbon capture and storage technology for the continued production of energy from coal is evident. The potential benefits from the use of renewables and from nuclear power are also clear.

The following factors need to be included in any overall analysis of the applicability of renewable energy sources:

12.3.1 Intermittency and Unpredictability

A serious problem with several of the renewable sources is their intermittency and low predictability. Photovoltaics will not generate electricity at night. Wind and wave power will be non-productive under still conditions.

Such constraints can be overcome by at least three methods. They are:

- a back-up supply available at the flick of a switch (probably a gas turbine)
- an interface with a large electricity grid, into which the energy would flow when generating and from which energy could be extracted when not
- reversible energy storage.

All would add costs and bring additional emissions and environmental impact.

As many of the most suitable sites for solar, wind and wave power generation are in relatively remote locations, the costs of transmission (and associated losses) from where the energy is generated to where it is used, while an engineering problem capable of solution, will nevertheless add further significant costs.

12.3.2 Geothermal

Heat at or near the Earth's surface arises from two sources in a ratio of *ca.* 4 : 1:

- that generated from radioactive decay in the crust (the top 40 km or so)
- that arising from conduction from the Earth's liquid (outer) core.[1a]

The liquid core is still hot from the planet's formation and is further heated by friction associated with tidal flows induced in the liquid core under the gravitational influence of the Moon and Sun. Such energy is, in principle, available perpetually (and continuously) and can be considered renewable. Shallow extraction using heat pumps is attractive for domestic space heating, and in some cases, regional heating—though this cannot provide a globally significant source of energy. Deep drilling (10–20 km) could access hot rock, though once the immediately available energy is extracted, longer-term supply would be governed by the (slow) rate of heat transfer through the Earth's rocks. Sources of geothermal energy closer to the surface are associated with regions of the world close to the edges of tectonic plates, such as Iceland.

12.3.3 Wind Power

Wind power derives from airflows in the atmosphere resulting from pressure gradients that can be traced back ultimately to the activity of the Sun. This form of power generation is associated with emissions of only 10 g CO_2e kWh^{-1}. It is a developing technology with rapidly increasing installed capacity (*ca.* 100 GW worldwide), both land-based (where its visual impact is controversial) and offshore (where winds are stronger, though connection, maintenance and operation are more costly[viii]).

A major disadvantage is the intermittent nature of the supply, as energy is not just unavailable if no wind blows but the turbines also have to be 'feathered' in high wind. This intermittency is also unpredictable, leading to low load factors (20–30%). This is illustrated (Figure 12.7) with time-resolved energy output data for a 1000 turbine wind farm in the Pacific Northwest of the USA during the month of January 2009.[17a] Maintaining continuity of electricity supply requires an immediately available standby back-up source (a 'spinning reserve') of equivalent capacity not susceptible to problems of intermittency. The costs and emissions associated with the provision of such back-up supplies need to be included in any economic or environmental assessment. It is calculated that 20% of the UK's generating capacity of 78 GW could be provided by about 20 000 turbines, nominally rated at 3 MW, but with a load factor of 25%. But because of the need to avoid a 'shadow', each turbine needs to be well-spaced from all the

[viii] Mackay[1a] tells us that corrosion from sea water caused all 80 turbines at the Danish Horns Reef windfarm to become inoperable after only 18 months. Such corrosion is a materials engineering and chemistry problem that should be soluble, at a cost, as part of the technological learning process. The learning from oil exploration and production in the North Sea (and elsewhere) suggests that some of the relevant experience already exists.

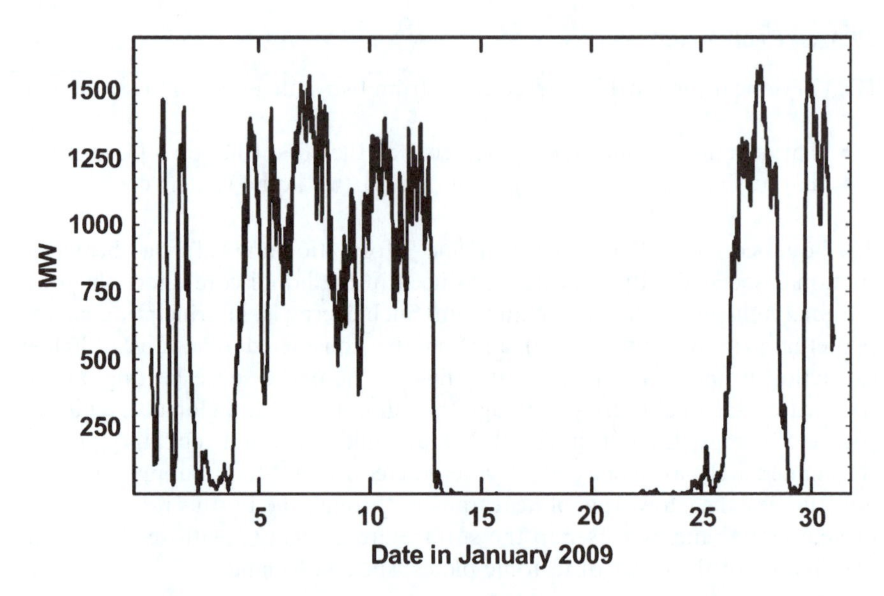

Figure 12.7 Wind power intermittency: time-resolved energy output (MW) from a thousand turbine windfarm during January 2009. Reproduced from ref. 17a with permission of the American Chemical Society.

others to achieve optimum performance. The area needed is 0.6–1 km^2 per turbine or 12–20 000 km^2 (an area the size of Yorkshire). The so-called 'power density' (MW km^{-2}) is 200–300 times less than for large conventional power stations.[17b]

The possibility of using tethered aerofoils to extract energy from the stronger and more constant winds at elevations >1000 m above the Earth's surface is currently under investigation.[17b]

12.3.4 Wave Power

Energy from waves is not a particularly well-developed technology,[18a,b] though over the years there has been a series of small-scale attempts with prototypes such as Salter's famous 'duck'[18c] to exploit it. The first commercially operated technology, the 'sea-snake' attenuator developed by Pelamis (Figure 12.8), was commissioned in Aguçadoura, Portugal, in September 2008. According to press reports (March 2009, Web 12a), technical challenges (inevitable in pioneering a technology) have been experienced, along with the collapse of the company providing most of the investment. However, Pelamis technology is now part of a recently announced major renewables project in Scotland (Web 12b).

Waves arise from the flow of air across water (in combination with the effect of gravity) (Figure 12.9). The energy extractable (per unit of wave-front length) from waves is related to wave height, wave speed, wavelength and sea depth according to a simplified empirical relationship (eqn 12.6).

Figure 12.8 The Pelamis 'sea-snake' attenuator for capturing wave energy.

$$P = \frac{1}{4}\rho \times h^2 \times v \quad (\rho = \text{density of water}) \qquad (12.6)$$

Wave size is determined by a combination of wind speed, and the depth and topography of sea shore. Being an island, the UK is considered well-suited to the exploitation of this source as it has long coastlines and appropriate prevailing winds. On the other hand, the quantity of energy available (even in principle) is small in relation to the country's total required capacity. This source is also intermittent and unpredictable, though less so than wind energy, and is subject to saltwater corrosion and to storm damage.

12.3.5 Tidal Power

Tidal power[19a] relies on the effect on the seas and oceans of the gravitational attraction of the Moon and the Sun, and the centrifugal force resulting from the Earth's rotation, allowing seawater to rise behind a dam or barrage, which is then released through turbines to generate electricity.

The approach is favoured by a large tidal range (the difference in the height of the water level measured between high and low tides). The River Severn, with a 14 m tidal range, has long been considered as a possible site. Recent proposals (Web 13) suggest that a scheme with 8 GW (peak) capacity would be able to supply *ca.* 5% UK needs (not necessarily continually, because of the reductions of flow at high and low tides), with a corresponding avoidance of up to $15\,\text{Mt}\,\text{y}^{-1}$ CO_2.[19b] Five different schemes are under consideration at the time of

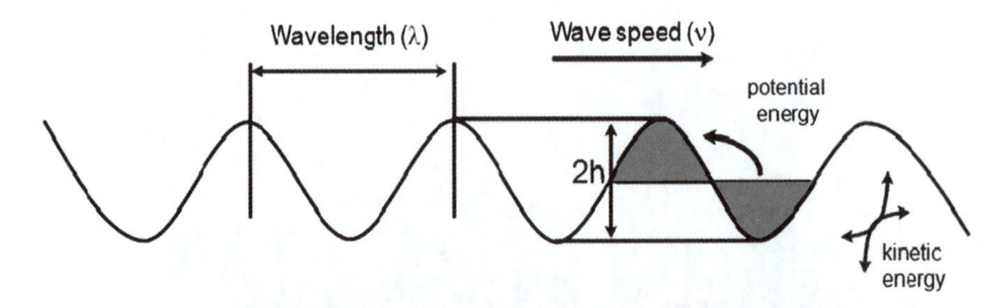

Figure 12.9 Idealised representation of waves and the factors that determine the energy that may be extracted.

writing. The costs are huge and there are concerns about habitat loss (though there would also be some habitat gain[19c]). This is not new technology with the La Rance Barrage on the estuary of the River Rance, Brittany, providing 240 MW (peak), having been in operation for 40 years (Web 14).

MacKay (ref. 1a, quoting Cartwright *et al.*) has suggested that the tidal flows into and out of the North Sea represent a *ca.* 250 GW power source, some of which could be captured by devices placed on the sea bed. Such tidal currents (which avoid the need for dams) are being exploited.

12.3.6 Solar Power: Photovoltaics

Capturing the energy from the sun directly to produce electricity, using the principle of photovoltaics (Figure 12.10) has been an area of scientific enquiry and engineering evaluation for many years.

A semiconductor is subjected to solar radiation to convert photons to electricity. If the photon energy is greater than the semiconductor bandgap, then an electron is promoted to the conduction band. Sunlight is made up a range of wavelengths: light with energy less than that of the band gap will not be captured; that with energy greater than the band gap may be captured with excess energy being lost. This describes a single junction cell; using a series of semiconductors with a range of bandgaps (multiple junction cell) would capture more energy. Photovoltaics (PV) is thus about the photogeneration of electrons and holes. The key requirement is to separate the charge carriers (and subsequently to invert the electricity from DC to AC if it is to be transmitted over long distances). The efficiency of such a device is given by eqn (12.7):[ix]

$$\text{Efficiency} = \frac{\text{Electric power output} \times 100}{\text{Light power input}} \tag{12.7}$$

[ix] This equation is one example of an empirical relationship used to estimate the effectiveness of an energy conversion process (the energy return on investment or EROI), with the term 'energy' used here in the practical rather than the rigorously thermodynamic sense.

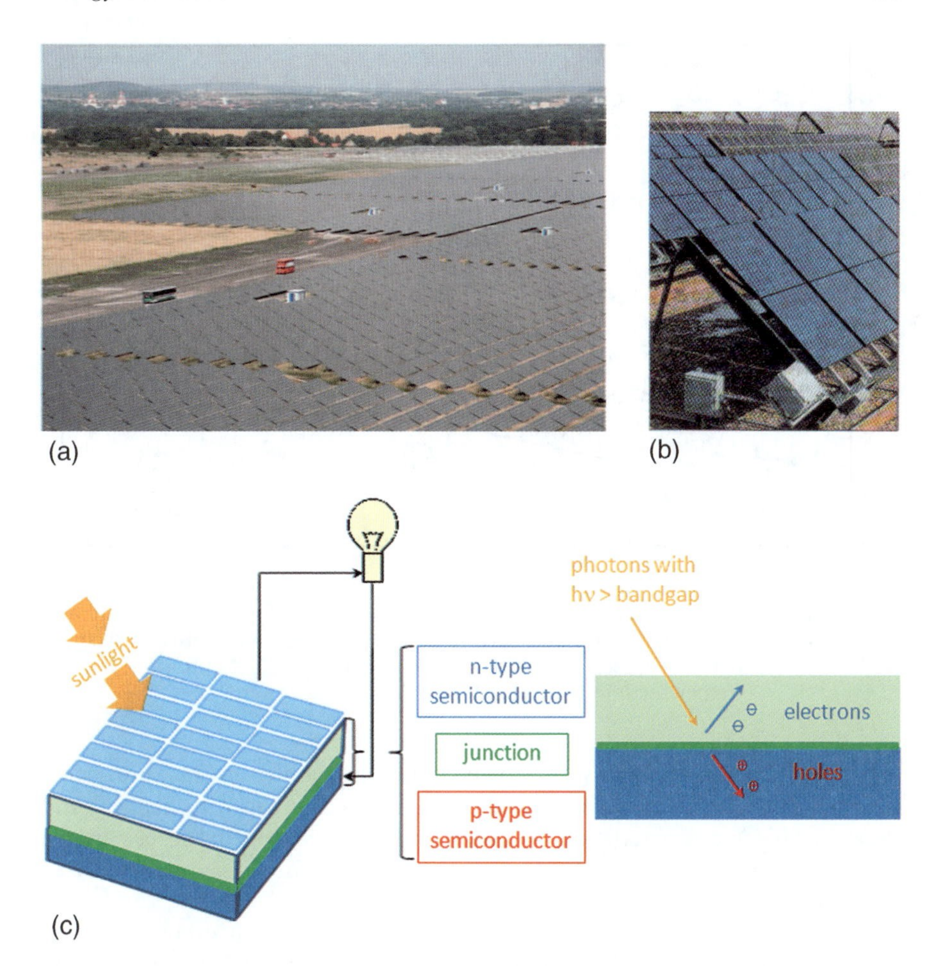

Figure 12.10 Solar energy captured by photovoltaics (PV): (a) Woldpolenz Solar Park, Germany, 40 MW capacity (photograph courtesy of JUWI Group); (b) detail of solar panel; and (c) principle of operation of a photovoltaic semiconductor device.

Some idea of the global capacity and more local practicality of photovoltaics is given from the following. Solar radiation is highly variable, depending on time of day, time of year, state of the weather and location. If solar radiation = 1000 W m^{-2} [assuming the following conditions: cloudless sky; noon (maximum height of the sun); equinox (length of day = the length of night); at the Equator (sun directly overhead)], then a 10% efficient PV system of area 1 m^2 would provide a power output of *ca.* 100 W. Clearly, this output will be significantly less on a cloudy late afternoon in winter in Liverpool. Commercially available PV cells/systems currently provide 10–20% efficiency,[20] though research (multiple junction) cells (Figure 12.11) deliver up to 40%.

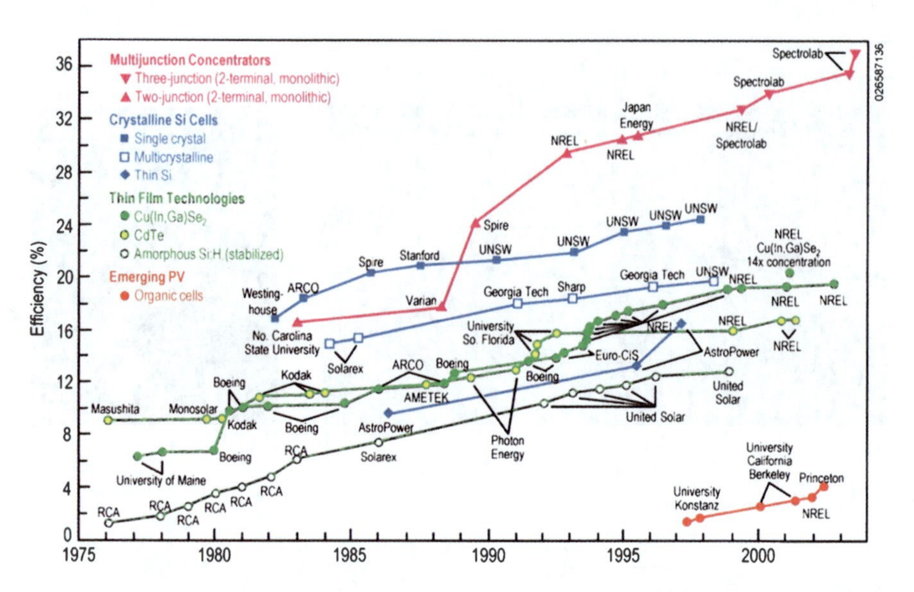

Figure 12.11 Improvements with time (1975–2004) in the efficiency (%) of photovoltaic light capture for research devices based on a range of technologies. Source: National Renewable Energy Laboratory, National Center for Photovoltaics (Web 15).

It is evident how relatively slowly efficiency improvements are being achieved. Many of the technical challenges have origins in the chemical and electrochemical characteristics of materials being developed for use in the various approaches to photovoltaic devices whose performance is summarised in Figure 12.11.

The problem of applicability is one of engineering (once a suitably efficient system has been identified) to optimise cost with efficiency and reliability. Any system installed at a suitable site must deliver at some minimum efficiency for an extended period to justify the large investment. A period of 15–20 years is the approximate minimum lifetime of a conventional power station. In any event, it will be necessary to provide an operating cost per unit of electricity low enough to justify the higher installation costs. Critical metrics that will be used to determine investment decisions are:

- the total installed system price per kilowatt hour delivered
- the number of years of productive operation till payback of the installation costs from the savings made.

The efficiency currently available is approaching that that would make PV electricity generation attractive. The technology is close to being used on the large scale. It is already used as a point-of-use power source and is increasingly being integrated into buildings. It is estimated that electricity produced this way is 4–10 times more expensive than today's retail electricity costs. Government subsidies are accelerating take up in some countries. The first major installations are in regions of high solar irradiation. Smaller local installations are suitable where there is no alternative electricity supply. At the end of 2008, the cumulative global PV installed capacity reached 15.2 GW (Web 16), though it has a very low load factor (*ca.* 15%[17]) associated with intermittency and variability (requiring back-up to ensure continuity of supply).

World desert covers *ca.* 2×10^9 ha; the world energy requirement in 2040 of 900–1000 EJ y^{-1} could, in principle, be supplied using PV solar energy requiring 10^7 ha or 10^{11} m^2 or 100 000 square kilometres (*cf.* UK land area = 243 000 km^2). New infrastructure will be needed for energy storage and transmission from currently remote sources (*e.g.* Sahara or Gobi deserts) to centres of population and industry. The issue of geopolitical security will be a major factor in deciding where the primary solar energy capture sites are located, with sites in regions of political stability being favoured.

12.3.7 Biomass[21]

The land area (and land type) needed to grow crops to provide for energy generation and conversion into transportation biofuels has already been discussed in the previous chapter. We have already seen from more detailed evaluation that the land requirement for growing crops for first generation biofuels appears unsustainable. Greater productivity may arise from other crops, including algae, though these would need to be grown close to a concentrated source of carbon dioxide. Biomass-derived products such as methanol, hydrogen and methane may provide a more sustainable secondary energy source, either as fuels for traditional power sources or for use in fuel cells.

12.3.8 Nuclear Fusion

Energy from nuclear fusion (the process occurring in the Sun, and other stars, which delivers solar energy) arises when two nuclei join to form a heavier nucleus in which a net loss of mass occurs. The energy generated, associated with the mass loss in this process, is given by Einstein's equation, $E = mc^2$, where m is the mass reduction resulting from the process and c is the speed of light. (1 g of mass is equivalent to *ca.* 25 GWh.) Fusion energy generation has been researched for over 50 years with slow progress towards solving some critical engineering solutions necessary for a reliable and productive technology. The front-running fusion process is the so-called D-T [deuterium (^2H)–tritium (^3H)] fuel cycle (eqn 12.8):

$$^2H + {}^3H \rightarrow {}^4He + n + 17.6\,\text{MeV} \tag{12.8}$$

Heavy water, D_2O, comprises 0.016% of natural water and can supply deuterium, ^2H, in quantities sufficient in practice for an inexhaustible source of electricity. Unfortunately, the other component of the reaction, the third hydrogen isotope, tritium, T or ^{3}H, is present naturally at only one part in 10^{17}. This must be generated synthetically and the most favoured process involves lithium (eqn 12.9):

$$^{6}\text{Li} + \text{n} \rightarrow {}^{3}\text{H} + {}^{4}\text{He};\ {}^{7}\text{Li} + \text{n} \rightarrow {}^{3}\text{H} + {}^{4}\text{He} + \text{n} \tag{12.9}$$

The challenges are significant, not least the need to generate, sustain and contain[x] a plasma (a mixture of electrons and ions) at a temperature of 200×10^6 K to ensure a self-sustaining reaction. Containment is achieved by magnetic confinement, though clearly there are problems of materials design in dealing with a high 14 MeV neutron flux and the handling of radioactive tritium.

Nevertheless, progress is being made (Web. 17) with fusion reactions sustained for sufficient time to generate 10 MW for 0.5 s. (The energies needed to initiate the process are themselves prodigious so a major challenge is for the reactor to become a net producer of energy and for this to be uninterrupted and continuous.) The project costs are enormous[xi] and can only be borne by an international consortium of countries. The latest development, the International Thermonuclear Experimental Reactor (ITER) (Web 18), with a current budget of 10 billion Euros, is undergoing site clearance in preparation for construction in Cadarache, France. Large-scale energy generation from nuclear fusion holds out the prospects for unlimited energy, but it can only be seen as a distant prospect. In addition, there are questions about the sufficiency of supplies of lithium in the long term necessary for tritium production.[23] While international collaborations such as ITER will certainly make important technical and scientific advances, nuclear fusion is unlikely to make a significant contribution to large-scale electricity generation over the next 20–30 years.

12.4 SECONDARY ENERGY SOURCES AND ENERGY STORAGE

Electricity cannot be stored and, for this reason, is converted into forms from which electricity can be recovered later. To avoid having to build generating capacity to meet peak demand (*e.g.* half-time in the final of the World Cup), electricity generated when the demand is low is converted into something that can be stored. This has the added benefit of running power stations at optimum efficiency, rather than turning them up and down as demand fluctuates. On the large scale, electricity is either converted into potential energy by pumping water from low level to high level (in pump-storage schemes associated with hydroelectric power generation) or into chemical storage in batteries. Alternatively, electricity can be chemically stored as hydrogen from which energy can be recovered in a fuel cell.

---

[x] The issue was put neatly by a leading figure in fusion research: '*We say we will put the Sun into a box. The problem is, we don't know how to make the box*'.

[xi] Whether or not at least a proportion of such funding might be more usefully directed towards areas of energy research that might deliver benefits more quickly is a particularly teasing question.[22]

12.4.1 Hydrogen

Hydrogen is a clean fuel as water is the sole product of its combustion (eqn 12.10).

$$H_2(g) + 0.5\,O_2(g) \rightarrow H_2O(l);\ \Delta G^\circ(298\,K) = -237\,kJ\,(mol\,H_2O)^{-1} \quad (12.10)$$

Hydrogen is currently neither a primary nor a renewable energy source, as 96% of its current production arises from fossil sources. Hydrogen is a co-product of sodium hydroxide production from the electrolysis of brine in mercury cells. Much is also produced by steam reforming at $>700\,^\circ C$ over a Ni on Al_2O_3 catalyst (eqn 12.11) and by the water-gas shift reaction (eqn 12.12), with the net process shown in eqn (12.13).

$$C_nH_m + n\,H_2O \rightarrow n\,CO + (n + m/2)\,H_2 \quad (12.11)$$

$$CO + H_2O \rightarrow CO_2 + H_2 \quad (12.12)$$

$$C_nH_m + 2n\,H_2O \rightarrow n\,CO_2 + (2n + m/2)\,H_2 \quad (12.13)$$

Importantly, 7.1 kg CO_2 are produced in such processes per kg H_2 produced.

If there is any future for the so-called hydrogen economy, there is an urgent need to find ways of producing (transporting and storing) dihydrogen on a large scale in a sustainable manner from renewable sources. Materials that permit the storage of hydrogen and other gases at temperatures above their boiling point (21 K) are a major research activity. A key characteristic of any technically acceptable storage material must be the ready reversibility of the process leading to storage. Interest includes the catalytic decomposition of ammonia,[24] a material that has excited attention because of its ready availability.

More sustainable hydrogen production may be achieved from biomass (Section 11.6.7), though this is a long way from being a developed technology. Electrolysis using electricity from PV or wind power may produce hydrogen, thereby storing solar energy for subsequent conversion to electricity in a fuel cell (Section 12.4.2) when the primary source is not available. However, this is likely to be applicable first in small-scale local free-standing systems.

PV can be used in two ways: to generate electricity that could then be used to electrolyse water; or photoelectrochemically to split water directly.[25a] Direct photocatalytic splitting of water (particularly using visible light) is an intense area of research in which chemists are playing a key part.[25b,c] A third (thermochemical) method of water splitting would require temperatures of $>2000\,K$ for useful conversions. However, using a series of linked chemical reactions it is possible to split water at the temperatures ($>1300\,K$) found in a nuclear reactor. Some 200 thermochemical cycles have been investigated, though none is currently considered economic. The so-called sulfur–iodine cycle (eqn 12.14–12.17, net reaction shown in eqn 12.18) is being studied[26] on the pilot scale:

$$H_2SO_4 \rightarrow H_2O + SO_3 \quad (12.14)$$

$$SO_3 \rightarrow SO_2 + 0.5\,O_2 \quad (12.15)$$

$$SO_2 + I_2 + 2\,H_2O \rightarrow 2\,HI + H_2SO_4 \quad (12.16)$$

$$2\,HI \rightarrow H_2 + I_2 \quad (12.17)$$

$$H_2O \rightarrow H_2 + 0.5\,O_2 \quad (12.18)$$

Even if the processes of hydrogen manufacture from renewable sources can be made suitably efficient, additional major challenges are evident, including those associated with hydrogen storage,[27] containment and transmission arising from its volatility, diffusivity and, interestingly, from its nuclear spin. Molecular hydrogen can exist in two forms, in which the two nuclei are parallel (*para*-hydrogen) or antiparallel (*ortho*-hydrogen). At room temperature, hydrogen is made up of 25% of the *para* form and 75% of the *ortho* form. However, at its boiling point (21 K) the equilibrium concentration of the *ortho* form reduces to only 0.2%. Unfortunately, the conversion of *ortho*-H_2 to *para*-H_2 is exothermic, with the heat of reaction increasing as the temperature is lowered, so that at temperatures $<77\,K$, it is greater than the latent heat of vaporisation. If unconverted liquid hydrogen is stored, then the heat released by the conversion will lead to its evaporation. These losses will be in addition to those arising from heat leakage (dependent in the surface-to-volume ratio of the container). Fortunately, it is possible to catalyse the equilibration of the two forms of H_2 with a paramagnetic oxide, an important additional processing step in the industrial production of liquid hydrogen. Additional losses from leakage during commercial transportation of hydrogen currently represent *ca.* 10% of production. It is difficult to see how industry will be able to reduce losses for automobile use to $0.05\,g\,(kg\,H_2)^{-1}\,h^{-1}$ by 2015 as specified (Web 19).

Using hydrogen as a transportation fuel for automobiles has the attraction that water is the only direct co-product. An analysis of the pros and cons of this application has been addressed.[28] A particular concern (highlighted for several alternatives in Figure 12.12) involves the storage capacity of hydrogen in various forms. For a fuel cell electric vehicle (FCEV) with a range of 300 miles (480 km), the necessary amount of H_2 (at NTP) occupies 3000 times the space occupied by petrol sufficient for the same range. A tank of compressed H_2 (70 MPa) occupies eight times the volume of a petrol tank for an equivalent amount of fuel, but as fuel cells are twice as efficient as the internal combustion engine, this reduces to four times on a range-adjusted basis. Fuel tanks must be heavily insulated thereby adding weight. Chilling H_2 to liquefy it requires *ca.* 30% of its energy content. In 2004, it cost US$5 to produce an amount of H_2 equivalent to 1 US gallon of petrol (costing *ca.* US$2).

However, while there have been suggestions that dihydrogen (which has a natural atmospheric abundance of $\sim$0.5 ppm[29]) may deplete stratospheric ozone,[30] a more significant impact is as an 'indirect' greenhouse gas[31] competing for tropospheric hydroxyl radicals (eqn 12.19) with methane, thereby reducing a decomposition route for the latter—and adding to its global warming potential.

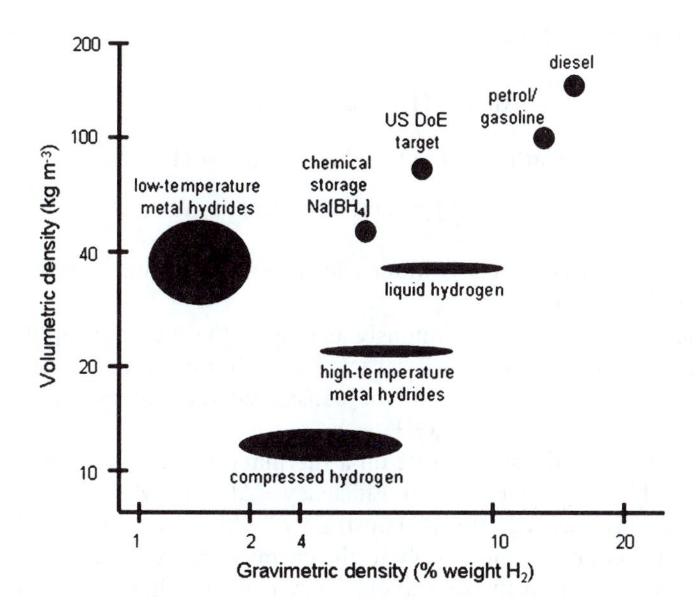

Figure 12.12 A comparison (based on ref. 28a) by volume and weight of the storage capacity for hydrogen in various forms using conventional petrochemical-derived transportation fuels as a benchmark.

$$H_2 + 2\,OH^{\bullet} \rightarrow 2\,H_2O \tag{12.19}$$

Finally, we should keep in mind not only the technical and the commercial aspects, but also the occupational and other hazards (represented by the fatalities per GW per year[32]) associated with different energy producing technologies. These are relatively high for coal and oil, but are much less for gas and nuclear. This may appear surprising, but this may simply be an indication of the extent of myth-making that exists.

12.4.2 Batteries and Fuel Cells

Fuel cells[33] and batteries[34] both convert chemical energy into electrical energy. A fuel cell differs from a battery in that reactants (one fed to the anode, the other to the cathode) are introduced from an external source. A battery, by contrast, is a device in which electrical energy is stored as chemical energy.

The development of fuel cells was given a major impetus through the requirements and investment of NASA's Apollo space programme, when the issue of cost was secondary. In a fuel cell, a fuel (H_2, CO, CH_4, CH_3OH) is supplied to the anode (the oxidation half-cell process, eqn (12.20)) and oxygen (or air) to the cathode (the reduction half cell process, eqn (12.21)), giving the net reaction eqn (12.22):

$$\text{Anode: } H_2 \rightarrow 2\,H^+ + 2e^- \tag{12.20}$$

$$\text{Cathode: } 0.5\,O_2 + 2\,H^+ + 2e^- \rightarrow H_2O \tag{12.21}$$

$$2\,H_2 + O_2 \rightarrow 2\,H_2O \tag{12.22}$$

The innocuous by-product is water which, in some circumstances such as in space, has significant value.

Electricity is generated continuously as long as the fuels are supplied to the electrodes. For reasons discussed earlier, fuel cells are secondary energy sources, and whether or not they can be considered renewable depends on the source of H_2, CO, CH_4 or CH_3OH.

Because fuel cells do not operate on a thermal cycle, they are not limited to the Carnot efficiency (Section 6.12). Efficiency $= \Delta G/\Delta H$, where $\Delta G^\circ = -n\,F\,E^\circ$. Having said this, fuel cells are still constrained by the laws of thermodynamics. The laws of thermodynamics apply to the chemical processes occurring in fuel cells, but the maximum theoretical efficiency is higher than the efficiency of a conventional combustion cycle. The efficiency of a fuel cell is usually normalised to the energy available by simply burning the fuel. Efficiency can be as high *ca.* 50%, though is reduced as current is extracted. While they are presently uneconomic compared with other power sources, they are attractive as distributed power sources (thereby avoiding transmission losses) and have found use in transport in fuel cell electric vehicles.

12.4.3 Energy Storage

Energy can be stored chemically, electrochemically, thermochemically or mechanically. We have already noted the use of hydrogen and methanol, of batteries and of pumped-storage. For completeness, we can add the use of phase-change materials and the storage of kinetic energy in flywheels, both requiring the development of appropriate materials.[35]

REFERENCES

1. (a) D. J. C. MacKay, *Sustainable Energy – without the hot air*, UIT Cambridge, Cambridge, 2009; (b) D. Abbott, *Proc. IEEE*, 2010, **98**, 42.
2. (a) J. Goldemberg, *Science*, 2007, **315**, 808; (b) J. Goldemberg, S. T. Coelho, P. M. Nastari and O. Lucon, *Biomass Bioenergy*, 2004, **26**, 301.
3. B. Lomborg, *The Skeptical Environmentalist: Measuring the Real State of the World*, Cambridge University Press, Cambridge, 2001, Figure 73, p. 133.
4. M. Jaccard, *Sustainable Fossil Fuels: The Unusual Suspect in the Quest for Clean and Enduring Energy*, Cambridge University Press, Cambridge, 2005.

5. (a) C. J. Campbell, Presentation to a House of Commons All-Party Committee, 7 July 1999 (Web 4); (b) M. K. Hubbert, presented at 'Nuclear Energy and the Fossil Fuels', Spring Meeting, Southern District, American Petroleum Institute, San Antonio, Texas, 7–9 March 1956; (c) S. Sorrell, J. Speirs, R. Bentley, A. Brandt and R. Miller, *An Assessment of the Evidence for a Near-Term Peak in Global Oil Production*, UK Energy Research Centre, London, 2009 (Web 5); (d) R. A. Kerr, *Science*, 2009, **326**, 1048; (e) I. S. Nashawi, A. Malallah and M. al-Bisharah, *Energy Fuels*, 2010, **24**, 1788.
6. G. J. Kramer and M. Haigh, *Nature*, 2009, **462**, 568.
7. Q. Schiermeier, J. Tollefson, T. Scully, A. Witze and O. Morton, *Nature*, 2008, **454**, 816.
8. (a) IPPC, *Special Report on Carbon Dioxide Capture and Storage*, ed. B. Metz, O. Davidson, H. C. De Coninck, M. Loos and L. A. Meyer, Cambridge University Press, Cambridge, 2005 (Web 6); (b) J. Lu, M. Wilkinson, R. S. Haszeldine and A. E. Fallick, *Geology*, 2009, **37**, 35; (c) B. Tohidi, J. Yang, M. Salehabadi, R. Anderson and A. Chapoy, *Environ. Sci. Technol.*, 2010, **44**, 1509.
9. (a) F. S. Zeman and D. W. Keith, *Phil. Trans. R. Soc. A*, 2008, **366**, 3901; (b) K. S. Lackner, *Eur. Phys. J. Spec. Top.*, 2009, **176**, 93; (c) N. Jones, *Nature*, 2009, **458**, 1094.
10. S. Pacala and R. Socolow, *Science*, 2004, **305**, 968.
11. B. K. Sovacool, *Energy Policy*, 2008, **36**, 2950.
12. D. Tullos, *J. Environ. Manage.*, 2009, **90**, 5208.
13. (a) R. W. Norman, *Science*, 1974, **186**, 350; (b) R. Labrecque, *Entropy*, 2009, **11**, 798.
14. E. Rebhan, *Eur. Phys. J. Spec. Top.*, 2009, **176**, 53.
15. Department for Business, Enterprise and Regulatory Reform, *Meeting the Energy Challenge: A White Paper on Nuclear Power*, HMSO, London, 2008, CM 7296 (Web 11a).
16. C. Okkerse and H. van Bekkum, *Green Chem.*, 1999, **1**, 107.
17. (a) W. Katzenstein and J. Apt, *Environ. Sci. Technol.*, 2009, **43**, 6108; (b) L. Fagiano, M. Milanese and D. Piga, *IEEE Trans. Energy Conv.*, 2010, **25**, 168.
18. (a) B. Drew, A. R. Plummer and M. N. Sahinkaya, *Proc. Inst. Mech. Eng. Part A.*, 2009, **223**, JPE782; (b) A. F. de O. Falcão, *Renew. Sustain. Energy Rev.*, 2010, **14**, 899; (c) S. H. Salter, *Nature*, 1974, **249**, 720.
19. (a) F. O'Rourke, F. Boyle and A. Reynolds, *Appl. Energy*, 2010, **87**, 398; (b) C. Woolcombe-Adams, M. Watson and T. Shaw, *Water Environ. J.*, 2009, **23**, 63; (c) J. Xia, R. A. Falconer and B. Lin, *Ocean Model.*, 2010, **32**, 86.
20. (a) D. Butler, *Nature*, 2008, **454**, 558; (b) D. M. Bagnall and M. Boreland, *Energy Policy*, 2008, **36**, 4390; (c) A. E. Curtright, M. G. Morgan and D. W. Keith, *Environ. Sci. Technol.*, 2008, **42**, 9031.
21. V. Dornburg, D. van Vuuren, G. van de Ven, H. Langeweld, M. Meeusen, M. Banse, M. van Oorschot, J. Ros, G. J. van den Born, H. Aiking,

M. Londo, H. Mozaffarian, P. Verweij, E. Lysen and A. Faaij, *Energy Environ. Sci.*, 2010, **3**, 258.
22. N. Winterton, *Clean Technol. Environ. Policy*, 2008, **10**, 309.
23. (a) D. Cohen, *New Sci.*, 23 May 2007, p. 34; (b) W. F. Pickard, *Global Planet. Change*, 2008, **61**, 285.
24. D. A. Hansgen, D. G. Vlachos and J. G. Chen, *Nat. Chem.*, 2010, **2**, 484.
25. (a) A. Kudo and Y. Miseki, *Chem. Soc. Rev.*, 2009, **38**, 253; (b) W. J. Youngblood, S.-H. A. Lee, K. Maeda and T. E. Mallouk, *Acc. Chem. Res.*, 2009, **42**, 1966; (c) A. J. Bard and M. A. Fox, *Acc. Chem. Res.*, 1995, **28**, 141.
26. K. Onuki, S. Kubo, A. Terada, N. Sakaba and R. Hino, *Energy Environ. Sci.*, 2009, **2**, 491.
27. A. Züttel, *Naturwissenschaften*, 2004, **91**, 157.
28. (a) R. F. Service, *Science*, 2004, **305**, 958; (b) R. F. Service, *Science*, 2009, **324**, 1257.
29. A. Grant, C. S. Witham, P. G. Simmonds, A. J. Manning and S. O'Doherty, *Atmos. Chem. Phys.*, 2010, **10**, 1203.
30. T. K. Tromp, R.-L. Shia, M. Allen and Y. L. Yung, *Science*, 2005, **300**, 1740.
31. M. J. Prather, *Science*, 2005, **302**, 581.
32. D. Elliot, in *Sustainability and Environmental Impact of Renewable Energy Resources, Issues in Environmental Science and Technology*, ed. R. E. Hester and R. M. Harrison, Royal Society of Chemistry, 2003, vol. 19, ch. 2, pp. 19–47.
33. B. Grievson, *Handbook of Green Chemistry and Technology*, ed., J. H. Clark and D. Macquarrie, Wiley-Blackwell, Oxford, 2002, ch. 20, pp. 466–481.
34. M. R. Palacín, *Chem. Soc. Rev.*, 2009, **38**, 2565.
35. (a) A. Gil, M. Medrano, I. Martorell, A. Lázaro, P. Dolado, B. Zelba and L. F. Cabeza, *Renew. Sustain. Energy Rev.*, 2010, **14**, 31; (b) M. Medrano, A. Gil, I. Martorell, X. Potau and L. F. Cabeza, *Renew. Sustain. Energy Rev.*, 2010, **14**, 56; (c) D. Lindley, *Nature*, 2010, **463**, 18; (d) M. Kenisarin, *Renew. Sustain. Energy Rev.*, 2010, **14**, 955.

BIBLIOGRAPHY[xii]

V. Smil, *Energy at the Crossroads: Global Perspectives and Uncertainties*, MIT Press, Cambridge, MA, 2005 [not examined by the author].
G. F. Hewitt and J. G. Collier, *Introduction to Nuclear Power*, Taylor & Francis, London, 2nd edn, 2000.
R. A. van Santen, in *Catalysis for Renewables*, ed. G. Centi and R. A. van Santen, Wiley-VCH, Weinheim, 2007, ch. 1, p. 11,18.

[xii] These texts supplement those already cited as references.

WEBLIOGRAPHY

1. www.eia.doe.gov/aer
2. https://publicaffairs.llnl.gov/news/energy/content/energy/energy_archive/energy_flow_2008/LLNL_US_EFC_2008_annotated1.pdf
3. www.bp.com/productlanding.do?categoryId = 6929&contentId = 7044622
4. www.hubbertpeak.com/Campbell/commons.htm
5. www.ukerc.ac.uk/support/Global%20Oil%20Depletion
6. www.ipcc.ch/pdf/special-reports/srccs/srccs_wholereport.pdf
7. www.rsc.org/chemistryworld/News/2008/September/15090801.asp
8. (a) http://earthobservatory.nasa.gov/images/imagerecords/6000/6574/3gorges_ast_2006135.jpg
   (b) www.ctgpc.com
9. (a) http://en.wikipedia.org/wiki/Glendoe_Hydro_Scheme
   (b) http://news.bbc.co.uk/2/hi/uk_news/scotland/highlands_and_islands/7171802.stm
   (c) www.nce.co.uk/news/structures/beleagured-glendoe-hydro-project-needs-16km-tunnel/8600364.article
10. www.statkraft.com/energy-sources/osmotic-power/
11. (a) www.berr.gov.uk/files/file43006.pdf
    (b) http://news.bbc.co.uk/1/hi/uk_politics/7179579.stm
    (c) www.world-nuclear.org/info/reactors.html
12. (a) www.guardian.co.uk/environment/2009/mar/19/pelamis-wave-power-recession
    (b) http://news.bbc.co.uk/1/hi/scotland/highlands_and_islands/8564662.stm
13. www.reuk.co.uk/Severn-Barrage-Tidal-Power.htm
14. www.reuk.co.uk/La-Rance-Tidal-Power-Plant.htm
15. www.nrel.gov/pv/thin_film/docs/kaz_best_research_cells.ppt
16. http://en.wikipedia.org/wiki/Photovoltaics
17. http://en.wikipedia.org/wiki/Fusion_energy
18. http://en.wikipedia.org/wiki/ITER
    www.iter.org/default.aspx
19. www1.eere.energy.gov/hydrogenandfuelcells/mypp/pdfs/storage.pdf

All the web pages listed in this Webliography were accessed in May 2010.

CHAPTER 13

The Chemist as Citizen

'Learning without thought is labour lost; thought without learning is perilous'

Confucius (551–479 BC)

My use of the term 'foundation' in the title of this book is meant to convey that there is more to defining and addressing (let alone solving) the challenges of sustainability that even this broadly drawn book could cover. In such a fast-moving area it will also, in some detailed respects, be out-of-date even before it is published. However, its purpose should be recalled: to equip the reader with some insights, starting points and tools to put their learning into a wider context, particularly relating to the challenges of sustainability. A foundation is meant to be built on.

To be long-lasting and fit-for-purpose, any structure needs a vision and a plan, investment and skill, commitment and effort. Most of all, it requires an optimism to believe that it is worth it. It is not too melodramatic to say that a new generation will, over the next 20–30 years, have both to meet an awesome responsibility and to grasp an epoch-defining opportunity. This will have to be done altruistically as the new generation will need to make sacrifices in the knowledge that they will not experience the benefits that will be their legacy. Bringing this about needs a dialogue of experts, the public and government, the success of which requires, on an unprecedented scale, informed, pragmatic, enlightened and disinterested decision-making. What role, therefore, should the chemist play?

The chemist, of course, has the same sort of individual and collective rights as any one else. He or she is free to think, to express and to act, subject to the law and individual conscience. However, does the chemist (or any other scientist,

Chemistry for Sustainable Technologies: A Foundation
By Neil Winterton

Published by the Royal Society of Chemistry, www.rsc.org

for that matter) have any particular responsibility when participating in the public discourse about sustainable development? It would be a presumption if this chapter were to become a manifesto or a prescription. It is not meant to be either. It simply raises questions and provides some illustrative instances that might stimulate the reader into a course of thought and reflection.

There are four definable (if overlapping) roles that cover most of the relevant circumstances that are likely to arise:

- the chemist as professional scientist working with other scientists;
- the chemist as a scientific expert in the public arena;
- the chemist as an advocate in the public arena and
- the chemist as a citizen in the public arena.

I have already introduced relevant matters (particularly Sections 2.5, 2.11 and 4.7), such as the role of scientists in their professional work and at times of public debate and controversy.

13.1 SCIENCE AND ETHICS

The discussion in Chapter 4 identified some of the key precepts that characterise the ways scientists contribute to the scientific enterprise. These norms will be the ones to which individuals would be expected to adhere when practising as a chemist. While scientists generally recognise that there are different ways in which individual scientists and scientific disciplines carry out their research, they recognise that there are certain core ethical principles[i] that govern the way they carry out research, produce and process data, and communicate their results and conclusions. The background to these principles is discussed in detail in the texts by Merton, Ziman and others, previously cited. The essence of this is trust, a characteristic that, by and large, still defines the relationship between science and society. However, as scientists are only human this trust can be undermined by individual failings or by conflicts of interest driven by either personal ambition or external pressure (whether political or commercial). The maintenance of the standards of science by each of its practitioners is the best guarantor that public trust will not irreparably be diminished.

The National Academies of Science and Engineering in the USA have published practical guidance to beginning researchers on some of the ethical dilemmas facing scientists. Their document is short and readable (63 pages with an extended bibliography of its own).[2] A similar checklist has been produced specifically for chemists by the Royal Society of Chemistry in a booklet entitled *Code of Conduct and Guidance on Professional Practice*[3] (Web 1).

---

[i] Seeking to provide an ethical framework for the relationship between science and technology and the environment in the era of sustainability includes the *Principia Ecologica* of Anton Moser.[1]

13.2 RHETORIC AND EVIDENCE

While it is clear what is expected of scientists working with other scientists in their professional capacity, it requires a little more analysis when we move into the domain of more public discussion of science and technology. The training of scientists, formally or informally, in seeking objectively to evaluate the evidence for or against a scientific proposition should enable them to apply the techniques of enquiry, analysis and the weighing of evidence to matters of broader public controversy. At the very least, they can use the techniques described in Appendix 1 to gather together a reasonable body of information and evidence, its source and provenance.

However, the positions which individuals (scientists or not) might take on particular questions will be determined (among a number of factors) by the way their own circumstances, experience, attitudes and values affect their assessment of information and opinion to which they are exposed or choose to seek out. However, much of the evidence is presented to us by the press, TV and radio, and the web and will have been selected[ii] and edited, sometimes to support a particular point of view. It will have been 'mediated'. In addition, we can be influenced one way or another simply by the way in which evidence and argument are presented. Does the person presenting an argument look to us trustworthy or not? How much of what he or she is saying chimes with our own views, attitudes or values?

That the persuasiveness or otherwise of someone presenting an argument can be very influential in the degree to which support for a position can be marshalled was well-known in the times of the ancient Greeks. The role of rhetoric (the art of using language and styles of delivery as a means of persuasion) was seen as an important component of public debate and the decision-making process. In ancient Greece, the debate usually took place in front of those who were also responsible for the decision-making and decisions could be affected by the quality of the advocacy for and against, sometimes overturning the factual basis of an argument. The contest between the rhetorician and the philosopher in the processes of decision making is the basis of one of Plato's Socratic dialogues, entitled *Gorgias*.[5] Socrates argues that the tools of persuasion can counter mere facts, the implication being that the rhetorician is more able to convince an uninformed audience than can an expert.

[ii] . . . and sometimes fabricated. Fictitious stories about so-called celebrities were given to journalists. These were printed in a number of daily newspapers in the UK (with some of the stories being picked up in the international press). The journalists who took these stories failed to carry out the elementary step of checking the facts (that would have immediately revealed the stories to be untrue). See the report in *The Guardian* of 15 October 2009 and in Web 2. It is hard to assess the degree to which such journalistic shortcomings extend to the treatment of more serious matters. Before we get too snooty about this, it is worth pointing out that the editors of learned academic journals can be caught out in a similar manner. As part of the 'science wars' (see footnote, p. 13) a physicist, Alan Sokal, submitted to (and, crucially, had published in) a spoof article made up of plausible nonsense entitled *Transgressing the Boundaries: Towards a Transformative Hermeneutics of Quantum Gravity* to a cultural studies journal called *Social Text*. Sokal discusses the issues this raised in his 2008 book.[4] It, therefore, always pays to check.

This is very much the issue that bears upon our responsibility as scientists to be on our guard in the face of seemingly persuasive arguments, particularly relating to issues of science, but more specifically relating to issues of the environment and sustainable development. This is not an arcane historical or academic point. The current problem with an increase in the incidence of measles in the UK, which may have serious consequences for some who contract the disease, arises from a very persuasive campaign, particularly in *The Daily Mail*, that sought to promote the suggestion that there was a link between the triple vaccine against measles, mumps and rubella (the 'MMR' vaccine) and the occurrence of autism, an idea initially proposed by a clinician, Dr Andrew Wakefield. It was at the time largely believed among experts that no such link could be demonstrated and that Dr Wakefield's initial assertions were flawed.[iii] Over the 10 or so years since this controversy began, and as a result of the campaign in the press, many parents decided not to have their children inoculated with the triple jab. As consequence the incidence of measles cases has risen sharply. It took a long time for the full scientific evidence (see, for instance, studies undertaken in Denmark that are particularly convincing[7]) to be taken note of but, by then, serious damage had been done.

One response from the lay person to the complexity of the issues raised in media controversies on scientific, technological and environmental issues is 'a plague on all your houses', declining to engage in consideration of the matter at all, or worse, begin to undermine a belief in the effectiveness of the methods of science.[8] This may be the case even when there is some important (if poorly defined) consequence that may affect that person, as is the case with climate change, when the impact on the individual appears ill-defined and remote. The use of 'icons' (symbolic representations to which an individual can relate) has been explored[9] to engage individuals in a consideration of climate change as a necessary precursor to the acceptance and implementation of associated societal and lifestyle changes. Icons were classified as 'non-expert' or 'expert' (ref. 9 provides an introduction to the deeper meanings of 'expertise'), with the polar bear being an example of the former and ocean acidification being an example of the latter. Scientists are likely to have an important role in the selection and use of both types of icons.

13.3 SCIENCE AND PUBLIC PERCEPTION

A day-to-day examination of newspapers, TV, radio and web-based news sources shows that science, technology and the environment continue to be matters of public concern and debate. As scientists, we should be qualified (within the limit of our own expertise, specific and general) to ask searching questions of those contributing to the debate and to make informed assessments of statements made about scientific and technological controversies. So, it is important for scientists, in particular, to understand what science is and

[iii] The editors of the journal, *The Lancet*, which published Dr Wakefield's original studies in 1998, have retracted the paper[6] following the judgement of The General Medical Council's Fitness to Practise Panel (Web 3).

how it works and to recognise that its methods may not be fully understood by the wider public.

As the work of scientists has to be broadly sanctioned by wider society, then it is important that any debate on matters of scientific controversy takes place in a way that misunderstandings can be explained, if not necessarily accepted. This is especially important when what is done by scientists and how it has been presented by the media (and thus perceived by the wider public) may be different.

The scientist is in a unique and privileged position, being able, through access to sources of information that may not be available to the general public, to explore the original sources of scientific controversies that are played out in the press and to check out stories in the press that have, are claimed or believed to have, a scientific basis.

Readers should be able, in most cases and using the methods outlined in Appendix 1, to track back to the original sources of material used by journalists and others so that they can, independently, establish what the facts, as determined and originally reported by scientists, actually are. While they may not be a specialist in the matter being discussed, as scientists they should certainly be better placed than a lay member of the public to frame appropriate questions. They should, at some rudimentary level, be able to understand the content, evaluate the experimental work and recognise the conclusions so they are in a position to test whether what has been highlighted in the media report is a fair and accurate presentation of what the authors have written. While experts in the area may be expected to challenge misleading or erroneous reports more publicly, the information the reader may gain access to can certainly help in putting the record straight if the matter is raised by family, friends and colleagues.

The process can be illustrated by the following real example,[iv] relating to an important topic, *viz.*, the degree to which human fertility may have been affected by exposure to chemicals in the environment. The graph shown in Figure 13.1 is taken from a full page article in the (then) broadsheet newspaper, *The Independent on Sunday* of 28 April 1996. The article by the paper's Environment Correspondent had as its headline: '*Sperm Warning: As male fertility declines, evidence is growing that chemical pollutants are playing hell with our hormones*'. This was a very arresting headline. The graph, showing a clear halving of sperm counts over 50 or so years, is likely to have evoked a degree of concern in the minds of many readers. It should also have evoked a desire to find out more. I am neither a fertility expert nor an endocrinologist;[v] however, I did spot that the graph included the legend, University of Copenhagen. By use of readily available websites and search engines, I was able to establish that the article was based on a paper[10] published in the very reputable scientific

[iv] Other examples arise regularly and frequently, and should not be too difficult to spot. Ben Goldacre's weekly column in the Saturday *Guardian* (Web 4) highlights some, often appalling, examples of scientific misrepresentation and should be required regular reading.

[v] Endocrinology is concerned with the study of hormones, their biosynthesis and physiological function, and with the glands and tissues that secrete them.

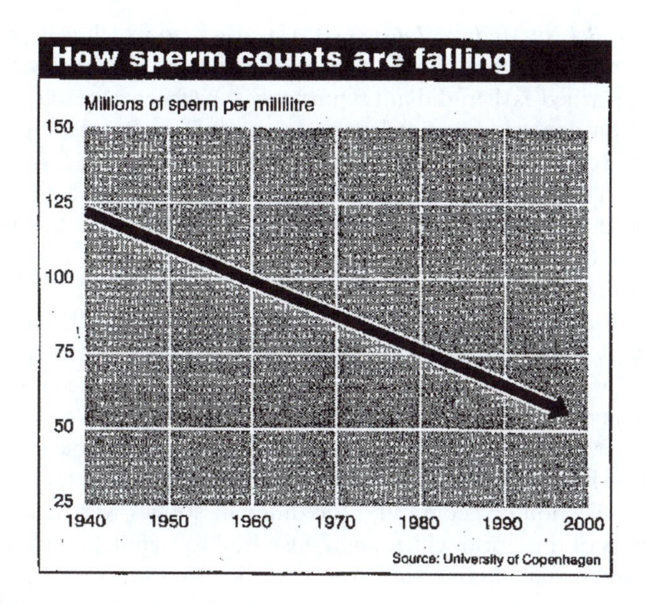

(a)

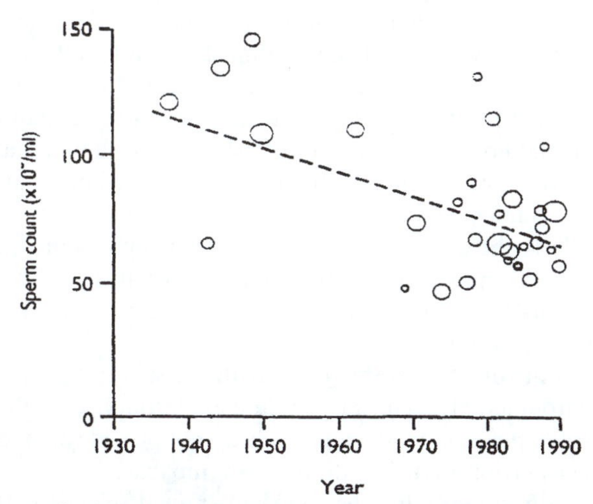

(b)

Figure 13.1 (a) Graph of changes in sperm counts with time used in *The Independent on Sunday*, 28 April 1996 (reproduced with permission from Independent News & Media plc) to suggest that these declines were associated with chemicals in the environment; and (b) original graph (reproduced from ref. 10 with permission from BMJ Publishing Group Ltd) on which (a) was based.

periodical, *British Medical Journal* (BMJ), access to which was possible *via* the University of Liverpool library website. The paper, peer-reviewed by the process we described earlier, did not report new experimental results of its own, but provided an analysis (a so-called 'meta' study) of 40 different investigations reported over the previous 60 years. Interestingly, a graph (Figure 13.1b) that bears some similarity to the one found in the newspaper article can be seen. However, there are also some fairly glaring differences: first, the data points are not clustered closely around the trend line. This becomes even clearer if one imagines Figure 13.1b without the trend line.

Even more of interest, if you read the original article itself, is to discover that the authors do indeed point to the possibility that environmental chemicals might be the origin of a decline. However, in contrast to the newspaper article, the authors rightly discuss the possibility that the data may not reveal a decline at all and, even if it did, then other causes may be responsible such as sampling methods (which were different at the beginning of the 60-year period from the end), self-selection or geography (the different studies were carried out in different parts of the world which might also lead to significant differences in diet, lifestyle and sexual habits). However, no conclusion was reached as to the most likely cause. The newspaper article, therefore, seriously misrepresented the scientific research and the expert conclusions based on it.

Further exploration of the literature using a technique called citation analysis (see Appendix 1) allows you, readily and online, to identify individual scientific papers that, subsequently, have referred to the earlier BMJ paper. An examination of these newer papers can show how the content of the paper stands up to later research, and how the authors and others may have undertaken further work that may add to or even change the conclusions of the original paper. One of these papers[11] undertook a review of studies carried out in the USA, allowing particularly the question of geography to be investigated. Figure 13.2 shows two graphs that plot, separately, studies undertaken in New York and those undertaken in other states of the USA. In neither of these cases is there an obvious or a marked trend.

It is worth pointing out that nothing can finally be said from the above about the role of environmental chemicals on human fertility, just that the data presented in the *Independent on Sunday* article do not provide an adequate basis for the headline assertion that they do have an impact.

I am not media-bashing: journalists are not trained scientists; they usually have exceedingly tight deadlines to meet and, sometimes, inevitably, errors may creep into a story in it preparation. So, when you read newspaper articles or see items on TV or the internet, it is a good idea to ask yourself some questions, such as:

- Who is saying what?
- Have the scientific data been published?
- Can I identify the scientist being quoted?
- Has the scientist been quoted correctly?
- What have they really said or written?

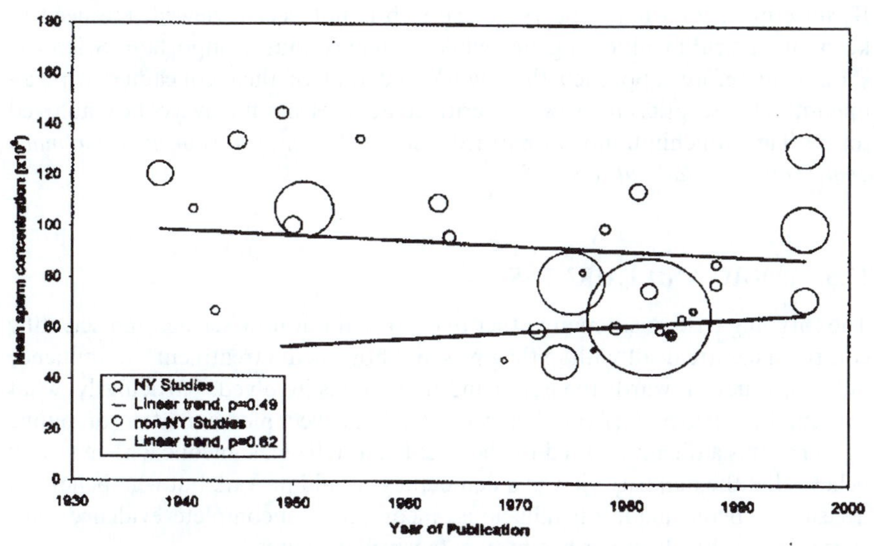

Figure 13.2 Further analysis of some US sperm-count data[11] showing weak trends with time of data from different geographical regions. Reproduced from ref. 11 with permission from Elsevier.

- Have they been quoted selectively?
- Is such selection fair and balanced, or not?
- Does this person (or the journalist) have an 'angle'?
- Does the quotation come from someone claiming to scientifically qualified?
- Can this be verified?
- Is their qualification relevant to the topic under discussion?

There is, of course, an argument that suggests that the public perception of science arises by default due to the absence of contributions from articulate, well-informed and credible scientists with the ability and motivation to engage publicly in debate. There is a variety of circumstances in which this might be changed: the first, and possibly most important, is to be able to present one's own research in a way that can be understood and to anticipate the sort of questions members of the public might want answers to. This is particularly the case, at the moment, at a time when the discussion of climate change and sustainable development is framed by scientific questions and the scientific evidence presented by scientists. These factors are discussed further below.

13.4 SCEPTICISM AND OPEN MINDEDNESS

It is important to be observant in how we read and retain material obtained from newspapers, TV and radio and the web: always keep our brains engaged.

Because much of what appears on the web is not peer-reviewed, the need to keep our critical faculties engaged while on-line is doubly important. Scientists should, therefore, approach their work and that of their colleagues with an open-minded scepticism. However, critical faculties should always be employed for, as Tim Minchin famously entitled a song: '*If you open your mind too much your brains will fall out*' (Web 5).

13.5 LOGIC AND FAIRNESS

The forgoing case raises many other questions about how science and scientific controversies are dealt with in the press and how media treatment can influence public attitudes towards the issues and individuals involved. Particularly, what role might an expert play when such an issue becomes a matter of public concern? It is a dilemma faced by the scientist as citizen, someone who may well believe that the majority view in a democracy should prevail, but who believes a decision is being made on fallacious, incorrect or incomplete evidence. Not surprisingly, this dilemma has been extensively discussed:

> '*Whose views should determine policy—the expert's or the public's? .... if the public is really misinformed we have a dilemma as to whether we ignore the public and follow the experts or blindly follow the public even though we know they're misinformed.*'[12]

Basing an approach to solving a political, social or environmental question solely on an objective assessment of the evidence might appear logical (setting aside for a moment whether or not it is possible to assemble such evidence when the problem is complex or 'wicked'). However, such a logical solution (one that may very well be proposed by an expert) may very negatively affect a section of society, some group or powerful interest. Any political process would seek to negotiate some compromise position to minimise conflict. Basing an approach on what might be fair to all those affected (and this is most likely to be that which is politically expedient or which may satisfy highly articulate, well organised or powerful interest groups) may not have an outcome that in anyway addresses the question initially raised. It is not surprising, therefore, that this dilemma which arises when seeking to reconcile the view of the expert and the democratic process has been expressed in more general form in Arrow's Impossibility Theorem, the stark suggestion that '*collective choice*[vi] *cannot, simultaneously, be both logical and fair*'.[13]

Bearing in mind the many factors involved in the political process and the associated process of collective decision-making, it is probably inevitably the case that there will always be winners and losers. The principles of economics suggest that the losers could, in some way, be compensated to persuade then to live with

[vi] This paradox was noted by Arrow in his analysis of different forms of voting[13b] from which he showed that no voting method can fairly aggregate individual ranked preferences for three or more choices.

their loss of rights or convenience. In the case of the longer-term needs of the environment and those likely to be affected by climate change, it is possible to identify some groups who will be adversely affected (*e.g.* those living in coastal regions close to sea level). One can but hope a combination of expert knowledge, political wisdom and social forbearance will allow an optimum value to emerge of the function that integrates, as far as sustainable development is concerned, logic and fairness over time and over all public and private decision-making.

We should nevertheless take note of the words of Freudenburg and of Kates that, in working towards more sustainable living, we should all develop a '*deeper kind of prudence*'[14] based on '*a capacity to worry intelligently*'.[15]

13.6 INDIVIDUAL ACTION

Our reaction to any threat can take a range of forms: to seek to blame someone else for it (particularly if we do not see ourselves directly responsible or we are immediately and negatively affected) or (more nobly!) to identify what action we can take (individually or collectively) to eliminate it or reduce it, depending on whether it is a threat to us alone or to a wider group.

We are exposed to many articles and programmes that suggest what individual actions we can take to reduce our own ecological or carbon footprint. By asking the sort of questions we have posed in this book, we may be better able to gauge the impact and benefit these actions might bring.

We may, for instance, be attracted to the idea to reduce our energy usage as this should both reduce the atmospheric burden of carbon dioxide and what we pay to our energy supplier. We might also be encouraged to buy fruit and vegetables at a local farmers' market rather than go to a supermarket and buy similar produce imported from distant lands.

These two examples raise some important (and quite general) questions: first, will such individual actions, on their own, have any beneficial effect? We need to go back to the idea of boundaries and examine the matter more widely. Certainly, if we switch off our TV rather than have it on standby or if we do not keep our telephone on recharge all the time we are not using it, then we will save energy. However, we need to ask: what do we do with the money we save? Will we, unwittingly, spend it on something that will, as a consequence of its manufacture, sale and use, consume more energy than we originally saved? We can only ensure the net effect is beneficial if we consciously ensure that the money saved is devoted totally to a non-carbon consuming product or activity. Preferably (and this may make a contribution towards sustainable development), we should donate the money saved to a project in the developing world designed demonstrably to have a minimum environmental impact (recognising that no activity will be without some sort of impact). When we buy our produce locally, we clearly avoid the costs and the emissions associated with its transport from where it is grown to where we buy it (and the costs of refrigeration that ensures that it arrives with us unspoiled, thereby avoiding waste). On the other hand, the farmer or grower, possibly in a country in the developing world, will no longer gain the economic benefit associated with our purchase.

In his excellent book, *Sustainable Energy—without the hot air*, David MacKay (in a section entitled 'Every BIG helps'[16]) goes into these questions in more detail, asking whether the 'every little helps' sentiment does anything more than the more objective statement that '*every little helps a little*'.

13.7 SCIENCE, ACTIVISM AND ADVOCACY

Probably the most difficult to balance is the role of a scientist as expert who also wishes to engage in activism and advocacy. We considered in Chapter 4 the importance of the characteristic of the scientist as being concerned with objective analysis and rational thought. It is unlikely that such objectivity and rationality can sit comfortably alongside the techniques of rhetoric and emotion that often are deployed in disputation. The serious issues raised are discussed in a recent paper.[17]

This is not to suggest that the scientist should disbar themselves from the democratic process; only that the difference between science and activism should be recognised when participating in the democratic process, placing a particular responsibility on the scientist to safeguard the ethical strength of science.

However, in seeking to raise awareness of environmental issues some may appear to have overstepped the mark a little. The late S. H. Schneider said in *Discover* magazine in 1989 (quoted in ref. 18):

> '*We need to capture the public's imagination. That entails getting loads of media coverage, so we have to offer up scary scenarios, make simplified and dramatic statements and make little mention of the doubts we might have.*'[vii]

This is not an argument against all advocacy and activism; more an argument for reason rather than rhetoric.

13.8 'CLIMATEGATE'

In the real world, the complex issues do not fall neatly into the separate sections I have used to present them in this chapter. Indeed, recent events have brought their complexity dramatically into sharp focus with the disclosure on 19

[vii] Schneider claimed vehemently (Web 6) he was selectively quoted out of context. I reproduce the quote here in full, so you can make up your own mind:

> '*On the one hand, as scientists we are ethically bound to the scientific method, in effect promising to tell the truth, the whole truth, and nothing but—which means that we must include all the doubts, the caveats, the ifs, ands, and buts. On the other hand, we are not just scientists but human beings as well. And like most people we'd like to see the world a better place, which in this context translates into our working to reduce the risk of potentially disastrous climatic change. To do that we need to get some broadbased support, to capture the public's imagination. That, of course, entails getting loads of media coverage. So we have to offer up scary scenarios, make simplified, dramatic statements, and make little mention of any doubts we might have. This 'double ethical bind' we frequently find ourselves in cannot be solved by any formula. Each of us has to decide what the right balance is between being effective and being honest. I hope that means being both.*'

November 2009 (and the subsequent torrent of comment and opinion) of private e-mails between scientists engaged in climate research at one of the world's foremost academic centres, the Climatic Research Unit (CRU) at the University of East Anglia. The speed with which the theft was picked up and volume of comment it generated, can be seen from the 31 700 000 hits (including its own Wikipedia entry, Web 7) arising from a search on 4 December 2009 on Google using the single term 'Climategate'.

The categorisation of these events as 'Climategate' at the very beginning of the debate is interesting. Commentators who form a view on events they believe suggest scandalous wrongdoing and want to convey this in a catchy fashion, frequently choose a word ending in '-gate'. The use of terms such as 'Contragate', 'Cheriegate', 'Bloodgate' and now 'Climategate' evokes the 'Watergate' scandal. Watergate is a one-word shorthand for illegal and unethical attempts to suborn the political process (and a high-level conspiracy to cover it up) in which those working for the then President of the USA, Richard Nixon, burgled the headquarters of their Democratic Party opponents at the Watergate complex in Washington DC in June 1972.

Reading some of the early news stories arising from the selective publication of the content of some of the CRU e-mails, it would seem that the scandal the journalists and commentators were seeking to highlight was not the theft of personal documents but an alleged conspiracy involving leading climate scientists. In that they have been spectacularly successful, despite countervailing attempts by those believing that the actual truth is not necessarily what it is being made out to seem. Unfortunately, those not fully engaged are unlikely to wait until the outcome of the deliberations of the various enquiries[viii] that have been set up to investigate the disclosures and their implications before coming to their view based on what they take the term 'Climategate' to represent. However, it ill-behoves those who have been responsible in the past for generating media coverage for their own views on climate change now to complain when their opponents use similar methods.

These disclosures and their aftermath represent a particularly challenging example of a public controversy that could be used, using the methods and ideas contained in this book, to practise how to seek out and assemble information and to use it to formulate your own views.

Should you read all the e-mails and other documents that the hackers published? Should you consider the timing of the disclosures and the possible motivations of those that made them? Do you wait until all the official investigations have been completed by those able to spend the time assembling all the evidence, assessing it and drawing conclusions from it? Or, do you buy into one or other of the responses from individuals? If so, how do you select them?

While there is no doubt that opinion on climate change will be further polarised, it is too early to form a view on the long-term implications for the

---

[viii] Note added in proof: An independent investigation of the allegations arising from the hacking of the CRU e-mails was undertaken by a team led by Sir Muir Russell. The 160 page report, its terms of reference and main findings can be found at www.cce-review.org. Compare the material contained in the report with any press coverage you care to choose.

efforts to mitigate changes in our climate arising from the activities of an ever-increasing number of humankind. However, the openness of the broader scientific community remains intact, despite whatever an individual scientist or group of scientists may do that gives the contrary impression and how damaging some unwise personal communications taken out of context can be made to seem. My own view (for what it is worth) is that the evidence (with or without that from the University of East Anglia) that humankind is having a measurable impact on climate (as represented by the reports of the IPCC and the many individual original contributions across a wide range of disciplines) will remain a sufficient basis for publicly supported political action. There remain, however, areas of genuine scientific uncertainty about the reliability of projections of changes that may arise in 50–100 years time, sufficient for the IPCC to define carefully (Web 8) the meaning to be attributed to terms such as 'high' or 'low' confidence or 'likely' or 'unlikely' when presenting its best estimates.

One highly publicised aspect of the controversy can be followed immediately using references in this book. The use of the term 'trick' in one e-mail has been widely seized upon as proof of skulduggery and the fiddling of results. We already have been alerted to the different meaning put on words in different contexts when used by different groups of people. In Chapter 9, I describe some new work on the synthesis of zeolites in which researchers were able to prepare sheets of ZSM-5 of single unit-cell thickness. In a commentary on this work, Avelino Corma categorises, in complimentary terms, the clever and innovative synthetic strategy used to prepare these new materials as a 'trick'. We should, therefore, not only be careful in our own use of words; we should also be careful in how we interpret the use of words by others.

A characteristic of a media controversy about supposed wrongdoing is the fact that the press becomes sensitised to further instances that are presented as being further examples of mischief. Such was the case of what has inevitably become known as 'Glaciergate' (Web 9), though this was more a case of science engaging in a process of self-correction, the proper role for climate 'sceptics' (as opposed to 'denialists'). The IPCC in their Fourth Assessment Report from Working Group II,[19] had claimed that all the Himalayan glaciers would melt by 2035. Not only was this date incorrect (the predicted date in the literature suggests 2350 not 2035), but the general conclusion appeared to be based, according to a letter published in the journal, *Science*,[20] on a non-peer-reviewed report from the World Wildlife Fund (WWF) that quoted a news story in the *New Scientist* about an unpublished study! The authors of the letter end with the following:

> '*These errors could have been avoided had the norms of scientific publication, including peer-review and concentration upon peer-reviewed work, been respected.*'[20,ix]

[ix] Note added in proof: A recently published review of the processes and procedures of the IPCC by the InterAcademy Council can be found at http://reviewipcc.interacademycouncil.net/report.html.

The foregoing underlines the fact that scientists are human too, and that their errors and personal rivalries can be made out to be evidence of conspiracy. It also reinforces the critical importance of reason rather than rhetoric in addressing the factual basis of societal problems and the responsibility of scientists to stay true to the scientific ethic.

Not least, the attention, which some might have found unwelcome, should have the benefit of improving the processes of consensus-forming and of focusing effort on fundamental areas where the science of climate change is less that robust (see, for instance, ref. 21a–h).

REFERENCES

1. A. Moser, *Sci. Eng. Ethics*, 1995, **1**, 241.
2. Committee on Science, Engineering and Public Policy, National Academy of Science, National Academy of Engineering, and Institute of Medicine of the National Academies, *On Being a Scientist: A Guide to Responsible Conduct in Research*, The National Academics Press, Washington DC, 3rd edn, 2009.
3. Royal Society of Chemistry, *Code of Conduct and Guidance on Professional Practice*, Royal Society of Chemistry, Cambridge, 2001 (Web 1).
4. A. Sokal, *Beyond the Hoax: Science, Philosophy and Culture*, Oxford University Press, Oxford, 2008.
5. Plato, *Gorgias*, translated by Walter Hamilton, Penguin Classics, London, 2004.
6. The Editors of the Lancet, *Lancet*, 2010, **375**, 445.
7. K. M. Madsen, A. Hviid, M. Vestergaard, D. Schendel, J. Wohlfahrt, P. Thorsen, J. Olsen and M. A. Melbye, *N. Engl. J. Med.*, 2002, **347**, 1477.
8. G. D. Munro, *J. Appl. Soc. Psychol.*, 2010, **40**, 579.
9. S. J. O'Neill and M. Hulme, *Global Environ. Change*, 2009, **19**, 402.
10. E. Carlsen, A. Giwercman, N. Keiding and N. E. Skakkebaek, *Br. Med. J.*, 1992, **305**, 609.
11. J. A. Saidi, D. T. Chang, E. T. Goluboff, E. Bagiella, G. Olsen and H. Fisch, *J. Urol.*, 1999, **161**, 460.
12. R. Raucher, quoted in R. A. Freeze, *The Environmental Pendulum*, University of California Press, Berkeley, CA, 2000, p. 141.
13. (a) T. Sager, *Prog. Plan.*, 1997, **50**, 75; (b) K. J. Arrow, *Social Choice and Individual Values*, Yale University Press, New Haven CT, 2nd edn, 1963.
14. W. R. Freudenburg, *Science*, 1988, **242**, 44.
15. R. W. Kates, *Ambio*, 1977, **6**, 247.
16. D. J. C. MacKay, *Sustainable Energy – without the hot air*, UIT Cambridge, Cambridge, 2009, ch. 19, p. 114.
17. M. P. Nelson and J. A. Vucetich, *Conserv. Biol.*, 2009, **23**, 1090.
18. M. Schrope, *Nature*, 2001, **412**, 112 quoting from J. Schell, *Discover*, Oct. 1989, 45 (see also Web 6).

19. R. V. Cruz, H. Harasawa, M. Lal and S. Wu et al., in *Climate Change 2007: Impacts, Adaptation and Vulnerability. Contribution of Working Group II to the Fourth Assessment Report of the Intergovernmental Panel on Climate Change*, ed. M. L. Parry, O. F. Canziani, J. P. Palutikof, P. J. van der Linden and C. E. Hanson, Cambridge University Press, Cambridge, 2007, ch. 10, pp. 469–506.
20. J. G. Cogley, J. S. Kargel, G. Kaser and C. J. van der Veen, *Science*, 2010, **327**, 522.
21. (a) O. Heffernan, *Nat. Rept.: Climate Change*, 2010, **4**, 15; (b) Q. Schiermeier, *Nature*, 2010, **463**, 284; (c) K. Kleiner, *Nat. Rept.: Climate Change*, 2010, **4**, 4; (d) M. Hulme, *Nature*, 2010, **463**, 730; (e) E. Zorita, *Nature*, 2010, **463**, 731; (f) T. F. Stocker, *Nature*, 2010, **463**, 731; (g) J. Price, *Nature*, 2010, **463**, 732; (h) J. R. Christy, *Nature*, 2010, **463**, 732.

WEBLIOGRAPHY

1. www.rsc.org/Membership/CodeofConduct.asp
2. www.guardian.co.uk/media/2009/oct/15/starsuckers-celebrity-cosmetic-surgery-hoax
3. http://news.bbc.co.uk/1/hi/health/8483865.stm
4. www.badscience.net
5. www.youtube.com/watch?v = RFO6ZhUW38w
6. http://stephenschneider.stanford.edu/Publications/PDF_Papers/DetroitNews.pdf
7. http://en.wikipedia.org/wiki/Climatic_Research_Unit_e-mail_hacking_incident
8. www.ipcc.ch/pdf/assessment-report/ar4/wg1/ar4-wg1-chapter1.pdf (Box 1.1)
9. http://news.bbc.co.uk/1/hi/world/south_asia/8387737.stm

All the web pages listed in this Webliography were accessed in May 2010.

APPENDIX 1

Finding Stuff Out

A1.1 LITERATURE SEARCHING

Below is a short introduction to the searching out of information. This is often called 'research', but should be distinguished from original research that is the prime focus of the scientific peer-reviewed literature on which the searching methods concentrate.

Much of the scientific literature is now accessible online. However, two key points must be made.

First, it is important to distinguish the scientific peer-reviewed literature from other web-based material that is not peer-reviewed. This applies particularly to very useful web sources of information (*e.g.* Wikipedia) that can be used as a starting point. (Am I being ironic in referring you to Web 1?) Generally, such material should always be checked out and independently verified in some way.

Secondly, older scientific material and important material to be found in books will not be available electronically. Ignore such material at your peril (displaying 'temporal chauvinism'), as it often contains information and explanations that you need to aid your own understanding that today's experts now take for granted. Know how to find your way round an academic library and understand how to obtain and use hard-copy material.

Like so many things, successful searching for information depends on careful planning, thorough execution and systematic analysis. With so much published information available and more coming available all the time, it is hardly surprising that we rely to a significant extent on computerised specialised databases and search engines. Chemists use SciFinder®,[1] (Web 2) Scopus™ (Web 3) and Web of Knowledge (Web 4) (within which Web of Science sits) among many others; other disciplines will use other databases appropriate to their speciality. Lists of such databases can be found (after a little effort) in the online indices of university libraries.

Chemistry for Sustainable Technologies: A Foundation
By Neil Winterton

Published by the Royal Society of Chemistry, www.rsc.org

The scale of the accessible information is daunting. For instance, the Web of Science database, within the Web of Knowledge, has around 45 million records, with approximately 40 000 articles being added to Web of Science each week.

This book assumes a basic understanding of database structure and organisation as well as an awareness of the principles of searching, accessing and retrieving the literature. (Those completely unfamiliar with what is involved could try the online tutorials provided by many database providers or may be able to obtain guidance from their university library helpdesk.)

Database searching is now taught in many undergraduate chemistry courses, along with an understanding of the Boolean logic employed so that search terms can be combined using the operators AND, NOT, OR, SAME.

An ideal design for a search strategy would collect all relevant material without missing anything. In practice, there is usually a trade-off. There is an art to constructing an effective, computerised literature search strategy such that it casts wide enough to collect a high proportion of the relevant material while maintaining a high relevance ratio, so that your search result doesn‘t contain too high a proportion of wholly irrelevant material ('dross'). This can only come with practice and experience.

In addition, to avoid duplication of effort and to generate one's own mini-databases, the records obtained from database searching (after sifting out the dross and the marginally relevant) can be exported to software such as EndNote (Web 5) to be stored for later use.

It is important to remember that the database records obtained from literature searches provide only a short summary of the content of the document (which is what you are usually after). The individual record generally includes (among other things):

- the title of the article
- the authors and their affiliation
- a short abstract
- a selection of key words
- links to the original document (and related documents).

These all help to establish the relevance of the paper and can provide an important means of improving and widening subsequent searches.

However, these database records are only **secondary** sources or, if the article found is a review of primary sources, **tertiary** sources. It is important always to go to the original or **primary** source, particularly if you want to establish the factual basis of scientific or technological stories in the press. Such original documents can now be obtained quite readily. This is crucial as important or highly relevant papers must be read fully and in detail.

Accessing original papers is now relatively straightforward, particularly via university libraries that link (if they pay a subscription) to many sources from which an electronic form of the document can be readily downloaded (*e.g.* from the publisher of a particular journal). If you do not have access to a library that subscribes, then you will have to pay for each download.

It is also important to be aware that not all journals are always this accessible: some journals are only available electronically back to a particular date. Access can also be governed by the period of the library's subscription. Information on this can be found in the library catalogue (or even by talking to a librarian).

If the library doesn't subscribe to the particular journal you are interested in, then the document you require can be obtained from another library. There is usually a charge for this service, though it can often be cheaper than buying a copy of the document of interest direct from the publishers.

There are a number of key messages.

Carrying out one search is never enough!

Simply carrying out the same search on a different database can be quite revealing as the coverage of databases can be different. The searching of patent databases can sometimes be worthwhile. Sometimes it may be a good idea to carry out further searches using the names of particular authors or a chemical compound or reaction type. The selection of relevant documents ('hits') obtained from such searches has to be analysed, important ones examined and read in detail. These may then provide further information (*e.g.* additional search terms, new references cited in the article and the names of authors of key papers) which will provide the basis of new searches.

Searching can be either retrospective or prospective.

Most searches are retrospective, *i.e.* dealing with material produced before a certain date (usually the point at which you are undertaking the search). This will include papers that are cited in the articles you have found (clearly they cannot be any later than the time when the article was written).

On the other hand, the technique of **citation searching** allows you to find articles appearing after the publication of a paper of interest which refers back to the original article. The later work 'cites' the paper of interest.

Keep your awareness current

Keeping abreast of current developments in a particular area can be done using a number of different means.

Examine the contents pages of the latest issue of particularly relevant journals: publishers are keen to send you these electronically as new issues of the journal appear and will do so for free. This can be important in a fast-moving area as some time may elapse between the publication of a research article in these journals and its appearance in a searchable, abstracting database.

It is also possible to access journal websites to view the contents of all published issues, though you or your institution will need a subscription to access fully articles of interest. An advantage of accessing such web pages is that most publishers now post articles that have been accepted for publication but have not yet appeared in a journal issue.

Regular (weekly or monthly) database searching is also useful to access material abstracted into databases during the immediate past (*i.e.* by limiting the search by time). As noted above, huge numbers of research papers are abstracted each week. Even so, searching (on 15 April 2010) 38 927 records added in the previous week instead of the 45 702 848 in the entire database has the advantage that the search can be cast somewhat wider in the expectation that a greater number of relevant hits will be found. Although the proportion of dross will be higher, the absolute number of useless records may be more manageable—making the tedious job of separating them out shorter.

Read a book (or two)

Those wishing to explore a research topic in more detail should seek out a recent review or monograph on the topic, preferably by someone involved in the topic of interest. Sometimes the area will have been subject to a long period of research and even old texts can be very readable, describing the fundamentals of an area of research. Although the treatment may not be fully up-to-date in these instances, it provides the degree of understanding that is often assumed in later treatments. The bibliographical lists at the end of each chapter of this book include a number of such classical texts. Occasionally, some books (and some research papers) are so influential and well-written that their usefulness, relevance and historical importance remain even after many years.

A1.2 SCIENCE IN THE MEDIA

A point I make in this book is the importance for those trained in science (whether still practising or not) to be able to establish something of the factual or evidential basis lying behind scientific, environmental or technological controversies that erupt in the media. At the very least, it should be possible to:

- check whether the original work has been accurately and fairly reported
- check whether the work quoted had been submitted to peer review
- look at who the key authors are (and what qualifications they have for saying what they have).

When put together with what you know or can access from other sources, this can help to develop an objective and critical view of the material and its implications (within the limits of your own expertise). For many topics for which new published research material is coming available, the subject matter, language and assumed level of understanding can be particularly daunting. Even so, the abstract, introduction and conclusions of a paper can often be more accessible. So may be a commentary or perspective piece written by a qualified scientific journalist working for a well-respected scientific periodical such as *Science* (Web 6) or *Nature* (Web 7).

This does require a familiarity with the techniques of information searching as described above, but also may require the additional use of other web-based sources.

The nature and difficulty of the searching depends on many factors, but key amongst them is the degree to which the journalist or publisher provides information in the news item about the source they are using or quoting. In the example I discuss in Chapter 13, the only link to the original material was the disclosure (see Figure 13.1a) of the source as the University of Copenhagen. This was sufficient using the University of Copenhagen's website and Web of Science to lead to the *British Medical Journal* article. This was possible as the newspaper story was using material that had been published four years earlier.

If the story relates to some very new work, then unfortunately plugging the name of the scientist into a database may not lead to the original article as insufficient time will have elapsed for the original work to be abstracted. If the web version of a news story has a link to the publisher of the material on which the article is based, then a search of the publisher's website using the name of the author as a search term can be successful. Even then, science publishers may have released the story early to obtain media exposure prior to its appearance even as a pre-publication article. Under such circumstances, it may be necessary to wait a short time and revisit the website.

The area of sustainable development is a complex area in which contributions from many scientific and other academic disciplines and intellectual domains are relevant. Some material is often not available in peer-reviewed form, appearing in periodicals, newspapers, websites and radio and TV programmes.

For example, an item from the BBC News website, 'New evidence on Antarctic warming' (Web 8), appeared as this book was being prepared. (You might want to track this down by typing the link into your internet browser.) How should a chemist, who may not be a specialist in the science being discussed, approach such a new item that deals with scientific and environmental matters? This example (others appear almost on a daily basis) concerns reports that the Antarctic has now been found to have undergone measurable warming over the last 50 years, something that had not, until the report appeared, been quite as apparent as temperature changes in other parts of the world. When faced with any press report of this type, it is a good idea to start by being sceptical about what is written and to approach it by following some fairly simple steps to avoid being persuaded to accept an idea or information that is erroneous or misleading. Open-minded scepticism is a necessary defence mechanism that should automatically be deployed.

The item highlighted a study to be published shortly in the highly respected scientific journal, *Nature*, by a group led by a named individual. This information encourages the belief that the story and its contents may be well-founded. Even so, journalists and the wider media have been known to report selectively, to highlight a particular part of the study that might be more newsworthy or might support a particular point of view. So, the next step should be to track down the original paper using the author's name and the journal in which it was due to appear. This should lead to the article entitled, 'Warming of the Antarctic ice-sheet surface since the 1957 International Geophysical Year'.[2] This paper can be accessed online or a hard copy can be scrutinised in most university libraries. While we may not be expert in the subject of the paper, it is possible with a knowledge of chemistry, science and the scientific method to gain a reasonable impression of its contents, particularly to evaluate the match between it (if only the abstract and conclusions) and the news report based on it. Does the news report give a fair representation of the material reported and properly reflect the main conclusions of the study? Does it give proper weight to any qualifications or reservations the authors of the paper may have made about these conclusions?

A second example comes from the BBC website, 'Global warming is "irreversible"' (Web 9). This brings to public attention a (then) forthcoming scientific paper with, as the lead author, the leading climate scientist, Susan Solomon. The headline in the web article suggests that climate change associated with anthropogenic releases of carbon dioxide is 'irreversible'. The scientific paper on which the item is based is to appear in the prestigious peer-reviewed scientific journal, the *Proceedings of the National Academy of Sciences of the United States of America* (PNAS for short). In passing, it is worth noting that, even though the journal is prestigious and Susan Solomon is a leading climate scientist, one's ingrained open-minded scepticism should always be in operation as an intellectual firewall against being misled. It is a straightforward exercise (for those with relevant access or those willing to pay for the download) to access the PNAS website, search for 'Solomon' as an author and look for papers published recently on the web. This leads straight to the article, 'Irreversible climate change due to carbon dioxide emissions', which was published online on 10 February 2009 (ref. 3 gives the full citation) and not forgetting the availability of supplementary information (Web 10). A consideration of the full article suggests that the BBC News headline fairly reflects the content of the paper.

These two stories highlight that we are at, or approaching (over a timescale that may be decades or centuries), a time in which change going on at the moment will have important consequences. One could argue that it is entirely appropriate that the fear of doom is put into us to get our attention. However, whether making judgements in a state of panic is necessarily evidence of wisdom, the urge to act needs to be tempered with an assessment of the outcomes that might arise and those that might arise from doing something different (*e.g.* concentrating on mitigation strategies rather than prevention strategies) or from doing nothing.

REFERENCES

1. D. D. Ridley, *Information Retrieval: SciFinder®*, John Wiley and Sons, Chichester, 2009.
2. E. J. Steig, D. P. Schneider, S. D. Rutherford, M. E. Mann, J. C. Comiso and D. T. Shindell, *Nature*, 2009, **457**, 459.
3. S. Solomon, G.-K. Plattner, R. Knutti and P. Friedlingstein, *Proc. Natl. Acad. Sci. U.S.A.*, 2009, **106**, 1704.

WEBLIOGRAPHY

1. http://en.wikipedia.org/wiki/Scientific_literature
2. www.cas.org/products/sfacad/index.html

3. http://info.scopus.com
4. http://wok.mimas.ac.uk
5. www.endnote.com
6. www.sciencemag.org
7. www.nature.com/nature/index.html
8. http://news.bbc.co.uk/1/hi/sci/tech/7843186.stm
9. http://news.bbc.co.uk/1/hi/sci/tech/7852628.stm
10. www.pnas.org/cgi/content/full/0812721106/DCSupplemental

All the web pages listed in this Webliography were accessed in May 2010.

APPENDIX 2
Units and Abbreviations

ε'	relative permittivity
ε''	microwave loss factor
λ	wavelength
4AR	Fourth Assessment Report (of the IPCC)
5-HMF	5-(hydroxymethyl)furfural
ACH	acetone cyanhydrin
ADMET	acyclic diene metathesis
ADP	adenosine diphosphate
AE	atom efficiency or atom economy
API	active pharmaceutical ingredient
APR	aqueous-phase reforming
ATP	adenosine triphosphate
AU	atom utilisation
BASF	Badische Anilin und Soda Fabrik
bbl	barrel
BMJ	British Medical Journal
bp	boiling point
BSE	bovine spongiform encephalopathy
BTU	British Thermal Unit
BTX	benzene, toluene, xylenes
CAS	Chemical Abstracts Service
CCN	cloud condensation nuclei
CCS	carbon capture and storage
CFC	chlorofluorocarbon
CHP	combined heat and power
CM	cross-metathesis
CO_2e	carbon dioxide equivalent
COSHH	Control of Substances Hazardous to Health

Chemistry for Sustainable Technologies: A Foundation
By Neil Winterton

Published by the Royal Society of Chemistry, www.rsc.org

CRU	Climatic Research Unit [University of East Anglia]
Da	Dalton
DDT	dichlorodiphenyltrichloroethane
DKR	dynamic kinetic resolution
DMF	2,5-dimethylfuran
DNA	deoxyribonucleic acid
DoE	Department of Energy [US]
D_p	microwave penetration depth
ee	enantiomeric excess
EM	ecological modernisation
[emim]Cl	1,3-ethylmethylimidazolium chloride
EPA	Environmental Protection Agency [US]
ESI	Environmental Sustainability Index
FAME	fatty acid methyl esters
FCC	fluid catalytic cracking
FCEV	fuel cell electric vehicle
FEP	fluorinated ethylene propylene
fl. oz	fluid ounce
F-T	Fischer–Tropsch
G	Gibbs free energy
G3P	glyceraldehyde-3-phosphate
GDP	gross domestic product
GHG	greenhouse gas
GM	genetic modification or genetically modified
GtC	gigatonne of carbon
GTL	gas-to-liquids
GtN	gigatonne of nitrogen
GtS	gigatonne of sulfur
GWP	global warming potential
H	enthalpy
ha	hectare [0.01 km^2]
HANPP	human appropriation of net primary production
HDS	hydrodesulfurisation
HFC	hydrofluorocarbon
HI	hypsicity index
HIV	human immunodeficiency virus
[hmim]BF_4	1,3-hexylmethylimidazolium tetrafluoroborate
Hz	Hertz
ICI	Imperial Chemical Industries
IE	industrial ecology
IPCC	Intergovernmental Panel on Climate Change
IPR	intellectual property rights
IR	infra-red
ITER	International Thermonuclear Experimental Reactor
J	Joule [unit of energy]
k_b	Boltzmann's constant

LCA	life- cycle analysis
L-DOPA	3,4-dihydroxy-L-phenylalanine
LGA	levoglucosan
mim	1-methylimidazole
MMA	methyl methacrylate
MMR	mumps, measles and rubella
mp	melting point
MRI	magnetic resonance imaging
MRP	material recovery parameter
MTG	methanol-to-gasoline
MTO	methanol-to-olefins
MWt	molecular weight
NADH	nicotinamide adenine dinucleotide (reduced form)
NADPH	nicotinamide adenine dinucleotide phosphate (reduced form)
NASA	National Aeronautics and Space Administration
NMR	nuclear magnetic resonance
NPP	net primary production
NTP	normal temperature and pressure
OPEC	Organisation of Petroleum Exporting Countries
Pa	Pascal
PDO	propanediol
PET	poly(ethylene terephthalate)
PHA	poly(hydroxyalkanoate)
PHB	poly(hydroxybutyrate)
PHBV	poly(hydroxybutyrate-co-hydroxyvalerate)
ppb	parts per billion
ppm	parts per million
ppmw	parts per million by weight
ppt	parts per trillion
PTT	poly(trimethylene terephthalate)
PV	photovoltaic
PVC	poly(vinyl chloride)
quad	10^{15} BTU
RCM	ring-closing metathesis
RME	reaction mass efficiency
RNA	ribonucleic acid
ROMP	ring-opening metathesis polymerisation
RSPB	Royal Society for the Protection of Birds
rubisco	ribulose bisphosphate carboxylase oxygenase
S	entropy
SCF	supercritical fluid
SD	standard deviation
SF	stoichiometric factor
SI	International System of Units
STY	space-time yield
t	tonne [metric ton] [(10^3 kg])

tof	turn-over frequency
ton	turn-over number
Troy oz	Troy ounce
UNESCO	United Nations Educational, Scientific and Cultural Organisation
UV	ultraviolet
W	Watt [unit of power]
ZSM	zeolite Socony Mobil [a zeolitic aluminosilicate mineral]

PREFIXES USED WITH SI UNITS

E	exa (10^{18})
P	peta (10^{15})
T	tera (10^{12}) 'trillion'
G	giga (10^{9}) 'billion'
M	mega (10^{6}) 'million'
k	kilo (10^{3})
μ	micro (10^{-6})
n	nano (10^{-9})
p	pico (10^{-12})

APPENDIX 3

Twelve More Green Chemistry Principles[i,ii]

Neil Winterton of the Leverhulme Centre for Innovative Catalysis[ii] suggests that laboratory and research chemists should consider twelve more green chemistry principles, objectives and requirements to help them plan and carry out their work and to help them assess the relative 'greenness' of a process.

INTRODUCTION

Chemical technology has achieved much in relation to waste minimisation even before sustainable development was ever conceived and is a major success story. Indeed, many of the major benefits in public health and our standard of life have arisen, in large measure, as a consequence of the application of chemistry, such as in the provision of clean drinking water, plentiful food and treatments for disease. There is, nevertheless, much still to do and the potential exists for chemical technology to continue to build on these earlier successes through the greater use of renewable resources and the development of cleaner products and processes. Some will arise by the continued application of existing methods and ways of working, whether consciously, coincidentally or unexpectedly—the potential benefits of serendipity in any scientific enterprise should not be underestimated! In addition, new approaches, consciously planned using the principles of sustainable development, industrial ecology, clean technology, life cycle analysis and green chemistry, are needed to bring about additional environmental and societal benefits.

Chemistry, being concerned with understanding the way the material world works, has become, to an extent greater than in the past, a central science

[i] Part of a presentation 'Sense and Sustainability: The Role of Chemistry, Green or Otherwise', given at the BA Festival of Science, University of Glasgow, September 2001.

[ii] Leverhulme Centre for Innovative Catalysis, Department of Chemistry, University of Liverpool, Liverpool L69 7ZD.

Chemistry for Sustainable Technologies: A Foundation
By Neil Winterton

Published by the Royal Society of Chemistry, www.rsc.org

relevant to technology, the environment and the understanding of human impact on the latter. Green chemistry, not simply being a subdivision of chemistry like organic or organometallic chemistry, has environmental, technological and societal goals and is linked to the wider sustainability movement. There is, thus, a wider context in which green chemistry is being developed and this requires awareness, clear thinking and pragmatism[1] to target the most critical problems as well as sound science to help resolve them.

Green chemistry has set itself the goal of making chemicals technology more environmentally benign by, in Roger Sheldon's words,[2] '*efficiently using (preferably renewable) raw materials, eliminating waste and avoiding the use of toxic and/or hazardous reagents and solvents in the manufacture and application of chemical products*'.

To help it achieve this Anastas and Warner[3] set down the 12 well-known principles of green chemistry. These are summarised in Table 1. They are difficult to fault.

Most chemists, including those working in industry, recognise these as worthwhile ideals and acknowledge that most of them already guide their work. The failure to adopt them perfectly in every case does not arise from innate perversity or from indifference to environmental considerations. It arises from technological, economic and other factors that chemists do not always address.

William H. Glaze, the editor of the American Chemical Society publication, *Environmental Science & Technology* has indeed suggested[4] that the 'greenness' of a chemical transformation can only be assessed in the context of its scale-up, its application and its practice. This will usually involve some trade-off between conflicting factors, often driven by valid technical, economic and commercial considerations.

Table 1 Anastas and Warner's green chemistry principles.

1. It is better to prevent waste formation than to treat it after it is formed
2. Design synthetic methods to maximise incorporation of all material used in the process into the final product
3. Synthetic methods should, where practicable, use or generate materials of low human toxicity and environmental impact
4. Chemical product design should aim to preserve efficacy whilst reducing toxicity
5. Auxiliary materials (solvents, extractants *etc.*) should be avoided if possible or otherwise made innocuous
6. Energy requirements should be minimised: syntheses should be conducted at ambient temperature/pressure
7. A raw material should, where practicable, be renewable
8. Unnecessary derivatisation (such as protection/deprotection) should be avoided, where possible
9. Selectively catalysed processes are superior to stoichiometric processes
10. Chemical products should be designed to be degradable to innocuous products when disposed of and not be environmentally persistent
11. Process monitoring should be used to avoid excursions leading to the formation of hazardous materials
12. Materials used in a chemical process should be chosen to minimise hazard and risk.

Table 2 Twelve more principles of green chemistry.

1.	Identify and quantify by-products
2.	Report conversions, selectivities and productivities
3.	Establish full mass-balance for process
4.	Measure catalyst and solvent losses in air and aqueous effluent
5.	Investigate basic thermochemistry
6.	Anticipate heat and mass transfer limitations
7.	Consult a chemical or process engineer
8.	Consider effect of overall process on choice of chemistry
9.	Help develop and apply sustainability measures
10.	Quantify and minimize use of utilities
11.	Recognise where safety and waste minimization are incompatible
12.	Monitor, report and minimize laboratory waste emitted.

To complement and build on those formulated by Anastas and Warner and to address Glaze's concerns, twelve more green chemistry principles, objectives and requirements are suggested (Table 2). They are proposed to aid laboratory and research chemists, interested in applying green chemistry, to plan and carry out their work to include the collection of data that are of particular use to those, usually process chemists, chemical engineers and chemical technologists, wishing to assess the potential for waste minimisation. These groups will recognise and already work to these principles. They are all intended to help demonstrate and assess the potential 'greenness' of a process and, just as important, its relative 'greenness' when compared with other processes; additional background is discussed in more detail elsewhere:[5–9]

1. ***Identify by-products and, if possible, quantify them***
   Separation and disposal of by-products can be both expensive and resource-consuming and indeed may determine the economic viability of a process or product. This may even be true of by-products formed in quite small amounts. Rarely do chemical processes give the targeted product with perfect specificity even after much effort. Because by-products may be recycled in scaled-up process, it will also be informative to examine the effect on reaction outcome of adding each by-product at the beginning of the reaction.
2. ***Report conversions, selectivities, productivities***
   A product formed in high yield may be produced in an overall very inefficient and wasteful process, requiring stoichiometric co-reagents producing a range of co-products requiring separation or post-treatment for recycle or disposal. To aid process design, therefore, more informative metrics than yields should be reported. These should include conversions, selectivities, productivities (and related measures, such as atom efficiency[10] or productivity[11] functions) and rates.
3. ***Establish a full mass balance for the process***
   All the material used in producing and isolating the desired material should be identified, quantified and account for. In addition, all

materials, including solvents, used in the recovery of the final product in its pure state should be specified and quantified.

4. ***Quantify catalyst and solvent losses***
A much more useful estimate of catalyst and solvent losses is obtained by measuring their concentration in the (measured volume of) solid waste, aqueous or gaseous effluent you dispose of, rather than by simply weighing the recovered catalyst or solvent.
5. ***Investigate basic thermochemistry to identify potentially hazardous exotherms***
A reaction operated on the small scale in the laboratory may be perfectly safe. However, surface-to-volume ratios decrease on scale-up leading to significant and unanticipated constraints on heat transfer. The design engineer would rather know about such effects before a pilot-plant or semi-tech unit is built.
6. ***Anticipate other potential mass and energy transfer limitations***
Check other factors affecting heat and mass transfer and the consequences for reaction outcome. These might include stirring rates, gas dispersion, solid–liquid contacting. Can these be managed on scale-up?
7. ***Consult a chemical or process engineer***
Get the perspective of someone involved in scale-up to comment on your reaction. Examine some of the published literature. Identify and understand further likely constraints on process chemistry and take account of these in your laboratory investigations, as far as possible.
8. ***Consider the effect of the overall process on choice of chemistry***
A range of options for the operation of process on the large scale will arise from the choice of raw material, feedstock, reactor, by-product separation, purity and purification, energy and utilities utilization, catalyst and solvent recovery and waste disposal. These should be considered pragmatically to establish the impact of your new chemistry on these options. This may be especially relevant to novel chemistry being considered for a well-established process. It makes sense to be aware of the current manufacturing technology. How many steps are used to obtain feedstock from a primary raw material, like oil? An alcohol may be made from the hydration of an olefin which itself is produced by cracking or dehydrogenation of a saturated hydrocarbon. Why not start using the hydrocarbon itself? Would your use of a very pure feedstock be ruled out on cost grounds? Would you create a waste-stream requiring disposal elsewhere in the product-chain in so doing? What would the effect be of using the commercial-grade rather than the research-grade material in your laboratory experiment?
9. ***Help develop and apply sustainability measures***
Attempt to evaluate the sustainability of your process using, for example, the parameters described by J. Dewulf *et al.*[12]
10. ***Quantify and minimize use of utilities and other inputs***
These are often neglected in laboratory studies. However, the use of water (e.g., for cooling and extraction), electricity or inerting gases may be a significant element in emissions. For example, compression and

recompression of CO_2 for use in supercritical fluid processing is very energy intensive.

11. ***Recognise where operator safety and waste minimization may be incompatible***
 Partial oxidation of hydrocarbons using dioxygen must avoid explosive regimes.
12. ***Monitor, report and minimise all waste emitted to air, water and as solids, from individual experiments or from laboratory overall***
 This would represent a real and direct demonstration of your green credentials.

The application of these principles, in itself, will not guarantee that clean chemical processes will be developed. However, it is hoped that they will aid in their earlier selection as well as directing the attention of green chemists into the most productive of areas of investigation. However, there is no substitute for an awareness of the wider technological factors, beyond the chemical, that affect the efficiency of chemical processing and the choices made whether or not to develop new processes arising from green chemistry. Having said all this, the wholly novel, arising unanticipated from research done for its own sake, may also be the basis of innovative and waste-eliminating chemical technology.

REFERENCES

1. *The Skeptical Environmentalist—Measuring the Real State of the World*, B. Lomborg, Cambridge University Press, 2001, ISBN 0 521 01068 3.
2. 'Atom utilization, E factors and the catalytic solution', R. A. Sheldon, *Comptes Rendus de l'Académie des Sciences—Series IIC-Chemistry*, 2000, **3**, 541.
3. *Green Chemistry: Theory and Practice*, P. T. Anastas and J. C. Warner, Oxford University Press, 1998.
4. 'Sustainability engineering and green chemistry', W. H. Glaze, *Environmental Science & Technology*, 2000, **34**, 449A.
5. 'So you think your process is green, how do you know?—Using principles of sustainability to determine what is green—a corporate perspective', A. D. Curzons, D. J. C. Constable, D. N. Mortimer and V. L. Cunningham, *Green Chem.*, 2001, **3**, 1.
6. Green chemistry measures for process research and development, D. J. C. Constable, A. D. Curzons, L. M. Freitos dos Santos, G. R. Green, R. E. Hannah, D. Hayler, J. Kitteringham, M. A. McGuire, J. E. Richardson, P. Smith, R. L. Webb and M. Yu, *Green Chem.*, 2001, **3**, 7.
7. 'Chemical process design using heuristics in the context of pollution prevention', D. W. Pennington, *Clean Prod. Process.*, 1999, **1**, 170.
8. 'A screening-level prototype for the synthesis or analysis of separation systems to support identification of inherently cleaner chemical processes', D. W. Pennington and P. L. Yue, *Clean Prod. Process.*, 2000, **2**, 82.

9. 'Process design for the environment: a multi-objective framework under uncertainty', Y. Fu, M. Diwekar, D. Young and H. Cabezas, *Clean Prod. Process*, 2000, **2**, 92.
10. R. Sheldon, *Chem. Ind.*, 7 December, 1992, 903.
11. 'Choosing Processes for their Productivity', A. Steinbach and R. Winkenbach, *Chem. Eng.*, April 2000, 94.
12. 'Illustrations towards quantifying the sustainability of technology', J. Dewulf, H. Vangenhove, J. Mulder, M. M. D. van den Berg, H. J. van der Kooi and J. de Swaan Arons, *Green Chem.*, 2000, **2**, 108.

Subject Index

References to figures are given in italic type; references to tables are given in bold type.